# Fundamentals of Electrochemical Corrosion

**E.E. Stansbury**
Professor Emeritus
Department of Materials Science and Engineering
The University of Tennessee

and

**R.A. Buchanan**
Robert M. Condra Professor
Department of Materials Science and Engineering
The University of Tennessee

The Materials Information Society

**ASM International®**
Materials Park, Ohio 44073-0002
www.asminternational.org

First printing, July 2000
Second printing, December 2004
Digital printing, August 2012



Comments, criticisms, and suggestions are invited, and should be forwarded to ASM International.

*ASM International staff who worked on this project included Veronica Flint, Manager of Book Acquisitions; Scott Henry, Assistant Director, Reference Publications; Bonnie Sanders, Manager of Production; Carol Terman, Copy Editor; Kathleen Dragolich, Production Supervisor; and Alexandru Popaz-Pauna, Book Production Coordinator.*

Library of Congress Cataloging-in-Publication Data

Stansbury, E.E.
Fundamentals of electrochemical corrosion / E.E. Stansbury and R.A. Buchanan
p. cm.
1. Electrolytic corrosion. 2. Corrosion and anticorrosives.
I. Buchanan, R.A. (Robert Angus), 1930- II. Title.
TA462.S714 2000 620.1'1223—dc21 99-058428

ISBN-13: 978-0-87170-676-8 ISBN-10: 0-87170-676-8
SAN: 204-7586

ASM International®
Materials Park, OH 44073-0002
www.asminternational.org

Printed in the United States of America

*Cover art represents autocatalytic processes occurring in a corrosion pit. The metal, M, is being pitted by an aerated NaCl solution. Rapid dissolution occurs within the pit, while oxygen reduction takes place on the adjacent surfaces. Source: U.R. Evans, Corrosion, Vol 7 (No. 238), 1951*

# Dedication

To my wife, Bernice; daughters, Ginny, Kate, and Barb; and son, Dave.

**Gene Stansbury**

To my wife, Billie; daughter, Karen; mother, Katherine; and in memory of my son, Mike.

**Ray Buchanan**

And to our graduate students who have extended our understanding of this fascinating field.

# ASM International Technical Books Committee (1999-2000)

# Contents

# Preface

The objective of this book is to provide a reasonably self-contained textbook covering the essential aspects of the corrosion behavior of metals in aqueous environments. It is designed to be used in courses for upper-level undergraduate and graduate students, for concentrated courses in industry, for individual study, and for reference. It has been our experience that both students and persons in industry come to a first course in corrosion with a wide diversity of backgrounds, both academically and in terms of experience in corrosion behavior. The usual pedagogical problem arises as to the minimum background for each participant allowing a useful understanding of the subject. This text has been designed to provide flexibility in meeting this need.

An introductory chapter, Chapter 1, provides an overview of aqueous corrosion. Emphasis is placed on the fact that corrosion is an interface phenomenon and, as such, is dependent on the variables defining the metal, the environment, and the physical aspects of the interface itself. Schematic electrochemical cell circuits are used to illustrate how these variables give rise to electrical potential differences across the interface and drive the corrosion process, resulting in current densities directly related to the corrosion rate. The fact that the current is also controlled by interface films allows emphasizing how passive-type alloys with their adherent oxide films have lower corrosion rates than the nonpassive alloys.

The essential electrochemical background is provided in Chapter 2 on electrode reactions and in Chapter 3 on electrode kinetics. These chapters contain the essential electrochemical concepts required for understanding the following chapters. Chapter 2 covers the principles governing the stability of metal/environment systems. Following an introduction to the classical thermodynamic criteria for stability, determination of stability based on electrochemical cell calculations allows an early introduction to the relative roles of the metal and the environment in corrosion. More than the usual emphasis is placed on the significance of environmental variables (pH, aeration, etc.), as is done throughout the text. Chapter 2 concludes with a rather detailed discussion of the so-called Pourbaix diagrams. While it is recognized that these diagrams must be used with caution in the analysis of corrosion problems, they are ready sources of information on the stability of metal/water systems and the corrosion products that can form. The somewhat more practical use of the diagrams is illustrated using Pourbaix's modified diagrams defining the conditions for immunity, passivity, and corrosion for several metals in aqueous environments.

Simple but pedagogically useful theories of electrode kinetics are presented in Chapter 3. This permits discussion of models for anodic and cathodic reactions at the metal/environment interface and for diffusion of species to and from the interface. Mathematical models of these theories lead to so-called kinetic parameters whose values govern the rate of the interface reaction. The range of values that these parameters can have and some of the variables that can influence the values are emphasized since these will relate to understanding the influence of such factors as surface conditions (roughness, corrosion product films, etc.), corrosion inhibitors and accelerators, and fluid velocity on corrosion rates. This chapter also introduces electrochemical measurements to determine values of the kinetic parameters.

The concepts in Chapters 2 and 3 are used in Chapter 4 to discuss the corrosion of so-called active metals. Chapter 5 continues with application to active/passive type alloys. Initial emphasis in Chapter 4 is placed on how the coupling of cathodic and anodic reactions establishes a mixed electrode or surface of corrosion cells. Emphasis is placed on how the corrosion rate is established by the kinetic parameters associated with both the anodic and cathodic reactions and by the physical variables such as anode/cathode area ratios, surface films, and fluid velocity. Polarization curves are used extensively to show how these variables determine the corrosion current density and corrosion potential and, conversely, to show how electrochemical measurements can provide information on the nature of a given corroding system. Polarization curves are also used to illustrate how corrosion rates are influenced by inhibitors, galvanic coupling, and external currents.

A separate chapter, Chapter 5, is used to introduce the corrosion behavior of active/passive type metals. This allows emphasis on the more complex anodic polarization behavior of these metals and the associated problems in interpreting their corrosion behavior. The chapter is introduced by discussing experimental observations on the anodic polarization of iron as a function of pH and how these observations can be related qualitatively to the iron-water Pourbaix diagram. Pedagogically, it would be desirable to analyze the corrosion behaviors of active/passive metals by relating their anodic polarization curves to curves for cathodic reactions as was done in Chapter 4 for nonpassive alloys. Because of the extreme sensitivity of an experimental curve to the environment, a reasonably complete curve usually can only be inferred. To do so requires understanding of the forms of experimental curves that can be derived from individual anodic and cathodic polarization curves. The basis for constructing such curves is discussed in some detail with ten cases analyzed showing the schematic construction of curves for an active/passive alloy with several environmental and alloy variables. The objective of the remainder of the chapter is to provide representative examples of (1) anodic polarization behaviors of commercial metals, (2) the effect of alloy composition on anodic polarization, and (3) the effect of several environmental variables on anodic polarization. Final sections illustrate the prediction of corrosion behavior of active/passive type alloys in specific environments.

Principles and procedures of electrochemical measurements used to investigate corrosion behavior are described in Chapter 6. Although some reference is made to subjects covered earlier in the book, the chapter is reasonably self contained and can be used as a condensed reference on electrochemical corrosion measurements and instrumentation. Also, the chapter is referenced in earlier chapters for readers wanting more information than accompanies an immediate discussion. Reference half cells and the use of electrometers for measuring electrochemical cell potentials are described in some detail including sources of error in measured values. This is followed by discussion of the potentiostat circuit and the use of potentiostats to determine the basic parameters of electrochemical reactions, and to measure corrosion potentials and current densities. Because of the more recent and expanding use of electrochemical impedance measurements to investigate many aspects of corrosion behavior, the theory and procedures underlying this technique are treated in some detail in the latter part of the chapter.

Chapter 7 describes localized corrosion phenomena and covers specific corrosion processes extending from pitting and crevice corrosion to stress corrosion cracking and corrosion fatigue. The discussion of each of these processes for several commercially important metals and alloys assumes familiarity with concepts covered in the earlier chapters. An objective of the chapter is to show that while there are general principles that can be brought to the investigation and understanding of corrosion behavior, identifying those that are applicable is frequently complicated because of conditions unique to each metal/environment system.

The material in Chapter 7 can be used in several ways: (1) it is a reasonably self-contained overview of localized corrosion and can be used as such for readers familiar with the principles developed in earlier chapters; (2) in covering the earlier chapters as a text, reference can be made to specific sections of Chapter 7 to illustrate the relevance of principles being developed to observations on real systems; (3) conversely, the chapter can be covered with emphasis on how knowledge of the principles of corrosion presented in earlier chapters is fundamental to understanding applied corrosion behavior; and (4) an outline of the major identifying features of each of the processes can be created as a guide to the reader in pursuing subjects for clarification or greater in-depth discussion.

The examples of localized corrosion in Chapter 7 are taken largely from the published literature, for which representative references are given. The major characteristics of each process are presented, followed by discussion of one or more mechanisms that have been proposed for the process. While generally a mechanism is discussed with reference to a specific metal and environment, application of the mechanism to other metal/environment systems should be recognized. The authors have used this chapter to emphasize that the range of corrosion phenomena directly involves a breadth of disciplines extending from electrochemistry and materials science to solid and fluid mechanics.

E.E. Stansbury
R.A. Buchanan

CHAPTER 1

# Introduction and Overview of Electrochemical Corrosion

## Definition and Examples of Corrosion

The deterioration of materials due to reactions with their environments is the currently accepted broad definition of corrosion. From a practical standpoint, the term *materials* refers to those substances used in the construction of machines, process equipment, and other manufactured products. These materials include metals, polymers, and ceramics. The environments are liquids or gases, although under special circumstances certain solid-solid reactions might be included as corrosion. The breadth of this definition can best be appreciated by considering examples, starting with cases that are usually recognized as corrosion and proceeding to those that are less obvious or are not generally recognized as corrosion:

- Rusting of steel and cast iron in water, including humid air, as occurs with domestic and industrial water tanks and supply piping, automobiles, and exposed steel structures
- Corrosion of copper, aluminum, and cast iron in automotive cooling systems

- Corrosion of iron-base, copper-base, nickel-base, etc. alloys in the chemical process industry
- Corrosion of automobile exhaust systems by direct reaction of the metal with high-temperature gases and by condensation of water and absorption of the oxides of sulfur and nitrogen to produce aqueous acid environments
- Corrosion of turbine blades in gas turbines by hot combustion gases
- Corrosion of metallic surgical implant materials used in orthopedic, cardiovascular, and dental devices resulting in the release of metal ions to tissues, and degradation of the physical properties of polymeric implant materials due to interactions with tissue fluids and/or blood
- Corrosion of iron-base and nickel-base alloys by liquid metals used as heat transfer agents (e.g., liquid sodium, potassium, and lithium)
- Enhanced deterioration of structural concrete and stone by interaction with condensed moisture and acidic contaminants in the air, such as the oxides of sulfur and nitrogen
- Stress-corrosion cracking (SCC) of gold and brass by mercury
- SCC and pitting of stainless steel in sea water

## The Need to Control Corrosion

The need to control corrosion almost always reduces to considerations of safety and economics. Machines, equipment, and functional products may fail due to corrosion in such a manner as to result in personal injury. Because the choice of materials, enforcement of manufacturing procedures, and control of products to minimize personal injury all involve economic considerations, implementation of safety measures not only involves humanitarian concerns but also economics. With all economic decisions, the basis for action is a compromise between the benefits generated by a certain level of corrosion control versus the costs that would result if that level of control were not maintained. Examples of economic decisions involving considerations of the consequences of corrosion include the following:

- Within limits of health and safety, materials should not be selected for individual products, or components of more complex products, if the corrosion resistance would permit the life of the part to be significantly longer than the life actually realized because of other factors. Thus, the muffler of an automobile could be made of materials that would permit it to outlast the use of some large fraction of all automobiles manufactured at a given time. Because driving habits have a major influence on muffler life, and reasonable performance and ease of replacement can be realized by using relatively inexpensive materials, it is not economical to use more highly corrosion-resistant materials. This choice also is favored by the fact that the muffler is not a critical component from the safety standpoint. For example, a different set of criteria would be required for critical components of the steering mechanism.

- Design for corrosion resistance may be almost exclusively for appearance when favorable appearance is an economic advantage. Stainless steel and aluminum are frequently used for architectural applications and in food service largely for appearance. They also are used for trim on automobiles for the same reason.
- On the other hand, materials exhibiting very low corrosion rates may be selected for reasons of both health and appearance in the processing of foods, pharmaceuticals, and cosmetics. Even if health is not involved, corrosion products producing objectionable color or particles of foreign material are not acceptable to the consumer. For example, such product contamination in paint obviously can lead to totally unacceptable products.
- In some cases, severely corrosive environments are contained by metals such as gold and platinum, which, in spite of high costs, are required because of their inertness. The initial cost, however, is countered by the ease of recovery of the metals following use and their high recycle value.
- A major economic factor in designing for corrosion resistance is the avoidance of interruption of plant production. Failure due to corrosion of critical components such as pumps and heat exchangers may necessitate large sections of a process or entire plants to become inoperative, leading to costs associated with lost production far in excess of the cost of replacement of the failed component. Process design and materials selection to minimize plant outage is a major engineering consideration.

## Corrosion Mechanisms

Particularly under the broad definition of corrosion as the deterioration of materials by reaction with the environment, the number of mechanisms whereby deterioration occurs is large. In general, a mechanism of corrosion is the actual atomic, molecular, or ionic transport process that takes place at the interface of a material. These processes usually involve more than one definable step, and the major interest is directed toward the slowest step that essentially controls the rate of the overall reaction. In corrosion, of course, this rate should be as slow as possible. Because these processes cannot be observed directly on an atomic scale, it is necessary to infer possible mechanisms from indirect measurements and observations. Examples are the rate of change in weight or dimensions, the rate of buildup of corrosion products in the environment, changes in surface appearance examined by optical or electron microscopy, or changes in mechanical or physical properties. When electrochemical corrosion is occurring, mechanisms may be inferred from measurements of electrical potential and current.

Considering engineering materials as metals, polymers, and ceramics, transport of mass across the interface to the environment may be broadly considered as electrochemical, chemical, or physical. Since electrochemical corrosion involves the release of ions to the environment and movement of electrons within the material, this mechanism can occur only if the environment can contain ions and the material can conduct electrons. The most important case of electrochemical mechanisms is the simple corrosion of metals in aqueous solutions, where atoms at the surface of the metal enter the solution as metal ions and electrons migrate through the metal to a site where, to sustain the reaction, they are consumed by species in contact with the metal. In more complicated cases, the metal ions move into solution by forming complex ions, or they combine with other species in the solution and precipitate compounds such as hydroxides, oxides, or sulfides. At sufficiently high temperatures, metals corrode in gases, particularly oxygen to form oxides. Whereas the mechanism in this case appears to be one of direct chemical attack, the mechanism may still be electrochemical in nature, with ions and electrons moving in the oxide which acts as the electrolyte supporting the electrochemical mechanism.

Polymeric and ceramic materials generally do not support electron conduction and hence corrode by either direct chemical or physical mechanisms. Chemical attack of polymers breaks bonds responsible for the properties of these materials, resulting in changes of molecular structure, possible transfer of material to the environment, and degradation of properties. In the case of chemical attack of ceramic materials, the composition of the environment may cause the ceramic or components in the ceramic to either become soluble or to be changed into soluble corrosion products. An example is the attack of sulfurous and sulfuric acid on limestone. Corrosion by direct chemical attack often results in the material being transported into the environment—polymers in certain organic solvents or metals in liquid metals. Direct physical attack often is the result of the mechanical action of the environment, which can remove protective films or actually disintegrate the material by intense local forces. Thus, cavitation corrosion results from the forces of collapsing vapor bubbles in a liquid impinging on the surface of the material. If the environment contains suspended matter, abrasive wear may cause a form of failure classified as erosion-corrosion.

In the present treatment, the fundamental mechanisms involved in aqueous electrochemical corrosion of metals and alloys and the effects of direct chemical and physical processes will be emphasized.

## Electrochemical Corrosion Processes and Variables

Before examining in detail the theories of aqueous corrosion processes and the bases for making quantitative calculations of corrosion rates, it will be useful to develop qualitatively the major phenomena involved. The following sections review several general types of metal/corrosive-environment combinations, the chemical reactions involved, idealized mechanisms for the transfer of metal ions to the environment, and the electrochemical processes occurring at the interface between the metal and the aqueous environment.

### *Uniform Corrosion with pH as the Major Variable*

For metals, M, that are thermodynamically unstable in water, the simplest corrosion reactions are:

$$M + mH^+ \rightarrow M^{m+} + \frac{m}{2} H_2 \qquad \text{at pH} < 7 \qquad \text{(Eq 1.1)}$$

$$M + mH_2O \rightarrow M^{m+} + mOH^- + \frac{m}{2} H_2 \qquad \text{at pH} \geq 7 \qquad \text{(Eq 1.2)}$$

Thus, the metal passes from the metallic state to ions of valence m in solution with the evolution of hydrogen. The reaction is considered to be directly with hydrogen ions in acid solution and progressively with water molecules as the pH increases to neutral and alkaline conditions. Two processes are involved in the reaction, with each involving a change in charge: M to $M^{m+}$ and $mH^+$ to m/2 $H_2$ (in acid solution). The changes in charge are accomplished by electron transfer from M to $H^+$. Because the metallic phase is an electron conductor, it supports the electron transfer, allowing the two processes to occur at separate sites on the metal surface. In limiting cases, these processes occur within a few atom diameters on the surface with the sites constantly changing with time, thus producing uniform corrosion. Otherwise, the corrosion is nonuniform. Uniform corrosion supported by pH is represented schematically in Fig. 1.1. In this example, oxygen is excluded by a nitrogen gas purge and overblanket.

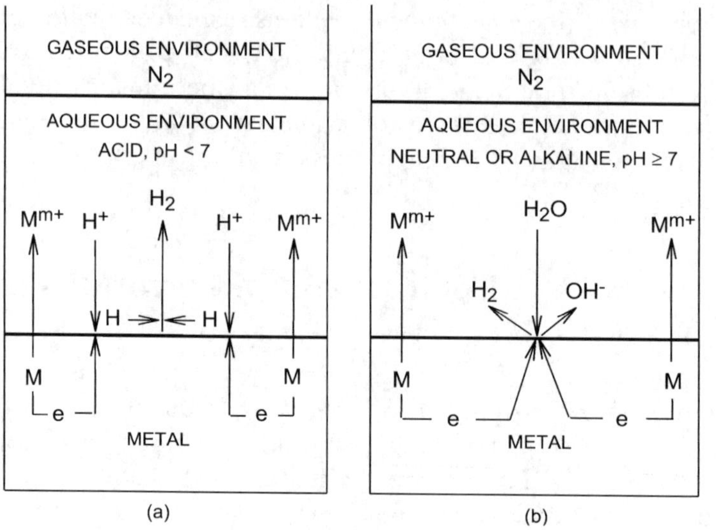


**Fig. 1.1** Uniform corrosion supported by controlled pH (oxygen excluded, deaerated). (a) Acid, pH < 7. (b) Neutral or alkaline, pH ≥ 7

## *Uniform Corrosion with pH and Dissolved Oxygen as Variables*

When dissolved oxygen is present in the solution, usually from contact with air (aerated environment), the following reactions apply *in addition* to those just considered:

$$M + \frac{m}{4} O_2 + mH^+ \rightarrow M^{m+} + \frac{m}{2} H_2O \quad \text{at pH} < 7 \qquad \text{(Eq 1.3)}$$

$$M + \frac{m}{4} O_2 + \frac{m}{2} H_2O \rightarrow M^{m+} + mOH^- \quad \text{at pH} \geq 7 \qquad \text{(Eq 1.4)}$$

Uniform corrosion supported by dissolved oxygen and pH is represented schematically in Fig. 1.2. Since electrons are now consumed by two reactions, the rate of corrosion of the metal increases. In the case of iron, dissolved oxygen is more important in supporting corrosion than the presence of hydrogen ions when the pH is greater than approximately 4. This is an initial illustration of the role of dissolved oxygen (aeration of solutions) in corrosion.

## *Uniform Corrosion with Corrosion Product Formation*

An example of corrosion product formation is the rusting of iron as illustrated in Fig. 1.3. When the pH is greater than approximately 4, and under aerated conditions, a layer of black $Fe_3O_4$, and possibly $Fe(OH)_2$, forms in contact with the iron substrate. In the presence of the dissolved oxygen, an outer layer of red $Fe_2O_3$ or FeOOH forms. The adherence

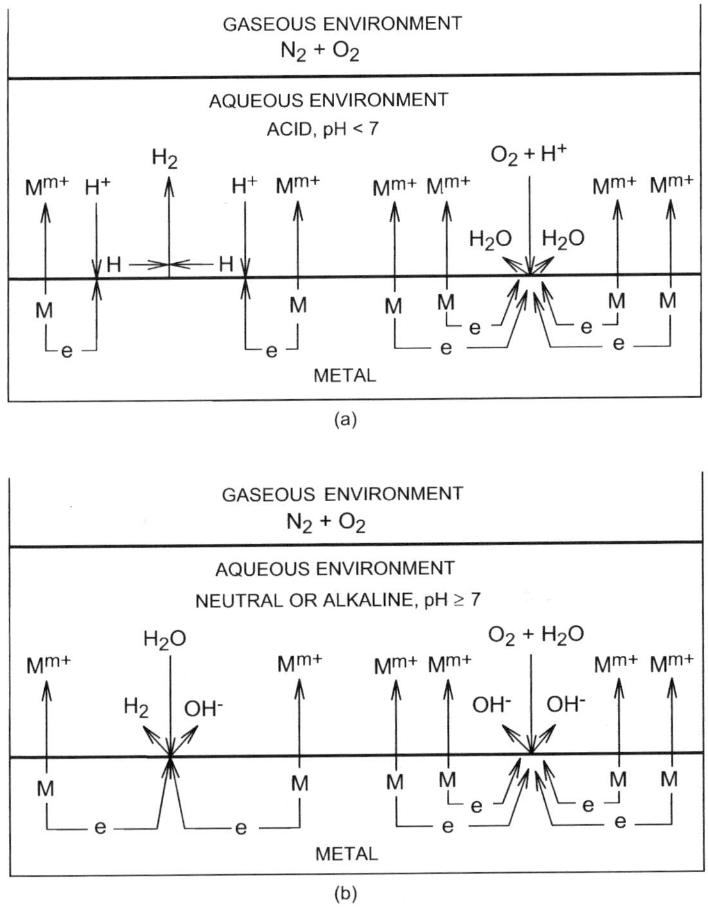


**Fig. 1.2** Uniform corrosion supported by pH and dissolved oxygen (aerated). (a) Acid, pH < 7. (b) Neutral or alkaline, pH ≥ 7

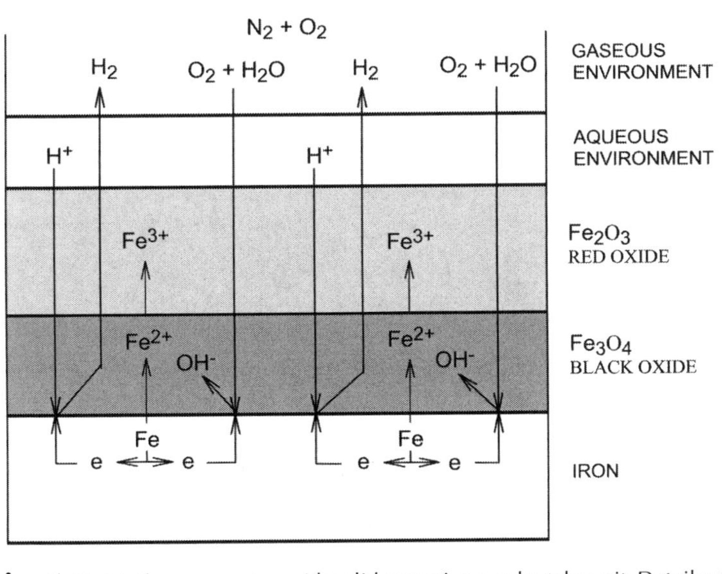


**Fig. 1.3** Uniform corrosion with solid corrosion product deposit. Details of the formation of oxide species are not considered at this point.

and porosity of these layers change with time and can be influenced by other chemical species in the environment, such as chloride and sulfate ions. In any case, the formation of the corrosion product layer influences the corrosion rate by introducing a barrier through which ions and oxygen must diffuse to sustain the corrosion process.

## *Some Basic Terminology, Reactions, and Variables in Aqueous Corrosion*

The basic corrosion process is represented in Fig. 1.4. In the simplest case, the corrosion reaction is the transfer of metal atoms from the solid to the solution where they exist as ions (i.e., $M \rightarrow M^{m+} + me$). Because there is a loss of electrons from the metal atom in this transfer, the metal has undergone oxidation. The oxidation is sustained by the consumption of the electrons by another reaction, generalized in this case as $X^{x+} + xe \rightarrow X$. The oxidation occurs at a site on the metal surface referred to as the anodic reaction site and is the location of the loss of metal by corrosion. The electrons are picked up at a cathodic reaction site. The areas over which the anodic and cathodic reactions occur individually vary greatly and may extend from positions a few atom distances apart on the surfaces to microscopic areas, and even to macroscopic areas extending to hundreds of square meters. When the sites are so close together that they cannot be distinguished, and when the sites undergo changes and reversals with time, uniform corrosion is said to occur. With resolvable areas and/or with anodic and cathodic sites that do not change with time, the corrosion will be largely identified by the anode areas only, and localized corrosion is said to occur. Obviously, there are large differences in interpretation of what is uniform corrosion and what is localized corrosion. It frequently depends on the scale of obser-

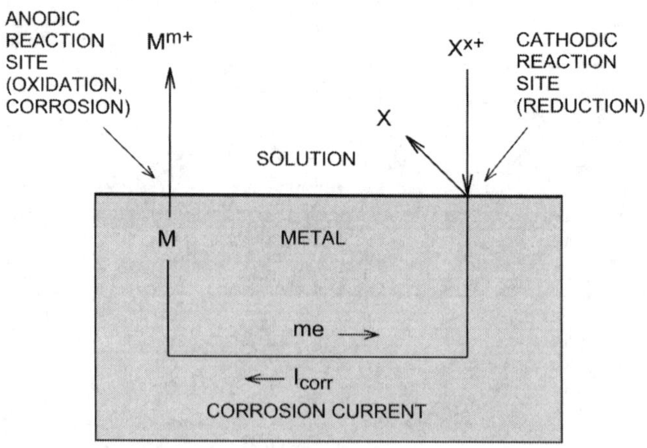


**Fig. 1.4** The basic corrosion process

vation, or the magnitude of the difference in corrosion rate between areas that are predominantly anodic and areas that are predominantly cathodic because both reactions often occur over the entire surface. If the two processes are occurring on a microscale, then the anodic and cathodic areas are considered the same and equal to the total area, A. If the two processes are occurring over separate areas, an anodic reaction area, $A_a$, is distinguished from a cathodic reaction area, $A_c$.

For a specific example, such as the corrosion of iron in an aerated acid solution, the net reaction due to acidity is:

Anodic reaction:

$$Fe \rightarrow Fe^{2+} + 2e \quad \text{(Eq 1.5)}$$

Cathodic reaction:

$$2H^+ + 2e \rightarrow H_2 \quad \text{(Eq 1.6)}$$

Overall reaction:

$$Fe + 2H^+ \rightarrow Fe^{2+} + H_2 \quad \text{(Eq 1.7)}$$

and the reaction due to dissolved oxygen is:

Anodic reaction:

$$Fe \rightarrow Fe^{2+} + 2e \quad \text{(Eq 1.8)}$$

Cathodic reaction:

$$\frac{1}{2}O_2 + 2H^+ + 2e \rightarrow H_2O \quad \text{(Eq 1.9)}$$

Overall reaction:

$$Fe + \frac{1}{2}O_2 + 2H^+ \rightarrow Fe^{2+} + H_2O \quad \text{(Eq 1.10)}$$

To show that these reactions actually proceed to the right (i.e., to show that corrosion actually occurs), it is necessary to calculate the Gibbs free-energy change and find that it is negative. To make this calculation requires quantitative information on the activity or effective concentration of iron ions ($a_{Fe^{2+}}$) in the solution, the acidity, or pH, and the concentration of dissolved oxygen that is related to the partial pressure of the oxygen, $P_{O_2}$, in contact with the solution. It is demonstrated in the following chapter that the change in the Gibbs free energy is negative for these reactions at all values of pH, and hence, iron tends to corrode at all pH values. The rate of corrosion, however, depends on factors influencing the kinetic mechanisms of the several processes involved in the transport of ions from metal to solution and in the supporting cathodic reactions. In addition to the species in solution relating directly to the above reactions ($Fe^{2+}$, $H^+$, and $O_2$), other species in solution can affect both the tendency to corrode in terms of thermodynamic driving forces and the kinetics of the several steps involved. For example, complexing agents reacting with metal ions in solution reduce the concentration of free metal ions and make it more favorable thermodynamically for metal ions to pass into solution, thereby increasing the corrosion rate. Conversely, if species in solution can form precipitates with metal ions and form protective diffusion barriers at the interface, corrosion rates may be decreased significantly.

The important processes, terminology, and variables associated with the anodic and cathodic reactions, and which characterize the environment, are summarized in Table 1.1.

**Table 1.1 Summary of processes, terminology, and variables associated with aqueous corrosion(a)**

**Anode**

- Area, $A_a$
- Reactions (oxidation)
  - General, $M \rightarrow M^{m+} + me$
  - Reduced state → oxidized state
  - Example, $Fe \rightarrow Fe^{2+} + 2e$

**Cathode**

- Area, $A_c$
- Reactions (reduction)
  - General, $X^{x+} + xe \rightarrow X$
  - Oxidized state → reduced state
  - Examples
    - Deaerated
      - Acid, $H^+ + e \rightarrow \frac{1}{2} H_2$
      - Neutral or alkaline
        - $H_2O + e \rightarrow \frac{1}{2} H_2 + OH^-$
    - Aerated (additive to above)
      - Acid, $O_2 + 4H^+ + 4e \rightarrow 2H_2O$
      - Neutral or alkaline
        - $O_2 + 2H_2O + 4e \rightarrow 4OH^-$

**Aqueous phase variables**

- Acidity
  - $H^+$ concentration
  - $C_{H^+}$, molal concentration
  - $a_{H^+}$, activity
  - $pH = -\log a_{H^+}$
  - $(a_{H^+})(a_{OH^-}) = 10^{-14}$
- Dissolved gases
  - $H_2$, $C_{H_2} \propto P_{H_2}$
  - $O_2$, $C_{O_2} \propto P_{O_2}$

**Other dissolved species**

$Fe^{2+}$, $Cl^-$, $SO_4^{2-}$, etc., with activities $a_{Fe^{2+}}$, etc.

Note: $C_Z$ = Molal concentration of species Z; $a_Z$ = Activity or effective concentration of species Z; $P_Z$ = Partial pressure of species Z. (a) Figure 1.4 shows a schematic representation of the interrelationships of the processes characterized in this table.

## The Elementary Electrochemical Corrosion Circuit*

Aqueous corrosion is most readily understood in terms of a "dead-shorted" battery or electrochemical cell consisting of two half cells (Fig. 1.5). In comparison with the battery, the solution or electrolyte above the corroding metal is the battery fluid, and the metallic path between the anodic site (exposed metal) and the cathodic site (for example, an area of adherent-conducting oxide) is the external circuit. At the anodic site, the net oxidation reaction is $M \rightarrow M^{m+} + me$, and at the cathodic site, the generalized net reduction reaction is $X^{x+} + xe \rightarrow X$. As a consequence of the transfer of ions and electrons at each interface, differences in electrical potential, $\Delta\phi_a$ and $\Delta\phi_c$, develop between the metal and the solution at the anodic and cathodic sites, respectively, where

$$\Delta\phi_a = \phi_{M,a} - \phi_{S,a} \quad \text{(Eq 1.11)}$$

$$\Delta\phi_c = \phi_{M,c} - \phi_{S,c} \quad \text{(Eq 1.12)}$$

The subscripts a and c designate the anodic and cathodic sites, and the subscripts M and S designate the metal and solution phases. These differences in potential, coupled as shown, constitute the electrochemical cell in which electrons are caused to flow from the anodic to the cathodic site in the metal; conventional electrical current (positive charge) flows in the opposite direction. In the solution, current flows from the anodic to the cathodic site as a consequence of the potential in the solution being

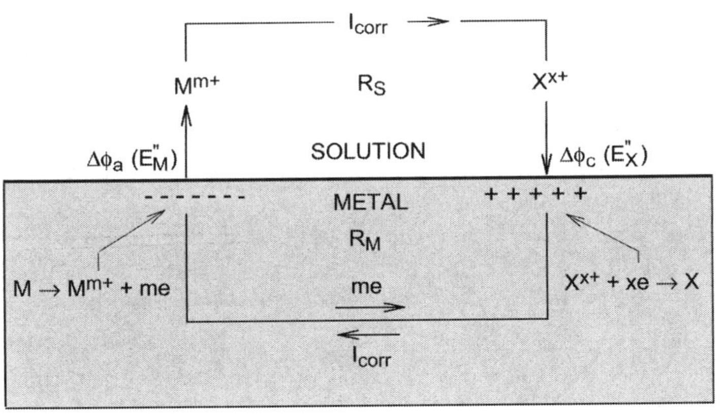


**Fig. 1.5** The elementary electrochemical corrosion circuit

* The following section provides a qualitative insight into the essentials of the corrosion process. Important factors such as current distributions, nonuniform metal and environment compositions, and finite resistance of the metal are considered later in the text.

higher above the anodic site than above the cathodic site; that is, $\phi_{S,a} > \phi_{S,c}$. This current is defined as a positive quantity for the spontaneous corrosion process represented in Fig. 1.5. In practice, individual interface differences in potential, $\Delta\phi$, are assigned values relative to the standard hydrogen electrode as discussed in the next chapter. In this text, these values are designated by E for the general case, by E′ for the case of no current passing, and by E″ for the case of a corrosion current passing the interface. If the potential of the standard reference electrode is taken as zero, then for the general case, $\Delta\phi_a = E_M$ and $\Delta\phi_c = E_X$.

The driving potential for the current in the solution, $\Delta\phi_S$, is:

$$\Delta\phi_S = \phi_{S,a} - \phi_{S,c} = (\phi_{M,a} - \Delta\phi_a) - (\phi_{M,c} - \Delta\phi_c) \quad \text{(Eq 1.13)}$$

If it is assumed that the metal path is a good conductor (as is the general case), then the potential difference in the metal will be small, and $\phi_{M,a} \approx \phi_{M,c}$. The driving potential for the current in the solution, using Eq 1.13, is then:

$$\Delta\phi_S = \Delta\phi_c - \Delta\phi_a = E''_X - E''_M \quad \text{(Eq 1.14)}$$

where the Es are now double primed to emphasize their values associated with the corrosion current. Recognizing that Ohm's law must apply, the corrosion current is given by:

$$I_{corr} = (E''_X - E''_M)/(R_S + R_M) \quad \text{(Eq 1.15)}$$

where $R_S$ and $R_M$ are the resistances of the solution and metal paths of the current. This current is called the corrosion current, $I_{corr}$, and when the area of the anode through which the current flows is taken into consideration, the corrosion penetration rate can be calculated, for example, in micrometers or mils (0.001 in.) per year. The total path resistance, $R_S + R_M$, is obviously an important variable in determining the corrosion rate. In addition, if high-resistance interface films form, the total circuit resistance, $R_S + R_M + R_{interface}$, increases, and the corrosion rate decreases.

The relative sizes and locations of anodic and cathodic areas are important variables affecting corrosion rates. As stated previously, these areas may vary from atomic dimensions to macroscopically large areas. In Fig. 1.6, areas have been depicted over which the anodic and cathodic reactions occur, designated as $A_a$ and $A_c$. If the current is uniformly distributed over these areas, then the *current densities*, $i_a = I_a/A_a$ and $i_c = I_c/A_c$, may be calculated.* The current density is fundamentally more

* Actually, the current will not be uniformly distributed. Rather, the current density near the anode/cathode junction will be higher, and hence, the corrosion rate will be higher because resistance of a current path is smaller here and increases with distance from the junction.

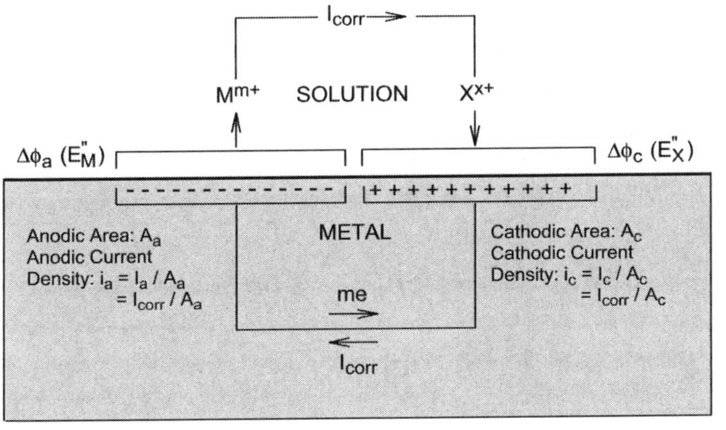


**Fig. 1.6** Relationships between anodic and cathodic areas, current densities, and potentials

important than the current for two reasons. First, through Faraday's law, the anodic current density, $i_a$, relates directly to *corrosion intensity* as mass loss per unit time per unit area, or to *corrosion penetration rate* as a linear dimension loss per unit time. Second, it is observed that interface potentials, E, are functions of current density, E(i), of the form:

$$E_X(i_c) = E'_X + \eta_X(i_c) = E'_X + \eta_X(I_c/A_c) \quad \text{(Eq 1.16)}$$

$$E_M(i_a) = E'_M + \eta_M(i_a) = E'_M + \eta_M(I_a/A_a) \quad \text{(Eq 1.17)}$$

In these expressions, $E_X$ and $E_M$ become the potentials $E'_x$ and $E'_M$ if the current is zero and, therefore, relate to the potential differences across the individual interfaces at equilibrium (i.e., no net transport of ions or electrons). These limiting potentials are referred to as *equilibrium half-cell potentials,* and when conditions of concentration and temperature are standardized, they characterize the *standard equilibrium half-cell reactions* to which they relate. Equations 1.16 and 1.17, therefore, indicate that the existing potential with current flow is the equilibrium value plus a term, $\eta(i)$, representing the shift in potential resulting from the current density. This shift is referred to as *overpotential* (or overvoltage) and increases in magnitude with increasing current density. During corrosion, the anodic current must equal the cathodic current, $I_a = I_c$, and this current is the corrosion current, $I_{corr}$. Thus, Ohm's law can be written as:

$$I_{corr} = \frac{E''_X - E''_M}{R_{total}} = \frac{[E'_X + \eta_X(I_{corr}/A_c)] - [E'_M + \eta_M(I_{corr}/A_a)]}{R_{total}} \quad \text{(Eq 1.18)}$$

where $E''_X$ and $E''_M$ are now the potentials when the cathodic and anodic reactions are coupled. If theoretically or experimentally based expressions for the polarized potentials, Eq 1.16 and 1.17, are available, the Ohm's law equation can be solved for the corrosion current, $I_{corr}$. $I_{corr}$ is a measure of the *total* loss of metal from the anode surface during corrosion. The anodic current density during corrosion, $i_{corr} = I_{corr}/A_a$, is a measure of the corrosion intensity from which the corrosion penetration rate can be calculated.

## Criteria for Metal/Aqueous-Environment Reactions: Corrosion

For the current to flow in the direction shown in Fig. 1.6, corresponding to the corrosion of M, $E''_X$ must be greater than $E''_M$. Because $\eta_X$ is always negative and $\eta_M$ always positive (as shown in Chapter 4), $E'_X$ must be greater than $E'_M$, and because these equilibrium potentials can be calculated from tables of standard equilibrium half-cell potentials, these tables are useful for establishing whether corrosion can occur. The corrosion rate, however, is also strongly dependent on both $\eta_X$ and $\eta_M$; $\eta_X$ is a function of the kinetic mechanisms of the physical, chemical, and electrochemical processes occurring at the cathode surface; $\eta_M$ relates to kinetic processes at the anode surface. It is essential, therefore, to realize that processes of corrosion, particularly the rate of corrosion, depend on both the anodic and cathodic reactions. In some cases, the anodic process will control, and in other cases, the cathodic process will control the corrosion rate. Conversely, in attempting to control corrosion by additives called corrosion inhibitors, control may be directed selectively to either the cathodic or anodic, or both, kinetic mechanisms. Obviously, it is important to understand the steps in each process as completely as possible.

## Comments on Cathodic Reactions

The corrosion of a metal, a process of oxidation or loss of electrons, is supported by a cathodic reactant or oxidizing agent, which is reduced in performing the cathodic reaction. In general, the stronger the oxidizing reaction is, thermodynamically and kinetically, the greater the induced corrosion rate will be.

The cathodic reaction has been generalized in the form $X^{X+} + xe \rightarrow X$. Representative specific cathodic reactions are classified in Table 1.2 along with the standard equilibrium half-cell potentials, $E^o$, relative to the standard hydrogen electrode (SHE), where $E^o_{H_2,H^+} \equiv 0$. The variables that must be set to correct the standard potentials, $E^o$, to values

**Table 1.2 Cathodic reactions and equilibrium potentials**

| Examples of cathodic reactions | Standard equilibrium half-cell potentials(a), $E^o$ (mV vs. SHE) | Variables required for correction of $E^o$ to $E'$ |
|---|---|---|
| **Oxidation due to $H^+$ ions or water** | | |
| $H^+ + e = \frac{1}{2}H_2$ pH < 7 | 0 | $a_{H^+}$ (pH), $P_{H2}$ |
| $H_2O + e = \frac{1}{2}H_2 + OH^-$ pH ≥ 7 | −820 | $a_{OH^-}$ (pH), $P_{H2}$ |
| **Oxidation due to dissolved oxygen** | | |
| $O_2 + 4H^+ + 4e = 2H_2O$ pH < 7 | +1,229 | $a_{H^+}$ (pH), $P_{O2}$ |
| $O_2 + 2H_2O + 4e = 4OH^-$ pH ≥ 7 | +401 | $a_{OH^-}$ (pH), $P_{O2}$ |
| **Oxidation due to change in valence of ionic species** | | |
| $Fe^{3+} + e = Fe^{2+}$ | +771 | $a_{Fe^{3+}}$, $a_{Fe^{2+}}$ |
| **Oxidation due to reaction to the metallic state** | | |
| $Cu^{2+} + 2e = Cu$ | +342 | $a_{Cu^{2+}}$ |
| **Oxidation due to "oxidizing" anion radical** | | |
| Dichromates $Cr_2O_7^{2-} + 14H^+ + 4e = 2Cr^{3+} + 7H_2O$ | +1,333 | $a_{Cr_2O_7^{2-}}$, $a_{Cr^{3+}}$, $a_{H^+}$ (pH) |
| Nitrites $NO_2^- + 8H^+ + 6e = NH_4^+ + 2H_2O$ | +890 | $a_{NO_2^-}$, $a_{NH_4^+}$, $a_{H^+}$ (pH) |
| Nitric acid: $2H^+ + NO_3^- + 2e = NO_2^- + H_2O$ | +940 | $a_{NO_3^-}$, $a_{NO_2^-}$, $a_{H^+}$ (pH) |

(a) It should be noted that all of these potentials, except for the reduction of water, are relatively positive, which reflects that they tend to be oxidizing and involve oxidizing agents that are reduced by the reaction. These standard values correspond to 25° C and to unit activity of the species and would need to be corrected for the actual temperature and activities.

that they would have under the actual equilibrium conditions, E′, are also given.

## Comments on Anodic Reactions

The anodic or corrosion half-cell reaction has been generalized as $M \rightarrow M^{m+} + me$. The previously presented schematic representations of anodic corrosion processes immediately raise three questions:

- What is the particular metal or alloy constituting the anode?
- What governs the positions on metal surfaces at which metal ions transfer from the metallic phase to the solution phase?
- What governs the rate at which the transfer occurs?

A pure metal can be anodic only if its equilibrium half-cell potential, $E'_M$, is less than the half-cell potential of some cathodic reaction, $E'_X$, such that the total cell potential ($E''_X - E''_M$) causes current to flow as in Fig. 1.6, that is, current away from the anode area as ions in the solution. A few representative anodic reactions are listed in Table 1.3 along with their standard equilibrium half-cell potentials.

For any specific pure metal, the physical state or condition may also influence the tendency for the metal to become anodic and corrode.

**Table 1.3 Anodic reactions and equilibrium potentials**

| Examples of anodic reactions | Standard equilibrium half-cell potentials(a), E° (mV vs. SHE) |
|---|---|
| $Zn = Zn^{2+} + 2e$ | −763 |
| $Fe = Fe^{2+} + 2e$ | −440 |
| $Pb = Pb^{2+} + 2e$ | −126 |
| $Cu = Cu^{2+} + 2e$ | +342 |
| $Ag = Ag^{+} + e$ | +799 |

(a) These standard values correspond to 25 °C and unit activity of the metal ions and would need to be corrected for the actual temperature and activity to determine E′.

These variables include the amount of general or localized cold working (e.g., scratches); the presence of imperfections such as dislocations and grain boundaries, the latter making grain size a variable; and crystal orientation. The latter becomes a variable because different crystal faces exposed to the environment have different arrangements of atoms and, hence, different tendencies to react with the environment.

When metals are combined to form alloys, it is no longer possible to define a unique half-cell potential, nor to calculate whether corrosion is possible, to the same extent that this calculation can be made for pure metals. Obviously, the response of an alloy to a corrosive environment depends on the kinds and amounts of alloying elements added to a given base metal. Solid-solution-type alloys tend to segregate alloying elements during solidification, and as a consequence, cast shapes, ingots, and even fabricated products, such as pipe and plates, may corrode in localized regions. Solidification segregation may be a particular problem leading to the corrosion of weldments. In most of these cases, heat treatments to remove the segregation are uneconomical. In multiphase alloys, different phases may act as relative anodes and cathodes. For all alloys, conditions affecting the physical state, such as cold work and grain boundaries, also may be significant.

## Corrosion Considerations Based on Relative Cathodic and Anodic Equilibrium Potentials

The initial consideration in analyzing an existing or proposed metal/environment combination for possible corrosion is determination of the stability of the system. According to Eq 1.18, the criterion is whether the equilibrium half-cell potential for an assumed cathodic reaction, $E'_X$, is greater than the equilibrium half-cell potential for the anodic reaction, $E'_M$. A convenient representation of relative positions of equilibrium half-cell potentials of several common metals and selected possible corrodent species is given in Fig. 1.7. To the left is the scale of potentials in millivolts relative to the standard hydrogen electrode (SHE). The solid vertical lines identified by the name of the metal give

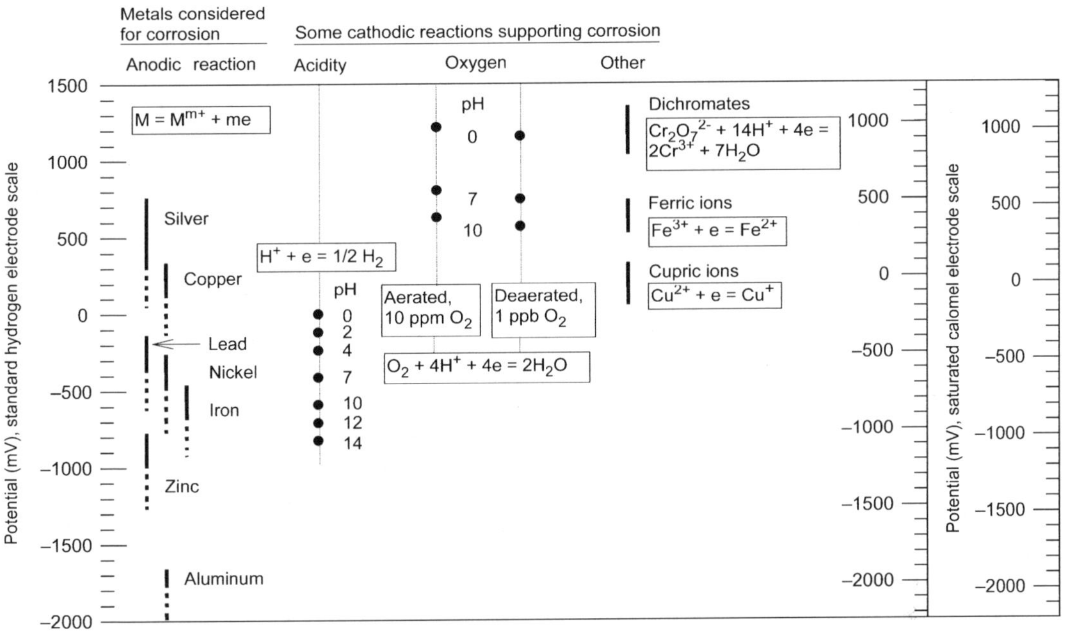


**Fig. 1.7** Ranges of half-cell potentials of some electrochemical reactions of importance in corrosion. Vertical bars represent metal ion concentration of 1 molal (approximately 10%) down to 1 ppm. Dashed extensions may apply with precipitated and complexing species. The hydrogen and oxygen reactions depend on both pH and pressure of the gases. Values for the hydrogen are at one atmosphere pressure. Values for oxygen are for water in contact with air (aerated) giving 10 ppm dissolved oxygen and for water deaerated to 1 ppb dissolved oxygen.

the range of half-cell potentials for the metal, extending from the potential at unit concentration of metal ions (1 mole per 1000 g of water) at the top to a concentration of about 1 ppm by weight at the bottom of the solid line. The dotted extensions to lower potentials apply when precipitating or complexing agents are present that reduce the metal ion concentration below 1 ppm. Reactions that might support corrosion involve hydrogen ions, dissolved oxygen, and ferric, cupric, and dichromate ions. The potential of the hydrogen ion reaction depends on pH and is given for the pH range of 0 to 14. The potential of the oxygen reaction depends on pH and dissolved oxygen concentration. Potentials are given for pH values of 0, 7, and 10 at 10 ppm dissolved oxygen, the approximate concentration of an aqueous solution in contact with air, and 1 ppb dissolved oxygen, an approximation to the deaerated condition. The other ions will have a range of potentials depending on concentration as shown by the solid vertical lines on the right.

The information in Fig. 1.7 allows quick estimation of the stability of a metal/environment combination. Thus, if the potential for a possible cathodic reaction is determined and found to be greater than that for the half-cell reaction of the metal being examined, then $[E'_X - E'_M]$ is positive, and according to Eq 1.18, the current flow induced will be positive and, therefore, corrosion will be expected. An example would be iron in contact with a completely deaerated aqueous environment at pH = 2 (all oxygen excluded; values can be found under the column "Acidity") and containing $Fe^{2+}$ ions at a concentration of 1 ppm. The difference in potential will be $[E'_X - E'_M] = -120 - (-670) = +550$ mV, and iron should undergo corrosion at pH = 2, as in fact it does.

It is emphasized that while following the above procedure to determine whether a metal/environment combination is susceptible to corrosion, no information is provided on the rate of corrosion, the physical nature of the attack (i.e., uniformity of attack), the influence of corrosion products, or factors relating to the environment, such as fluid velocity and uniformity of fluid composition.

## Importance of Solid Corrosion-Product Formation: Corrosion Acceleration Versus Passivation

The formation of solid corrosion products may be a dominant factor in controlling corrosion. These products form when the metal ions passing into solution (corrosion) reach a critical concentration, causing precipitation with some species in the environment. Since the metal-ion concentration is greatest at the surface where transfer is occurring across the metal-solution interface, the precipitate tends to form at or near the surface of the metal. Common solid corrosion products are hydroxides, oxides, sulfides, or complex mixtures of these. If the precipitate does not adhere to the surface, and the solubility is very small, the precipitation process will maintain the metal-ion concentration at a low value, and the corrosion rate will be high due to the continual removal of metal ions from solution and the resulting driving force to compensate for this removal by transfer of ions from the metal to the solution.

In contrast to the above, precipitates that adhere to the metal surface as continuous, nonporous films greatly reduce corrosion rates because the controlling mechanism becomes the slow solid-state diffusion of ions through the films. Further, if the film is a poor conductor of electrons, then the oxidation (corrosion) reaction is retarded because electrons have difficulty reaching the solution interface to enter into the cathodic reaction.

As discussed at some length in this introduction, metals corrode as a consequence of species in solution supporting a cathodic reaction (i.e., accepting electrons released at the corrosion sites where metal ions are discharged into the solution). The cathodic reactant is acting as an oxidizing agent oxidizing the metal from $M^o$ to $M^{m+}$ with the transfer of electrons to the cathodic reactant, which is reduced. The more positive the cathodic-reactant half-cell potential (Fig. 1.7) and the greater the concentration, the greater is the oxidizing power of the environment and, therefore, the tendency for corrosion to occur. However, for those metals capable of forming protective corrosion-product films, such films are observed to form at critical oxidizing conditions, and once formed, the corrosion rate may decrease by several orders of magnitude. When this occurs, the metal is described as having undergone *passivation.* That is, it becomes passive to its environment rather than, as might be expected, progressively more active with increasingly aggressive properties of the environment. The phenomenon can be represented by a schematic plot of corrosion rate as a function of oxidizing power of the environment as shown in Fig. 1.8. The shape and position of the curve depends on the particular metal or alloy and a number of environmental factors, such as acidity (pH), temperature, and the presence of a number of nonoxidizing anions, particularly the chloride ion. Obviously, a metal or alloy should be selected that will form a passive protective film in the environment in which it is used. Consideration also should be given to adjustments in the environmental conditions to provide oxidizing conditions that will form the passive film on the metal surface.

For some materials in some environments, it is not possible to form passive films for corrosion protection. In this case, the corrosion rate continues to increase with increasing oxidizing conditions, and satisfactory use of materials of this type depends upon maintaining acceptably low oxidizing conditions and, therefore, acceptably low corrosion

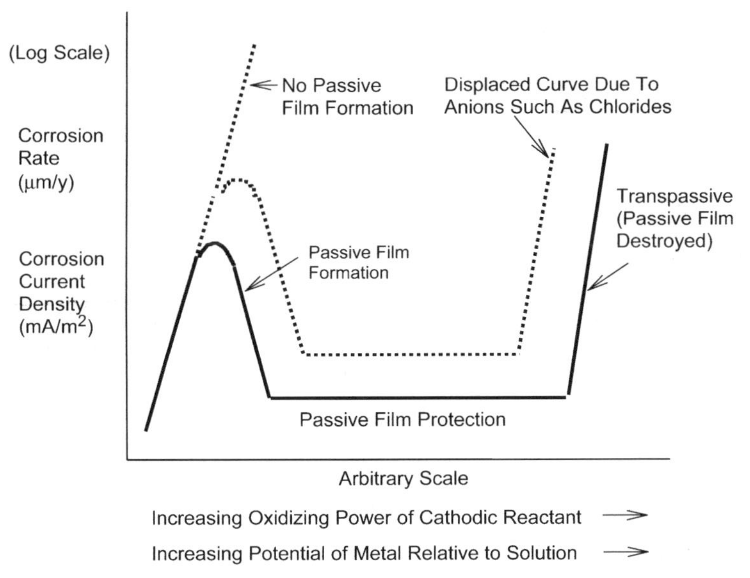


**Fig. 1.8** Schematic representation of the effect of increasing oxidizing power of the environment on the corrosion of an active-passive type alloy such as stainless steel

rates. The best example of corrosion control based on these general observations is the deaeration of water in heat transfer loops to reduce the dissolved oxygen, which is the principal cathodic reactant.

Iron does not passivate in most environments and, therefore, performs best when the oxidizing power of the environment is as low as possible, for example, by deaeration as mentioned above. In contrast, a large class of industrially important alloys depend upon sufficiently oxidizing conditions to produce a protective passive film if they are to perform satisfactorily. These alloys include stainless steels, nickel-base alloys, titanium and its alloys, and many others.

## Chapter 1 Review Questions

1. Give four examples of the economic significance of the control of corrosion.
2. Show schematically the processes involved in the corrosion of a metal, M, in a simple acid ($pH < 7$) and in a neutral or alkaline ($pH \geq 7$) environment in both deaerated and aerated conditions.
3. For the case of an aerated alkaline environment, list the reasonably possible electrochemical, chemical, and physical (diffusion, electron conduction) steps in the total corrosion process.
4. Under what circumstances can the formation of insoluble corrosion products (a) increase corrosion and (b) decrease corrosion?
5. The current given by the Ohm's law expression (Eq 1.18) is the total current referred to as $I_{corr}$. Later in the course, considerable significance is given to the fact that $I_{corr} = I(\text{cathode}) = I(\text{anode})$. Why will it always be necessary to equate $I_c = I_a$?
6. In calculating corrosion rates, the anodic current density should be evaluated as $i_a = I_{corr}/A_a$. Why?
7. Relative to question 6, give another reason why current density is fundamentally more important than current.
8. In a corroding system involving distinguishable anodic and cathodic areas, which is more desirable, (a) a large $A_a/A_c$ area ratio or (b) a small $A_a/A_c$ area ratio? Explain.
9. In Eq 1.18, for corrosion to occur, $I_{corr}$ must be positive, or $E'_X$ must be greater than $E'_M$. On this basis, which of the cathodic reactions listed in Table 1.2 should support the corrosion of copper (see Table 1.3)? Assume standard conditions such that $E' = E^o$.
10. As discussed in the text, in reacting electrochemical systems (corroding), the values of $E''_X$ and $E''_M$ depend upon current density (Eq 1.18).
    a. When corrosion is occurring, is it desirable for $\eta_M$ and $\eta_X$ to be weak or strong functions of the current density? Explain.

b. Comment on "a" for electrochemical reactions in a battery.

11. List at least eight conditions relating to a metal or alloy and/or its environment that could cause localized regions on the surface to become anodic and result in localized corrosion.
12. Plain carbon steels may be heat treated to have dispersions of small, round, isolated iron carbides in the continuous iron matrix. The amount of carbide is usually less than 10% of the structure. With two-phase alloys such as this, the carbide may become anodic in some environments and cathodic in others. Predict the progress of corrosion if the carbide is (a) anodic and (b) cathodic. Be reasonably specific in describing changes at the surface.
13. With reference to question 12, predict the corrosion behavior if the carbide is in the form of a continuous thin film between the grains.
14. If an alloy can be passivated, is it generally desirable to have oxidizing conditions in the environment? Explain.
15. If an alloy does not form passive films, is it generally desirable to have minimum oxidizing conditions in the environment? Explain.

CHAPTER **2**

# Electrochemical Thermodynamics: The Gibbs Function, Electrochemical Reactions, and Equilibrium Potentials

## Decrease in the Gibbs Function as a Condition for Spontaneous Reaction*

The first law of thermodynamics may be written as:

$$dU = q - w \quad \text{(Eq 2.1)}$$

where dU is an incremental change in the internal energy during a process associated with heat absorbed, q, and work done, w, by the system. If the process is conducted reversibly, the heat absorbed is TdS, where T is the absolute temperature, and dS is the change in entropy associated

*A general introduction to chemical thermodynamics, including electrochemical cells, can be found in Ref 1.

with the process. It is useful to consider the work term as divided between PdV work, associated with volume changes of the system doing work on or receiving work from the surrounding atmosphere, and other work, considered here as electrical, which will be designated as w′.

The reversible case is then written as:

$$dU = TdS - PdV - w'_r \qquad \text{(Eq 2.2)}$$

which on rearrangement becomes:

$$dU - TdS + PdV = -w'_r \qquad \text{(Eq 2.3)}$$

The left-hand side of this expression is the differential of the function (U – TS + PV) taken at constant T and P; thus, d(U – TS + PV) = dU – TdS + PdV. Thus:

$$d(U - TS + PV) = -w'_r \quad \text{(constant T and P)} \qquad \text{(Eq 2.4)}$$

The expression U + PV naturally arises in thermodynamics and is called the enthalpy, H; thus, H = U + PV. The entire expression U – TS + PV, which was shown to naturally develop by this argument, is called the Gibbs free energy, G:* thus, G = U + PV – TS = H – TS. Thus:

$$dG = -w'_r \qquad \text{(Eq 2.5)}$$

or:

$$-dG = w'_r \quad \text{(reversible, constant T and P)} \qquad \text{(Eq 2.6)}$$

This is true for a reversible process (essentially at equilibrium) carried out at constant T and P. Therefore, under these conditions, the maximum work over and above that associated with the volume change is given by the decrease in the Gibbs free energy (GFE).

In the reversible process, the heat effect, q = TdS, and work against the environment, PdV, are inherently associated with the process. However, the heat effect, q, will be equal to TdS only if the process is reversible. Strictly speaking, PdV will be the work effect against the environment only if the process is reversibly carried out, although from a practical standpoint, reversibility is not as critical for this term as for the heat term. When a process is considered, whether it represents a small (incremental) or large change, definite (definable) initial and final states exist. For each of these states, the thermodynamic variables have definite values characteristic of the state. Thus G, S, V, etc. each undergo specific changes for the system regardless of whether the change

*For convenience, the Gibbs free energy or Gibbs function is indicated by GFE.

is brought about reversibly or irreversibly. The product, TdS, however, which in principle can be calculated, will equal the experimentally observed heat effect only if the process occurs reversibly.

Since, for a given increment of a process, dU is the same whether it is brought about reversibly or irreversibly:

$$dU = q_i - w_i = q_r - w_r \qquad \text{(Eq 2.7)}$$

where the subscripts indicate irreversible and reversible cases. Consider an irreversible process in which no w′ work is actually done (again, it can be done if the process is conducted reversibly).*
Then:

$$dU = q_i - PdV = q_r - PdV - w'_r \qquad \text{(Eq 2.8)}$$

or

$$q_i - q_r = -w'_r \qquad \text{(Eq 2.9)}$$

when work $w'_r$ is done by the system, $w'_r$ will be positive and thus:

$$q_r > q_i \qquad \text{(Eq 2.10)}$$

or

$$q_r > q_i \quad \text{(system)} \qquad \text{(Eq 2.11)}$$

where the second form is used to emphasize that the q's refer to the system or process involved; also, it should be remembered that q is taken as positive for heat absorbed by the system and negative for heat rejected to the environment. Therefore, for systems undergoing reactions that liberate heat (negative q values) (e.g., chemical or electrochemical reactions):

$$|q_i| > |q_r| \quad \text{(system)} \qquad \text{(Eq 2.12)}$$

The conclusion is that more heat is rejected by the system and hence absorbed by the surroundings in the irreversible case. Specifically, the magnitude of the extra heat is that of the work $w'_r$, which could have been realized in a reversible process. Hence, since $-dG = w'_r$, (at constant T and P), $-dG$ is the energy available from the process and represents either useful work if the process is permitted to occur reversibly or extra heat rejected to the environment (which, importantly, can never be used isothermally to do the work otherwise possible). Since this energy,

---

*Many processes, particularly under the conditions of constant T and P, do not involve doing w′ work under reversible or irreversible conditions. The present argument is made under the conditions of constant T and P and also that the process is a chemical or electrochemical reaction.

dG, is not a part of the energy change associated with TdS or PdV, both of which are fixed inherently by the process, it can be said that this energy is spontaneously available and that, from a thermodynamic viewpoint, the process can occur spontaneously. Thus, the condition for a spontaneous process is that $-dG > 0$, or

$$dG < 0 \quad \text{(constant T and P)} \qquad \text{(Eq 2.13)}$$

## Standard Gibbs Free-Energy Change for Chemical Reactions

In chemical thermodynamics, the process of frequent interest is the chemical reaction, abbreviated as:

$$aA + bB \rightarrow cC + dD \qquad \text{(Eq 2.14)}$$

The change in the GFE for a finite amount of reaction at constant T and P may be written as:

$$\Delta G = \Delta U + P\Delta V - T\Delta S = \Delta H - T\Delta S \qquad \text{(Eq 2.15)}$$

In principle, values of U, H, and S, from which G may be calculated, exist for each chemical species. If these values could be determined, then the change in the GFE could be calculated for the reaction as follows:

$$\Delta G_{react} = G_{products} - G_{reactants} \qquad \text{(Eq 2.16)}$$

$$\Delta G_{react} = cG_C + dG_D - (aG_A + bG_B) \qquad \text{(Eq 2.17)}$$

where $G_A$ etc. are the GFEs per mole for each species indicated by the subscript. If the calculation leads to $\Delta G < 0$, then the reaction as written (left to right, reactants to products) is capable of occurring spontaneously.

Although Eq 2.17 suggests that absolute values of the GFEs of the chemical species can be obtained and that these values can be used to calculate the change in the GFE for the reaction, such absolute values cannot be determined. This is due to the fact that the GFE is derived from the internal energy, U, or the enthalpy, H, neither of which can be assigned absolute values. As a consequence, the GFE can be assigned a numerical value only relative to its value in some reference state. The usual reference state is the stable form of the substance at the reference conditions, these usually being one atmosphere pressure and either 0 K or 298 K. Since the absolute values of G in the reference state cannot be determined, an arbitrary value must be assigned. A consistent basis for calculations results if the GFEs of the *elements* in their stable forms at the reference conditions of one atmosphere pressure and 298 K are assigned the value of zero. The pure elements at other conditions will have definite values, for example, $\Delta G(T) = G(T) - G(T_{ref})$. When the GFE of a chemical compound is needed as a function of temperature, its value at the reference temperature also could be assigned a value of zero and values at other temperatures calculated relative to $T_{ref}$ just as for pure elements. However, as discussed next, the GFE of a compound can be referenced to the elements that compose it. This reference method is used in most calculations involving chemical reactions.

Any reaction between elements to form compounds has associated with it a change in the GFE between the compound and the reactant elements. Thus, for the oxidation of iron at T:

$$\frac{4}{3} Fe + O_2 \rightarrow \frac{2}{3} Fe_2O_3 \qquad \text{(Eq 2.18)}$$

for which

$$\Delta G_f(T) = \frac{2}{3} G_{Fe_2O_3} - \frac{4}{3} G_{Fe} - G_{O_2} \qquad \text{(Eq 2.19)}$$

$\Delta G_f(T)$ is the GFE of formation of $Fe_2O_3$ at temperature T and at the particular conditions of the reaction. In this case, the only important variable other than the temperature is the pressure of the oxygen since the other two species are solids of fixed composition whose GFE is essentially independent of pressure. If the reactant elements and the product oxide are in their stable forms at one atmosphere, the symbol $\Delta G_f^o(T)$ is used to indicate the *standard* GFE of formation at temperature T. Standard values are usually reported for reference temperatures of 0 K and/or 298 K.

In general, $\Delta G_f^o(T_{ref})$ values are based on calculations from direct experimental reactions of the elements to form the compound and from specific heat and related calorimetric measurements on each species, which allows correction of the data from $T_{exp}$ (the experimental reaction temperature) to $T_{ref}$. Tabulations of $\Delta G_f^o$ (0 K) or $\Delta G_f^o$ (298 K) for reactions and of the specific heats of reactants and products to allow temperature corrections form the source from which many chemical thermodynamic calculations are made (Ref 2).

## Calculation of Standard Change of Gibbs Free Energy for Chemical Reactions from Gibbs Free Energy of Formation

For chemical reactions in which all of the reactants and products are in their standard states, the change in the GFE for the reaction is given by:

$$\Delta G^o_{react} = \Sigma \Delta G_f^o(\text{products}) - \Sigma \Delta G_f^o(\text{reactants}) \qquad \text{(Eq 2.20)}$$

The free energy of formation of a pure element is zero because no change in the element's state is involved (e.g., $O_2 \rightarrow O_2$, $\Delta G_f^o = 0$). Therefore, in implementing Eq 2.20:

$$\Delta G_f^o \text{ (pure element)} = 0 \qquad \text{(Eq 2.21)}$$

For example, in considering the oxidation of $Fe_3O_4$ to $Fe_2O_3$ by $H_2O$:

$$Fe_3O_4 + \frac{1}{2}H_2O \rightarrow \frac{3}{2}Fe_2O_3 + \frac{1}{2}H_2 \qquad \text{(Eq 2.22)}$$

for which, fundamentally:

$$\Delta G^o_{react} = \frac{3}{2}G^o_{Fe_2O_3} + \frac{1}{2}G^o_{H_2} - \left(G_{Fe_3O_4} + \frac{1}{2}G^o_{H_2O}\right) \qquad \text{(Eq 2.23)}$$

As stated previously, absolute values of G are not available, and the above calculations cannot be made directly. If Eq 2.20 is correct, it is necessary to show that the following equation based on Eq 2.20 is equivalent to Eq 2.23:

$$\Delta G^o_{react} = \frac{3}{2}\Delta G^o_{f_{Fe_2O_3}} + \frac{1}{2}\Delta G^o_{f_{H_2}} - \left(\Delta G^o_{f_{Fe_3O_4}} + \frac{1}{2}\Delta G^o_{f_{H_2O}}\right) \qquad \text{(Eq 2.24)}$$

The reactions for the formations of the compounds, and expressions for the standard free energies of formation, are:

$$3Fe + 2O_2 \rightarrow Fe_3O_4 \qquad \text{(Eq 2.25)}$$

$$\Delta G^o_{f_{Fe_3O_4}} = G^o_{Fe_3O_4} - \left(3G^o_{Fe} + 2G^o_{O_2}\right) \qquad \text{(Eq 2.26)}$$

$$H_2 + \frac{1}{2}O_2 \rightarrow H_2O \qquad \text{(Eq 2.27)}$$

$$\Delta G^o_{f_{H_2O}} = G^o_{H_2O} - \left(G^o_{H_2} + \frac{1}{2}G^o_{O_2}\right) \qquad \text{(Eq 2.28)}$$

$$2Fe + \frac{3}{2}O_2 \rightarrow Fe_2O_3 \qquad \text{(Eq 2.29)}$$

$$\Delta G^o_{f_{Fe_2O_3}} = G^o_{Fe_2O_3} - \left(2G^o_{Fe} + \frac{3}{2}G^o_{O_2}\right) \qquad \text{(Eq 2.30)}$$

$$\Delta G^o_{f_{H_2}} = 0 \qquad \text{(Eq 2.31)}$$

When the $\Delta G_f^o$ expressions, Eq 2.26, 2.28, 2.30, and 2.31 are substituted into Eq 2.24, Eq 2.23 is produced. Thus, $\Delta G_f^o$ data can be used to calculate $\Delta G^o_{react}$ through Eq 2.20.

Equation 2.20 gives the GFE of reaction when reactants in their standard states are converted to products in their standard states, an initial calculation usually applying to the reference temperature for which data are tabulated. From specific heat data, the change in the $\Delta G_f^o$ of each reactant and product with temperature may be calculated. The values of $\Delta G_f^o(298\ K)$ can then be corrected to $\Delta G_f^o(T)$, where T is the reaction temperature of interest. The new set of values of $\Delta G_f^o(T)$ is then appropriately combined to give $\Delta G_{react}^o(T)$ for any reaction. *The condition for a reaction to occur spontaneously, however, is not that the standard GFE of reaction,* $\Delta G_{react}^o(T)$, *is negative, but rather that the change for the actual conditions of reaction,* $\Delta G_{react}(T)$, *is negative.* $\Delta G_{react}(T)$ is calculated from $\Delta G_{react}^o(T)$ by correcting the latter for the differences in concentrations of reactants and products from those of the standard state to those of the state corresponding to the actual conditions of reaction. Then, if $\Delta G_{react}(T) < 0$, the reaction will occur spontaneously.

## Electrochemical Reactions, the Electrochemical Cell, and the Gibbs Free-Energy Change

Many chemical reactions may be divided into two half reactions, each reaction involving loss or gain of electrons by chemical species, which, as a result, undergo valence changes. Frequently, the half reactions involve metal surfaces at which metal ions either pass into or are deposited from solution or at which the valence state of another species is changed. If the half reactions occur on physically separated metals in an appropriately conducting medium (usually an aqueous solution), then a difference in electrical potential is generally observed to exist between them. For example, consider the reaction:

$$Fe + 2HCl \rightarrow FeCl_2 + H_2 \qquad \text{(Eq 2.32)}$$

or, if the ionized states of the HCl and $FeCl_2$ are taken into account, the equivalent reactions are written as:

$$Fe + 2H^+ + 2Cl^- \rightarrow Fe^{2+} + 2Cl^- + H_2 \qquad \text{(Eq 2.33)}$$

and

$$Fe + 2H^+ \rightarrow Fe^{2+} + H_2 \qquad \text{(Eq 2.34)}$$

Reaction 2.34 is the sum of the following half reactions:

$$Fe \rightarrow Fe^{2+} + 2e \qquad \text{(Eq 2.35)}$$

$$2H^+ + 2e \rightarrow H_2 \qquad \text{(Eq 2.36)}$$

in which the iron, having lost electrons to form ferrous ions, is oxidized, and the hydrogen ions are reduced to hydrogen gas. These reactions are generally observed to take place from left to right as written. Conceptually, the two half reactions may be caused to occur at physically distinct surfaces by placing iron into a solution of ferrous ions and platinum, which is chemically inert, in a solution of hydrogen ions into which hydrogen gas is bubbled. The arrangement is shown in Fig. 2.1. A porous barrier is indicated between the two electrodes, across which electrical conduction can occur but with minimum mixing of solutions. There is a potential difference at this liquid/liquid junction, but it is generally small compared to other potential differences and will not be considered in the present discussion.

The electrochemical cell, or battery, that results will have a difference in electrical potential between the metal electrodes (Ref 3, 4). This potential difference is a function of the concentration of $Fe^{2+}$ ions, the $H^+$ ions, and the pressure of the hydrogen gas at a given temperature. If these variables are adjusted to unit activity (essentially unit molality, or moles per 1000 g of solvent, for the ions in dilute solution, and 1 atm pressure for the hydrogen), the potential difference in the limiting idealized case at 25 °C, with the electrodes not electrically connected, is 440 mV, with the platinum on which the hydrogen reaction occurs being positive. It is important to note that measurement of the potential difference with an electrometer does not constitute electrical connection since the internal resistance is extremely high (>$10^{14}$ ohms), and essentially no current is allowed to flow. Also, the assumption is made here that the spontaneous hydrogen reaction on iron (Fe) is negligible compared to that on platinum (Pt). The overall reaction, Eq 2.34, will not occur until the two electrodes are connected externally either directly or through some device using the current to perform work. For example, upon connection of an electrical motor (Fig. 2.1), electrons will flow from the iron electrode (at which net oxidation occurs, $Fe \rightarrow Fe^{2+} + 2e$), through the motor, to the platinum electrode (at which net reduction occurs, $2H^+ + 2e \rightarrow H_2$). (Unfortunately, it is customary to consider electrical current as a flow of positive charge from the positive to the negative terminal—just the opposite of the electron flow direction.) If the

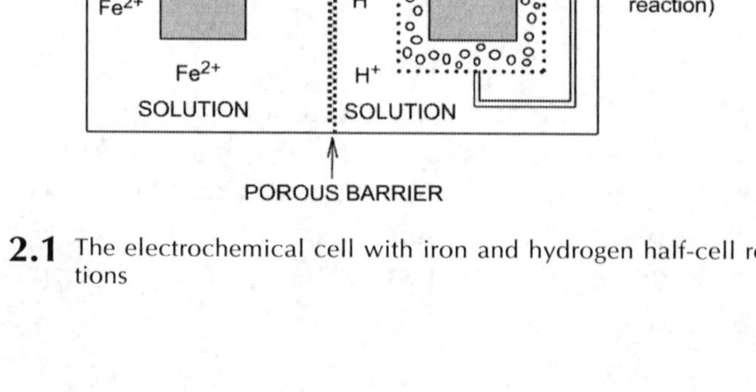


**Fig. 2.1** The electrochemical cell with iron and hydrogen half-cell reactions

motor is mechanically and electrically perfect, then the electrochemical energy released by the cell reaction results in an equivalent amount of work; otherwise, part or all of this energy may be dissipated as heat.

The maximum amount of work that can be obtained per unit of reaction (here, per mole of iron) is that of the reversible transfer of the electrons (electrical charge) through the potential difference between the electrodes. This is also the $w'_r$ work represented by the change in the GFE at constant pressure and temperature (Eq 2.6). Conventional electrical circuit analysis considers that positive electricity (positive charge) flows as a consequence of the difference in potential. If unit positive charge (with magnitude equal to that of the electron charge) is designated as $e^+$ and c charges are transferred per unit of reaction, then the *reversible* electrical work is given by:

$$w'_r = ce^+E_{cell} \qquad \text{(Eq 2.37)}$$

where $E_{cell}$ is defined such as to be positive when $w'$ work is done as a consequence of the spontaneous reaction (i.e., work done by the system). If each symbol for a chemical species in a reaction is interpreted to represent a mole of the species, then in the present example, the unit of reaction involves 1 mol, or Avogadro's number ($N_o$) of iron atoms, which produces $2N_o$ charges upon reaction. In general then, $c = nN_o$, where n is the number of mols of unit charges (electrons) transferred per unit of reaction. The reversible electrical work is therefore:

$$w'_r = nN_oe^+E_{cell} \qquad \text{(Eq 2.38)}$$

$$w'_r = nFE_{cell} \qquad \text{(Eq 2.39)}$$

where $N_oe^+ = F$ is Faraday's constant or the absolute value of the charge of $N_o$ electrons. Substitution of Eq 2.39 into Eq 2.5 gives:

$$\Delta G_{react} = -nFE_{cell} \qquad \text{(Eq 2.40)}$$

Since $E_{cell}$ is defined to be positive for a spontaneous reaction, this equation correctly expresses a decrease in Gibbs function, which is the thermodynamic criterion for a spontaneous reaction at constant T and P. It is evident that if $\Delta G_{react}$ can be calculated from $\Delta G_f^o$ data, the potential of a cell arranged for reversible operation can be determined; conversely, experimental measurements of $E_{cell}$ permit calculation of $\Delta G_{react}$. Both types of calculations are useful in electrochemical work and, thus, in the analysis of corrosion.

The calculation of $E_{cell}$ for the reaction of Eq 2.32 can be used as an example. The reaction is rewritten as follows to show the activities in aqueous solution, $a_{HCl}$ and $a_{FeCl_2}$:

$$Fe + 2HCl(aq., a_{HCl} = 1) \rightarrow FeCl_2(aq., a_{FeCl_2} = 1) + H_2 \qquad \text{(Eq 2.41)}$$

This reaction can be derived from the following four reactions, thermodynamic data for which may be found tabulated in handbooks:

$$Fe + Cl_2(\text{gas, 1 atm}) \rightarrow FeCl_2(\text{solid}) \qquad \Delta G_f^o(298) = -302{,}200 \text{ joules(J)} \qquad \text{(Eq 2.42)}$$

$$FeCl_2(\text{solid}) \rightarrow FeCl_2(\text{aq., } a_{FeCl_2} = 1) \qquad \Delta G^o_{soln}(298) = -45{,}200 \text{ J} \qquad \text{(Eq 2.43)}$$

$$\frac{1}{2}H_2 + \frac{1}{2}Cl_2 \rightarrow HCl(\text{gas, 1 atm}) \qquad \Delta G_f^o(298) = -95{,}300 \text{ J} \qquad \text{(Eq 2.44)}$$

$$HCl(\text{gas, 1 atm}) \rightarrow HCl(\text{aq., } a_{HCl} = 1) \quad \Delta G^o_{soln} = -35{,}900 \text{ J} \qquad \text{(Eq 2.45)}$$

If reactions 2.44 and 2.45 are multiplied by two, reversed, and added to the sum of reactions 2.42 and 2.43, reaction 2.41 results. Then, for this reaction:

$$\Delta G^o_{react}(298) = -85{,}000 \text{ J per mol of Fe} \qquad \text{(Eq 2.46)}$$

Solving for $E_{cell}$ from Eq 2.40 gives:

$$E^o_{cell} = -\frac{\Delta G^o_{react}}{nF} = -\frac{-85{,}000}{2(96{,}485)} = +0.44 \text{ V} = +440 \text{ mV} \qquad \text{(Eq 2.47)}$$

In this calculation, n is 2 because two moles of charges are transferred per mole of iron reaction (or per unit of this reaction); this is usually referred to as two electrochemical equivalents, 1 electrochemical equivalent (ee) being defined as moles of material that will produce 1 mol or Avogadro's number of electrons (i.e., for iron in this example, 1 ee = 0.5 mol, and 1 mol of iron reacting represents 2 ees). The Faraday constant, F, is 96,485 coulombs (joule/volt) per electrochemical equivalent (Ref 2).

An electrochemical cell such as that represented in Fig. 2.1 will have a difference in potential, $E_{cell}$, between the metallic conductors extending out of the solution (i.e., Fe and Pt). This difference in potential is a consequence of the electrochemical reaction at each metal/solution interface and the accompanying potential difference established across each interface (discussed further in the next section). If these individual-interface potential differences could be measured, the cell potential for any combination of electrochemical reactions could be calculated. Unfortunately, a single metal/solution interface potential difference cannot be measured directly because the metal probe from an electrometer used to measure the potential difference will, on contacting the solution, introduce another metal/solution interface. Therefore, the electrometer will indicate only the difference in potential between the metal under investigation and the metal probe in contact with the same solution. A practical solution to this dilemma is provided by selecting one of several specific metal/aqueous-environment combinations that will give a highly reproducible interface potential difference and, therefore, function as a standard reference electrode. More specifically, these combinations are referred to as standard reference electrodes or half cells because they must be used in conjunction with the metal under investigation to produce a complete electrochemical cell, with metal contacts between which a difference in potential can be determined.

The accepted primary reference electrode is the hydrogen half cell described in association with Fig.2.1 (Ref 5). It consists of platinum (which serves as an inert conductor) in contact with a solution at 25 °C, saturated with hydrogen gas at one atmosphere pressure, and containing hydrogen ions at pH = 0 ($a_{H^+} = 1$). In practice, the major use of the standard hydrogen electrode (SHE) is for calibration of secondary reference electrodes, which are more convenient to use. Two common reference electrodes are the calomel or mercury/saturated-mercurous-chloride half cell with a potential of +241 mV relative to the SHE and the silver/saturated-silver-chloride half cell with a relative potential of +196 mV. Both of these electrodes are saturated with potassium chloride to maintain a constant chloride and hence metal-ion concentration.

## Interface Potential Difference and Half-Cell Potential (Ref 3, 6)

It is useful to consider a metal as an array of ions, $M^{m+}$, the valence electrons of each atom having been transferred to the crystal as a whole. These "free" electrons account for the electrical conductivity of the metal and other electronic properties. The metal in aqueous solution also exists as an ion, and thus, the relative tendency for the ion to exist in the metal or in the solution depends, along with other factors such as the concentration, on the relative *electrochemical free-energy* of the ion in these two phases. The electrochemical free energy is used in this application rather than the Gibbs free energy because charged phases are involved. The electrochemical free energy per ion, $g_{el}$, is composed of a chemical contribution, g, and a charge contribution, $q\phi$, such that:

$$g_{el} = g + q\phi \quad \text{(Eq 2.48)}$$

where q is the charge on the ion, and $\phi$ is the electrical potential at the ion in the phase (solid or liquid). The electrical potential at the ion is defined by the work required to move unit positive charge from an infinite reference state to the position of the ion. The difference in electrical potential between two points is therefore directly related to the work required to move unit positive charge between the points; this difference of potential is the more important concept for the present discussion.

Just as the condition for chemical equilibrium is $\Delta g = 0$, the condition for electrochemical equilibrium is $\Delta g_{el} = 0$. This condition is now applied to the transfer of ions across the metal/electrolyte interface. For convenience, the symbols $g_{M^0}$ and $\phi_{M^0}$ are used to indicate the GFE and electrical potential of the ion in the metal; the symbols $g_{M^+}$ and $\phi_{M^+}$ apply to the ion in solution. The change in electrochemical free energy on going from an ion in the solid to an ion in solution is given by:

$$\Delta g_{el} = (g_{M^+} - g_{M^0}) + q(\phi_{M^+} - \phi_{M^0}) \quad \text{(Eq 2.49)}$$

At equilibrium, $\Delta g_{el} = 0$, and therefore:

$$(g_{M^+} - g_{M^0}) = -q(\phi'_{M^+} - \phi'_{M^0}) \quad \text{(Eq 2.50)}$$

where the primed $\phi$s indicate equilibrium values.

The charge transferred per ion is $q = me^+$, where m is the valence and $e^+$ the unit positive charge. Therefore, per ion:

$$(g_{M^+} - g_{M^0}) = -me^+(\phi'_{M^+} - \phi'_{M^0}) \quad \text{(Eq 2.51)}$$

Multiplying by $N_o$, and with $G = N_o g$ and $F = N_o e^+$, the change in GFE per mole is:

$$(G_{M^+} - G_{M^0}) = -mF(\phi'_{M^+} - \phi'_{M^0}) \quad \text{(Eq 2.52)}$$

These equations imply that metal ions tend to transfer from the solid across the interface to the solution due to a decrease in the GFE (i.e., $G_{M^+} < G_{M^0}$). They tend to transfer in the opposite direction as a consequence of the difference in potential between the two phases (i.e., $(\phi'_{M^+} > \phi'_{M^0})$. These concepts are summarized in Fig. 2.2. This result leads to the brief generalization: *At equilibrium, the GFE driving force to transfer ions from the metal to the solution is exactly balanced by the electrical potential difference attracting the ions back to the metal.*

It is not possible to calculate or experimentally measure absolute values for $G_{M^+}$, $G_{M^0}$, $\phi'_{M^+}$, or $\phi'_{M^0}$. However, relative potential differences can be measured by connecting two electrode systems as indicated in the electrochemical cell of Figure 2.1, and also as indicated by the abbreviated cell representation of Fig. 2.3. In Fig 2.3, the right-hand electrode (RHE) is shown as the hydrogen reaction, $2H^+ + 2e = H_2$, occurring on platinum as an inert conductor. When the activity (effective concentration) of the hydrogen ions is unity (molality, $m_{H^+} \approx 1$), the pressure of the hydrogen gas is one atmosphere, and the temperature is 25 °C, this electrode is called the standard hydrogen electrode (SHE). Its interface potential difference may be indicated as $(\phi'_{H_2} - \phi'_{H^+})_s$, with the s subscript indicating standard conditions. This combination of electrodes is an electrochemical cell, the potential difference between the electrodes being defined as:

$$E'_{M,M^{m+}} = \left(\phi'_{M^0} - \phi'_{M^{m+}}\right) - \left(\phi'_{H_2} - \phi'_{H^+}\right)_S \quad \text{(Eq 2.53)}$$

$E'_{M,M^{m+}}$ is called the single *electrode* or *half-cell potential* of the $M,M^{m+}$ electrode on the standard hydrogen scale. It should be recalled that in this text, E denotes the potential in the general case, E′ the potential at equilibrium, and E° the potential at equilibrium under standard

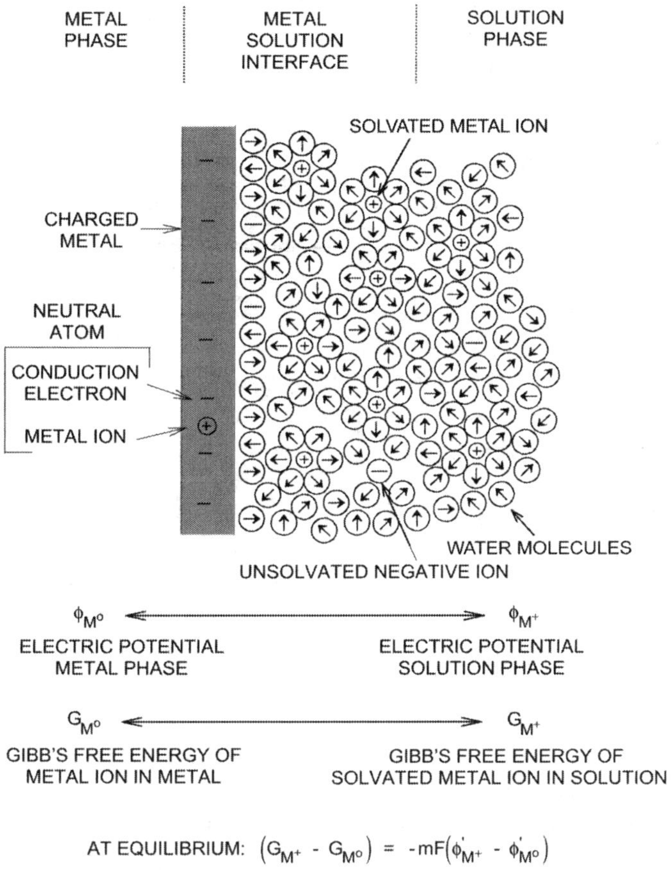


**Fig. 2.2** The metal/solution interface. Based on Ref 3

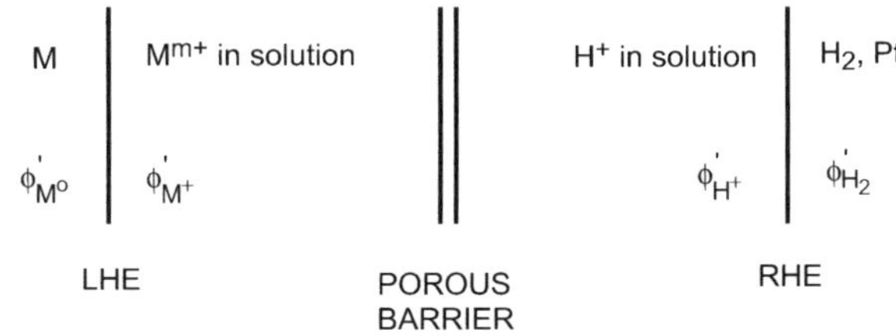


**Fig. 2.3** Abbreviated cell representation showing absolute potentials

conditions, all relative to the standard hydrogen electrode (SHE). It is to be noted, based on Eq 2.53, that the half-cell potential of the hydrogen reaction under standard conditions is zero (i.e., $E^o_{H_2,H^+} = 0$).

The sign or polarity of the electrode ($M,M^{m+}$) is determined basically by the difference in the work required to move unit positive charge from infinity to the metal, M, less the work required for transport to the SHE. The electrode requiring the greater amount of work in moving the unit positive charge from infinity will be at a higher potential and is said to be positive relative to the second electrode, which is called the negative electrode. If the electrodes are connected externally through a conductor, conventional positive current, I, will flow from the positive to the negative electrode, although the actual carriers are electrons flowing in the opposite direction. Practically, the polarity of the electrode whose potential is being measured relative to the SHE is given by the polarity of the terminal of a high-impedance voltmeter or electrometer that must be attached to the electrode to obtain a positive meter reading. Thus, if M spontaneously oxidizes to $M^{m+}$ when coupled to the SHE, the M electrode will be negative relative to the SHE, and $E'_{M,M^{m+}}$ will be negative for the half-cell reaction, $M = M^{m+} + me$.

It is important to realize that the standard half-cell potential, $E^o$, or the half-cell potential at other than standard conditions, $E'$, is sign invariant with respect to how the equilibrium reaction is written or considered, for example, $E^o_{Fe,Fe^{2+}} = -440$ mV (SHE) for both $Fe = Fe^{2+} + 2e$ and $Fe^{2+} + 2e = Fe$. This point can be appreciated by examining the measurement of the difference in electrical potential of the cell in Fig. 2.1.* Although these measurements are usually made with an electrometer ($>10^{14}$ ohms internal resistance), it is helpful to examine measurements with a potentiometer. The potentiometer is a variable potential device that is attached to the cell and adjusted until the current flow is zero. At this condition, the potentiometer is applying a potential to the cell that just equals the cell potential, for example, 440 mV for $Fe = Fe^{2+} + 2e$ with the negative terminal of the potentiometer connected to the Fe electrode, that is, $E^o_{Fe,Fe^{2+}} = -440$ mV (SHE). If the potentiometer is adjusted to slightly increase the potential of the Fe electrode relative to the SHE, for example, −430 mV (SHE), equilibrium no longer exists, the cell reaction occurs as it would spontaneously (but at a reduced rate), and net oxidation occurs (i.e., $Fe \rightarrow Fe^{2+} + 2e$). Thus, for the M electrode in general, very slight increasing or decreasing of the potential of M relative to the SHE by the potentiometer upsets the equilibrium and causes net oxidation, $M \rightarrow M^{m+} + me$, or net reduction, $M^{m+} + me \rightarrow M$, but with only a very small change relative to $E^o_{M,M^{m+}}$.

*The assumption is still made here as previously that the spontaneous hydrogen reaction on iron is negligible compared to that on platinum.

# The Generalized Cell Reaction

It is useful to establish a more generalized representation for the electrochemical cell reaction as follows:

$$xM + mX^{x+} \leftrightarrow xM^{m+} + mX \quad \text{(Eq 2.54)}$$

which is the sum of the following two half-cell reactions:

$$xM \leftrightarrow xM^{m+} + (xm)e \quad \text{(Eq 2.55)}$$

$$mX^{x+} + (xm)e \leftrightarrow mX \quad \text{(Eq 2.56)}$$

or

$$x(M \leftrightarrow M^{m+} + me) \quad \text{(Eq 2.57)}$$

$$m(X^{x+} + xe \leftrightarrow X) \quad \text{(Eq 2.58)}$$

where the parentheses above contain the usual representations of the half-cell reactions (except for the $\leftrightarrow$ symbol) that are tabulated in reference tables for the equilibrium condition, for example, $M = M^{m+} + me$. The standard half-cell potentials for many of the reactions are contained in Table 2.1. The $\leftrightarrow$ symbol is used in this text to denote the stoichiometric relationship between reactive species. It is specifically employed to indicate that no assumption is being made regarding the spontaneous direction of the overall reaction, reaction 2.54 (i.e., it could be either left to right or right to left). If, for example, the spontaneous direction for reaction 2.54 is left to right, the spontaneous direction for the half reactions, Eq 2.55 to 2.58, will also be left to right.

The abbreviated cell representation for the generalized reaction is shown in Fig. 2.4. The reduced species on the left side of the overall reaction (M) and its associated ion ($M^{m+}$) are identified as the left-hand

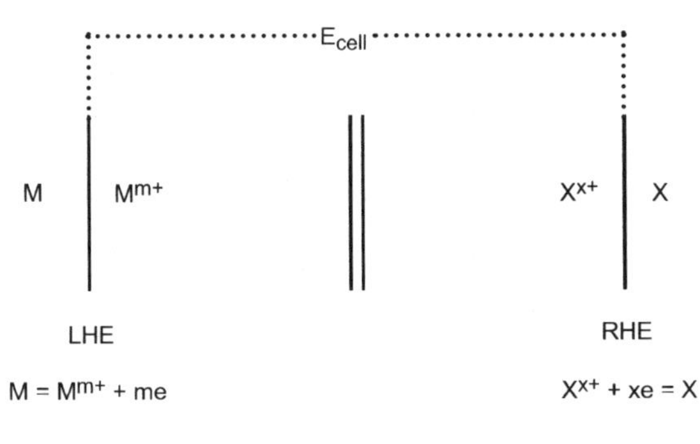


**Fig. 2.4** Abbreviated cell representation showing $E_{cell}$ and half-cell reactions

**Table 2.1 Standard aqueous half-cell potentials at 25 °C (also known as standard electrode, redox, or oxidation potentials, and as the standard emf series)(a)**

| Electrode reaction | $E^0$, mV (SHE) | Electrode reaction | $E^0$, mV (SHE) |
|---|---|---|---|
| **Acid solutions** | | **Acid solutions (continued)** | |
| $Li = Li^+ + e$ | −3040 | $H_2O_2(aq) = O_2(g) + 2H^+ + 2e$ | +695 |
| $K = K^+ + e$ | −2931 | $3NH_4^+ = NH_3(aq) + 11H^+ + 8e$ | +695 |
| $Ca = Ca^{2+} + 2e$ | −2868 | $Fe^{2+} = Fe^{3+} + e$ | +771 |
| $Na = Na^+ + e$ | −2714 | $2Hg = Hg_2^{2+} + 2e$ | +797 |
| $Mg = Mg^{2+} + 2e$ | −2356 | $Ag = Ag^+ + e$ | +799 |
| $H(g) = H^+ + e$ | −2106 | $N_2O_4(g) + 2H_2O = 2NO_3^- + 4H^+ + 2e$ | +803 |
| $Al + 6F^- = AlF_6^{3-} + 3e$ | −2069 | $HNO_2 + H_2O = NO_3^- + 3H^+ + 2e$ | +940 |
| $U = U^{3+} + 3e$ | −1798 | $NO + 2H_2O = NO_3^- + 4H^+ + 3e$ | +957 |
| $Al = Al^{3+} + 3e$ | −1662 | $NO + H_2O = HNO_2 + H^+ + e$ | +983 |
| $Ti = Ti^{2+} + 2e$ | −1630 | $2NO + 2H_2O = N_2O_4 + 4H^+ + 4e$ | +1035 |
| $Zr = Zr^{4+} + 4e$ | −1550 | $2HNO_2 = N_2O_4 + 2H^+ + 2e$ | +1065 |
| $Mn = Mn^{2+} + 2e$ | −1185 | $Pt = Pt^{2+} + 2e$ | ca +1200 |
| $Zn = Zn^{2+} + 2e$ | −762 | $2H_2O(liq.) = O_2 + 4H^+ + 4e$ | +1229 |
| $Cr = Cr^{3+} + 3e$ | −744 | $2Cr^{3+} + 7H_2O = Cr_2O_7^{2-} + 14H^+ + 6e$ | +1232 |
| $U^{3+} = U^{4+} + e$ | −607 | $2Cl^- = Cl_2 + 2e$ | +1360 |
| $Fe = Fe^{2+} + 2e$ | −440 | $Mn^{2+} + 4H_2O = MnO_4^- + 8H^+ + Se$ | +1507 |
| $Cr^{2+} = Cr^{3+} + e$ | −408 | $2H_2O = H_2O_2 + 2H^+ + 2e$ | +1776 |
| $Cd = Cd^{2+} + 2e$ | −403 | $Fe^{3+} + 4H_2O = FeO_4^{2-} + 8H^+ + 3e$ | +2200 |
| $Pb + SO_4^{2-} = PbSO_4 + 2e$ | −359 | $2F^- = F_2(g) + 2e$ | +2866 |
| $Sn + 6F^- = SnF_6^{2-} + 4e$ | −250 | | |
| $Ni = Ni^{2+} + 2e$ | −257 | **Basic solutions** | |
| $Mo = Mo^{3+} + 3e$ | −200 | $Mg + 2OH^- = Mg(OH)_2 + 2e$ | −2690 |
| $Sn(white) = Sn^{2+} + 2e$ | −136 | $Zn + S^{2-} = ZnS(wurtzite) + 2e$ | −1405 |
| $Pb = Pb^{2+} + 2e$ | −126 | $Zn + 4CN^- = Zn(CN)_4^{2-} + 2e$ | −1260 |
| $H_2 = 2H^+ + 2e$ (SHE) | 0 | $Zn + 2OH^- = Zn(OH)_2 + 2e$ | −1249 |
| $Ag + 2S_2O_3^{2-} = Ag(S_2O_3)_2^{3-} + e$ | +17 | $Fe + S^{2-} = FeS(\alpha) + 2e$ | −950 |
| $Ag + Br^- = AgBr + e$ | +71 | $Fe + 2OH^- = Fe(OH)_2 + 2e$ | −877 |
| $Sn^{2+} = Sn^{4+} + 2e$ | +150 | $H_2 + 2OH^- = 2H_2O + 2e$ | −828 |
| $Cu^+ = Cu^{2+} + e$ | +153 | $Fe + CO_3^{2-} = FeCO_3 + 2e$ | −756 |
| $Ag + Cl^- = AgCl + e$ | +222 | $Ni + 2OH^- = Ni(OH)_2 + 2e$ | −720 |
| $2Hg + 2Cl^- = Hg_2Cl_2 + 2e$ | +268 | $Cu + 2CN^- = Cu(CN)_2^- + e$ | −429 |
| $Cu = Cu^{2+} + 2e$ | +342 | $Ag + 2CN^- = Ag(CN)_2^- + e$ | −310 |
| $Fe(CN)_6^{4-} = Fe(CN)_6^{3-} + e$ | +358 | $Cu + 2NH_3 = Cu(NH_3)_2^+ + e$ | −120 |
| $2Ag + CrO_4^{2-} = Ag_2CrO_4 + 2e$ | +447 | $Ag + CN^- = AgCN + e$ | −17 |
| $Cu = Cu^+ + e$ | +521 | $4OH^- = O_2 + 2H_2O + 4e$ | +401 |
| $2H_2SO_3 = S_2O_6^{2-} + 4H^+ + 2e$ | +570 | $Cu(CN)_2^- = Cu^{2+} + 2CN^- + e$ | +1103 |
| $2Hg + SO_4^{2-} = Hg_2SO_4 + 2e$ | +613 | | |

(a) Selected values from Ref 2, 7, and 8.

electrode (LHE); the reduced species on the right side (X) and its associated ion ($X^{x+}$) are identified as the right-hand electrode (RHE).

If reaction 2.54 occurs spontaneously from left to right, then:

$$\Delta G_{react} < 0 \quad \text{(Eq 2.59)}$$

where $\Delta G_{react}$ always applies to the left-to-right direction of reaction 2.54. For this condition, if the electrochemical cell reaction is allowed

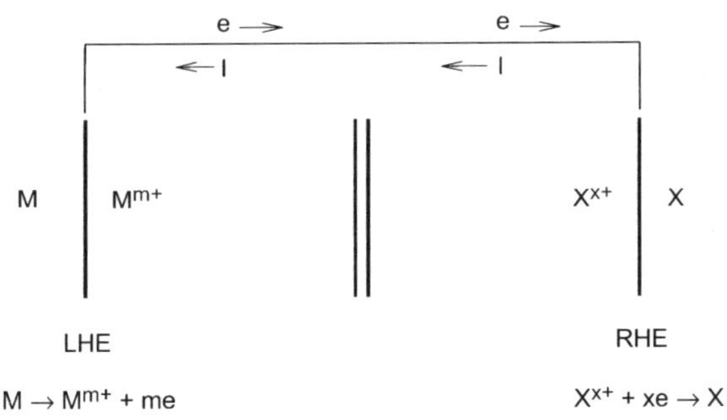


**Fig. 2.5** Abbreviated cell representation showing current flow when the half-cell reactions are coupled

to occur, the electron-flow and conventional-current-flow directions will be as shown in Fig. 2.5. According to electrical circuit convention, X (in this case) is at a higher potential than M, and the flow of current from X to M provides electrical energy capable of doing work. As discussed previously, this work is related to the change in GFE through Eq 2.40, namely:

$$\Delta G_{react} = -nFE_{cell} \qquad (n = xm) \qquad \text{(Eq 2.60)}$$

In this relationship, n is the number of moles of electrons transferred per unit of the reaction (i.e., per x moles of M etc.).

Care must be exercised in assigning a sign to $E_{cell}$ such that the cell potential and the change in the GFE for the reaction are consistent with Eq 2.60. This is one of the most critical points with respect to notation in electrochemistry. If reaction 2.54 occurs spontaneously from left to right, $\Delta G_{react}$ must be negative (Eq 2.59). Then, in order to be consistent with Eq 2.60, $E_{cell}$ must be positive. For these conditions, as shown in Fig. 2.5, the half-cell potential of the RHE is greater than that of the LHE. Therefore, a positive $E_{cell}$ value is accomplished by defining $E_{cell} = E'_{RHE} - E'_{LHE}$. Indeed, for all conditions, $E_{cell}$ will have the proper sign if the following convention is adopted:

$$E_{cell} = E'_{RHE} - E'_{LHE} \qquad \text{(Eq 2.61)}$$

This convention and additional terminology and relationships are summarized in Table 2.2.

It follows from the above discussion that if calculations result in $E'_{RHE} < E'_{LHE}$, $E_{cell}$ will be negative. A negative value of $E_{cell}$ results in $\Delta G_{react} > 0$ and, hence, the conclusion that the reaction will not proceed from left to right, but rather that the spontaneous direction is from right to left.

The significant points of the foregoing discussion may be summarized as follows:

**Table 2.2 Summary of electrochemical cell conventions, terminology, and relationships**

| Comment | Representation | |
|---|---|---|
| Cell reaction | $xM + mX^{x+} \leftrightarrow xM^{m+} + mX$ | |
| Cell representation | $M \mid M^{m+}(a_{M^{m+}}) \parallel X^{x+}(a_{X^{x+}}) \mid X$ | |
| Electrode identification | LHE | RHE |
| Electrode potential | $E'_{M,M^{m+}}$ | $E'_{X,X^{x+}}$ |
| Cell potential | $E_{cell} = E'_{RHE} - E'_{LHE}$ | |
| **If reaction is spontaneous from left-to-right, $\Delta G_{react} < 0$, which results in:** | | |
| Electrode designation | Anode | Cathode |
| Terminal polarity | Negative | Positive |
| Electrode reaction | Oxidation (corrosion) | Reduction |
| Current flow in external circuit | $\leftarrow I$ | |
| Electron flow in external circuit | $e \rightarrow$ | |

- The electrochemical reaction is written in the form:

$$xM + mX^{x+} \leftrightarrow xM^{m+} + mX$$

- The cell is represented with the reduced species on the left side of the reaction (M) and its associated ion ($M^{m+}$) as the LHE (i.e., $M,M^{m+}$ or $M = M^{m+} + me$), and the reduced species on the right of the reaction (X) and its associated ion ($X^{x+}$) as the RHE (i.e., $X,X^{x+}$ or $X = X^{x+} + xe$).

$$\underset{\text{LHE}}{M \mid M^{m+}} \parallel \underset{\text{RHE}}{X^{x+} \mid X}$$

- If the reaction proceeds spontaneously from left to right:

$$\Delta G_{react} < 0$$

- For the relationship $\Delta G_{react} = -nFE_{cell}$ to be consistent with the previous three statements, $E_{cell}$ must be positive, which follows when $E_{cell}$ is defined as:

$$E_{cell} = E'_{RHE} - E'_{LHE}$$

- $E'_{RHE}$ and $E'_{LHE}$ are equilibrium half-cell potentials, or electrode potentials, which depend in sign on the definitions of positive and negative electricity and assignment of $E'_{H_2,H^+} = 0$ at standard conditions. They do not depend on the direction in which the half-cell reaction is written (i.e., $M = M^{m+} + me$ versus $M^{m+} + me = M$).
- It follows that:

a. If $E'_{RHE} > E'_{LHE}$, $E_{cell} > 0$, $\Delta G < 0$, and the reaction is spontaneous from left to right.

b. $E'_{RHE} < E'_{LHE}$, $E_{cell} < 0$, $\Delta G > 0$, and the reaction is spontaneous from right to left.

c. If $E'_{RHE} = E'_{LHE}$, $E_{cell} = 0$, $\Delta G = 0$, and the reaction is at equilibrium.

It is useful to include as much information as reasonable with the cell representation. In particular, it is important to specify all variables, including the nature of the phase or phases associated with each side of the electrode across which electron transport occurs. Using the LHE to illustrate:

| *Electron Transporting Phase* (e.g., *M*) | *Solution Phase or Ion Transporting Phase*<br>Identify reacting species and activities or concentrations.<br>Identify any precipitated phase.<br>Identify any reacting dissolved nonionized species (e.g., $O_2$) |
|---|---|

LHE

In these representations, the electron-transporting phase is usually a metal; however, in certain cases it can be an electron-conducting oxide, other compound, or other material, such as graphite. Furthermore, two categories of electron-transporting phases may be encountered:

- Active electron-conducting electrodes, for example:

$$M \mid M^{m+} (a_{M^{m+}})$$

LHE

- Inert electron-conducting electrodes (Pt, Au, graphite, etc., in certain solutions), for example:

$$H^+ (a_{H^+} \text{ or pH}, P_{H_2}) \mid \text{on Pt}$$

RHE

It is important to define the solution phase with respect to variables establishing the half-cell potentials. The identified species will be ionic with ionic activities, for example, $H^+$ ($a_{H^+}$ or pH), $NO_2^-$($a_{NO_2^-}$), and $Fe^{2+}$ ($a_{Fe^{2+}}$), or neutral, such as oxygen ($a_{O_2}$ or $P_{O_2}$), and hydrogen ($a_{H_2}$ or $P_{H_2}$).

During the corrosion process, it is important to realize that both the anodic reaction (oxidation, for example, $M \rightarrow M^{m+} + me$) and the cathodic reaction (reduction, for example, $2H^+ + 2e \rightarrow H_2$) occur on the same metal; in this case, therefore, the electron-conducting phase for both the LHE and the RHE would be the metal, M.

# The Nernst Equation: Effect of Concentration on Half-Cell Potential (Ref 3, 6)

Consider again the generalized electrochemical reaction:

$$xM + mX^{x+} \leftrightarrow xM^{m+} + mX \quad \text{(Eq 2.62)}$$

One of the most significant equations derived from chemical thermodynamics permits calculation of the change in the GFE for this reaction at constant total pressure and temperature as a function of $\Delta G_f^o$ of the reactant and product species in their standard states and the concentrations of those species with concentrations that can be varied. The equation is:

$$\Delta G_{react} = \Delta G_{react}^o + RT \ln \frac{a_X^m \cdot a_{M^{m+}}^x}{a_M^x \cdot a_{X^{x+}}^m} \quad \text{(Eq 2.63)}$$

In this equation, $\Delta G_{react}^o$ is the change in the GFE for the reaction as written for reactants and products in their standard states; it is calculated from Eq 2.20. The a's are the activities of the species indicated by the subscripts: each activity is raised to a power equal to the stoichiometric coefficient of the species as it appears in the reaction. The activity is frequently called the effective concentration of the species because it naturally arises as a function of the concentration, that is necessary to satisfy the changes in the thermodynamic functions (here, the GFE). In electrochemical systems, the activity is usually related to the molality of the species (moles per 1000 g of solvent) by the following equation:

$$a = \gamma m \quad \text{(Eq 2.64)}$$

where $\gamma$ is the activity coefficient and m, the molality. Although in principle the activity of a single ionic species has meaning, and theoretically, expressions have been developed for it, direct experimental measurement is not possible. The reason for the latter limitation is not discussed in detail here; it is sufficient to state that the problem relates to the fact that writing a single activity for an ionic species implies that this species can be added to a solution independent of other species. This is not possible because of the necessity of simultaneously adding or having present in the solution ions of opposite charge in amounts to satisfy electrical neutrality. Although Eq 2.62 is frequently written with the ions of opposite charge present, as in Eq 2.32 or 2.33, and Eq. 2.63 can be modified to include the activities of the actual species dissolved to give the solution ($FeCl_2$, for example), this is not done in the present treatment. The primary reason for using individual ion activities in the present treatment is that it allows focus of attention on the ions involved in the individual electrode reactions, the influence of which is important in controlling corrosion rates. In many corrosion calculations, it is sufficient to use estimates of the individual ion activities, or to use the molality directly. Some reasons and justifications for this often-necessary approach are as follows. Measurement or calculation of accurate activities in concentrated electrolytes and in electrolytes of complex mixtures is generally not possible. Also, a tenfold change in the concentration results in a change of less than 100 mV in the electrode potential, which is frequently small compared to the potentials involved in cell reactions (i.e., $E_{cell}$ values). And, finally, metal ion concentrations in many corrosive environments are usually small ($<10^{-4}$), in which case the activity coefficient is essentially unity and, therefore, $a \approx m$.

The standard state for reactants and products in reaction 2.62 is pure solid for solid species, one atmosphere pressure for gas species, and unit activity (approximately unit molality) for ionic species. The activity is unity in each of these standard states, and if these conditions are substituted into Eq 2.63, $\Delta G_{react}$ will equal $\Delta G^{o}_{react}$; this must follow if the derivation of this equation is examined. If, under the actual conditions of the reaction, one or more species are solids, or a gas exists at one atmosphere pressure, then unit activity for each of these species is substituted in Eq 2.63, which effectively removes these activities from the log term. Also, the activity of water can usually be set equal to unity because its concentration changes insignificantly in most reactions in aqueous solution. Thus, taking M and X as solids, Eq 2.63 reduces to:

$$\Delta G_{react} = \Delta G^{o}_{react} + RT \ln \frac{a^{x}_{M^{m+}}}{a^{m}_{X^{x+}}} \qquad \text{(Eq 2.65)}$$

but

$$\Delta G_{react} = -(xm)FE_{cell} \qquad \text{(Eq 2.66)}$$

and

$$\Delta G^{o}_{react} = -(xm)FE^{o}_{cell} \qquad \text{(Eq 2.67)}$$

Therefore,

$$E_{cell} = E^{o}_{cell} - \frac{RT}{xmF} \ln \frac{a^{x}_{M^{m+}}}{a^{m}_{X^{x+}}} \qquad \text{(Eq 2.68)}$$

From the convention relating the cell reaction to the cell representation (Table 2.2), the cell potentials are written as:

$$E_{cell} = E'_{RHE} - E'_{LHE} = E'_{X,X^{x+}} - E'_{M,M^{m+}} \qquad \text{(Eq 2.69)}$$

and

$$E^{o}_{cell} = E^{o}_{RHE} - E^{o}_{LHE} = E^{o}_{X,X^{x+}} - E^{o}_{M,M^{m+}} \quad \text{(Eq 2.70)}$$

Substitution into Eq 2.68 yields:

$$E'_{X,X^{x+}} - E'_{M,M^{m+}} = \left(E^{o}_{X,X^{x+}} + \frac{RT}{xF}\ln a_{X^{x+}}\right) - \left(E^{o}_{M,M^{m+}} + \frac{RT}{mF}\ln a_{M^{m+}}\right) \quad \text{(Eq 2.71)}$$

or

$$E'_{X,X^{x+}} = E^{o}_{X,X^{x+}} + \frac{RT}{xF}\ln a_{X^{x+}} \quad \text{(Eq 2.72)}$$

$$E'_{M,M^{m+}} = E^{o}_{M,M^{m+}} + \frac{RT}{mF}\ln a_{M^{m+}} \quad \text{(Eq 2.73)}$$

Equations 2.72 and 2.73 are Nernst half-cell equations. For example, with Eq 2.73, when $a_{M^{m+}} = 1$, $E'_{M,M^{m+}} = E^{o}_{M,M^{m+}}$. Hence, $E^{o}_{M,M^{m+}}$ is the half-cell potential at unit activity of the ions (i.e., the standard electrode half-cell potential). Values of the standard potentials of many electrode reactions are available in the literature, some of which are given in Table 2.1 (Ref 2, 7, 8). All values are given in sign and magnitude relative to the standard hydrogen electrode as previously discussed.

Many half-cell reactions involve species on both sides of the reaction that have variable concentrations in solution. These circumstances are handled by using the half-cell equation in the following more general form:

$$E'_{X,Y,Z} = E^{o}_{X,Y,Z} = \frac{RT}{nF}\ln\frac{\Pi[Ox_i]^{\upsilon_i}}{\Pi[Red_i]^{\upsilon_i}} \quad \text{(Eq 2.74)}$$

In this equation, X, Y, and Z are symbolic representatives of the important species involved in the reaction; $\Pi[Ox_i]^{\upsilon_i}$ is the product of the activities of the species on the "oxidized side" of the reaction (the side showing electrons produced), each raised to its stoichiometric coefficient ($\upsilon_i$); $\Pi[Red_i]^{\upsilon_i}$ has similar meaning for the "reduced side" of the reaction; and n is the number of mols of electrons produced (or consumed) per unit of the half reaction. Application of Eq 2.74 is illustrated in the following examples:

**Example 1.**

$$\frac{OH^-}{\text{reduced}} = \frac{1/2\,H_2O + 1/4\,O_2 + e}{\text{oxidized}} \quad \text{(Eq 2.75)}$$

$$E'_{OH^-,O_2} = E^{o}_{OH^-,O_2} + \frac{RT}{1F}\ln\frac{P^{1/4}_{O_2}}{a_{OH^-}} \quad \text{(Eq 2.76)}$$

**Example 2.**

$$\frac{HNO_2 + H_2O}{\text{reduced}} = \frac{NO_3^- + 3H^+ + 2e}{\text{oxidized}} \quad \text{(Eq 2.77)}$$

$$E'_{HNO_2,NO_3^-,H^+} = E^o_{HNO_2,NO_3^-,H^+} + \frac{RT}{2F} \ln \frac{a_{NO_3^-} a^3_{H^+}}{a_{HNO_2}} \quad \text{(Eq 2.78)}$$

## Half-Cell Reactions and Nernst-Equation Calculations

Five examples are given of the application of the Nernst equation to half-cell reactions. These examples illustrate the influence of ion concentration, pH, precipitate phases, and complex-ion formation on the electrode potential. All of these variables have significance in aqueous corrosion:

**Example 1: Metal/Metal-Ion Half-Cells.**

Reaction: $M = M^{m+} + me$ (Eq 2.79)

Nernst Equation: $E'_{M,M^{m+}}\,[\text{mV (SHE)}] = E^o_{M,M^{m+}}\,[\text{mV (SHE)}] +$

$$\frac{(8.314)(298)(1000)}{m(96{,}485)} 2.303 \log\left(\gamma_{M^{m+}} m_{M^{m+}}\right) \quad \text{(Eq 2.80)}$$

$$E'_{M,M^{m+}}\,[\text{mV (SHE)}] = E^o_{M,M^{m+}}\,[\text{mV (SHE)}] + \frac{59}{m} \log a_{M^{m+}} \quad \text{(Eq 2.81)}$$

Example:

$Zn = Zn^{2+} + 2e$

$$E'_{Zn,Zn^{2+}} = -763 + \frac{59}{2} \log a_{Zn^{2+}} \quad \text{Eq 2.82)}$$

The half-cell potential will be –763 mV (SHE) when the activity is unity. An increase in the activity causes the potential to become more positive. The change is shown graphically in Fig. 2.6.

**Example 2: Hydrogen Electrode.**

Reaction: $\frac{1}{2} H_2 = H^+ + e$ (Eq 2.83)

Nernst Equation: $E'_{H_2,H^+} = E^o_{H_2,H^+} + 59 \log \frac{a_{H^+}}{(P_{H_2})^{1/2}}$ (Eq 2.84)

$$E'_{H_2,H^+} = 0 + 59 \log a_{H^+} - \frac{59}{2} \log P_{H_2} \quad \text{(Eq 2.85)}$$

It is very convenient to introduce the pH as a measure of the hydrogen ion activity since the acidity of solutions is usually expressed in these terms and is measured with pH meters, indicator papers, or other measuring devices. The definition is pH = $-\log a_{H^+}$. Thus:

$$E'_{H_2,H^+} = -59\,pH - 29.5\log P_{H_2} \qquad \text{(Eq 2.86)}$$

This equation is plotted in Fig. 2.7 in terms of pH as the variable with $P_{H_2}$ as a parameter, and then in Fig. 2.8 in terms of $P_{H_2}$ with pH as the

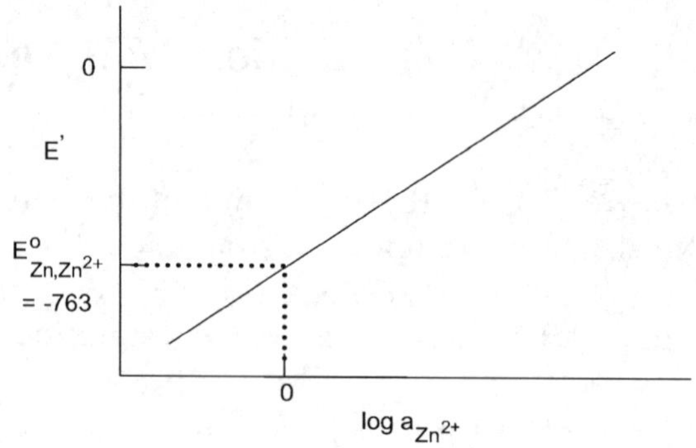


**Fig. 2.6** Dependence of metal-reaction equilibrium potential on metal-ion activity

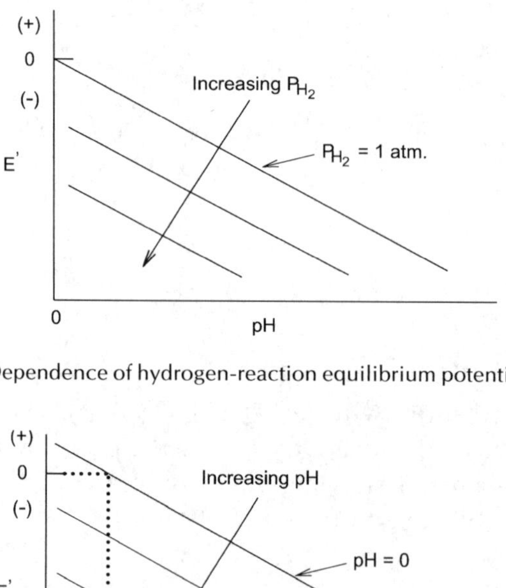


**Fig. 2.7** Dependence of hydrogen-reaction equilibrium potential on pH

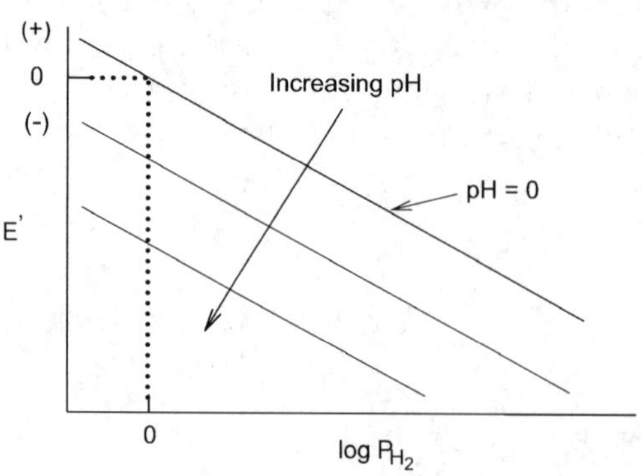


**Fig. 2.8** Dependence of hydrogen-reaction equilibrium potential on hydrogen-gas partial pressure

parameter. Increasing pH and $P_{H_2}$ causes the potential to move in the negative direction.

**Example 3: Oxygen Electrode.**
Reaction:

$$OH^- = \frac{1}{2}H_2O + \frac{1}{4}O_2 + e \quad \text{(Eq 2.87)}$$

Nernst Equation:

$$E'_{OH^-,O_2} = E^o_{OH^-,O_2} + 59\log\frac{P_{O_2}^{1/4}}{a_{OH^-}} \quad \text{(Eq 2.88)}$$

$$E'_{OH^-,O_2} = +401 + \frac{59}{4}\log P_{O_2} - 59\log a_{OH^-} \quad \text{(Eq 2.89)}$$

In aqueous solution, the $a_{OH^-}$ is related to $a_{H^+}$ through the activity product for water, $(a_{OH^-})(a_{H^+}) = 10^{-14}$. Hence,

$$E'_{OH^-,O_2} = +401 + 15\log P_{O_2} - 59\log\frac{10^{-14}}{a_{H^+}} \quad \text{(Eq 2.90)}$$

$$E'_{OH^-,O_2} = 1229 - 59\,pH + 15\log P_{O_2} \quad \text{(Eq 2.91)}$$

This equation is plotted in Fig. 2.9 in terms of pH as the variable with $P_{O_2}$ as a parameter, and then in Fig. 2.10 in terms of $P_{O_2}$ with pH as a parameter. Note that the potential decreases with increasing pH but increases with increasing $P_{O_2}$. The latter effect of increasing gas pressure is opposite that observed on increasing the hydrogen pressure for the hydrogen half cell. This follows from the fact that the gaseous species occur on opposite sides of their half-cell reactions (i.e., the oxidized side for oxygen and the reduced side for hydrogen). The last expression above may be written as:

$$E' = 1229 + 59\log P_{O_2}^{1/4} a_{H^+} \quad \text{(Eq 2.92)}$$

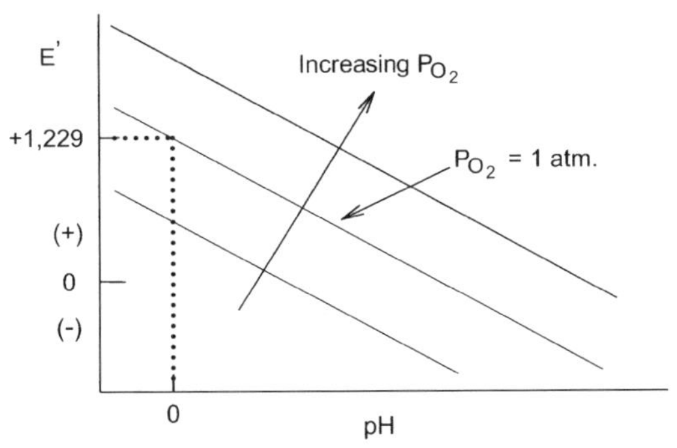


**Fig. 2.9** Dependence of oxygen-reaction equilibrium potential on pH

which corresponds to the reaction:

$$\frac{1}{2}H_2O = \frac{1}{4}O_2 + H^+ + e \qquad \text{(Eq 2.93)}$$

with $E^o_{H_2O,O_2} = +1229\ \text{mV (SHE)}$.

**Example 4: Metal/Insoluble-Metal-Salt Electrodes.** These electrodes are particularly important from two standpoints:

- Reference cells for electrochemical measurements are usually metals in contact with solutions containing precipitated metal salt.
- If ions in the corrosive solution can form an insoluble salt with the metal ions that enter the solution as a result of corrosion, the formation of the insoluble salt may have a controlling influence on the corrosion. If the insoluble salt forms as an adherent nonporous film on the metal, corrosion may essentially stop. On the other hand, if the precipitate forms irregularly on the surface, pitting may be introduced either by exposure of the metal between precipitate patches or by exclusion of oxygen from the regions covered by the precipitate.

In these electrodes, the solubility product for the salt is an important consideration. According to this principle, the precipitation reaction:

$$bA^{a+} + aB^{b-} = A_bB_a \quad \text{(solid ppt)} \qquad \text{(Eq 2.94)}$$

actually occurs if the following condition is met in the solution:

$$(a_{A^{a+}})^b (a_{B^{b-}})^a = K_{sp} \qquad \text{(Eq 2.95)}$$

where $K_{sp}$ is a constant known as the solubility product constant. In the usual application of this equation, the concentration of $B^{b-}$ ions is either known in the solution or can be estimated. This concentration is then

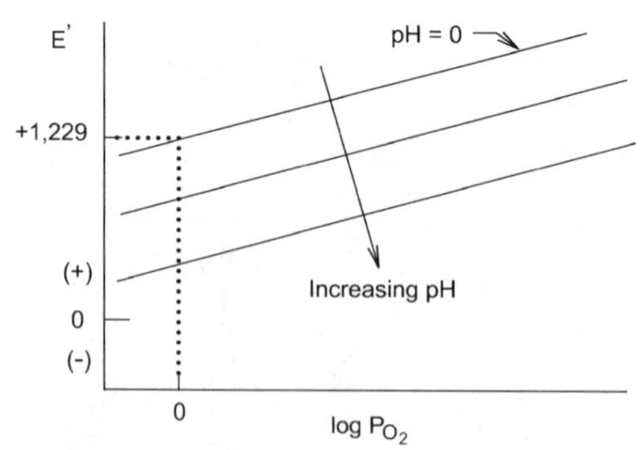


**Fig. 2.10** Dependence of oxygen-reaction equilibrium potential on oxygen-gas partial pressure

substituted in the previous equation, and the metal-ion concentration is calculated, for example:

$$Ag \mid Ag^{+} (a_{Ag^{+}} \text{ established by } 1.0 \text{ m } Cl^{-} (KCl), \gamma = 0.66) \qquad \text{(Eq 2.96)}$$

$$Ag = Ag^{+} + e \qquad E^{o}_{Ag,Ag^{+}} = +799 \text{mV (SHE)} \qquad \text{(Eq 2.97)}$$

$$E'_{Ag,Ag^{+}} = 799 + 59 \log a_{Ag^{+}} \qquad \text{(Eq 2.98)}$$

but

$$(a_{Ag^{+}})(a_{Cl^{-}}) = K_{sp} = 1.56 \times 10^{-10} \qquad \text{(Eq 2.99)}$$

from which

$$a_{Ag^{+}} = (1.56 \times 10^{-10}) / (a_{Cl^{-}}) = (1.56 \times 10^{-10}) / (0.66 \times 1) \qquad \text{(Eq 2.100)}$$

Therefore:

$$E'_{Ag,Ag^{+}} = 799 + 59 \log \frac{1.56 \times 10^{-10}}{0.66 \times 1} = +231 \text{ mV (SHE)} \qquad \text{(Eq 2.101)}$$

This silver/silver-chloride electrode is sufficiently important as a reference electrode to find it tabulated in tables of half cells. In these tables, the standard half-cell value is given, which is the potential when the ion functioning as the variable controlling the potential is at unit activity. In the present example, the $Cl^{-}$ ion is the variable and the standard half-cell potential is that which results for $a_{Cl^{-}} = 1.0$:

$$E'_{Ag,Ag^{+}} = 799 + 59 \log \frac{1.56 \times 10^{-10}}{1} = +222 \text{ mV (SHE)} = E^{o}_{Ag,AgCl,Cl^{-}} \qquad \text{(Eq 2.102)}$$

This standard half-cell potential, $E^{o}_{Ag,AgCl,Cl^{-}}$, is seen to apply to the reaction:

$$Ag + Cl^{-} = AgCl + e \qquad \text{(Eq 2.103)}$$

If the chloride ion activity is now recognized as the variable controlling the potential, then the half-cell potential is written as:

$$E'_{Ag,AgCl,Cl^{-}} = E^{o}_{Ag,AgCl,Cl^{-}} + 59 \log \frac{1}{a_{Cl^{-}}} \qquad \text{(Eq 2.104)}$$

Many of the reference half cells for electrochemical measurements are metals in contact with insoluble salts suspended in solutions very dilute in the metal ion and containing the anion causing precipitation at specified concentrations. Some important reference electrodes and their potentials are listed in Table 2.3.

**Example 5: Cells with Complexing Agents.** Complexing agents are soluble species that combine with metal ions in solution to form soluble complexes, thus reducing the effective concentration of metal ions to extremely low values. This effect causes the electrode potential to move in the negative direction (half-cell potential is decreased) and increases the susceptibility to corrosion. Complexing agents are both inorganic and organic chemical species and may be found in many chemical process solutions. Naturally occurring complexing agents are frequently found in foods and may alter the predicted corrosion behavior of metal food containers.

The example to be considered is the analysis of the tendency for copper to corrode in a deaerated solution of pH = 8 with no complexing agents as compared to the corrosion tendency if cyanide ions are added to the same solution.

If copper is placed in the solution of pH = 8, then calculation of the half-cell potential of the copper will depend on the copper-ion concentration of the solution. Generally, this will be small and unknown. For purposes of estimation, it may be assumed that $a_{Cu^{2+}} = 10^{-4} = m_{Cu^{2+}}$, and that if corrosion occurs, the cathodic reaction is the release of hydrogen gas at one atmosphere pressure (solution is deaerated). The reaction under consideration is then:

$$Cu + 2H^+ \leftrightarrow Cu^{2+} + H_2 \qquad \text{(Eq 2.105)}$$

$$Cu \mid Cu^{2+}(m_{Cu^{2+}} = 10^{-4}) \parallel H^+(pH = 8) \mid H_2(1\ atm)\ on\ Cu$$

$$E'_{LHE} = 342 + \frac{59}{2}\log 10^{-4} = 224\ mV\ (SHE) \qquad \text{(Eq 2.106)}$$

**Table 2.3 Potentials of some reference electrodes or half cells**

| | | | |
|---|---|---|---|
| $H_2/H^+$ (Standard Hydrogen Electrode, SHE) | Reaction: | $H_2 = 2H^+ + 2e$ | |
| | | $H_2/H^+$, pH = 0, $P_{H_2}$ = 1.0 atm | 0 mV (SHE) |
| Ag/AgCl | Reaction: | $Ag = Ag^+ + e$ | |
| | or | $Ag + Cl^- = AgCl + e$ | |
| | | Ag/AgCl, saturated KCl | +196 mV (SHE) |
| | | Ag/AgCl, 1 N KCl | +234 mV (SHE) |
| | | Ag/AgCl, 0.1 N KCl | +289 mV (SHE) |
| $Hg/Hg_2Cl_2$ (Calomel) | Reaction: | $2Hg = Hg_2^{++} + 2e$ | |
| | or | $2Hg + 2Cl^- = Hg_2Cl_2 + 2e$ | |
| | | $Hg/Hg_2Cl_2$, saturated KCl | +241 mV (SHE) |
| | | $Hg/Hg_2Cl_2$, 1 N KCl | +280 mV (SHE) |
| | | $Hg/Hg_2Cl_2$, 0.1 N KCl | +334 mV (SHE) |
| $Cu/CuSO_4$ | | $Cu/CuSO_4$, saturated $CuSO_4$ | +316 mV (SHE) |

$$E'_{RHE} = -59\ pH = -59(8) = -472\ mV\ (SHE) \qquad \text{(Eq 2.107)}$$

$$E_{cell} = E'_{RHE} - E'_{LHE} = -472 - 224 = -696\ mV\ (SHE) \qquad \text{(Eq 2.108)}$$

$$\Delta G_{react} = -nFE_{cell} = -2F(-696) > 0 \qquad \text{(Eq 2.109)}$$

Therefore, copper does not corrode with the evolution of hydrogen at pH = 8.

KCN is now added to the solution until $a_{CN^-} = 0.5$. $CN^-$ complexes $Cu^+$ to $Cu(CN)_2^-$. The equilibrium reaction is:

$$Cu(CN)_2^- = Cu^+ + 2CN^- \qquad \text{(Eq 2.110)}$$

and the equilibrium constant is:

$$\frac{(a_{Cu^+})(a_{CN^-})^2}{a_{Cu(CN)_2^-}} = K = 10^{-16} \qquad \text{(Eq 2.111)}$$

It should be noted that the complex is formed with cuprous ions ($Cu^+$) and not cupric ions ($Cu^{2+}$). Since $Cu^{2+}$ is usually considered to be the corrosion product of copper, it is necessary to calculate the relationship of these two ions in solution. This can be done from electrode potential data. For the reaction:

$$Cu = Cu^{2+} + 2e \qquad \text{(Eq 2.112)}$$

$$E'_{Cu,Cu^{2+}} = E^o_{Cu,Cu^{2+}} + \frac{59}{2}\log a_{Cu^{2+}} \qquad \text{(Eq 2.113)}$$

where $E^o$ = +342 mV (SHE). For the reaction:

$$Cu^+ = Cu^{2+} + e \qquad \text{(Eq 2.114)}$$

$$E'_{Cu^+,Cu^{2+}} = E^o_{Cu^+,Cu^{2+}} + 59\log\frac{a_{Cu^{2+}}}{a_{Cu^+}} \qquad \text{(Eq 2.115)}$$

where $E^o$ = +153 mV (SHE). Consider a cell made up of these two electrodes with the cell at equilibrium (i.e., $E_{cell} = 0$). Then:

$$E'_{Cu,Cu^{2+}} = E'_{Cu^+,Cu^{2+}} \qquad \text{(Eq 2.116)}$$

$$342 + 29.5\log a_{Cu^{2+}} = 153 + 59\log\frac{a_{Cu^{2+}}}{a_{Cu^+}} \qquad \text{(Eq 2.117)}$$

Solving for $a_{Cu^+}$:

$$a_{Cu^+} = 6.3\times10^{-4}\,a^{1/2}_{Cu^{2+}} \qquad \text{(Eq 2.118)}$$

and still taking $a_{Cu^{2+}} = 10^{-4}$ gives

$$a_{Cu^+} = 6.3 \times 10^{-4} (10^{-4})^{1/2} = 6.3 \times 10^{-6} \quad \text{(Eq 2.119)}$$

With $CN^-$ present, the $a_{Cu^+}$ is further reduced. Also, the activity of the complex ion will be the initial activity of the $Cu^+$ ($6.3 \times 10^{-6}$) less the existing $a_{Cu^+}$:

$$a_{Cu(CN)_2^-} = 6.3 \times 10^{-6} - a_{Cu^+} \quad \text{(Eq 2.120)}$$

Since the amount of $CN^-$ consumed is negligibly small ($<10^{-4}$), $a_{CN^-} \approx 0.5$. Substituting into Eq 2.111 and solving for $a_{Cu^+}$ yields:

$$\frac{a_{Cu^+}(0.5)^2}{(6.3 \times 10^{-6} - a_{Cu^+})} = 10^{-16} \quad \text{(Eq 2.121)}$$

$$a_{Cu^+} = 2.5 \times 10^{-21} \quad \text{(Eq 2.122)}$$

Equation 2.118 is now used to determine the $a_{Cu^{2+}}$ with $CN^-$ present:

$$2.5 \times 10^{-21} = 6.3 \times 10^{-4}\, a_{Cu^{2+}}^{1/2} \quad \text{(Eq 2.123)}$$

$$a_{Cu^{2+}} = 1.6 \times 10^{-35} \quad \text{(Eq 2.124)}$$

In summary, in order to reach the very low equilibrium concentration of $a_{Cu^{2+}}$ required in the presence of $CN^-$, a combination of reactions 2.112 and 2.114 occurs, $Cu + Cu^{2+} \rightarrow 2Cu^+$, until the $a_{Cu^{2+}}$ goes from $10^{-4}$ to $1.6 \times 10^{-35}$ (it should be noted that during this process metallic copper is being corroded), with the resulting $Cu^+$ ions being complexed by $CN^-$ until the $a_{Cu^+}$ reaches $2.5 \times 10^{-21}$. The net result is that virtually all of the $Cu^{2+}$ is taken out of solution. With the $CN^-$ present, the copper half-cell potential is now given by:

$$E'_{Cu,Cu^{2+}} = 342 + 29.5 \log(1.6 \times 10^{-35}) \quad \text{(Eq 2.125)}$$

$$E'_{Cu,Cu^{2+}} = -684 \text{ mV (SHE)} \quad \text{(Eq 2.126)}$$

The cell under consideration is now

$$Cu \mid Cu^{2+}(a_{Cu^{2+}} = 1.6 \times 10^{-35}) \parallel a_{H^+} (pH = 8) \mid H_2 \text{ (1 atm) on Cu} \quad \text{(Eq 2.127)}$$

$$E_{cell} = E'_{RHE} - E'_{LHE} = -59(8) - (-684) = +212 \text{ mV (SHE)} \quad \text{(Eq 2.128)}$$

The cell potential is now positive. The change in the GFE is negative and the reaction will occur. Thus, the copper that could not corrode in the deareated solution of pH = 8 in the absence of $CN^-$ now corrodes because the complexing of the $CN^-$ to form $Cu(CN)_2^-$ has reduced the $Cu^{2+}$ concentration to the extremely low value of $1.6 \times 10^{-35}$.

Another interpretation, more reasonable in terms of mechanism, can be placed on the effect of a complexing agent on the electrode potential. In the example just given, the activity of the $Cu^{2+}$ ion was so small ($1.6 \times 10^{-35}$) that these ions could not play a significant role in the electrode process. Rather, the complex ion could be looked upon as a chemical species in the solution to which the metal ion was attached at a lower energy than for attachment to water molecules. More metal ions could therefore exist in solution before a state of equilibrium was reached; this corresponded to a lower (more negative or active) electrode potential.

## Electrochemical Cell Calculations in Relationship to Corrosion

In most corrosion calculations, the metal-ion concentration in the environment is usually unknown. In the absence of specific values of activity, a reasonably low activity of $10^{-6}$ is usually assumed. This corresponds to less than 1 ppm (parts per million) by weight. Also, most environments will not contain hydrogen, and the question arises as to the value of $P_{H_2}$ to use in calculations on cathodic reactions involving hydrogen evolution. Since hydrogen bubbles cannot form unless the hydrogen pressure is about 1 atm, the usual approximate pressure of the surroundings, it is common practice to assume $P_{H_2} = 1$ atm. Assuming zero for either the metal-ion concentration or the pressure of the hydrogen leads to an infinite potential because the activity appears in the log term of the Nernst half-cell equation. This implies that some corrosion should always occur initially because $E_{cell}$ would be infinite corresponding to an infinite decrease in ΔG. Therefore, it is reasonable to assume that activities of the above magnitude are quickly established on contact of a metal with an aqueous environment if corrosion is thermodynamically possible at all.

**Example 1.** Determine the thermodynamic tendency for silver to corrode in a deaerated acid solution of pH = 1.0. Assume: $a_{Ag^+} = 10^{-6}$ and $P_{H_2} = 1$ atm. Cell reaction:

$$Ag + H^+ \leftrightarrow Ag^+ + \frac{1}{2}H_2$$

Cell representation and calculations:

$$\underbrace{Ag \mid Ag^+(a_{Ag^+} = 10^{-6})}_{\text{LHE}} \Bigg\| \underbrace{\begin{matrix} H^+ \ (pH = 1) \\ H_2 \text{ dissolved} \end{matrix} \Bigg| \begin{matrix} H_2 \text{ on Ag} \\ P_{H_2} = 1 \text{ atm} \end{matrix}}_{\text{RHE}}$$

At LHE:

$$Ag \leftrightarrow Ag^{+} + e$$

$$E'_{LHE} = E'_{Ag,Ag^+} = 799 + \frac{59}{1}\log 10^{-6}$$

$$E'_{LHE} = 445\ \text{mV (SHE)}$$

At RHE:

$$H^{+} + e \leftrightarrow \frac{1}{2}H_2$$

$$E'_{RHE} = E'_{H_2,H^+} = 0 + \frac{59}{1}\log\left(\frac{a_{H^+}}{1}\right)$$

$$E'_{RHE} = -59\,\text{pH} = -59\ \text{mV (SHE)}$$

$$E_{cell} = E'_{RHE} - E'_{LHE} = -59 - (445) = -504\ \text{mV}$$

$E_{cell}$ is found to be negative, which means that $\Delta G = -nFE_{cell}$ is positive. Therefore, the spontaneous direction for the cell reaction is right to left; consequently, silver will not corrode due to the acidity represented by pH = 1.0.

**Example 2.** Determine the thermodynamic tendency for silver to corrode in an aerated acid solution at pH = 1.0. Assume: $a_{Ag^+} = 10^{-6}$, $P_{H_2} = 1.0$ atm, and $P_{O_2} = 0.2$ atm. Compare the result to that of Example 1 (deaerated solution). Cell reaction:

$$4Ag + O_2 + 4H^+ \leftrightarrow 4Ag^+ + 2H_2O$$

Cell representation and calculations:

$$Ag \mid Ag^{+}(a_{Ag^+} = 10^{-6}) \left\| \begin{matrix} H^{+}\ (pH = 1) \\ O_2\ \text{dissolved} \end{matrix} \right| \begin{matrix} O_2\ \text{on Ag} \\ P_{O_2} = 0.2\ \text{atm} \end{matrix}$$

**LHE** **RHE**

At LHE: Same as Example 1

$$E'_{LHE} = 445\ \text{mV (SHE)}$$

At RHE:

$$O_2 + 4H^{+} + 4e \leftrightarrow 2H_2O$$

$$E'_{RHE} = E'_{O_2,H^+} = 1229 + \frac{59}{4}\log P_{O_2}\ a^4_{H+} = 1229 - 59\,\text{pH} + 15\log 0.2$$

$$E'_{RHE} = 1160\ \text{mV (SHE)}$$

$$E_{cell} = E'_{RHE} - E'_{LHE} = 1160 - (445)$$

$$E_{cell} = 715 \text{ mV}$$

$E_{cell}$ is found to be positive, which means that ΔG is negative. Therefore, the spontaneous direction for the cell reaction is left to right; consequently, silver will corrode in this aerated acid solution (pH = 1) due to the dissolved oxygen.

**Example 3.** Determine the pH at which silver will not corrode in an aerated aqueous solution. Refer to Example 2 and set $E_{cell} = 0$ with the pH as the unknown variable.

$$E_{cell} = E'_{RHE} - E'_{LHE}$$

$$0 = (1229 - 59 \text{ pH} + 15 \log 0.2) - 445$$

$$\text{pH} = 13.1$$

**Example 4.** Determine the thermodynamic tendency for tin to corrode in deaerated sulfuric acid at pH = 2. Assume: $a_{Sn^{2+}} = 10^{-6}$ and $P_{H_2} = 1.0$ atm. Cell reaction:

$$Sn + 2H^+ \leftrightarrow Sn^2 + H_2$$

Cell representation and calculations:

$$Sn|Sn^{2+}(a_{Sn^{2+}} = 10^{-6}) \left\| \begin{matrix} H^+ \text{ (pH} = 2) \\ H_2 \text{ dissolved} \end{matrix} \right| \begin{matrix} H_2 \text{ on Sn} \\ (P_{H_2} = 1 \text{ atm}) \end{matrix}$$

LHE RHE

At LHE:

$$Sn \leftrightarrow Sn^{2+} + 2e$$

$$E'_{LHE} = -136 + \frac{59}{2} \log 10^{-6}$$

$$E'_{LHE} = -313 \text{ mV (SHE)}$$

At RHE:

$$2H^+ + 2e \leftrightarrow H_2$$

$$E'_{RHE} = 0 + \frac{59}{2} \log \frac{a^2_{H^+}}{P_{H^2}} = -59 \text{ pH} = -59(2)$$

$$E'_{RHE} = -118 \text{ mV (SHE)}$$

$$E_{cell} = E'_{RHE} - E'_{LHE} = -118 - (-313) = +195\,mV$$

$E_{cell}$ is positive, $\Delta G$ is negative, and the spontaneous direction for the cell reaction is left to right; therefore, corrosion will occur.

**Example 5.** Determine the tendency for iron to corrode in deaerated water. Assume $a_{Fe^{2+}} = 10^{-5}$, pH = 7, and $P_{H_2} = 1.0$ atm. Cell reaction:

$$Fe + 2H^+ \leftrightarrow Fe^{2+} + H_2$$

Cell representation and calculations:

$$Fe \,|\, Fe^{2+}(a_{Fe^{2+}} = 10^{-5}) \,\|\, \begin{matrix} H^+\ (pH = 7) \\ H_2\ \text{dissolved} \end{matrix} \,\Big|\, \begin{matrix} H_2\ \text{on Fe} \\ (P_{H_2} = 1\ \text{atm}) \end{matrix}$$

LHE RHE

At LHE:

$$Fe \leftrightarrow Fe^{2+} + 2e$$

$$E'_{LHE} = E'_{Fe,Fe^{2+}} = -440 + \frac{59}{2}\log 10^{-5}$$

$$E'_{LHE} = -558\ mV\ (SHE)$$

At RHE:

$$E'_{RHE} = E'_{H_2,H^+} = -59\ pH$$

$$E'_{RHE} = -59(7) = -413\ mV\ (SHE)$$

$$E_{cell} = E'_{RHE} - E'_{LHE} = -413 - (-558)$$

$$E_{cell} = +175\ mV$$

$E_{cell}$ is positive, and the spontaneous direction for the cell reaction is left to right; therefore, iron will corrode.

**Example 6.** Determine the tendency for iron to corrode in deaerated water contaminated with dissolved $H_2S$. Assume $a_{S^{2-}} = 10^{-12}$, pH = 4, and $P_{H_2} = 1.0$ atm. Compare the result with that of Example 5.

Hydrogen sulfide dissolved in water to give the indicated sulfide ion activity will produce acidity at about pH = 4. Concern also arises from knowing that FeS is a relatively insoluble substance and therefore may influence the corrosion behavior. Either of two approaches may be taken.

*Solution I.* Assume that the iron corrodes as $Fe^{2+}$ and then precipitates as FeS. Cell reaction:

$$Fe + 2H^{+} \leftrightarrow Fe^{2+} + H_2$$

Cell representation and calculations:

$$Fe \mid Fe^{2+}(a_{Fe^{2+}} = ?) \parallel \begin{matrix} H^{+}\ (pH = 4) \\ H_2\ \text{dissolved} \end{matrix} \mid \begin{matrix} H_2\ \text{on Fe} \\ (P_{H_2} = 1\ \text{atm}) \end{matrix}$$

LHE RHE

At LHE:

$$Fe \leftrightarrow Fe^{2+} + 2e$$

$$E'_{LHE} = E'_{Fe,Fe^{2+}}$$

$$E'_{LHE} = -440 + \frac{59}{2}\log\ a_{Fe^{2+}}$$

$a_{Fe^{2+}}$ is determined from the solubility product for the reaction:

$$Fe^{2+} + S^{2-} = FeS\ (ppt)$$

$$a_{Fe^{2+}} \cdot a_{S^{2-}} = K_{sp} = 3.7 \times 10^{-19}$$

Therefore,

$$a_{Fe^{2+}} = \frac{3.7 \times 10^{-19}}{10^{-12}} = 3.7 \times 10^{-7}$$

Then,

$$E'_{LHE} = -440 + \frac{59}{2}\log(3.7 \times 10^{-7})$$

$$E'_{LHE} = -630$$

At RHE:

$$2H^{+} + 2e \leftrightarrow H_2$$

$$E'_{RHE} = -59\ pH = -59(4)$$

$$E'_{RHE} = -236\ mV\ (SHE)$$

$$E_{cell} = E'_{RHE} - E'_{LHE} = -236 - (-630)$$

$$E_{cell} = +394\ mV$$

$E_{cell}$ is positive, and the spontaneous direction for the cell reaction is left to right; therefore, iron will corrode. Note that the driving potential for corrosion ($E_{cell}$) in the $H_2S$ contaminated water is about twice that in the uncontaminated water (a comparison can be made with Example 5).

*Solution II.* Assume the following direct half-cell reaction, which is listed in Table 2.1:

$$Fe + S^{2-} \leftrightarrow FeS + 2e$$

$$E^{o}_{Fe,S^{2-},FeS} = -950 \text{ mV (SHE)}$$

Cell reaction:

$$Fe + S^{2-} + 2H^{+} \leftrightarrow FeS + H_2$$

Cell representation and calculations:

$$\begin{array}{c|c||c|c} Fe & S^{2-} & H^{+}\ (pH = 4) & H_2 \text{ on Fe} \\ FeS & (a_{S^{2-}} = 10^{-12}) & H_2 \text{ dissolved} & (P_{H_2} = 1 \text{ atm}) \end{array}$$

LHE RHE

At LHE:

$$Fe + S^{2-} \leftrightarrow FeS + 2e$$

$$E'_{LHE} = -950 + \frac{59}{2}\log\frac{1}{10^{-12}}$$

$$E'_{LHE} = -596 \text{ mV (SHE)}$$

At RHE:

$$2H^{+} + 2e \leftrightarrow H_2$$

$$E'_{RHE} = -59(4)$$

$$E'_{RHE} = -236 \text{ mV (SHE)}$$

$$E_{cell} = E'_{RHE} - E'_{LHE} = -236 - (-596)$$

$$E_{cell} = +360 \text{ mV}$$

As in solution I, $E_{cell}$ is positive, and therefore, corrosion occurs. Note that the $E'_{LHE}$ values obtained in solutions I and II are slightly different (–630 versus –596). This difference results from using two different sources of data and shows that some uncertainty exists for the precise value of $K_{sp}$.

**Example 7.** Copper is generally considered to be corrosion resistant in nonoxidizing, deaerated acids. However, a recent publication reported measurable corrosion in HCl ($m = 12$, $a_{H^+} = a_{Cl^-} = 5$). Consider this apparent dilemma.

First, consider that the only cathodic reaction is the evolution of hydrogen due to reduction of hydrogen ions, and show that copper should not corrode by calculating $E_{cell}$. Assume $a_{Cu^{2+}} = 10^{-6}$ and $P_{H_2} = 1$ atm. Cell reaction:

$$Cu + 2H^+ \leftrightarrow Cu^{2+} + H_2$$

Cell representation and calculations:

$$\underset{\textbf{LHE}}{Cu \left| \begin{array}{l} Cu^{2+} \\ (a_{Cu^{2+}} = 10^{-6}) \end{array} \right\|} \begin{array}{l} H^+ \, (a_{H^+} = 5) \\ H_2 \text{ dissolved} \end{array} \underset{\textbf{RHE}}{\left| \begin{array}{l} H_2 \text{ on Cu} \\ (P_{H_2} = 1 \text{ atm}) \end{array} \right.}$$

At LHE:

$$Cu \leftrightarrow Cu^{2+} + 2e$$

$$E'_{LHE} = 342 + \frac{59}{2} \log 10^{-6}$$

$$E'_{LHE} = +165 \text{ mV (SHE)}$$

At RHE:

$$2H^+ + 2e \leftrightarrow H_2$$

$$E'_{RHE} = -59 \text{ pH} = +59 \log a_{H^+} = 59 \log 5$$

$$E'_{RHE} = +41 \text{ mV (SHE)}$$

$$E_{cell} = E'_{RHE} - E'_{LHE} = 41 - (165)$$

$$E_{cell} = -124 \text{ mV}$$

$E_{cell}$ is negative, $\Delta G$ is positive, and the spontaneous direction for the cell reaction is right to left; therefore, copper will not corrode due to the acidity.

Next, consider the suggestion that copper corrodes in the concentrated HCl because of the formation of a soluble chloride complex with an equilibrium constant for the reaction $Cu^{2+} + 4Cl^- = (CuCl_4)^{2-}$ of $K = 10^{+6}$. If $a_{(CuCl_4)^{2-}} = 10^{-4}$, and the activity of the $Cl^-$ is that given above in the concentrated acid ($a_{Cl^-} = 5$), calculate $E_{cell}$ and determine whether corrosion will occur due to the formation of the complex ion. Cell reaction:

$$Cu + 2H^+ \leftrightarrow Cu^{2+} + H_2$$

Cell representation and calculations:

$$\underset{\text{LHE}}{Cu \,\Big|\, \begin{array}{l} Cu^{2+}(a_{Cl^-} \text{ in equilibrium} \\ \text{with } (CuCl_4)^{2-}) \end{array}} \Bigg\| \underset{\text{RHE}}{\begin{array}{l} H^+ (a_{H^+} = 5) \\ H_2 \text{ dissolved} \end{array} \Bigg| \begin{array}{l} H_2 \text{ on Cu} \\ (P_{H_2} = 1 \text{ atm}) \end{array}}$$

At LHE:

$$Cu \leftrightarrow Cu^{2+} + 2e$$

$$E'_{LHE} = 342 + \frac{59}{2} \log a_{Cu^{2+}}$$

$$K = 10^{+6} = \frac{a_{(CuCl_4)^{2-}}}{(a_{Cu^{2+}} \cdot a^4_{Cl^-})}$$

$$a_{Cu^{2+}} = \frac{10^{-4}}{(10^6 \cdot (5)^4)}$$

$$a_{Cu^{2+}} = 1.6 \times 10^{-13}$$

$$E'_{LHE} = 342 + \frac{59}{2} \log (1.6 \times 10^{-13})$$

$$E'_{LHE} = -35 \text{ mV (SHE)}$$

At RHE: Same as in previous calculation

$$E'_{RHE} = +41 \text{ mV (SHE)}$$

$$E_{cell} = E'_{RHE} - E'_{LHE} = 41 - (-35)$$

$$E_{cell} = +76 \text{ mV}$$

$E_{cell}$ is positive, and the spontaneous direction for the cell reaction is left to right; therefore, corrosion of copper will occur. The effect of the $Cl^-$ in the HCl is to complex the $Cu^{2+}$ and reduce its activity to the point that corrosion by hydrogen ions is thermodynamically possible.

## Graphical Representation of Electrochemical Equilibrium: Pourbaix Diagrams

### *Origin and Interpretation of Pourbaix Diagrams (Ref 9, 10)*

The equilibrium electrochemistry of an element in aqueous solution can be represented graphically using coordinates of equilibrium half-cell potential, E′, and pH. These graphical representations, known as Pourbaix diagrams, are essentially phase diagrams from which the conditions for thermodynamic stability of a single aqueous phase, or equilibrium of this phase with one or more solid phases, may be determined. For purposes of the present discussion, the solid phases will be restricted to the pure metal and to the metal oxides or hydroxides that may form under appropriate conditions. The aqueous phase is characterized by the activities of simple and/or complex metal ions and the pH that is established by adding acid or base that ideally has no other effect. The objective of these diagrams is to provide a large amount of information in a convenient form for quick reference. Unfortunately, the data available do not always permit construction of the diagram with acceptable accuracy for some purposes. For example, there may be uncertainties as to exactly which phase is the stable one at a given pH and potential. Also, different phases may appear depending on the immediate past sequence of pH and potential changes, and although such phases are not truly equilibrium phases, they may persist and for practical purposes represent the steady state of the system.

A somewhat simplified Pourbaix diagram for the iron/water system is shown in Fig. 2.11. In this case, the possible solid phases are restricted to metallic iron, $Fe_3O_4$, and $Fe_2O_3$. A more detailed diagram and a diagram with $Fe(OH)_2$ and $Fe(OH)_3$ are shown subsequently.

Interpretation of the Pourbaix diagram in Fig. 2.11 requires discussion of the experimental conditions under which, at least in principle, it would be determined. The coordinates are pH and electrode potential, and it is implied that each of these may be established experimentally. Their values will locate a point on the diagram, and from this point the equilibrium state of the system is determined. It is assumed that the pH may be established by appropriate additions of an acid or base.

To establish any predetermined electrode potential, the experimental arrangement shown in Fig. 2.12 is used. The components and their functions include:

- The aqueous solution of controlled pH. This solution may contain dissolved oxygen, or the container may be closed and an inert gas, such as $N_2$ or He, bubbled through the solution to remove the oxygen present from contact with air.
- The working electrode, which is the electrode under study. It may be an active metal such as iron, with iron ions being exchanged between the electrode and the solution. This electrode may also be an inert metal, such as platinum, which supplies a conducting surface through which electrons pass to oxidize or reduce species in solution.
- The auxiliary or counter electrode, usually platinum, against which the potential of the working electrode is established.

- The reference electrode, against whose known half-cell potential the electrode potential of the working electrode is measured.
- The electrometer or high impedance voltmeter, which is used to measure the potential of the working electrode relative to the reference electrode. The impedance of these instruments should be approximately $10^{14}$ ohms or greater, such that the current required to allow measurement will have a negligible effect on the working electrode.
- The potentiostat, which establishes the potential of the working electrode. The potential between the working and auxiliary elec-

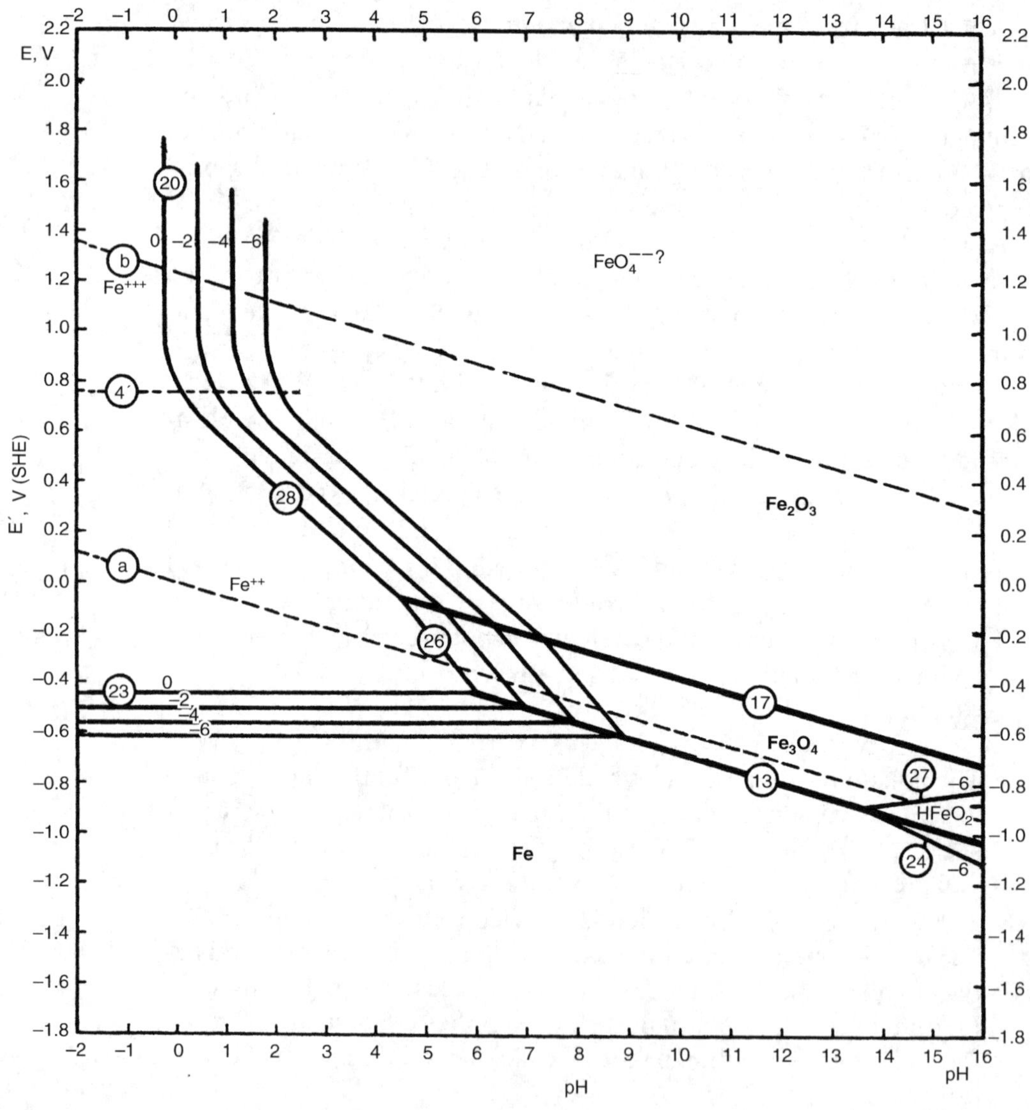


**Fig. 2.11** Simplified Pourbaix diagram for the iron/water system (iron/iron-oxides). Source: Ref 9

trodes is changed until the electrometer indicates the desired potential for the working electrode relative to the reference electrode. Potentiostats are usually electronic instruments that may be set to the desired potential, and this potential is maintained by feedback control from the reference electrode.

In the following discussion of the Pourbaix diagram for the system iron/water (Fig. 2.11), it is convenient to consider that the potentials represented along the ordinate axis have been established by a potentiostat. It should be emphasized that the potentiostat is a device that is useful for studying electrode behavior. In the majority of instances, however, the potential will not be established by an external device. Rather, one or more electrochemical reactions on the metal will establish the potential, and reference to the Pourbaix diagram at this potential gives information on the state of the system. If the potential is established at –0.44 V (SHE) on an iron-working electrode in contact with an aqueous solution at pH $\leq$ 6.0, then the equilibrium condition is that of line 23(0) on the diagram with the 0 representing the $Fe^{2+}$ activity of $10^0$. Further interpretations of this line, and other lines and areas (all labeled in accordance with Pourbaix's published diagrams), are as follows:

- Lines 23, that is, 23(0), 23(–2), etc., represent the equilibrium half-cell or electrode potential of iron as a function of $Fe^{2+}$ activity.

$$E'_{Fe,Fe^{2+}} = -440 + \frac{59}{2} \log a_{Fe^{2+}} \qquad \text{(Eq 2.129)}$$

The parallel lines are identified by the exponent of 10 of the activity of $Fe^{2+}$ ions in solution (i.e., $a_{Fe^{2+}} = 10^0, 10^{-2}, 10^{-4}, 10^{-6}$, and others, which are not shown, at greater dilution). The lines are horizontal because the half-cell potential is independent of the pH at lower

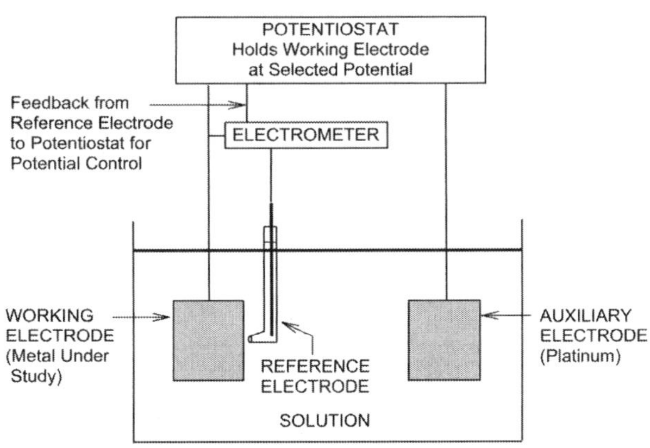


**Fig. 2.12** The potentiostatic-circuit/polarization-cell arrangement

values of pH. If the potential that is applied to the iron is below that corresponding to the $a_{Fe^{2+}}$ in contact with iron, the iron will be stable and will not corrode; rather, iron will tend to be deposited from solution. If $E_{applied}$ is above E′ for a given ion concentration, iron will tend to pass into solution, increasing the concentration of iron ions up to the equilibrium value corresponding to the applied potential.

- At a given $a_{Fe^{2+}}$, increasing the pH eventually results in the reaction:

$$3Fe^{2+} + 4H_2O = Fe_3O_4 + 8H^+ + 2e \quad \text{(Eq 2.130)}$$

$$E' = +980 - 236\ pH - 89 \log a_{Fe^{2+}} \quad \text{(Eq 2.131)}$$

Lines 26, therefore, represent the equilibrium of $Fe^{2+}$ ions with $Fe_3O_4$ at various $Fe^{2+}$ activities (i.e., $10^0$, $10^{-2}$, etc.).

- Conditions along line 13 correspond to a film of $Fe_3O_4$ on Fe. That is, Fe and $Fe_3O_4$ coexist at equilibrium with water containing $Fe^{2+}$ ions at an activity given by the appropriate line 23. Actually, line 13 is the locus of intersections of lines 23 and 26.
- Above lines 23, the stable state of the system is virtually all iron in solution (i.e., $a_{Fe^{2+}} > 10^0$) with $a_{Fe^{2+}} > a_{Fe^{3+}}$. A platinum working electrode must be used to establish these potentials.
- Line 4′ corresponds to $a_{Fe^{2+}} = a_{Fe^{3+}}$ and is located at the half-cell potential for the $Fe^{2+}$ | $Fe^{3+}$ half cell.

$$E^o_{Fe^{2+},Fe^{3+}} = 770\ mV\ (SHE) \quad \text{(Eq 2.132)}$$

- Below Line 4′:

$$a_{Fe^{2+}} > a_{Fe^{3+}} \quad \text{(Eq 2.133)}$$

Above Line 4′:

$$a_{Fe^{2+}} < a_{Fe^{3+}} \quad \text{(Eq 2.134)}$$

- Lines 28 correspond to the reaction:

$$2Fe^{2+} + 3H_2O = Fe_2O_3 + 6H^+ + 2e \quad \text{(Eq 2.135)}$$

$$E' = -728 - 177\ pH - 59 \log a_{Fe^{2+}} \quad \text{(Eq 2.136)}$$

These lines give the conditions for precipitation of $Fe_2O_3$ from solution. Again, the lines are identified by the exponent of 10 for the $a_{Fe^{2+}}$.

- Lines 20 correspond to the formation of $Fe_2O_3$ from solutions of $a_{Fe^{3+}} > a_{Fe^{2+}}$. Here, the curves identified as 0, –2, –4, and –6 correspond to $a_{Fe^{3+}} = 10^0, 10^{-2}, 10^{-4}, 10^{-6}$.
- Line 17 corresponds to the equilibrium of $Fe_3O_4$, $Fe_2O_3$, and solutions of indicated $a_{Fe^{2+}}$ as a function of potential and pH. With increasing potential, $Fe_3O_4$ is oxidized to $Fe_2O_3$.
- Lines a and b correspond to the following equilibrium reactions:

Line a:

$$H^+ + e = \frac{1}{2}H_2$$

or

$$H_2O + e = \frac{1}{2}H_2 + OH^- \qquad \text{(Eq 2.137)}$$

Line b:

$$2H_2O = O_2 + 4H^+ + 4e$$

or

$$4OH^- = O_2 + 2H_2O + 4e \qquad \text{(Eq 2.138)}$$

Therefore, below line a, $H_2$ is produced by reduction of $H^+$ or $H_2O$, and above line b, $O_2$ is produced by oxidation of $H_2O$ or $OH^-$. Between lines a and b, water is stable (i.e., it is neither reduced to $H_2$ nor oxidized to $O_2$).

In 1966, Pourbaix published his *Atlas of Electrochemical Equilibria in Aqueous Solutions,* which contains electrode-potential/pH diagrams for many elements and a critical analysis of the data on which the diagrams are based (Ref 9). Figures 2.13 and 2.14 are from this publication and represent the iron/water system, assuming the solid phases to be iron and iron oxides in the first case and iron and iron hydroxides in the second case. It should be noted that the two diagrams differ only in relatively small detail, which results from the relatively small difference between the GFEs of a hydroxide and the oxide related to it. This can be demonstrated by writing:

$$2Fe(OH)_3 \rightarrow Fe_2O_3 \cdot 3H_2O \rightarrow Fe_2O_3 + 3H_2O \rightarrow Fe_2O_3 \qquad \text{(Eq 2.139)}$$

as the sequence of changes in the conversion of ferric hydroxide to the red dehydrated rust ($Fe_2O_3$). The short, dashed lines in Fig. 2.13 and

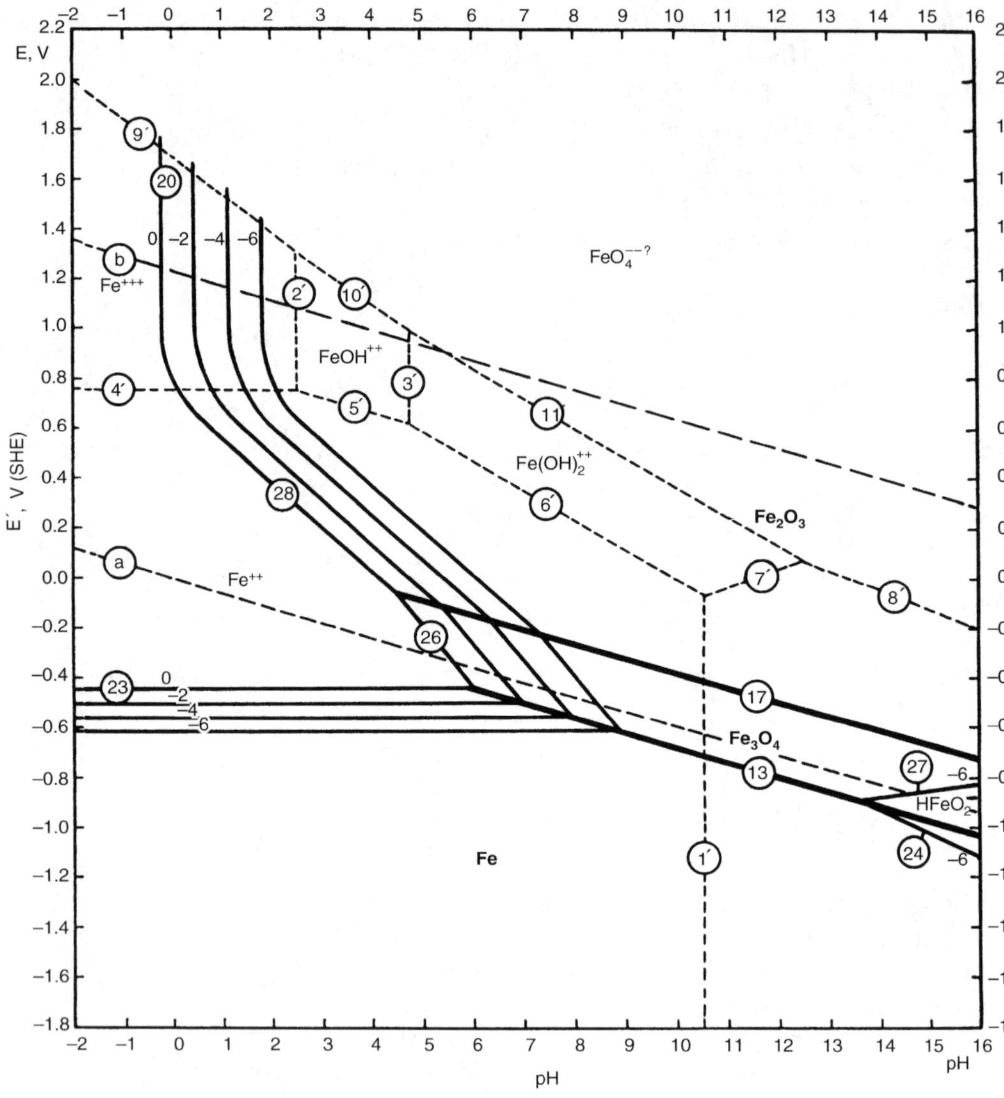


**Fig. 2.13** Pourbaix diagram for the iron/water system (iron/iron-oxides). Source: Ref 9

2.14 separate regions in which the indicated iron-bearing ionic species are observed as the major species in solution. For example, there is experimental evidence that at positive electrode potentials above 1000 mV (SHE) and in all alkaline solutions, the iron exists in solution as $FeO_4^{2-}$ ions.

The Pourbaix diagram for the copper/water system is shown in Fig. 2.15. The more positive standard electrode potential of copper (+337 mV (SHE)) as compared to iron (–440 mV (SHE)) is evident. This greater nobility results in copper being thermodynamically stable in water; that is, line 14 (–6) representing $a_{Cu^{2+}} = 10^{-6}$ lies above line a.

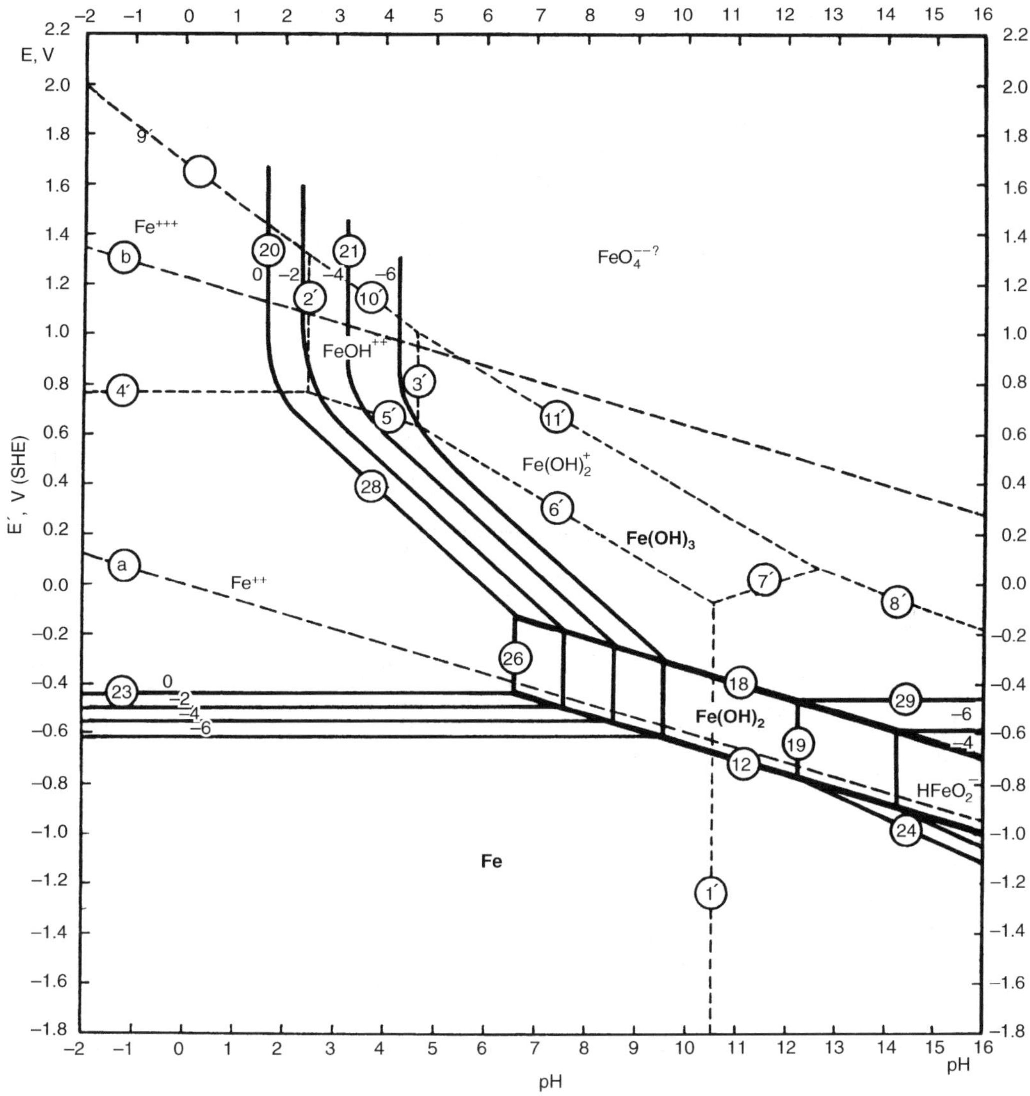


**Fig. 2.14** Pourbaix diagram for the iron/water system (iron/iron-hydroxides). Source: Ref 9

## *Use of Pourbaix Diagrams to "Predict" Corrosion*

The Pourbaix diagram can be used to make preliminary predictions of the corrosion of metals as a function of electrode potential and pH. It is emphasized that the predictions are very general, and the method has been criticized in leading to incorrect conclusions because reference only to the diagram does not recognize the generally controlling factors of rate and nonequilibrium. Figure 2.11 is reproduced in Fig. 2.16(a) with Pourbaix's areas of corrosion, immunity, and passivation indicated (Ref 9). Figure 2.16(b) shows the form frequently used to represent these areas assuming that the activity of reacting ions is $10^{-6}$. The terms are defined as follows:

- *Immunity:* If the potential and pH are in this region, the iron is thermodynamically immune from corrosion. At a point, such as X in Fig. 2.16(a), it is estimated that the $Fe^{2+}$ activity should adjust to about $10^{-10}$, and no corrosion should occur. $H_2$ would be evolved. In the case of iron, an external current source (i.e., a potentiostat) would be required to hold the system at this potential.

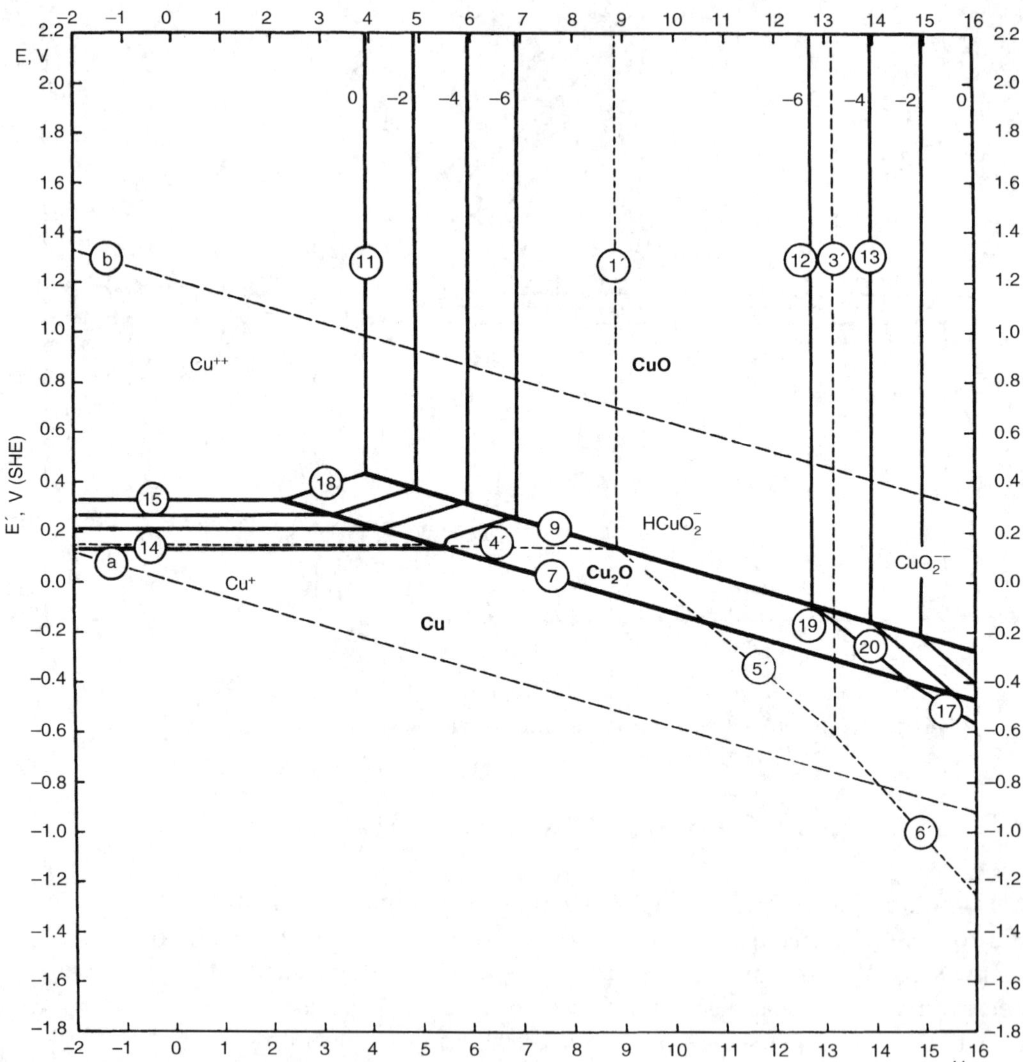


**Fig. 2.15** Pourbaix diagram for the copper/water system. Source: Ref 9

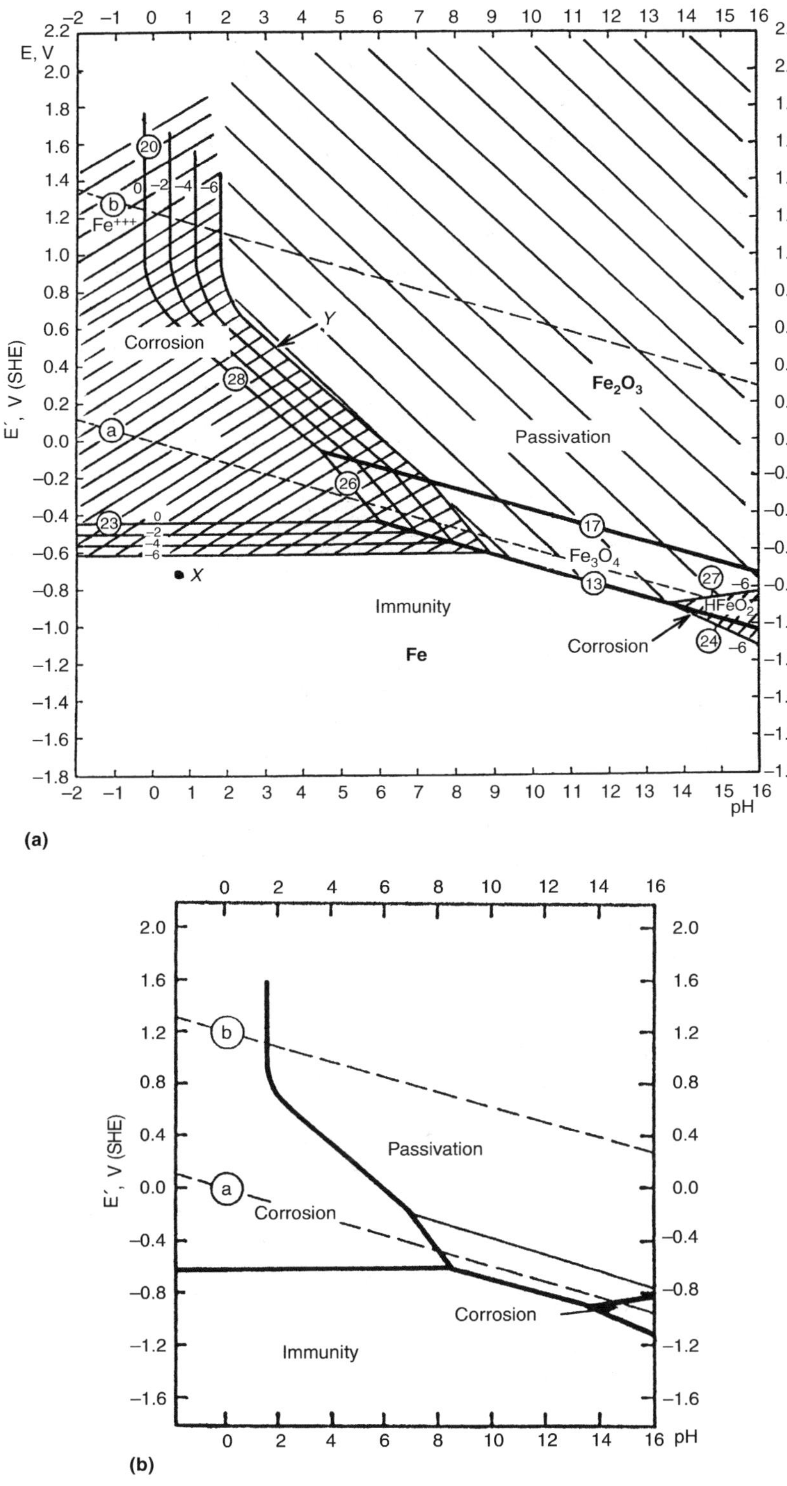


**Fig. 2.16** Pourbaix diagrams for the iron/water system. (a) Reproduction of Fig. 2.11 showing regions of corrosion, immunity, and possible passivation. (b) Form of the diagram frequently employed. Source: Ref 9, 10

- *Corrosion:* In these regions of potential and pH, the iron should ultimately become virtually all ions in solution, and therefore, iron exposed at these conditions should corrode.
- *Passivation:* In this region, the equilibrium state is one of oxide plus solution, meaningful only along a boundary such as Y in Fig. 2.16(a). If iron is placed in potential-pH environments along one of these boundaries, oxide will form on the surface. If this oxide is adequately adherent, nonporous, and has high resistance to ion and/or electron transport, it will significantly decrease the rate of corrosion. Under these conditions, the iron is said to have undergone passivation. These regions in Pourbaix diagrams would be more accurately identified as regions of "possible passivation."

The diagrams in Fig. 2.17 are taken from Pourbaix's *Atlas of Electrochemical Equilibria in Aqueous Solutions* as representative of how regions of immunity, corrosion, and passivation can be identified (Ref 9). Lines a (lower diagonal line) and b (upper diagonal line) are indicated for the possible cathodic reactions involving hydrogen ions and dissolved oxygen as discussed previously with respect to the corrosion of iron and copper. Relative to these diagrams, the regions for immunity, corrosion, and passivation for iron, copper, platinum, and tantalum should be compared. Platinum is corrosion resistant because its region of immunity extends over the entire pH range and to high potentials. Tantalum is corrosion resistant because its region of passivation extends over the entire pH range, and the oxide film that forms is adherent and nonporous; that is, the metal passivates even though the upper limit of the region of immunity is below line a, indicating that spontaneous corrosion should occur with evolution of hydrogen.

## Pourbaix Diagram Interpretations in Relationship to Corrosion

The following examples are with reference to the Pourbaix diagram for the lead/water system (Fig. 2.18) (Ref 9).

**Example 1a.** Use the Nernst half-cell equation for the hydrogen reaction at $P_{H_2} = 1$ atm and the Pb = $Pb^{2+}$ + 2e reaction at $a_{Pb^{2+}} = 10^{-6}$ to confirm the value of the pH at which line a intersects line 16 (–6).

The point of intersection of line 16 (–6) and line a corresponds to the equilibrium of lead at $a_{Pb^{2+}} = 10^{-6}$ with the hydrogen reaction. Since the intersection point represents equilibrium, imagine the following cell and set $E_{cell} = 0$.

$$\text{Pb}\left|\begin{matrix}\text{Pb}^{2+}\\ (a_{Pb^{2+}} = 10^{-6})\end{matrix}\right\|\begin{matrix}\text{H}^{+}\ (\text{pH} = ?)\\ \ \end{matrix}\left|\begin{matrix}\text{H}_2 \text{ on Pb}\\ (P_{H_2} = 1 \text{ atm})\end{matrix}\right.$$

LHE RHE

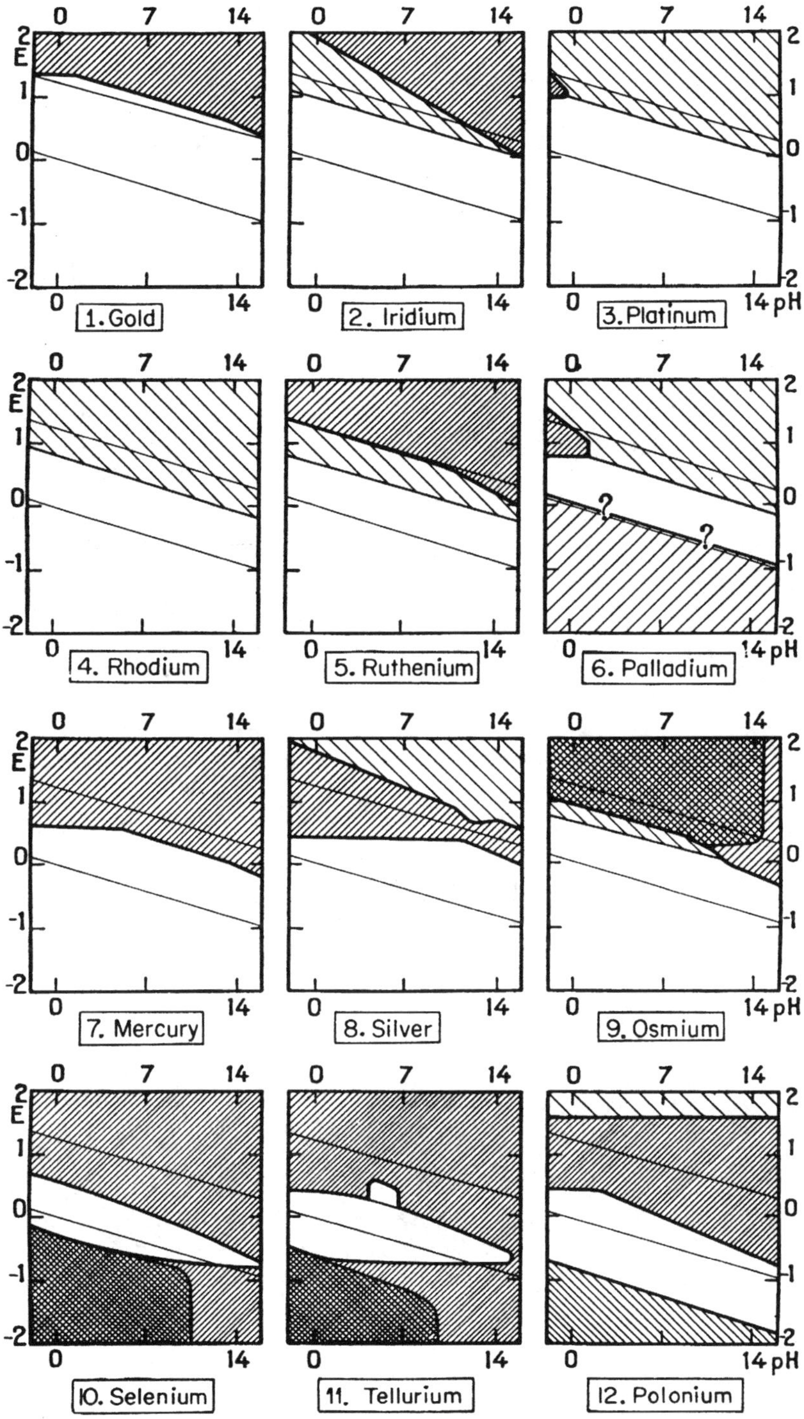


**Fig. 2.17** Pourbaix diagrams for selected metals showing regions of corrosion, immunity, and possible passivation. Source: Ref 9

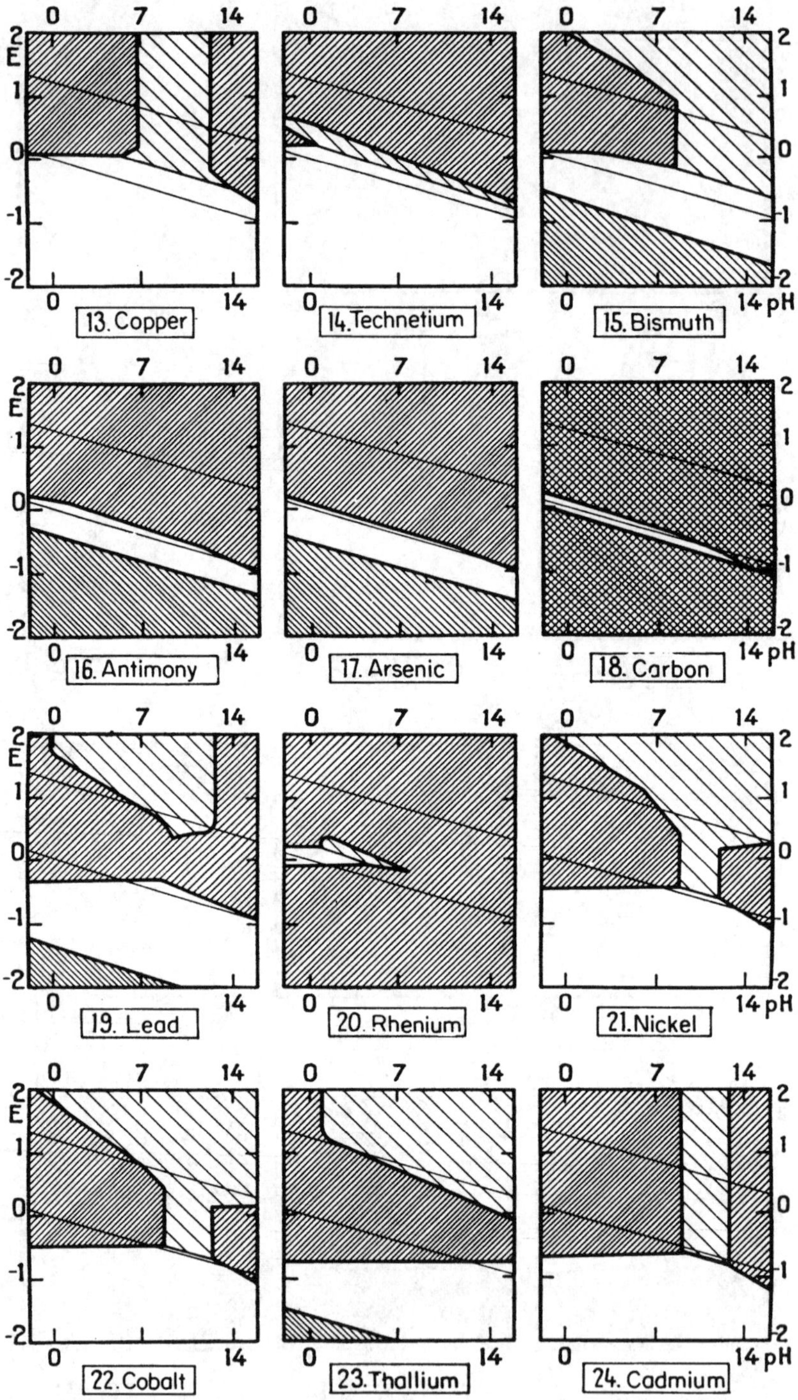


**Fig. 2.17 (continued)** Pourbaix diagrams for selected metals showing regions of corrosion, immunity, and possible passivation. Source: Ref 9

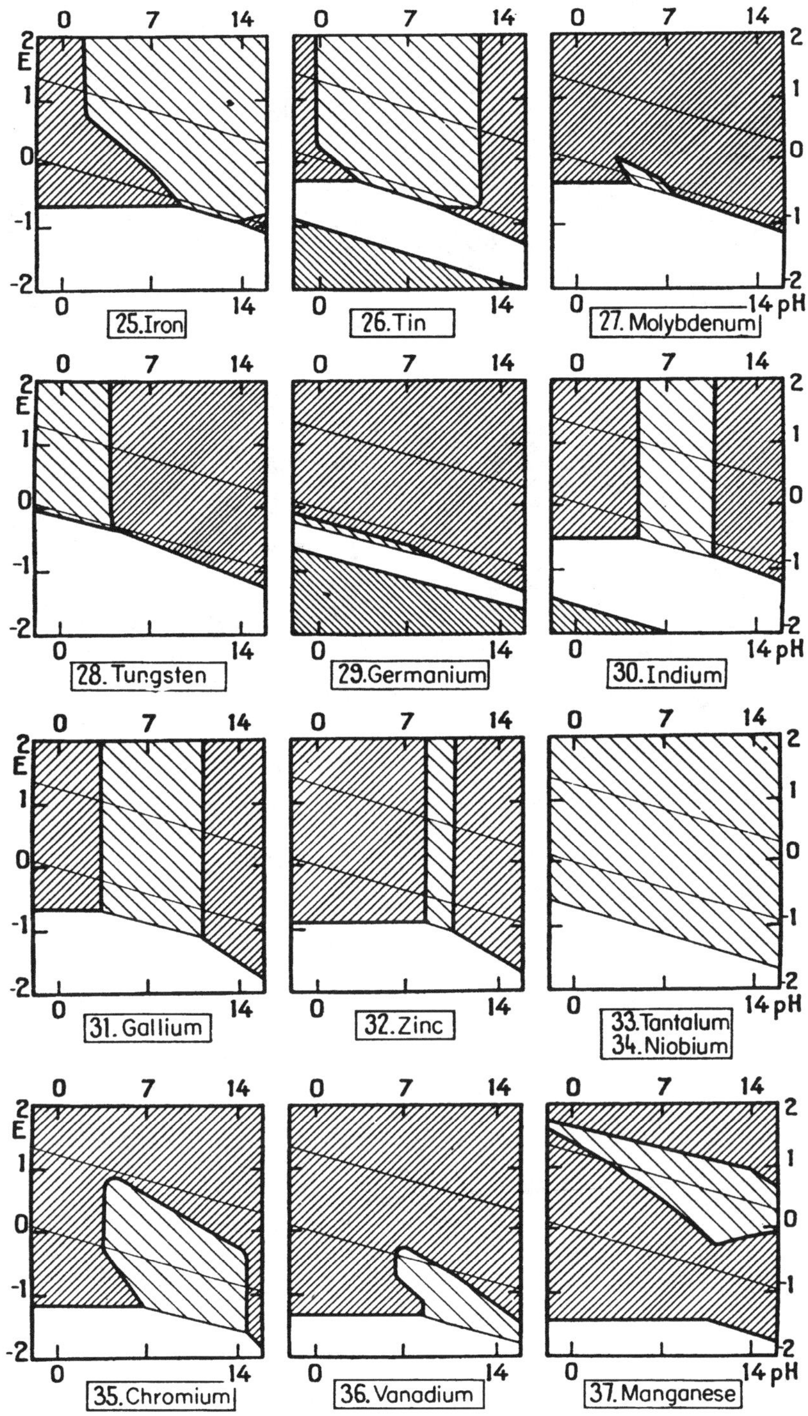


**Fig. 2.17 (continued)** Pourbaix diagrams for selected metals showing regions of corrosion, immunity, and possible passivation. Source: Ref 9

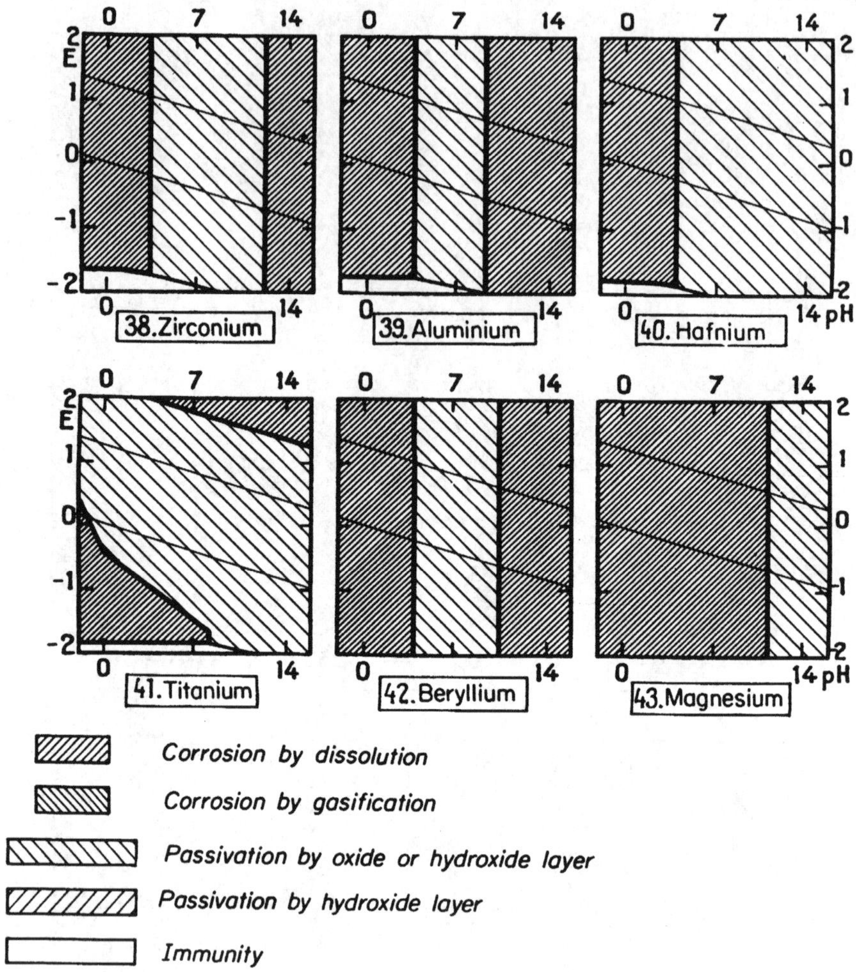


**Fig. 2.17 (continued)** Pourbaix diagrams for selected metals showing regions of corrosion, immunity, and possible passivation. Source: Ref 9

At LHE:

$$Pb \leftrightarrow Pb^{2+} + 2e$$

$$E'_{LHE} = E'_{Pb,Pb^{2+}} = -126 + \frac{59}{2}\log 10^{-6}$$

$$E'_{LHE} = -303 \text{ mV (SHE)}$$

At RHE:

$$2H^{+} + 2e \leftrightarrow H_2$$

$$E'_{RHE} = E'_{H_2,H^{+}} = 0 + \frac{59}{2}\log\left(\frac{a^2_{H^+}}{P_{H_2}}\right)$$

$$E'_{RHE} = 59 \log a_{H^+} = -59 \text{ pH}$$

$$E_{cell} = E'_{RHE} - E'_{LHE} = -59 \text{ pH} - (-303) = 0$$

$$\text{pH} = 5.13$$

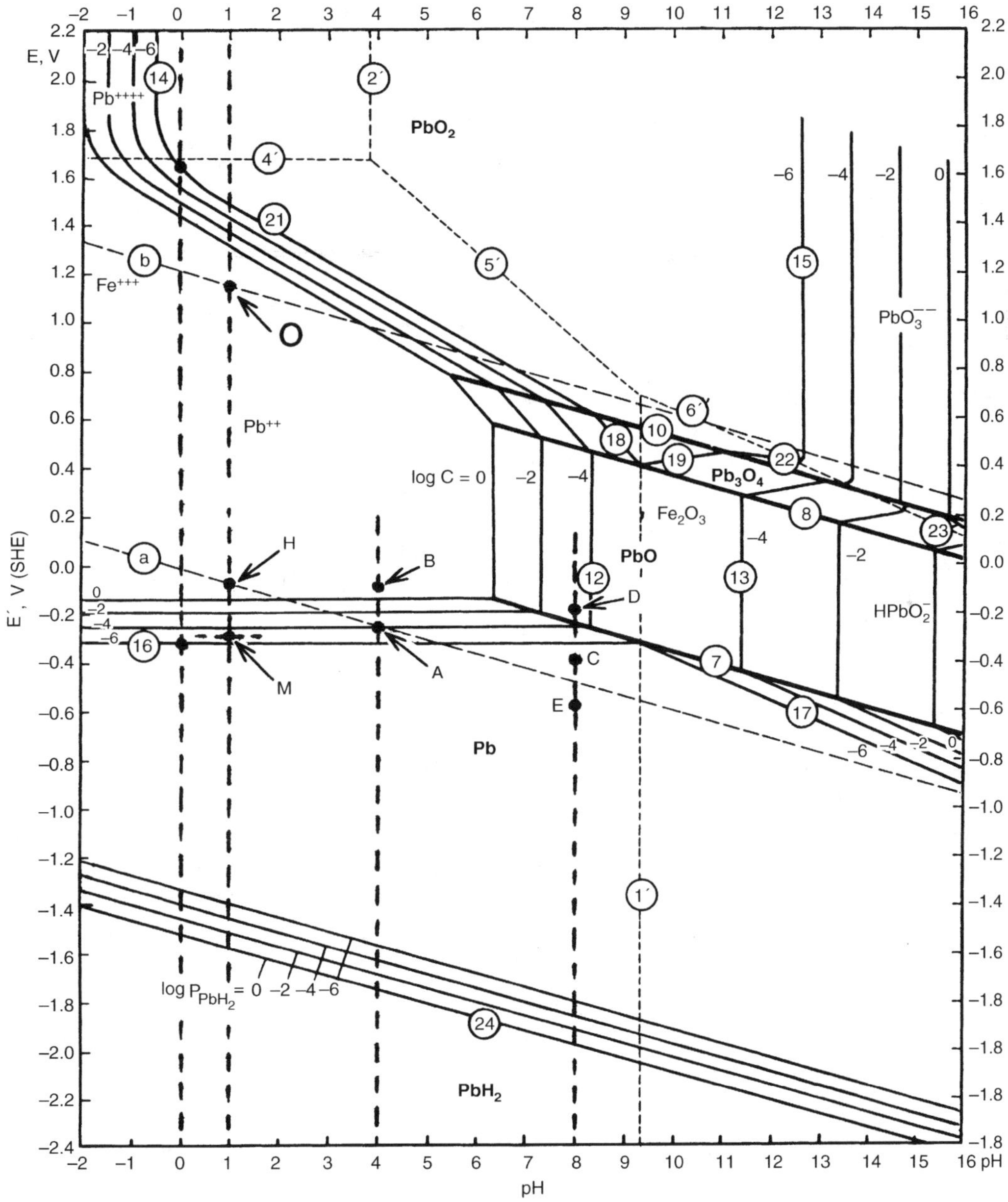


Fig. 2.18 Pourbaix diagram for the lead/water system. Based on Ref 9

These calculations lead to intersection coordinates of $E'_{Pb,Pb^{2+}} = -303$ and pH = 5.13, which agree with values read from the Pourbaix diagram.

**Example 1b.** Use the Pourbaix diagram to estimate $E_{cell}$ for the proposed corrosion of lead in deaerated solution at pH = 1 and $a_{Pb^{2+}} = 10^{-5}$.

Draw a vertical line at pH = 1 as shown. Draw a horizontal (constant potential) line midway between lines 16(–4) and 16(–6) to represent Pb in equilibrium with $Pb^{2+}$ at $a_{Pb^{2+}} = 10^{-5}$. Estimate the values of the potentials at which the vertical line at pH = 1 intersects lines a and 16(–5). The intersection with line a is approximately –60 mV (SHE) and with line 16(–5), approximately –270 mV (SHE). The former is more positive than the latter, which means that the lead tends to corrode. $E_{cell} = -60 - (-270) = +210$ mV.

**Example 1c.** Estimate $E_{cell}$ for the proposed corrosion of lead in contact with a solution at pH = 1, $a_{Pb^{2+}} = 10^{-5}$, and in equilibrium with oxygen at $P_{O_2} = 1$ atm.

From the intersection of the pH = 1 line and line b, $E'_{O_2,H^+} = 1170$ mV (SHE). $E_{cell} = 1170 - (-270) = +1440$ mV. Note that this value is much larger than the $E_{cell}$ with hydrogen as the cathodic reaction and, therefore, indicates a greater driving potential for corrosion.

In both Example 1b and Example 1c, the Pb in contact with $Pb^{2+}$ ions is the more negative or active of the half-cell pair and, hence, in the electrochemical cell would be the anode with a potential corresponding to point M in Fig. 2.18. The cathodic reaction for deaerated conditions is the hydrogen reaction at a potential corresponding to point H. The additional cathodic reaction under aerated conditions is the oxygen reaction at point O. If a potentiostat holds the lead at point M, the metal will not corrode, although hydrogen will be evolved and oxygen consumed because the potential is below lines a and b. If the potentiostat is removed, the lead will spontaneously corrode with the evolution of hydrogen in the deaerated case and with the consumption of dissolved oxygen in the aerated case. With the freely corroding metal, the question arises as to what a reference electrode measures relative to the metal when placed some distance from the surface. This can be determined quantitatively only if the kinetics of the electrode processes are known. Applications of electrode-kinetics principles for estimating corrosion potentials and rates are covered in later chapters. It is sufficient here to state that in the case of the deaerated solution, the measured potential cannot be more negative than the potential of point M nor more positive than point H. In the aerated case, theoretically, the potential could be as high as the potential of point O. The potential determined for a corroding surface is called the corrosion potential, $E_{corr}$, and is an important quantity in the analysis of corrosion behavior. If the corrosion potential is between points M and H, and the solution is aerated, the corrosion will be supported by electrons consumed by both the hydrogen and oxygen reactions. If the corrosion potential is above point H, this indicates that the dissolved oxygen has raised the potential into this region and is the single reaction consuming electrons and supporting corrosion. Depending on the electrode kinetics, either of these conditions could occur.

**Example 2.** According to the Pourbaix diagram for the lead/water system, if the lead is in contact with a solution of $Pb^{2+}$ ions at $a_{Pb^{2+}} = 10^{-6}$, line 16(–6), increasing the pH will eventually cause the lead to be coated with PbO, which can possibly decrease the corrosion rate. The reaction is:

$$Pb^{2+} + 2OH^- = PbO + H_2O$$

for which the equilibrium constant is:

$$K = \frac{1}{a_{Pb^{2+}} \cdot a^2_{OH^-}} = 2.22 \times 10^{15}$$

This equation can be used to calculate the pH at which PbO forms if $Pb^{2+}$ ions at $a_{Pb^{2+}} = 10^{-6}$ are in contact with metallic lead. From the equilibrium constant expression:

$$a^2_{OH^-} = \frac{1}{a_{Pb^{2+}} \cdot 2.22 \times 10^{15}} = \frac{1}{10^{-6} \cdot 2.22 \times 10^{15}} = 4.5 \times 10^{-10}$$

$$a_{OH^-} = 2.12 \times 10^{-5}$$

$$a_{H^+} = \frac{10^{-14}}{2.12 \times 10^{-5}} = 4.71 \times 10^{-10}$$

$$pH = -\log a_{H^+} = -\log(4.71 \times 10^{-10}) = 9.3$$

This value agrees with the coordinates of the three-phase equilibrium: Pb, $Pb^{2+}$ (a = $10^{-6}$), and PbO.

**Example 3.** The voltage of the common lead storage battery can be easily estimated from the diagram. The *negative electrode* consists of Pb in contact with solid $PbSO_4$ in $H_2SO_4$ at a pH of approximately zero. Under these conditions, Pb is in contact with $Pb^{2+}$ at approximately $a_{Pb^{2+}} = 10^{-6}$. The *positive electrode* is $PbO_2$ in contact with the same solution but $PbO_2$ is the only solid phase.

Draw a vertical line at pH = 0. The intersection with line 16(–6) is $E'_{Pb,Pb^{2+}} = -300$ mV (SHE). The intersection with line 21(–6) is $E'_{PbO_2,Pb^{2+}} = +1650$ mV (SHE). The former is the anode (negative electrode), and the latter is the cathode (positive electrode). The cell potential is:

$$E_{cell} = 1650 - (-300) = 1950 \text{ mV}$$

which is in close agreement with the accepted value of 2 V.

**Example 4.** Refer to points A through E as indicated on the Pourbaix diagram (Fig. 2.18). The state of the system at each point and the change in state when going from one point to another are to be interpreted:

- *What interpretation is given to point A?* Two-phase state consisting of metallic Pb at equilibrium with $Pb^{2+}$ in solution at $a_{Pb^{2+}} = 10^{-4}$ and pH = 4. Since point A is on line a, a potentiostat is not required to maintain the equilibrium in a deaerated solution, and hydrogen gas is not evolved.
- *What should happen on changing from point A to point B?* The pH remains unchanged at 4. Pb goes into solution as $Pb^{2+}$ and no metallic lead remains (this assumes that in attempting to reach the very high $Pb^{2+}$ activity of approximately $10^{+2}$, all the available metallic Pb undergoes dissolution). The potential is –50 mV (SHE). To achieve this potential in a deaerated solution, it would be necessary to use an inert electron-conducting material (e.g., platinum) as an auxiliary electrode and to hold the potential of the Pb with a potentiostat. Hydrogen gas is not evolved.
- *What interpretation is given to point C?* At pH = 8, the system is in a two-phase state with metallic lead in equilibrium with $Pb^{2+}$ ions in solution at $a_{Pb^{2+}}$ estimated to be $10^{-8}$. The potential to hold this equilibrium would be approximately –400 mV (SHE) and, certainly in a deaerated solution, would be held with a potentiostat. No hydrogen gas is evolved.
- *What should happen on changing from point C to point D?* Metallic Pb goes into solution. The activity of $Pb^{2+}$ becomes approximately $10^{-3}$, and PbO precipitates. The final state is two phase with $Pb^{2+}$ ions in equilibrium with PbO at pH = 8 and $E' = -200$ mV (SHE). In a deaerated solution, the state would be held with an inert electrode connected to a potentiostat. No $H_2$ is evolved.
- *What would happen on changing from point C to point E?* The pH remains at 8. Metallic Pb deposits on the existing Pb. The $Pb^{2+}$ ion concentration decreases to a very low value estimated to be $10^{-12}$. A potentiostat would be used to hold the potential at –600 mV (SHE). Hydrogen would be evolved because point E is below line a.
- *What interpretation is given to line 4′?* This is the condition at which the ratio of activities of $Pb^{2+}$ to $Pb^{4+}$ is equal to unity.

From the standpoint of corrosion, there could be a significant difference in behavior between a potential change from A to B as compared to a change from C to D. In both cases, there is a driving force to corrode, and in fact, at equilibrium, all metallic lead will disappear. However, on changing from C to D, PbO will first form on the metallic lead surface, and if this coating is adherent and nonporous, the corrosion rate may be very small since the continuation of the process will depend on the solid-state diffusion of ions through the oxide coating. This mechanism of material transport will generally result in low corrosion rates.

If Pb is in contact with a strongly alkaline, aerated solution, at, for example, pH = 14, corrosion is thermodynamically possible with the formation of $HPbO_2^-$ ions. For example, if the activity of $HPbO_2^-$ is $10^{-6}$, the equilibrium potential is –800 mV (SHE). Higher potentials caused by dissolved oxygen would result in corrosion of Pb to $HPbO_2^-$.

## Chapter 2 Review Questions

1. The difference in electrical potential of a cell made up of a Zn electrode (anode) and $H_2$ electrode (cathode) immersed in 0.5 m $ZnCl_2$ is +590 mV (i.e., with Zn as the LHE, $E_{cell}$ = +590 mV). What is the pH of the solution? $\gamma_{Zn^{2+}}$ at this concentration is estimated to be 0.38.
2. Tin cans are made from tin-coated steel. At breaks in the tin coating, both tin and iron are in contact with the contained solution. If tin ions ($Sn^{2+}$) and iron ions ($Fe^{2+}$) are in the solution, then the following reaction is to be considered:

$$Fe^{2+} + Sn \leftrightarrow Fe + Sn^{2+}$$

   a. In estimating the tendency for this reaction to occur in either direction, approximate values of $a_{Fe^{2+}}$ and $a_{Sn^{2+}}$ are required. Assume initially that $a_{Fe^{2+}} = a_{Sn^{2+}} = 10^{-5}$. Determine $E_{cell}$ and $\Delta G$ for the above reaction and conclude whether the iron is protected from corrosion by the tin.
   b. If a complexing agent is in the solution that reduces the $a_{Sn^{2+}}$ to very low values, determine what this value must be to bring the above reaction to equilibrium if $a_{Fe^{2+}} = 10^{-5}$.
3. Calculate the theoretical tendency for nickel to corrode ($E_{cell}$) in deaerated water (pH = 7). Assume the corrosion product is $Ni(OH)_2$, the solubility product of which is $1.6 \times 10^{-16}$. (Ref 11)
4. Determine whether silver will corrode with $H_2$ gas evolution (1 atm) in deaerated KCN solution under the conditions: pH = 8, $a_{CN^-} = 1.0$, $a_{Ag(CN)_2^-} = 0.001$. (Ref 11)
5. Determine the pressure of hydrogen required to stop corrosion of iron immersed in a deaerated 0.1 m $FeCl_2$ solution at pH = 3. Assume $\gamma_{Fe^{2+}} = 1.0$. (Ref 11)

6. Determine the pressure of hydrogen required to stop corrosion of iron in deaerated water with $Fe(OH)_2$ as the corrosion product. The solubility product for $Fe(OH)_2$ is $1.6 \times 10^{-14}$. Assume pH = 7.0. (Ref 11)
7. The rate of corrosion of many metals is greatly influenced by oxygen dissolved in the solution from air. The presence of oxygen is responsible for the following important cathodic reaction:

$$\frac{1}{4}O_2 + \frac{1}{2}H_2O + e \rightarrow OH^- \qquad E^o = +401 \text{ mV (SHE)}$$

Another important cathodic reaction is:

$$H^+ + e \rightarrow \frac{1}{2}H_2 \qquad E^o = 0 \text{ mV (SHE)}$$

which is frequently the most important reaction in deaerated solutions. Calculate the pressure of oxygen in equilibrium with a solution that is required to make these two cathodic reactions equally possible. The conditions are to be taken as pH = 3 and $P_{H_2} = 1$ atm.
8. Lead is used as a construction material to contain sulfuric acid because of the formation of an adherent coating of $PbSO_4$. Calculate the driving potential, $E_{cell}$, for the corrosion of lead in 1 m $H_2SO_4$. The pH is –0.3, $a_{SO_4^{2-}} = 1.0$, and $K_{sp}$ is $1.3 \times 10^{-8}$ for $PbSO_4$.
9. In considering the corrosion of iron in deaerated solutions, the reaction $H^+ + e \rightarrow \frac{1}{2} H_2$ is the usual cathodic reaction in acid solution, this reaction becoming less favorable as the acidity is decreased.

   a. Calculate the pH at which the hydrogen reaction is no longer thermodynamically possible as a cathodic reaction if the solution contains $10^{-3}$ m $FeCl_2$ in contact with iron.
   b. Show, however, that Fe cannot be at equilibrium with $a_{Fe^{2+}} = 10^{-3}$ at this pH because of the formation of $Fe(OH)_2$, which was not considered in part a. For $Fe(OH)_2$, $K_{sp} = 1.95 \times 10^{-15}$.
   c. Does the Pourbaix diagram for iron, which considers $Fe(OH)_2$ and $Fe(OH)_3$ as possible additional solid phases, indicate that the hydrogen evolution reaction is thermodynamically possible at all values of pH? That is, will iron tend to corrode at all values of pH in deaerated solutions?
10. a. What conclusion is made if the same calculation as in part (a) of problem 9 is made for Cu in contact with deaerated $10^{-3}$ m $CuCl_2$? Assume that the maximum reasonable acidity corresponds to pH = –1.0.

b. What conclusion is made concerning the possibility of copper corrosion in aerated acid of pH = –1.0 containing $10^{-3}$ m $CuCl_2$ if the cathodic reaction is:

$$O_2 + 4H^+ + 4e \rightarrow 2H_2O \qquad E^o = 1229 \text{ mV (SHE)}$$

and $P_{O_2} = 0.2$ atm?

11. Compare the tendencies for nickel to corrode under the following conditions:

a. Deaerated water:

$$Ni \leftrightarrow Ni^{2+} + 2e,\ a_{Ni^{2+}} = 10^{-4},\ pH = 7,$$

$$P_{H_2} = 1 \text{ atm}$$

b. Deaerated water contaminated with $H_2S$:

$$pH = 4,\ a_{S^{2-}} = 10^{-12},\ P_{H_2} = 1 \text{ atm}$$

$$Ni + S^{2-} \leftrightarrow NiS \text{ (ppt)} + 2e,\ E^o = -1040 \text{ mV (SHE)}$$

12. A copper storage tank containing dilute $H_2SO_4$ at pH = 0.1 is blanketed with $H_2$ gas at 1 atm. Calculate the maximum $Cu^{2+}$ contamination of the acid expressed as $a_{Cu^{2+}}$. (Ref 11)

13. Consider that you are required to find a method for removing by selective corrosion the tin coating from tinned copper wire. It is proposed to dip the tinned wire into a solution containing $Fe^{3+}$ and $Fe^{2+}$ ions.

a. Discuss why it is reasonable to consider that a solution of these ions might be used for this purpose.

b. Determine the ratio, $a_{Fe^{3+}} / a_{Fe^{2+}}$, which would remove the tin without corroding the copper. Is this a reasonable ratio to attempt to control for the practical removal of the tin? Explain. Assume the allowable $a_{Cu^{2+}} = 10^{-4}$.

14. An alternative suggestion for the removal of tin from "tinned" copper wire (see problem 13) is by dipping the wire into a solution containing $Sn^{2+}$ ions at an activity of $10^{-2}$ and $Sn^{4+}$ ions at the same activity of $10^{-2}$. Assume a large amount of solution relative to the amount of tin to be removed. Also assume that if the copper corrodes, it does so as $Cu^{2+}$ and that the solution contains $Cu^{2+}$ at $a_{Cu^{2+}} = 10^{-4}$.

a. Determine whether this solution will remove the tin as $Sn^{2+}$.

b. Will the solution corrode the copper if the tin is removed?

15. In making printed electric circuit boards, ferric chloride ($FeCl_3$) is used to corrode exposed copper on a plastic substrate.

    a. From the following data, calculate the potential of the cell inducing corrosion:

$$a_{Cu^{2+}} = 10^{-3},\ a_{Fe^{2+}} = 10^{-3},\ a_{Fe^{3+}} = 1$$

    b. As the corrosion solution is continually used, the $Cu^{2+}$ and $Fe^{2+}$ activities will increase, and the activity of the $Fe^{3+}$ will decrease until corrosion no longer occurs. Calculate the value of the $a_{Fe^{3+}}/a_{Fe^{2+}}$ ratio when the $Cu^{2+}$ activity has increased to 1.0.

16. Silver is usually assumed to be chemically inert and therefore might be considered for use in photographic processing equipment, for example, to contain acid "hyposolution" (sodium thiosulphate). From the following data, determine whether silver is satisfactory for this solution, considering that silver forms the complex ion, $Ag(S_2O_3)_2^{3-}$. The half-cell reactions are:

$$Ag + 2S_2O_3^{2-} \leftrightarrow Ag(S_2O_3)_2^{3-} + e$$

$$\frac{1}{2}H_2 \leftrightarrow H^+ + e$$

    Assume the maximum allowed values:

$$a_{Ag(S_2O_3)_2^{3-}} = 10^{-5},\ a_{S_2O_3^{2-}} = 1.0,\quad pH = 3.5$$

17. From the following data, calculate the potential of the calomel half cell in 0.1 m KCl. Half-cell reaction wanted:

$$2Hg + 2Cl^- = Hg_2Cl_2 + 2e, \qquad E' = ?$$

    Other data:

$$2Hg = Hg_2^{2+} + 2e, \qquad E^o = +796\ mV\ (SHE)$$

$$\left(a_{Hg_2^{2+}}\right)\left(a_{Cl^-}\right)^2 = 2\times10^{-18}, \qquad \gamma_{Cl^-} = 0.77\ at\ 0.1\ m$$

The following questions refer to the Pourbaix diagram for the nickel/water system as shown in Fig. 2.19 (Ref 9):

18. From half-cell data and the Nernst half-cell equation for the $Ni,Ni^{2+}$ and $H_2,H^+$ reactions, confirm the point of intersection of lines 9(–6) and a.

19. Over what range of pH is nickel thermodynamically stable in deaerated water if $a_{Ni^{2+}} = 10^{-6}$?

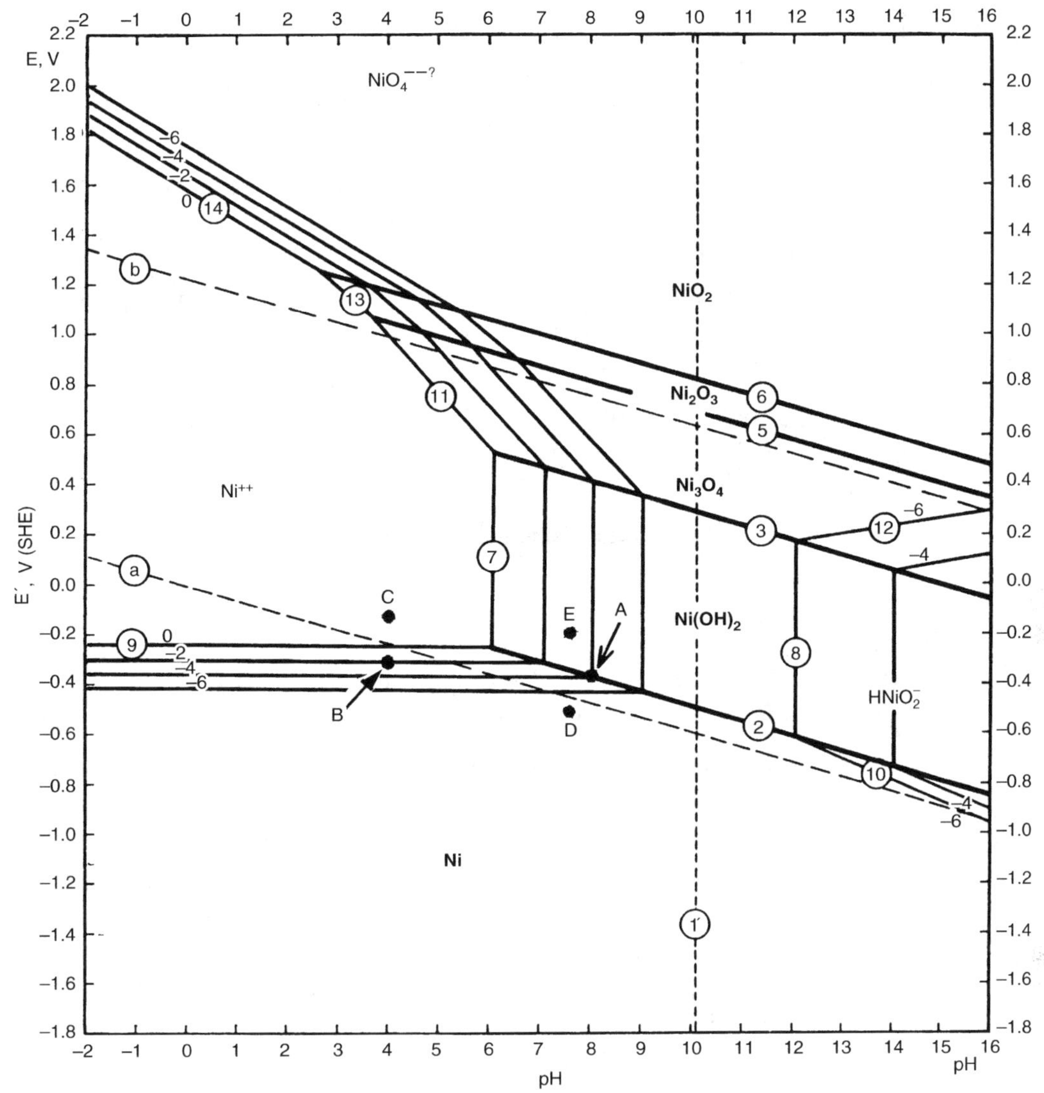


**Fig. 2.19** Pourbaix diagram for the nickel/water system. Based on Ref 9

20. In determining the conditions for the 3-phase equilibrium involving Ni, $Ni(OH)_2$, and $Ni^{2+}$, Pourbaix used $1.0 \times 10^{-16}$ as the solubility product for $Ni(OH)_2$. Using this value, confirm the conditions represented by point A.
21. At what potential should Ni be held in order to not corrode to form $HNiO_2^-$ in concentrated caustic of pH = 14? Assume that the allowed activity of $HNiO_2^-$ is $10^{-6}$.
22. Interpret point B. What should happen on changing from point B to C?
23. Interpret point D. What should happen on changing from point D to E?

## Answers to Chapter 2 Review Questions

1. pH = 3.3
2. a. With Sn|$Sn^{2+}$ as the LHE, $E_{cell}$ = −304 mV, ΔG = +58,700 J/mol of Fe; no, Fe is not protected from corrosion.
   b. $a_{Sn^{2+}} = 5.0 \times 10^{-16}$
3. With Ni|$Ni^{2+}$ as the LHE, $E_{cell}$ = −105 mV; Ni will not corrode.
4. With Ag, $CN^-$|$Ag(CN)_2^-$ as the LHE, $E_{cell}$ = +15 mV; yes, Ag will corrode.
5. $P_{H_2} = 8.3 \times 10^9$ atm
6. $P_{H_2} = 5.14$ atm
7. $P_{O_2} = 6.5 \times 10^{-84}$ atm
8. With Pb|$Pb^{2+}$ as the LHE, $E_{cell}$ = +376 mV
9. a. pH = 9.0
   b. $a_{Fe^{2+}} = 1.95 \times 10^{-5}$
   c. Yes
10. a. pH = −4.30, an impossibly high $H^+$ concentration
    b. With Cu|$Cu^{2+}$ as the LHE, $E_{cell}$ = +1024 mV; copper will corrode.
11. a. With Ni|$Ni^{2+}$ as the LHE, $E_{cell}$ = −40 mV; Ni will not corrode.
    b. With Ni, $S^{2-}$|NiS as the LHE, $E_{cell}$ = +450 mV; Ni will corrode.
12. $a_{Cu^{2+}} = 1.62 \times 10^{-12}$
13. a. Could possibly adjust the equilibrium potential of the $Fe^{3+} + e = Fe^{2+}$ reaction so that it would be higher than $E'_{Sn,Sn^{2+}}$ but lower than $E'_{Cu,Cu^{2+}}$.
    b. $a_{Fe^{3+}} / a_{Fe^{2+}} = 5.37 \times 10^{-10}$; no, not a practical ratio to control.
14. a. Yes; with Sn|$Sn^{2+}$ as the LHE, $E_{cell}$ = +345 mV.
    b. No; with Cu|$Cu^{2+}$ as the LHE, $E_{cell}$ = −74 mV.
15. a. With Cu|$Cu^{2+}$ as the LHE, $E_{cell}$ = +694 mV.
    b. $a_{Fe^{3+}} / a_{Fe^{2+}} = 5.37 \times 10^{-8}$
16. With Ag|$Ag^+$ as the LFE, $E_{cell}$ = +71 mV; Ag will corrode.
17. E′ = +339 mV (SHE)
19. pH ≥ 7.2
21. E ≤ −820 mV (SHE)

## References

1. D.R. Gaskell, *Introduction to Metallurgical Thermodynamics*, Taylor and Francis, 1981

2. D.R. Lide., Ed., *CRC Handbook of Physics and Chemistry,* CRC Press, 1997
3. J.O. Bockris and A.K.N. Reddy, *Modern Electrochemistry,* Plenum Press, 1973
4. J.M. West, *Electrodeposition and Corrosion Processes,* D. Van Nostrand Co., New York, 1965
5. D.J.G. Ives and G.J. Janz, *Reference Electrodes,* NACE International, Reprint, 1996
6. J.M. West, *Basic Corrosion and Oxidation,* Halsted Press, 1980
7. A.J. Bard, R. Parsons, and J. Jordan, *Standard Potentials in Aqueous Solutions,* Marcel Dekker, Inc., 1985
8. A.J. de Bethune and N.A.S. Loud, *Standard Aqueous Electrode Potentials and Temperature Coefficients at 25 °C,* Hampel, Skokie, IL, 1964
9. M. Pourbaix, *Atlas of Electrochemical Equilibria,* Pergamon Press, 1974
10. M. Pourbaix, *Lectures on Electrochemical Corrosion,* Plenum Press, 1973
11. H.H. Uhlig, *Corrosion and Corrosion Control,* John Wiley & Sons, 1971

CHAPTER 3

# Kinetics of Single Half-Cell Reactions

Electrochemical cells associated with corrosion obviously are not at equilibrium. Net anodic and cathodic currents flow to and from the surface over areas that can vary in size from atomic dimensions to large macroscopically identifiable areas. Any local region of the metal/solution interface is either consuming electrons from the metal, appearing as a local cathodic reaction, or releasing electrons to the metal, appearing as a local anodic reaction. For example, although a given region may be consuming electrons, a single cathodic reaction over this region may not be responsible. Rather, corrosion may be occurring over the region but with the cathodic reaction rate exceeding the anodic reaction rate, the imbalance being supplied by electrons from regions external to the immediate area. With respect to either of these regions, neither the source of the electrons at the cathodic region, nor the sink for the electrons at the anodic region, is important other than how they determine the current density. These sources or sinks may, therefore, be considered as external to the local region and may be due to nearby or remote half-cell reactions or sources of current either purposefully or accidentally introduced into the regions from remote batteries, power supplies, or electrical equipment. The latter are frequently referred to as stray, or leakage, current sources unless imposed intentionally by potentiostats or galvanostats.

Regardless of the cause of the electron flow at the interface, deviations of the half-cell potentials along the interface from their equilibrium values are functions of the current density. These deviations reflect the polarization behavior of the reaction, a phenomenon of fundamental importance in all electrochemical processes, including corrosion. In this text, the term *polarization* is used in a general sense, referring to either a change in the potential relative to the equilibrium half-cell potential, E′ (as used in the present chapter), or relative to the open-circuit corrosion potential, $E_{corr}$ (as used in later chapters).

Polarization behavior relates to the kinetics of electrochemical processes. Study of the phenomenon requires techniques for simultaneously measuring electrode potentials and current densities and developing empirical and theoretical relationships between the two. Before examining some of the simple theories, experimental techniques, and interpretations of the observed relationships, it is useful to characterize the polarization behavior of several of the important electrochemical reactions involved in corrosion processes.

Historically, Faraday observed that single-electrode half-cell potentials shifted from their equilibrium values when current passed through electrochemical cells. This deviation is referred to as *overpotential* or *overvoltage*. It is generally designated as η and is defined by the relationship:

$$\eta = E(i) - E' \qquad \text{(Eq 3.1)}$$

where E(i) is the potential represented as a function of current density, i, and E′ is the equilibrium half-cell potential, which would exist with no current and can be calculated from the Nernst half-cell equation. In 1905, Tafel observed that for a number of electrode reactions, η could be expressed in the form:

$$\eta = A + B \log i \qquad \text{(Eq 3.2)}$$

where A and B are constants (Ref 1).

It is shown subsequently that simple electrode-kinetics theory leads to the following equations for the oxidation and reduction half-cell reactions, respectively:

$$\eta = \beta_{ox} \log\left(\frac{i_{ox}}{i_o}\right) = -\beta_{ox} \log i_o + \beta_{ox} \log i_{ox} \qquad \text{(Eq 3.3a)}$$

$$\eta = -\beta_{red} \log\left(\frac{i_{red}}{i_o}\right) = \beta_{red} \log i_o - \beta_{red} \log i_{red} \qquad \text{(Eq 3.3b)}$$

where $\beta_{ox}$ and $-\beta_{red}$ are constants equal to the slopes of the straight lines produced by plots of η versus log $i_{ox}$ and η versus log $i_{red}$, respectively, and $i_o$ is a characteristic parameter of the half-cell reaction called the *exchange current density*. It is evident that Eq 3.3(a) and 3.3(b) are in the form of Eq 3.2. From Eq 3.1, the potential can be written as:

$$E(i) = E' + \eta \quad \text{(Eq 3.4)}$$

Substituting Eq 3.3(a) and the Nernst half-cell equation into Eq 3.4 gives an expression for the potential as a function of current density for the oxidation of a metal, M ($M \rightarrow M^{m+} + me$):

$$E_{ox,M} = E'_{M,M^{m+}} + \beta_{ox,M} \log \frac{i_{ox,M}}{i_{o,M}}$$

$$= E^{o}_{M,M^{m+}} + \frac{59}{m} \log a_{M^{m+}} + \beta_{ox,M} \log \frac{i_{ox,M}}{i_{o,M}} \quad \text{(Eq 3.5)}$$

where $E_{ox,M}$, $E'_{M,M^{m+}}$, and $E^{o}_{M,M^{m+}}$ are expressed in millivolts relative to the standard hydrogen electrode, mV (SHE), and

$i_{ox,M}$ = oxidation or anodic current density, mA/m$^2$
$i_{o,M}$ = exchange current density for the reaction $M = M^{m+} + m$, mA/m$^2$
$\beta_{ox,M}$ = slope of the oxidation overpotential curve, mV/(log decade)

This single equation is plotted in Fig. 3.1(a). Note that when $i_{ox,M}$ is equal to $i_{o,M}$, the last term is zero, and $E_{ox,M}$ becomes equal to the equilibrium potential, $E'_{M,M^{m+}}$. For the oxidation reaction, the slope of the curve, $\beta_{ox,M}$, is positive. Hence, as the current density is increased, the potential moves in the positive direction. For the reduction reaction, as shown in Fig. 3.1(b), the slope of the curve, $-\beta_{red,M}$, is negative, although the curve must go through the same $i_{o,M}$. The potential for the reduction reaction ($M^{m+} + me \rightarrow M$) is expressed as:

$$E_{red,M} = E'_{M,M^{m+}} - \beta_{red,M} \log \frac{i_{red,M}}{i_{o,M}}$$

$$= E^{o}_{M,M^{m+}} + \frac{59}{m} \log a_{M^{m+}} - \beta_{red,M} \log \frac{i_{red,M}}{i_{o,M}} \quad \text{(Eq 3.6)}$$

The two curves usually are plotted on the same coordinates as shown in Fig. 3.1(c), which more clearly emphasizes that they cross at $i_{o,M} = i_{ox,M} = i_{red,M}$.

The linear relationships shown for E as a function of log i are frequently observed for only small deviations from equilibrium. It is shown subsequently that the linear relationship corresponds to an upset in the mechanism of transfer of the ions between the metal and the solution and is termed *charge-transfer polarization*. As the potential is changed progressively from E′, the curves deviate from linearity (Fig. 3.2). Along the reduction branch, $E_{red,M}$ becomes more negative than the linear relationship would indicate. This additional deviation is due to removal of metal ions from the solution in the vicinity of the interface at a rate such that diffusion of the ions in the solution toward the inter-

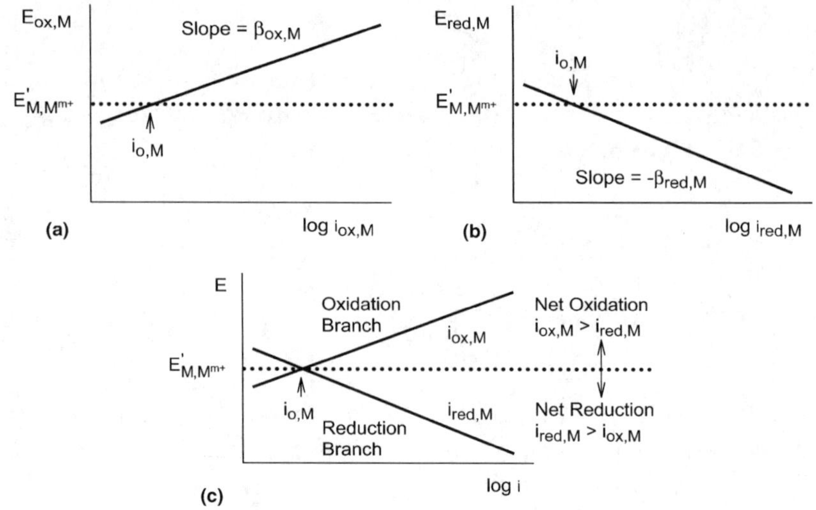


**Fig. 3.1** Polarization curves illustrating charge-transfer polarization (Tafel behavior) for a single half-cell reaction. (a) Anodic polarization. (b) Cathodic polarization. (c) Both anodic and cathodic polarization

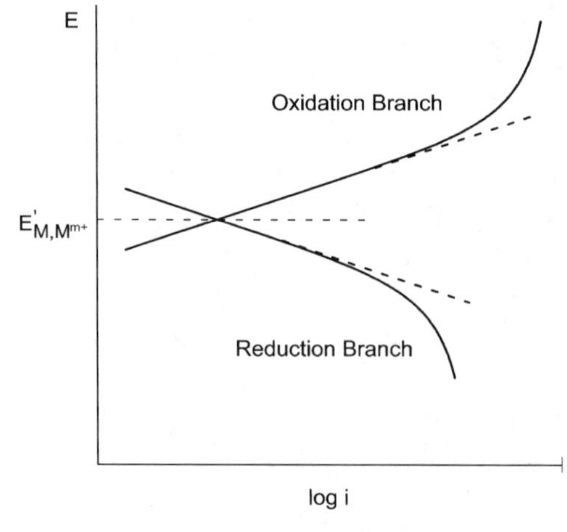


**Fig. 3.2** Deviation in the linear Tafel relationships at higher current densities due to diffusion or other current limiting processes

face becomes a rate-determining factor. Along the oxidation branch, $E_{ox,M}$ becomes more positive than the linear relationship would indicate. In this region, ions are passing into solution faster than they can diffuse into the bulk of the solution, and this diffusion process becomes rate determining.

As the potential is progressively increased to produce oxidation of a metal electrode, a critical potential may be reached at which the current density decreases significantly as indicated by the $E_{ox,M}$ versus log $i_{ox,M}$ relationship shown in Fig. 3.3. For systems showing this behavior, the decrease in current density from a to b is associated with formation of a

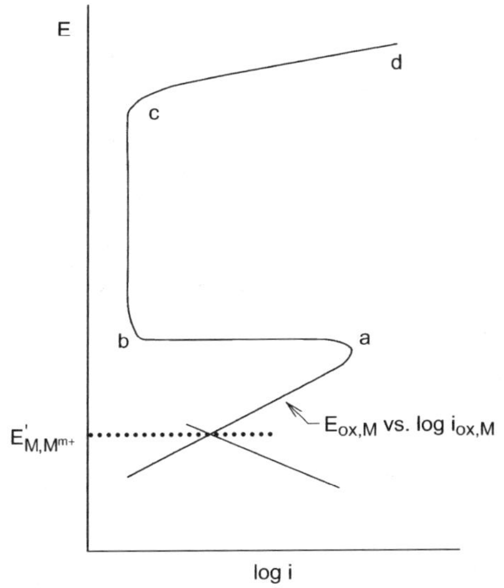


**Fig. 3.3** Anodic polarization curve representative of active/passive alloys. Oxide films forming in the potential range a to c cause a decrease in current density.

precipitate phase along the interface, usually an oxide. The more adherent and nonporous the precipitate film is, the greater the decrease in current density will be. From b to c, the film remains protective and grows in thickness holding log i to small values. This occurs even though the oxidizing conditions are increasing (i.e., to make the metal progressively more positive requires removal of electrons, which can only occur by oxidizing the metal at the interface). A potential may be reached, near c, at which new ionic species may form, and if these are soluble, the current density may increase along c to d. In this region of the polarization curve, the protective film formed at lower potentials is observed to disappear, and corrosion rates may become very large.

Curves of the types described are observed for all electrochemical reactions. The curves differ greatly in shape and position, which reflects differences in electrode processes and, in particular, the kinetic mechanisms of the electrode processes.

## The Exchange Current Density

The linear E versus log i curve, reflecting Tafel-type behavior, is referred to as *charge-transfer polarization* because it is associated with the actual separation of charge at the electrode interface. In the case of a metal, charge transfer involves either transfer of a metal ion into the solution and an electron(s) into the metal (oxidation or corrosion) or the combination of a metal ion in solution with an electron(s) to form an effectively neutral atom (reduction or electroplating). In the case of a reaction involving species in solution only (referred to as a redox reaction), such as the $H_2 \leftrightarrow 2H^+ + 2e$ reaction, electrons are transferred to or from the metal phase with either the formation of $H^+$ ions from $H_2$ molecules or the formation of $H_2$ molecules from $H^+$ ions; the metal substrate itself does not enter into the reaction.

For a single half-cell reaction at equilibrium, a dynamic state exists in which charges move in equal numbers in each direction across the interface, as represented in Fig. 3.4. The kinetic activity of this dynamic equilibrium may be expressed as the number of ions transferring in either direction per unit area per unit time. Since ions are transferred, the movement also may be expressed as charges transferred per unit area per unit time, or equivalently, as the current density, i (millicoulombs/(s · m$^2$) or mA/m$^2$). Positive ions passing into solution constitute an oxidation component of the current density, $i_{ox,M}$; positive ions passing from the solution account for the reduction component of the current density, $i_{red,M}$. At equilibrium:

$$i_{ox,M} = i_{red,M} = i_{o,M} \qquad \text{(Eq 3.7)}$$

where $i_{o,M}$ is called the *exchange current density*; it is a measure of the kinetic activity of the half-cell reaction at equilibrium and is an important parameter in the analysis of corrosion. Values of $i_o$ vary from the order of $10^{-7}$ to $10^{+5}$ mA/m$^2$.

Theoretical electrochemistry is concerned with developing models for these charge transfer processes and with deriving mathematical expressions based on these models from which values of the exchange current density may be calculated. It is sufficient for present purposes to examine one particularly simple model and derive, semiquantitatively, expressions for the exchange current density. Details of the model and the derivations are open to argument, but the result is of a mathematical form that is observed experimentally for a number of half-cell reactions.

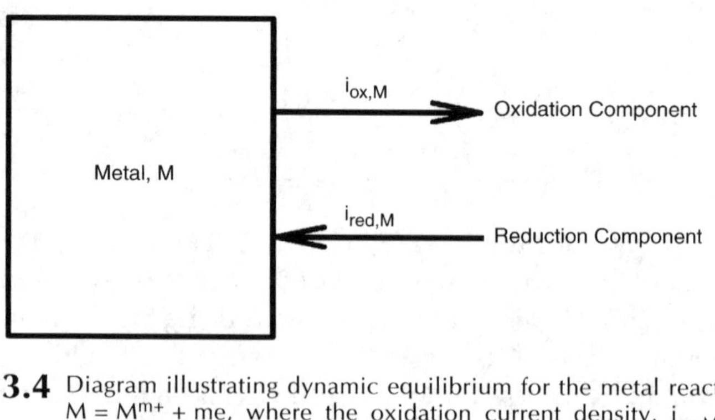


**Fig. 3.4** Diagram illustrating dynamic equilibrium for the metal reaction $M = M^{m+} + me$, where the oxidation current density, $i_{ox,M}$, is equal to the reduction current density, $i_{red,M}$.

As illustrated in Fig. 3.5, the ion in the metal, $M^{m+}$, is surrounded by other metal ions and by electrons (Ref 2). In the solution, the metal ion, being positively charged, is surrounded by oriented polar water molecules, this configuration lowering the energy of electrostatic attraction between the negative poles of the water molecules and the positive metal ion. In transferring between metal and solution, the ion must pass through configurations of higher energy than exist in either end state. For the condition of dynamic equilibrium, the electrochemical free energy, $G_{el}$, will be a function of the path between the two minimum energy positions but will be the same in the end positions as a consequence of the equilibrium. The electrochemical free energy as a function of distance from metal to solution through the interface is shown schematically in Fig. 3.6 (Ref 3, 4).

The chemical and electrical components of the electrochemical free energy ($G_{el} = G + mF\phi$ as discussed in Chapter 2) also are represented in Fig. 3.6. The shape of the electrical-component curve is defined by $\alpha'$, the fractional change in the potential as a function of position; $\alpha$, the transfer coefficient, is the fractional change at the maximum of the $G_{el}$ curve. $G^*$ and $G^*_{el}$ are the GFE and the electrochemical free energy, respectively, of the ion at the position of the maximum of these energies on traversing the interface. Ions in this state are frequently referred to as

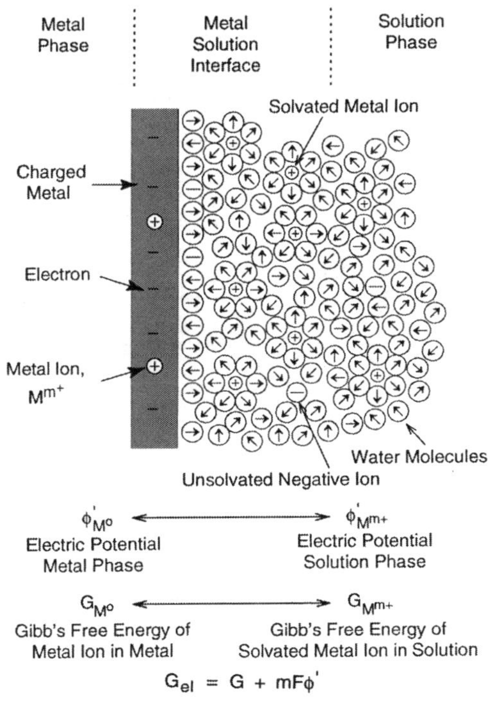


**Fig. 3.5** Representation of the environment of metal ions in the metal and aqueous phases at the interface. Based on Ref 2

being in the activated state. $\Delta G^*_{el}$ is the electrochemical free energy of activation and is the energy that statistical fluctuations of energies in the metal or solution must supply to cause the ion to move across the interface.

A simple model for the oxidation reaction, $M \rightarrow M^{m+} + me$, is based on the assumptions that the metal ions are detached from the metal at selected sites, such as dislocations, grain boundaries, or steps in the surface, pass through the interface, and reside in selected sites within the aqueous phase, such as within a sheath of water molecules as depicted in Fig. 3.5. The rate of the reaction is then assumed to be proportional to the concentration of sites from which the metal ions can jump from the surface, the concentration of sites in the solution to which they can jump, and to an exponential term involving the electrochemical free energy of activation. The latter term is equivalent to assuming that the rate is proportional to the probability that an energy fluctuation will occur of sufficient magnitude to allow the ion to pass through the interface. The resulting rate equation is of the form:

$$i_{ox,M} = K'_{ox,M} C_M C_R \exp\left(\frac{-\Delta G^*_{el,ox}}{RT}\right) \qquad \text{(Eq 3.8)}$$

where $C_M$ is the number of metal atoms per unit area at critical reaction sites on the metal surface; $C_R$ is the concentration of species in the

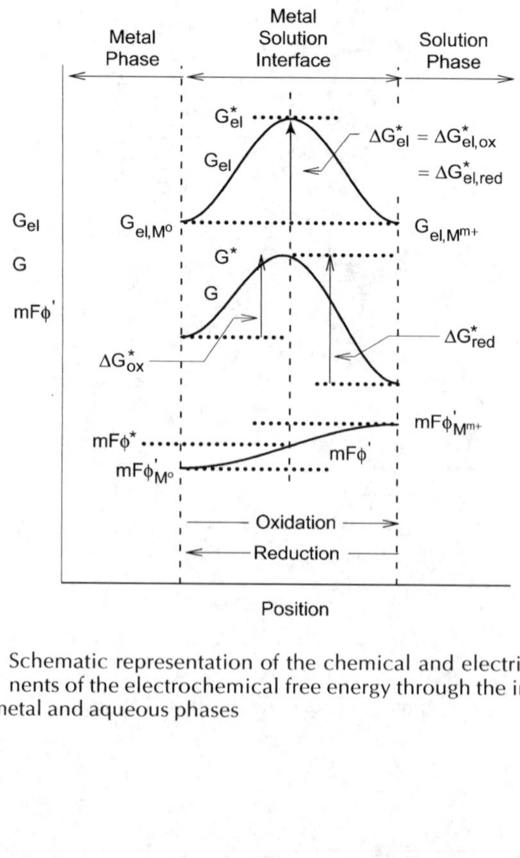


**Fig. 3.6** Schematic representation of the chemical and electrical components of the electrochemical free energy through the interface between the metal and aqueous phases

solution to which $M^{m+}$ becomes bonded, for example, solvated ions ($(H_2O)_qM^{m+}$), complex ions ($Cu(NH_3)_4^{2+}$), etc.; $K'_{ox,M}$ is a constant including terms resulting in the reaction rate expressed as current density; and $\Delta G^*_{el,ox} = G^*_{el} - G_{el,M^o}$ = electrochemical free energy of activation for the oxidation reaction.

Since $G_{el} = G + mF\phi$, $G^*_{el} = G^* + mF\phi^*$, and from Fig. 3.6 it is seen that in the oxidation direction (i.e., on going from the metal to the activated state):

$$\Delta G^*_{el,ox} = \Delta G^*_{ox} + mF(\phi^* - \phi'_{M^o}) \qquad \text{(Eq 3.9)}$$

where, as in Chapter 2, $\phi'_{M^o}$ represents the potential of the positive ion in the metal at equilibrium. Upon introducing the transfer coefficient, $\alpha = (\phi^* - \phi'_{M^o})/(\phi'_{M^{m+}} - \phi'_{M^o})$, where $\phi'_{M^{m+}}$ represents the potential of the ion in the solution at equilibrium, Eq 3.9 becomes:

$$\Delta G^*_{el,ox} = \Delta G^*_{ox} + \alpha mF(\phi'_{M^{m+}} - \phi'_{M^o}) \qquad \text{(Eq 3.10)}$$

Therefore:

$$i_{ox,M} = K'_{ox,M} C_M C_R \exp\left(\frac{-\Delta G^*_{ox} - \alpha mF(\phi'_{M^{m+}} - \phi'_{M^o})}{RT}\right) \qquad \text{(Eq 3.11)}$$

While Eq 3.11 is useful in form, it is limited in direct application because, by arguments given in Chapter 2, absolute values or even differences in the potentials, $\phi$, cannot be obtained. Rather, relative values are referenced to the standard hydrogen electrode (SHE). From Chapter 2, at equilibrium:

$$E'_{M,M^{m+}} = (\phi'_{M^o} - \phi'_{M^{m+}}) - (\phi'_{H_2} - \phi'_{H^+})_s \qquad \text{(Eq 3.12)}$$

$$E'_M = (\phi'_{M^o} - \phi'_{M^{m+}}) - \Delta\phi_{SHE} \qquad \text{(Eq 3.13)}$$

Substitution into Eq 3.11 yields:

$$i_{ox,M} = K_{ox,M} C_M C_R \exp\left(\frac{\alpha mFE'_M}{RT}\right) \qquad \text{(Eq 3.14)}$$

where the new constant, $K_{ox,M}$, is:

$$K_{ox,M} = K'_{ox,M} \exp\left(\frac{-\Delta G^*_{ox}}{RT}\right) \exp\left(\frac{\alpha mF\Delta\phi_{SHE}}{RT}\right) \qquad \text{(Eq 3.15)}$$

A corresponding model for the reduction reaction, $M^{m+} + me \rightarrow M$, is expressed as a reduction current density in the form:

$$i_{red,M} = K'_{red,M} C_{M^{m+}} \exp\left(\frac{-\Delta G^*_{el,red}}{RT}\right) \quad \text{(Eq 3.16)}$$

where $K'_{red,M}$ is a constant, $C_{M^{m+}}$ is the concentration of metal ions in solution, and $\Delta G^*_{el,red}$ is the electrochemical free energy of activation for the reduction reaction.

This model assumes that the rate of deposition of metal ions on the surface is proportional to the concentration of ions in solution and the probability of an ion overcoming the free energy barrier on jumping from solution to metal. It also assumes that ions hitting the surface attach at any position. From Fig. 3.6, it is seen that in the reduction direction (i.e., on going from the solution to the activated state):

$$\Delta G^*_{el,red} = \Delta G^*_{red} + mF(\phi^* - \phi'_{M^{m+}}) \quad \text{(Eq 3.17)}$$

and since it can be shown that $(\phi^* - \phi'_{M^{m+}}) = -(1-\alpha)(\phi'_{M^{m+}} - \phi'_{M^o})$:

$$\Delta G^*_{el,red} = \Delta G^*_{red} - (1-\alpha)\, mF\, (\phi'_{M^{m+}} - \phi'_{M^o}) \quad \text{(Eq 3.18)}$$

Upon substituting into Eq 3.16 and employing the relationship $E'_M = (\phi'_{M^o} - \phi'_{M^{m+}}) - \Delta\phi_{SHE}$:

$$i_{red,M} = K_{red,M} C_{M^{m+}} \exp\left(\frac{-(1-\alpha)\, mFE'_M}{RT}\right) \quad \text{(Eq 3.19)}$$

where the new constant, $K_{red,M}$, is:

$$K_{red,M} = K'_{red,M} \exp\left(\frac{-\Delta G^*_{red}}{RT}\right) \exp\left(\frac{-(1-\alpha) mF\Delta\phi_{SHE}}{RT}\right) \quad \text{(Eq 3.20)}$$

Since Eq 3.14 and 3.19 have been derived on the basis of equilibrium, the oxidation and reduction current densities must be equal and equal to the exchange current density:

$$i_{ox,M} = i_{red,M} = i_{o,M} \quad \text{(Eq 3.21)}$$

Therefore:

$$i_{o,M} = K_{ox,M} C_M C_R \exp\left(\frac{\alpha mFE'_M}{RT}\right) \quad \text{(Eq 3.22)}$$

and

$$i_{o,M} = K_{red,M} C_{M^{m+}} \exp\left(\frac{-(1-\alpha) mFE'_M}{RT}\right) \quad \text{(Eq 3.23)}$$

These equations can be interpreted as a kinetic basis for establishing the equilibrium electrode potential since, in principle, all terms in the right-hand expressions can be determined, thus allowing solution for $E'_M$. Practically, many of the terms cannot be accurately determined, and hence, it is necessary to measure $E'_M$ experimentally. The form of the equation, however, is quite useful. It is used here to derive a concentration dependence of $i_o$. From Eq 3.22 and 3.23:

$$\frac{K_{ox,M}C_M C_R}{K_{red,M}C_{M^{m+}}} = \exp\left(\frac{-(1-\alpha)mFE'_M - \alpha mFE'_M}{RT}\right) \quad \text{(Eq 3.24)}$$

$$\frac{K_{ox,M}\, C_M C_R}{K_{red,M}\, C_{M^{m+}}} = \exp\left(\frac{-mFE'_M}{RT}\right) \quad \text{(Eq 3.25)}$$

Substituting back into a rearranged form of Eq 3.23:

$$i_{o,M} = K_{red,M}\, C_{M^{m+}}\left(\exp\left(\frac{-mFE'_M}{RT}\right)\right)^{(1-\alpha)} \quad \text{(Eq 3.26)}$$

$$i_{o,M} = K_{red,M}C_{M^{m+}}\left(\frac{K_{ox,M}C_M C_R}{K_{red,M}C_{M^{m+}}}\right)^{(1-\alpha)} \quad \text{(Eq 3.27)}$$

$$i_{o,M} = K(C_{M^{m+}})^{\alpha} \quad \text{(Eq 3.28)}$$

where K is a constant independent of solution composition and electrode potential. K is equal to $i_{o,M}$ at unit concentration of $M^{m+}$ (essentially unit activity). Although experimental values of $\alpha$ of approximately 0.5 have been reported for several electrode systems, the value may vary over wide limits; also, more involved expressions for $i_o$ that take into account other species in the solution and the state of the electrode surface have been reported (Ref 3).

The foregoing equations have resulted from the stated assumptions of the models employed. These models are examples of many models that have been proposed for electrochemical reactions. The equations are accepted here because of their simplicity of form and the fact that they do predict relationships between exchange current density, equilibrium half-cell potential, and concentration, which are frequently observed experimentally. The theories and resulting equations are obviously more complicated for surface reactions, such as the reduction of dissolved oxygen, $O_2 + 4H^+ + 4e \rightarrow 2H_2O$. Theories for this reaction have proposed as many as eight individual steps.

At this point, and as somewhat of a digression, it is useful to consider a simple derivation of the Nernst half-cell equation from the kinetics principles that have been introduced. Thus, using activity rather than concentration, Eq 3.25 becomes:

$$\exp\left(\frac{-mFE'_M}{RT}\right) = \frac{K_{ox,M}C_M C_R}{K_{red,M}a_{M^{m+}}} \qquad \text{(Eq 3.29)}$$

But under standard conditions, $a_{M^{m+}} = 1$ and $E'_M = E^o_M$; therefore:

$$\exp\left(\frac{-mFE^o_M}{RT}\right) = \frac{K_{ox,M}C_M C_R}{K_{red,M}} \qquad \text{(Eq 3.30)}$$

Upon substitution of Eq 3.30 into Eq 3.29 and rearranging, the Nernst half-cell equation for the metal reaction $M = M^{m+} + me$ is produced:

$$E'_M = E^o_M + \frac{RT}{mF}\ln a_{M^{m+}} \qquad \text{(Eq 3.31)}$$

# Charge-Transfer Polarization

In the derivations of Eq 3.14 and 3.19 for the metal oxidation current density, $i_{ox,M}$, and the metal-ion reduction current density, $i_{red,M}$, it was not necessary to restrict the half-cell potential to its equilibrium value. Deviation from $E'_M$ will occur if the potential of either the metal or the solution is changed, resulting in an overpotential defined in general by Eq 3.1. More specifically, small deviations are associated with charge-transfer polarization, and the overpotential is designated as:

$$\eta = E - E' = \eta_{charge\ transfer} = \eta_{CT} \qquad \text{(Eq 3.32)}$$

With reference to a metal, M, the equilibrium potential, $E'_M$, was defined in terms of $\phi$ values by Eq 3.13. At 25 °C, it is given by the Nernst half-cell equation, $E'_M = E^o_M + (59/m)\log a_{M^{m+}}$. The *polarized* potential, $E_M$, is defined in terms of $\phi$ values as:

$$E_M = (\phi_{M^o} - \phi_{M^{m+}}) - \Delta\phi_{SHE} \qquad \text{(Eq 3.33)}$$

where, upon comparison with Eq 3.13, it is seen that primes are not used with $\phi_{M^o}$ and $\phi_{M^{m+}}$ because generally $E_M$ is not equal to the equilibrium half-cell potential, $E'_M$.

Oxidation overpotential is said to occur if the potential of the metal is increased, relative to $E'_M$, as would be accomplished by attaching the positive terminal of a battery to the metal, thus raising the potential by removal of electrons. (This also induces metal ions to pass into solution.) A somewhat more descriptive statement is that for oxidation overpotential *the metal is attached to an electron "sink," such as a more noble half cell or the positive terminal of a battery*, the negative terminal of which is attached to an inert electrode such as platinum, which completes the circuit in the solution. The experimental arrangement for such polarization measurements is shown in Fig. 3.7. An external power source is connected between the metal to be studied and the auxiliary (or counter) electrode. If the power source controls the current, a galvanostatic polarization measurement is made. If the power source supplies current to the specimen for a series of fixed working-electrode potentials, a potentiostatic polarization measurement is made. The potential of the electrode under study is determined by measuring the potential of the electrode relative to a reference electrode such as the Ag/AgCl (saturated) half cell. The electrical characteristics of this system are discussed in greater detail subsequently.

With an oxidation overpotential, the removal of electrons from the electrode makes it more positive relative to the solution, an effect that the electrode attempts to counteract by increasing the rate of transfer of ions from metal to solution (i.e., $i_{ox,M}$ is increased and $i_{red,M}$ is decreased relative to their equilibrium value, $i_{o,M}$), giving a net oxidation current density.

With a reduction overpotential, the supply of electrons to the electrode makes it more negative relative to the solution, an effect that the electrode attempts to counteract by increasing the rate of transfer of electrons to the metal ions in solution (i.e., $i_{red,M}$ is increased and $i_{ox,M}$ is decreased). These concepts are considered in more detail in the following discussion.

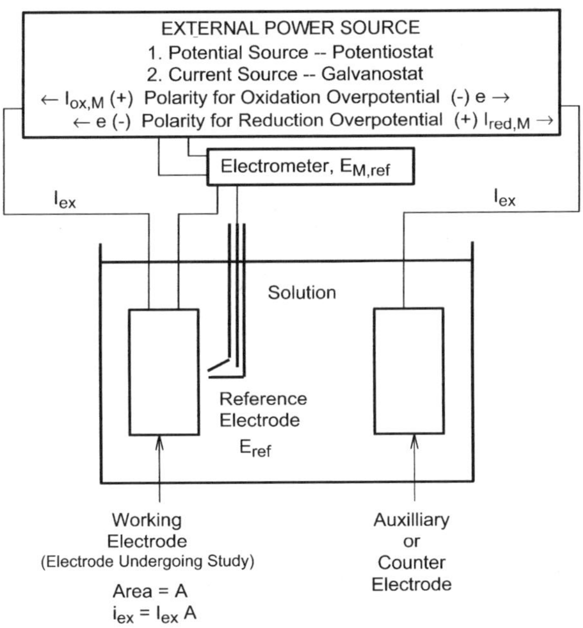


**Fig. 3.7** Components for the experimental determination of polarization of electrochemical reactions

From Eq 3.32, the polarized potential is the equilibrium potential plus the overpotential:

$$E_M = E'_M + \eta_{CT} \qquad \text{(Eq 3.34)}$$

On the basis that the model introduced to obtain expressions for the kinetics of the forward and reverse interface reactions at equilibrium is also valid when an overpotential exists, the polarized potential given by Eq 3.34 replaces the equilibrium potential in the exponential term. For the oxidation component of the reaction, Eq 3.14 becomes:

$$i_{ox,M} = K_{ox,M}\, C_M C_R \exp\left(\frac{\alpha m F(E'_M + \eta_{CT})}{RT}\right) \qquad \text{(Eq 3.35)}$$

and Eq 3.19 for the reduction reaction becomes:

$$i_{red,M} = K_{red,M} C_{M^{m+}} \exp\left(\frac{-(1-\alpha) m F(E'_M + \eta_{CT})}{RT}\right) \qquad \text{(Eq 3.36)}$$

These equations are now written in more compact form by defining $\beta'_{ox,M}$ and $\beta'_{red,M}$ as follows:

$$\beta'_{ox,M} \equiv \frac{RT}{\alpha m F} \qquad \text{(Eq 3.37)}$$

$$\beta'_{red,M} \equiv \frac{RT}{(1-\alpha) m F} \qquad \text{(Eq 3.38)}$$

Upon substitution, Eq 3.35 becomes:

$$i_{ox,M} = K_{ox,M} C_M C_R \exp\left(\frac{(E'_M + \eta_{CT})}{\beta'_{ox,M}}\right) \qquad \text{(Eq 3.39)}$$

$$i_{ox,M} = \left\{K_{ox,M} C_M C_R \exp\left(\frac{E'_M}{\beta'_{ox,M}}\right)\right\} \exp\left(\frac{+\eta_{CT}}{\beta'_{ox,M}}\right) \qquad \text{(Eq 3.40)}$$

On noting that the { } term is just the expression for the exchange current density given by Eq 3.22, Eq 3.40 can be written as:

$$i_{ox,M} = i_{o,M} \exp\left(\frac{+\eta_{CT}}{\beta'_{ox,M}}\right) \qquad \text{(Eq 3.41)}$$

By similar reasoning, Eq 3.36 for the current density for the reduction reaction becomes:

$$i_{red,M} = i_{o,M} \exp\left(\frac{-\eta_{CT}}{\beta'_{red,M}}\right) \qquad \text{(Eq 3.42)}$$

It should be noted that Eq 3.41 and 3.42 have the relatively simple form of an exponential term involving the overpotential, $\eta_{CT}$, multiplying the exchange current density to give the current densities of the oxidation and reduction components of the polarized half-cell reaction. When an overpotential exists, the oxidation and reduction current densities are no longer equal: When $\eta_{CT} > 0$, then $i_{ox,M} > i_{red,M}$, and when $\eta_{CT} < 0$, then $i_{red,M} > i_{ox,M}$.

In terms of the reaction-rate model, the influence of the sign of the overpotential, $\eta_{CT}$, on the dominance of the reaction components is illustrated by the curves of Fig. 3.8. When $\eta_{CT} = 0$, the $G_{el}$ versus distance curve represents the equilibrium condition and corresponds to the curve in Fig. 3.6. The activation energies for the oxidation and reduction components are equal, the oxidation and reduction rates are therefore equal, and the interface reaction is at equilibrium. If $\eta_{CT}$ is made positive by, for example, connecting the metal to the positive terminal of an external source as in Fig. 3.7, $G_{el,M^o}$ is raised relative to $G_{el,M^{m+}}$, and net oxidation occurs. That is, the activation energy for the oxidation component has been reduced relative to the reduction component of the

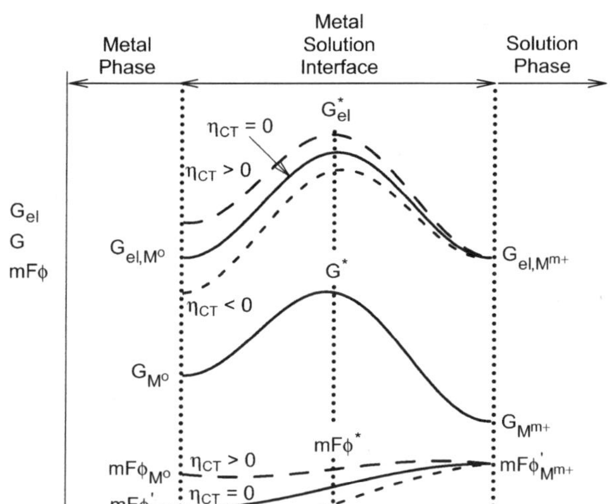


| Equilibrium | Oxidation Overpotential | Reduction Overpotential |
|---|---|---|
| $\eta_{CT} = 0$ | $\eta_{CT} > 0$ | $\eta_{CT} < 0$ |
| $\Delta G^*_{el,ox} = \Delta G^*_{el,red}$ | $\Delta G^*_{el,ox} < \Delta G^*_{el,red}$ | $\Delta G^*_{el,red} < \Delta G^*_{el,ox}$ |
| $i_{ex,M} = i_{red,M} = i_{o,M}$ | $i_{ox,M} > i_{red,M}$ | $i_{red,M} > i_{ox,M}$ |

**Fig. 3.8** Representation of the shifts in electrical and electrochemical free energies when conditions are imposed producing oxidation and reduction overpotentials

reaction. Conversely, if $\eta_{CT}$ is made negative, the activation energies are unbalanced in the opposite sense, and net reduction occurs. These effects are summarized in the table accompanying Fig. 3.8.

Equations 3.41 and 3.42 give the current density of the oxidation and reduction components of the interface electrochemical reaction as a function of the overpotential, $\eta_{CT}$, with $i_{o,M}$ and the β's as kinetic parameters characterizing the reaction mechanism. To obtain the Tafel relationship (Eq 3.2),which expresses the overpotential as a function of the current density, Eq 3.41 and 3.42 are changed to make the current density the independent variable:

$$\eta_{CT} = +\beta_{ox,M} \log\left(\frac{i_{ox,M}}{i_{o,M}}\right) \quad \text{(Eq 3.43)}$$

$$\eta_{CT} = -\beta_{red,M} \log\left(\frac{i_{red,M}}{i_{o,M}}\right) \quad \text{(Eq 3.44)}$$

where

$$\beta_{ox,M} = 2.303\beta'_{ox,M} = \frac{2.303RT}{\alpha mF} \quad \text{(Eq 3.45)}$$

$$\beta_{red,M} = 2.303\beta'_{red,M} = \frac{2.303RT}{(1-\alpha)mF} \quad \text{(Eq 3.46)}$$

An expression for the polarized potential, $E_M$, is now obtained by substituting the overpotential given by Eq 3.43 and 3.44 into $E_M = E'_M + \eta_{CT}$, or

$$E_M = E'_M + \beta_{ox,M} \log\left(\frac{i_{ox,M}}{i_{o,M}}\right) \quad \text{(Eq 3.47)}$$

$$E_M = E'_M - \beta_{red,M} \log\left(\frac{i_{red,M}}{i_{o,M}}\right) \quad \text{(Eq 3.48)}$$

These equations are frequently called the Tafel equations for the oxidation and reduction components of the half-cell reaction (Ref 3). Thus, the polarized potentials should plot as linear functions of the logarithm of current density as shown in Fig. 3.9(a). Note that the lines cross when $i_{ox,M} = i_{red,M} = i_{o,M}$ at the equilibrium half-cell potential, $E'_M$.

At any overpotential, $i_{ox,M}$ will not equal $i_{red,M}$, and the difference is a net current density defined as:

$$i_{net} = i_{ox,M} - i_{red,M} \quad \text{(Eq 3.49)}$$

Conservation of electrons requires that this net current density, to or from the polarized interface, relate to an external current ($I_{ex}$), from or to a source of the overpotential. This source, in corroding systems, will be other metal/solution interfaces acting as cathodic sites (accepting electrons) or as anodic sites (supplying electrons). The net current density also can be related to stray or leakage currents from electrical devices in contact with the system under study. When electrochemical measurements are being made, the net current can be measured as that flowing between a working electrode and a potentiostat or galvanostat as shown in Fig. 3.7. In any case, $I_{ex} = i_{net}(A_{polarized})$, where $A_{polarized}$ is the metal/solution interface area under analysis. It may be visualized as a small element of a larger surface or as an entire electrode as in Fig. 3.7.

The net current density as a function of the overpotential is obtained by substituting Eq 3.41 and 3.42 into Eq 3.49 to give:

$$i_{net} = i_{o,M}\left[\exp\left(\frac{+\eta_{CT}}{\beta'_{ox,M}}\right) - \exp\left(\frac{-\eta_{CT}}{\beta'_{red,M}}\right)\right] \quad \text{(Eq 3.50)}$$

The net current density in terms of the polarized potential is then obtained by substituting $\eta_{CT} = E_M - E'_M$:

$$i_{net} = i_{o,M}\left[\exp\left(\frac{+(E_M - E'_M)}{\beta'_{ox,M}}\right) - \exp\left(\frac{-(E_M - E'_M)}{\beta'_{red,M}}\right)\right] \quad \text{(Eq 3.51)}$$

Thus, if a positive overpotential is applied, the first exponential will be larger than the second, a positive $i_{net}$ results, and the net reaction is oxidation or anodic. Conversely, a negative overpotential will lead to a negative $i_{net}$, and the net reaction is reduction or cathodic.

Since Eq 3.51 cannot be solved explicitly for the polarized potential, $E_M$, it is not possible to express $E_M$ as a function of log $|i_{net}|$ for comparison to the Tafel equations for the individual anodic and cathodic com-

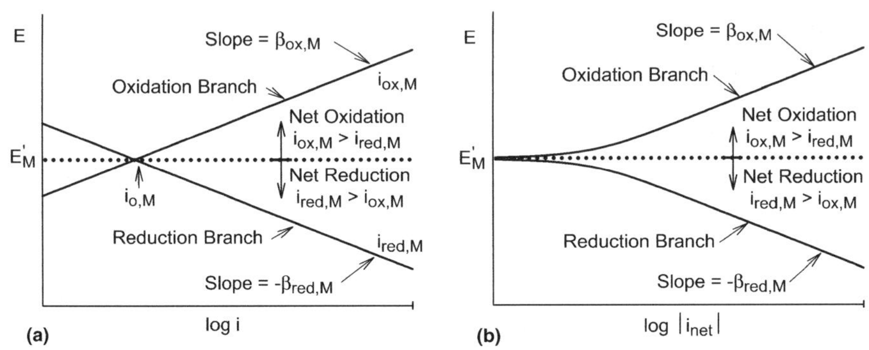


**Fig. 3.9** (a) Tafel relationships for the individual anodic and cathodic components of the interface reaction. (b) Net oxidation and reduction polarization curves derived from (a) by taking the difference between the oxidation and reduction components at each potential

ponents of the interface reaction (Fig. 3.9a). However, pairs of values of $i_{net}$ and $E_M$ can be obtained that satisfy Eq 3.51 and when plotted as $E_M$ versus log $|i_{net}|$ produce the curves shown in Fig. 3.9(b). Since $i_{net}$ is positive for $E_M > E'_M$ and negative for $E_M < E'_M$, the logarithm of the absolute value of $i_{net}$ is plotted. Comparison with Fig. 3.9(a) shows that at higher values of log $|i_{net}|$, the branches become linear and correspond to the Tafel lines of Fig. 3.9(a). This follows by noting in Eq 3.50 that as the positive overpotential increases, the first exponential term becomes larger as the second exponential term becomes smaller. At sufficiently large positive values of $\eta_{CT}$, the second term becomes negligible and the equation reduces to Eq 3.41 for the single oxidation reaction. At sufficiently negative values of $\eta_{CT}$, the change in relative values of the exponential terms reverses, and Eq 3.50 reduces to Eq 3.42. In the opposite limit, as $\eta_{CT} = E - E'_M$ approaches zero, $|i_{net}|$ approaches zero, and the curves in Fig. 3.9(b) become asymptotic to the equilibrium potential, $E'_M$.

## Interpretation of Charge-Transfer Polarization from Experiment

An objective in performing electrochemical measurements on a half-cell reaction is determination of the three kinetic parameters, $i_o$, $\beta_{ox}$, and $\beta_{red}$. With these parameters determined, the individual polarization curves can be drawn for the oxidation and reduction reactions using Eq 3.47 and 3.48. In the experimental measurement of overpotential, the external-circuit current, $I_{ex}$, and the potential of the metal (frequently called the working electrode) relative to a reference electrode are measured (Fig. 3.7). For experimental convenience, the reference electrode is most often not the standard hydrogen electrode (SHE) but rather, for example, the saturated calomel electrode (SCE) or saturated Ag/AgCl reference electrode. The metal electrode potential relative to a reference electrode will be designated as $E_{M,ref}$ and is assigned the polarity of the attached electrometer terminal when the electrometer indicates a positive reading. The working electrode potential, $E_M$, relative to the SHE is then calculated as:

$$E_M = E_{M,ref} + E_{ref} \quad \text{(Eq 3.52)}$$

where $E_{ref}$ is the potential of the reference electrode (e.g., saturated calomel) relative to the SHE (Table 2.3 provides selected $E_{ref}$ values).

In the previous section, emphasis is placed on the fact that the externally measured current relates only to the difference of the currents of the oxidation and reduction components of the reaction, neither of which are known initially at a given potential. It is useful to visualize $I_{ex}$, and $i_{ex} = I_{ex}/A$ at any small area over which the imbalance of oxidation and reduction currents occur, as shown in Fig. 3.10. More specifically, the area is representative of a working electrode in the experimental arrangement of Fig. 3.7. The example is for positive overpotential, $\eta_{CT} > 0$, resulting in:

$$i_{ex,ox} = i_{ox,M} - i_{red,M} \qquad \text{(Eq 3.53)}$$

From this:

$$i_{ox,M} = i_{ex,ox} + i_{red,M} \qquad \text{(Eq 3.54)}$$

which gives the current density of the oxidation component of the reaction in terms of the experimentally measured current density and the reduction component current density, the latter at the moment not known. Similarly, for negative overpotential, $\eta_{CT} < 0$, the external current density will be:

$$i_{ex,red} = i_{ox,M} - i_{red,M} \qquad \text{(Eq 3.55)}$$

which is negative since now $i_{red,M} > i_{ox,M}$. From this:

$$i_{red,M} = i_{ox,M} - i_{ex,red} \qquad \text{(Eq 3.56)}$$

The theoretical Tafel expression for polarization of the oxidation reaction was given as Eq 3.47, into which $i_{ox,M}$ from Eq 3.54 is now substituted to give:

$$E_M = E'_M + \beta_{ox,M} \log\left(\frac{i_{ex,ox} + i_{red,M}}{i_{o,M}}\right) \qquad \text{(Eq 3.57)}$$

or

$$E_M = E^o_M + 2.303\frac{RT}{mF}\log a_{M^{m+}} + \beta_{ox,M} \log\left(\frac{i_{ex,ox} + i_{red,M}}{i_{o,M}}\right) \qquad \text{(Eq 3.58)}$$

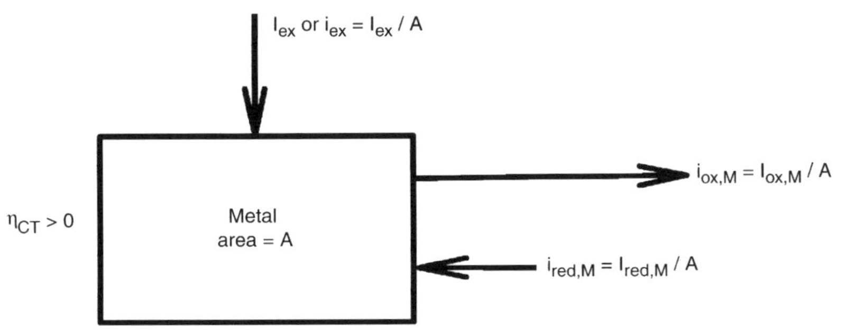


**Fig. 3.10** Illustration that an external current (measurable externally) is the difference in the oxidation and reduction currents at the interface, neither of which can be directly measured.

Equation 3.58 provides a theoretical expression for $E_M$ as a function of a measurable external current density. For a specific half cell, $E^o_M$ and $a_{M^{m+}}$ would be known, and $i_{o,M}$ and $\beta_{ox,M}$ are constants to be determined for the particular reaction. However, a plot of $E_M$ as a function of $i_{ex,ox} = I_{ex,ox}/A$ cannot be made because values of the reduction component, $i_{red,M}$, are not known. This problem is circumvented by examining the behavior of a plot of Eq 3.58 (or equivalently, Eq 3.57) in the limits of very small and very large values of $i_{ex,ox}$. It can be seen in the following analysis that Eq 3.57 has the form of the upper solid curve in Fig. 3.11. Qualitatively, the initial part of the curve has a small slope because in this current density range, $i_{ex,ox}$ is small relative to the exchange current density, and $i_{red,M}$ is close in magnitude to the exchange current density. In the limit of $i_{ex,ox} = 0$, Eq 3.54 leads to $i_{ox,M} = i_{red,M}$. When these are equal, a state of dynamic equilibrium exists, and both components are equal to the exchange current density, $i_{o,M}$. Substituting $i_{ex,ox} = 0$ and $i_{red,M} = i_{o,M}$ into Eq 3.57 results in the last term becoming zero, and therefore, $E_M = E'_M$ (i.e., the experimental curve is asymptotic to the equilibrium half-cell potential, $E'_M$, as $i_{ex,ox} \to 0$). In the limit of large external current densities where $i_{ex,ox} >> i_{red,M}$, Eq 3.54 indicates that $i_{ex,ox} \cong i_{ox,M}$; therefore, the last term of Eq 3.57, which is the overpotential term, becomes:

$$\eta_{CT} = \beta_{ox,M} \log\left(\frac{i_{ex,ox}}{i_{o,M}}\right) = \beta_{ox,M} \log\left(\frac{i_{ox,M}}{i_{o,M}}\right) \qquad \text{(Eq 3.59)}$$

This is equivalent to the Tafel equation (Eq 3.2, 3.3, 3.5, 3.43, and 3.47). If Eq 3.59 is used as the last term in Eq 3.57, the potential will be linear as a function of log $i_{ex,ox}$. This equation would plot exactly as the

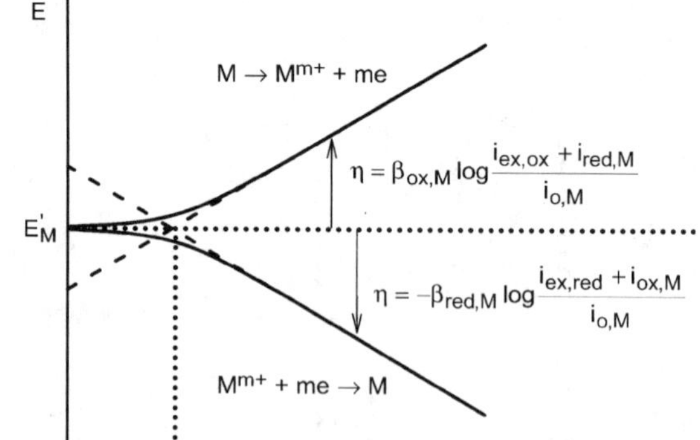


**Fig. 3.11** Experimental charge-transfer polarization curves, E vs. log $|i_{ex}|$, for positive and negative overpotentials

linear portion of Eq 3.57 in Fig. 3.11 but would extend as shown by the dashed portion of the line. The intersection of the dashed extension with the ordinate value corresponding to the equilibrium half-cell potential, $E'_M$, gives the exchange current density, $i_{o,M}$. That is, at $\eta_{CT} = 0$, $i_{ox,M} = i_{o,M}$, and $E_M = E'_M$ mathematically.

If this analysis is carried through for negative overpotentials, the following equation results:

$$E_M = E'_M - \beta_{red,M} \log\left(\frac{i_{ox,M} - i_{ex,red}}{i_{o,M}}\right) \quad \text{(Eq 3.60)}$$

The $E_M$ versus log $(-i_{ex\text{-}red})$ or log $|i_{ex,red}|$ (remember that $i_{ex,red}$ is a negative quantity) behavior is shown as the lower solid curve in Fig. 3.11. In the initial part of the curve, $-i_{ex,red}$ is small relative to $i_{o,M}$, and $i_{ox,M}$ is close in magnitude to $i_{o,M}$. In the limit, when $i_{ex,red} = 0$, $i_{ox,M} = i_{o,M}$ and therefore, $E_M = E'_M$. Consequently, this experimental curve also asymptotically approaches the equilibrium half-cell potential as $i_{ex,red} \rightarrow 0$. In the limit of large $|i_{ex,red}|$ values (i.e., large negative values of $i_{ex,red}$, $-i_{ex,red} >> i_{ox,M}$, and from Eq 3.56), $-i_{ex,red} \cong i_{red,M}$. Therefore, the last term of Eq 3.60, which is the overpotential term, becomes:

$$\eta_{CT} = -\beta_{red,M} \log\left(\frac{-i_{ex,red}}{i_{o,M}}\right) = -\beta_{red,M} \log\left(\frac{i_{red,M}}{i_{o,M}}\right) \quad \text{(Eq 3.61)}$$

This is equivalent to the Tafel equation (Eq 3.2, 3.3a, 3.6, 3.44, and 3.48). If Eq 3.61 is used as the last term in Eq 3.60, the potential will be linear as a function of log $(-i_{ex,red})$ (i.e., log $|i_{ex,red}|$). This equation would plot exactly as the linear portion of Eq 3.60 in Fig. 3.11 but would extend as shown by the dashed portion of the line. The intersection of the dashed extension with the ordinate value corresponding to the equilibrium potential, $E'_M$, again gives the exchange current density, $i_{o,M}$. That is, at $\eta_{CT} = 0$, $i_{red,M} = i_{o,M}$ and $E_M = E'_M$ mathematically.

If the mechanisms of the oxidation and reduction reactions are the same, the values of the Tafel constants, $\beta$, in Eq 3.57 and 3.60 should be the same; otherwise, they should be distinguished as $\beta_{ox,M}$ and $\beta_{red,M}$.

The previous concepts may be summarized by briefly reviewing the experimental procedures for determining the kinetic parameters, $i_o$, $\beta_{ox}$, and $\beta_{red}$. If a single half-cell reaction is involved, the equilibrium half-cell potential will be measured against some reference electrode. If the electrode is now connected to a potentiostat and the potential increased in the positive or oxidation direction, the upper solid curve of Fig. 3.11 will be plotted. If the potential is decreased, the lower solid curve will be plotted. The higher current-density linear sections of each curve are then extrapolated through the value of the equilibrium potential, E′. Their intersection gives a value for $i_o$, and the slopes of the curves give values for $\beta_{ox}$ and $-\beta_{red}$.

It should be noted that $i_{ex}$ has been consistently defined as $i_{ex} \equiv i_{ox,M} - i_{red,M}$, where $i_{ox,M}$ and $i_{red,M}$ are always positive quantities. Therefore, the sign of $i_{ex}$ will reveal whether the net reaction is oxidation ($i_{ex} > 0$) or reduction ($i_{ex} < 0$). This convention is consistent with external current measurements, wherein positive values reflect net oxidation at the working electrode and negative values net reduction.

## Diffusion Polarization

A net oxidation or reduction current at a local electrode will result in a change in the concentration at the interface of ions, or neutral species such as dissolved oxygen, involved in the electrode reaction. These changes in concentration at and near the interface result in concentration gradients causing diffusion of these species to or away from the interface. If the current density is great enough to cause significant concentration changes in the vicinity of the interface, the electrode potential will change in accordance with the Nernst half-cell equation, which for a simple metal/metal-ion reaction is:

$$E'_M = E^o_M + \frac{RT}{mF} \ln a_{M^{m+}} \qquad \text{(Eq 3.62)}$$

Oxidation currents will increase $a_{M^{m+}}$, causing the equilibrium electrode potential, $E'_M$, to become more positive. For reduction currents, the change in potential is in the opposite direction; the potential becomes more negative.

The change in electrode potential due to local concentration change is called *diffusion polarization*. A relationship between the magnitude of the change in potential and the external current density will be derived by considering the Nernst equation and relating the change in ion concentration to the rate of diffusion of ions under concentration and potential gradients.

Following a simple model, a theoretical expression for the *diffusion overpotential*, $\eta_D$, is derived as follows (Ref 2, 3). Consider a single interacting ion, $A^{n+}$, with activity, a, undergoing a net reduction reaction, $A^{n+} + ne \rightarrow A$ (i.e., $\eta < 0$ such that $i_{red,A} > i_{ox,A}$). Assume the activity of the ion to be equal to the molal concentration, a = c. The Nernst equation is (from Eq 2.72):

$$E' = E^o + \frac{RT}{nF} \ln a \qquad \text{(Eq 3.63)}$$

or

$$E' = E^{o} + \frac{59}{n}\log a \qquad \text{(Eq 3.64)}$$

at 25 °C with E′ and E° in mV (SHE).

Let $a_s = c_s$ = concentration in the bulk solution and $a_i = c_i$ = concentration at the interface. Applying the Nernst equation to the conditions of the bulk-solution concentration and the diffusion-depleted interface concentration:

$$E'_s = E^{o} + \frac{59}{n}\log c_s \qquad \text{(Eq 3.65)}$$

$$E'_i = E^{o} + \frac{59}{n}\log c_i \qquad \text{(Eq 3.66)}$$

The difference is:

$$E'_i - E'_s = \frac{59}{n}\log\frac{c_i}{c_s} \equiv \eta_D \qquad \text{(Eq 3.67)}$$

This would be the change in potential on establishing a diffusion layer, reducing the interface concentration from $c_s$ to $c_i$.

Let $J_D$ (mol/(s · cm$^2$)) represent the net flux of positive ions through the interface by diffusion. Fick's first law applied at the interface is:

$$J_D = -D\left(\frac{dC}{dx}\right)_{x=0} \qquad \text{(Eq 3.68)}$$

where x is the distance from the interface into the solution, and C is the concentration of ions (mol/cm$^3$). Concentration profiles are shown in Fig. 3.12 for zero time, an intermediate time, and a long time sufficient to reduce the interface concentration to zero ($C_i = 0$). The form of the plot is approximately linear near the interface, and the slope is approximately:

$$\left(\frac{dC}{dx}\right)_{x=0} \cong \frac{C_s - C_i}{\delta_i} \qquad \text{(Eq 3.69)}$$

Therefore:

$$J_D \cong -D\left(\frac{C_s - C_i}{\delta_i}\right) \qquad \text{(Eq 3.70)}$$

For the limiting case of $C_i = 0$ at the interface:

$$\left(\frac{dC}{dX}\right)_{x=0} \cong \frac{C_s}{\delta} \qquad \text{(Eq 3.71)}$$

and

$$J_D \cong -D\left(\frac{C_s}{\delta}\right) \qquad \text{(Eq 3.72)}$$

where δ is the *diffusion boundary-layer thickness*. Since the ions carry a charge $ne^+$ (n = number of unit charges per ion, $e^+$ = unit positive charge in coulombs), the flux is also associated with the net reduction current density, $i_{net,red}$, which is equivalent to the external reduction current density, $i_{ex,red}$. The charge in coulombs (C) per mole of ions is $N_o ne^+ = n(N_o e^+) = nF$ ($N_o$ = Avogadro's number, F = Faraday's constant). In terms of these quantities, the net flux is:

$$J_D\,(\text{mol}/\text{s}\cdot\text{cm}^2) = \frac{i_{ex,red}\,(\text{C}/\text{s}\cdot\text{cm}^2)}{nF(\text{C}/\text{mol})} \qquad \text{(Eq 3.73)}$$

$$J_D = \frac{i_{ex,red}}{nF} \qquad \text{(Eq 3.74)}$$

where $i_{ex,red}$ is a negative quantity; therefore, $J_D$ is a negative quantity (i.e., the flux in Fig. 3.12 is in the negative x direction). Equating the two flux expressions, Eq 3.70 and 3.74, gives:

$$\frac{i_{ex,red}}{nF} = -D\frac{(C_s - C_i)}{\delta_i} \qquad \text{(Eq 3.75)}$$

or

$$C_i = C_s + \frac{i_{ex,red}\,\delta_i}{nFD} \qquad \text{(Eq 3.76)}$$

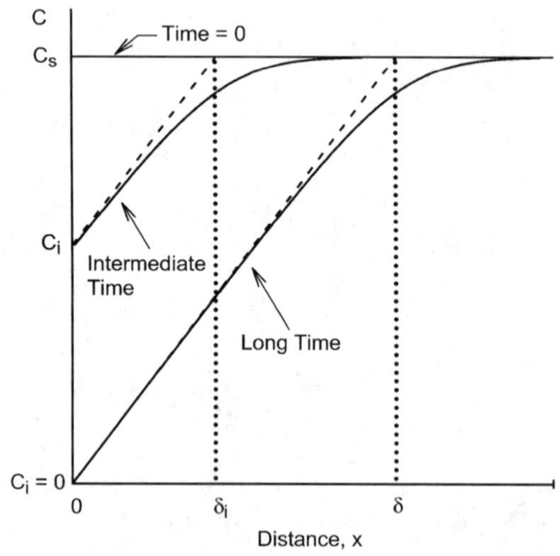


**Fig. 3.12** Reactive ion concentration profile in solution at the metal interface at initial, intermediate, and long times following initiation of current. The example corresponds to the deposit of reactive ions at the interface where ion concentration is depleted. δ is the diffusion boundary layer thickness.

Equation 3.76 is to be substituted into Eq 3.67; but before making the substitution, it should be recognized that for dilute solutions, the ratio of molal concentration, $c_i/c_s$ (each in moles per 1000 g of solvent), is approximately the same as the ratio of volumetric concentrations, $C_i/C_s$ (each in moles per cm$^3$). Making the substitution:

$$\eta_{D,red} = \frac{59}{n} \log \left( \frac{C_s + (i_{ex,red}\, \delta_i)/nFD}{C_s} \right) \quad \text{(Eq 3.77)}$$

or

$$\eta_{D,red} = \frac{59}{n} \log \left( \frac{(nFDC_s/\delta_i) + i_{ex,red}}{(nFDC_s/\delta_i)} \right) \quad \text{(Eq 3.78)}$$

With reference to Eq 3.75, in the limiting case when the concentration at the interface is reduced to zero ($C_i = 0$), $\delta_i$ becomes equal to $\delta$, and the absolute magnitude of the resultant limiting current density is identified as the positive quantity $i_{D,red}$, the *limiting diffusion current density* (i.e., $|i_{ex,red}| = -i_{ex,red} \equiv i_{D,red}$). Thus, under net reduction conditions:

$$\frac{i_{D,red}}{nF} = \frac{DC_s}{\delta} \quad \text{(Eq 3.79)}$$

or

$$\frac{nFDC_s}{\delta} = i_{D,red} \quad \text{(Eq 3.80)}$$

Therefore, substituting into Eq 3.78:

$$\eta_{D,red} = \frac{59}{n} \log \frac{(i_{D,red} + i_{ex,red})}{i_{D,red}} \quad \text{(Eq 3.81)}$$

or

$$\eta_{D,red} = -\frac{59}{n} \log \frac{i_{D,red}}{(i_{D,red} + i_{ex,red})} \quad \text{(Eq 3.82)}$$

where it should be recalled that $i_{ex,red}$ is a negative quantity.

The derivation carried out for the oxidation reaction, $A \rightarrow A^{n+} + ne$, leads to:

$$\eta_{D,ox} = +\frac{59}{n} \log \frac{i_{D,ox}}{(i_{D,ox} - i_{ex,ox})} \quad \text{(Eq 3.83)}$$

where $i_{ex,ox}$ and $i_{D,ox}$ are positive quantities. Furthermore, since diffusion control occurs at higher current densities, where at negative overpotentials, $i_{ex,red} \cong -i_{ex,red}$, and at positive overpotentials, $i_{ex,ox} \cong i_{ox}$, Eq 3.82 and 3.83 may be written as:

$$\eta_{D,red} = -\frac{59}{n}\log\frac{i_{D,red}}{(i_{D,red} - i_{red})} \quad \text{(Eq 3.84)}$$

$$\eta_{D,ox} = +\frac{59}{n}\log\frac{i_{D,ox}}{(i_{D,ox} - i_{ox})} \quad \text{(Eq 3.85)}$$

The limiting diffusion current density for a negative overpotential corresponds to a rate of reduction of species at the surface, which reduces the interface concentration to essentially zero. According to Eq 3.84, as $i_{red} \rightarrow i_D$, $\eta_D$ approaches $-\infty$. This effect also is deduced by inspecting Eq 3.63 and noting that as $a \rightarrow 0$, $E' \rightarrow -\infty$. The corresponding condition for a positive overpotential according to Eq 3.85 is that as $i_{ox} \rightarrow i_D$, $\eta_D \rightarrow +\infty$, which implies the buildup of ions to an infinite concentration at the interface. This, of course, does not have physical meaning. Practically, the ionic concentration is limited by the precipitation of some chemical species, which then controls the concentration through the solubility product. Frequently, this limiting concentration is relatively small and not significantly different from the initial concentration, in which case, $\eta_D$ is small (i.e., diffusion effects generally have little effect on polarization behavior for positive overpotentials). Schematic representations of positive and negative diffusion overpotentials are shown in Fig. 3.13.

The importance of diffusion polarization in corrosion results from the observation that in many situations, the current density of the reduction reaction is large enough to place it under diffusion control. Two important examples are the depletion of hydrogen ions in the solution adja-

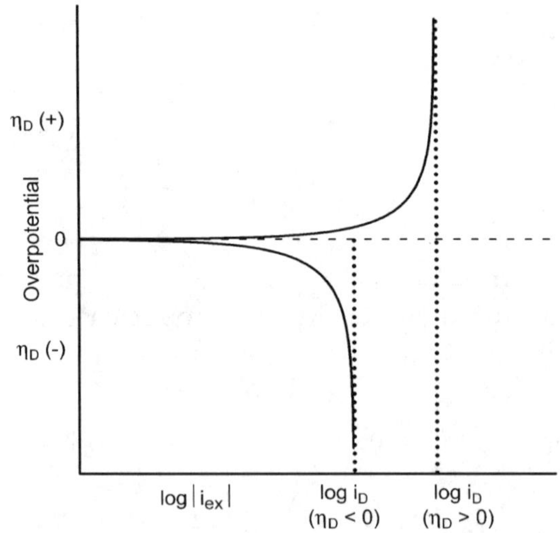


**Fig. 3.13** Diffusion overpotentials as a function of current density. Overpotentials become very large as the current density approaches the limiting current density.

cent to the interface as the reaction $H^+ + e \rightarrow 1/2H_2$ supports corrosion and the depletion of dissolved oxygen resulting from the reaction $O_2 + 2H_2O + 4e \rightarrow 4OH^-$. Diffusion control of the latter reaction is largely the consequence of the small solubility of oxygen in water (10 ppm at $P_{O_2} = 0.2$ atm). This diffusion limitation frequently becomes the controlling factor in the corrosion of many metals in aerated solutions.

## *Effect of Solution Velocity on Diffusion Polarization.*

For a specific solution and temperature, reference to Eq 3.80 indicates that the diffusion layer thickness, $\delta$, is the only variable that might be changed by a change in fluid velocity, and thus cause changes in the value of $i_D$. The upper limit for $\delta$ and, hence, the lower limit for $i_D$ occurs for a stagnant (zero velocity) solution, in which case $\delta$ is strictly determined by the properties of the solution. If the solution is flowing relative to the interface, the diffusion layer thickness is decreased, and hence, $i_D$ is increased. The effect on the diffusion-overpotential reduction curve is shown schematically in Fig. 3.14. The magnitude of the limiting diffusion current density, $i_{D,red}$, increases one log decade for each tenfold decrease in diffusion layer thickness. The change in $\delta$ with increased velocity (V), however, will depend on the fluid dynamics, increasing with $V^{-0.5}$ for laminar flow and with $V^{-0.9}$ for turbulent flow.

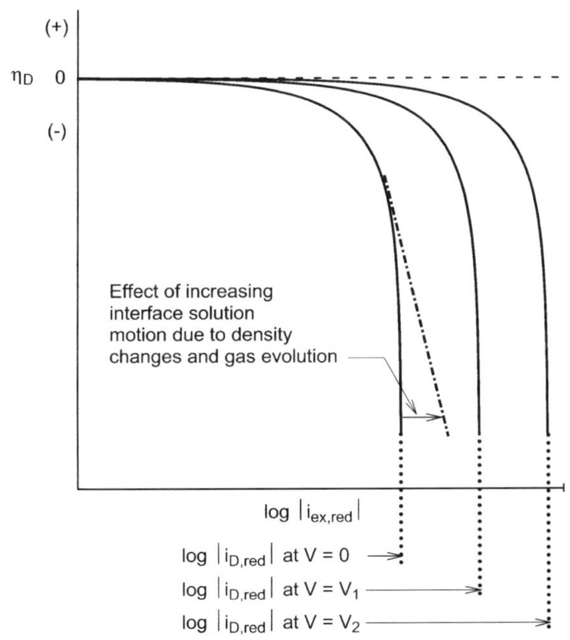


**Fig. 3.14** Effect of increasing solution velocity in increasing the limiting diffusion current density. At zero bulk fluid velocity, density changes and gas evolution can produce interface turbulence, which increases the current density.

It is emphasized that this brief discussion of diffusion-controlled polarization is based on simple diffusion and velocity-dependent models. Experimental determination of behavior at current densities causing a reaction to be diffusion controlled reveals more complex phenomena occurring at the interface. In part, completely stagnant conditions are seldom realized because the depletion of diffusing species near the interface results in changes in solution density, which then causes fluid motion under gravitational forces. This effect is more significant along vertical surfaces where flow parallel to the surfaces is induced. This fluid motion gives rise to overpotential curves of the form shown by the dashed curve in Fig. 3.14. The shift of the polarization curve from that shown by the solid curve is due to increased velocity induced by progressively larger changes in fluid density and, therefore, the velocity. Similar deviations may result from mixing at the interface, resulting from gas evolution, particularly $H_2$, at the interface. The greater the current density is, the greater the rate of gas generation will be and, hence, the larger the effect of turbulence in reducing the diffusion layer thickness.

## Complete Polarization Curves for a Single Half-Cell Reaction

By combining the Nernst equation with the expressions for charge-transfer overpotential ($\eta_{CT}$) and diffusion overpotential ($\eta_D$), equations can be written for the total experimental polarization behavior, $E(i_{ex,ox})$ and $E(i_{ex,red})$, of a single half-cell reaction:

$$E = E' + \eta_{CT} + \eta_D \qquad \text{(Eq 3.86)}$$

Using the $M \leftrightarrow M^{m+} + me$ reaction as an example, at positive overpotentials (net oxidation):

$$E_M = E_M^o + \frac{59}{m}\log a_{M^{m+}} + \beta_{ox,M}\log\frac{(i_{ex,ox} + i_{red,M})}{i_{o,M}} + \frac{59}{m}\log\frac{i_{D,ox,M}}{(i_{D,ox,M} - i_{ex,ox})} \qquad \text{(Eq 3.87)}$$

and at negative overpotentials (net reduction):

$$E_M = E_M^o + \frac{59}{m}\log a_{M^{m+}} - \beta_{red,M}\log\frac{(i_{ox,M} - i_{ex,red})}{i_{o,M}} - \frac{59}{m}\log\frac{i_{D,red,M}}{(i_{D,red,M} + i_{ex,red})} \qquad \text{(Eq 3.88)}$$

Similarly, and for reference and comparison, the following equations can be written for the total polarization behavior, $E(i_{ox})$ and $E(i_{red})$, for the single half-cell reaction, for the oxidation reaction:

$$E_M = E_M^o + \frac{59}{m}\log a_{M^{m+}} + \beta_{ox,M}\log\frac{i_{ox,M}}{i_{o,M}} + \frac{59}{m}\log\frac{i_{D,ox,M}}{i_{D,ox,M} - i_{ox,M}}$$

(Eq 3.89)

and for the reduction reaction:

$$E_M = E_M^o + \frac{59}{m}\log a_{M^{m+}} - \beta_{red,M}\log\frac{i_{red,M}}{i_{o,M}} - \frac{59}{m}\log\frac{i_{D,red,M}}{i_{D,red,M} - i_{red,M}}$$

(Eq 3.90)

In Eq 3.88, it should be recalled that $i_{ex,red}$ is a negative quantity; all other current densities in Eq 3.87 to 3.90 are positive quantities.

Curves representative of positive and negative overpotentials are shown in Fig. 3.15 for two electrodes. Electrode $X,X^{X+}$ has a more noble equilibrium potential, $E'_X$, and is shown with a higher exchange

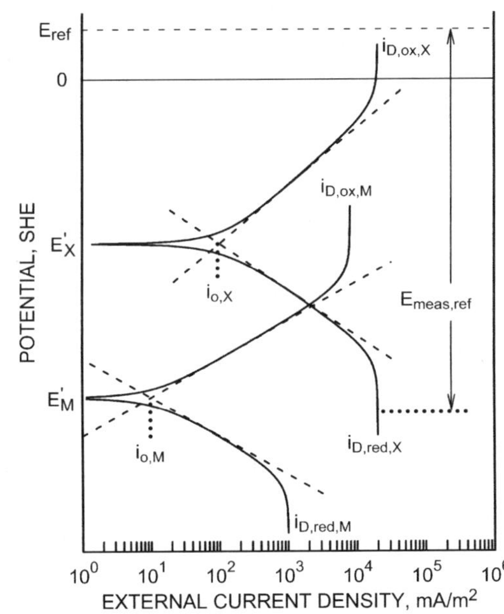


**Fig. 3.15** Example of overpotential curves for two electrochemical reactions illustrating that the thermodynamic and kinetic parameters place each reaction in different regions of the range of potentials and log $|i_{ex}|$

current density, $i_{o,X}$, than the $M,M^{m+}$ electrode with values of $E'_M$ and $i_{o,M}$. The solid curves are plotted as a function of external current density, $i_{ex}$, since this quantity can be measured experimentally and expresses the intensity or flux of ion transfer at the interface, which is the fundamentally correct basis for representing the characteristic behavior of the electrode. The significance of the linear portions of these curves and their extensions through the exchange current density for each electrode was previously discussed, and reference should be made to that discussion.

## *Polarization Behavior of the Hydrogen-Ion and Oxygen Reduction Reactions*

These reactions ($2H^+ + 2e \rightarrow H_2$ and $O_2 + 2H_2O + 4e \rightarrow 4OH^-$), occurring either independently or simultaneously, are, in many respects, the two most important reactions supporting corrosion. Both reactions have been studied extensively as a function of the pH and the metal surface on which the reactions occur (Ref 3, 5, 6). The data on, and mechanisms for, the hydrogen evolution reaction are reasonably well established; in contrast, the oxygen reduction reaction is poorly understood, particularly with respect to the values of the exchange current density. Also, in the potential range near +600 mV (SHE), electrode reactions involving hydrogen peroxide may make measurable contributions to the current density.

From polarization measurements on platinum and iron in 4% NaCl solution with the pH controlled by HCl additions, values of $i_o$, $\beta_{red}$, and $i_{D,red}$ for the hydrogen reaction have been approximated and used to construct the idealized E versus log $i_{red}$ polarization curves shown in Fig. 3.16 (Ref 5). In constructing these curves, the equilibrium potential was calculated from $E' = -59$ pH, $i_{o,H_2}$ on Fe was taken to be 1 mA/m$^2$ and independent of the pH, and the slope of the linear region ($-\beta_{red}$) was taken to be −100 mV per log decade. From the diffusion coefficient of hydrogen ions, $i_{D,red}$ was calculated to be $6 \times 10^4$ mA/m$^2$ at pH = 1. These parameters lead, for example, at pH = 1, to a curve starting at the equilibrium potential $E' = -59$ mV (SHE) and 1 mA/m$^2$ and ending as a vertical line at the limiting diffusion current density of $6 \times 10^4$ mA/m$^2$. The curves shift regularly with pH as shown. Corresponding to the vertical (diffusion control) sections of these curves, the interface hydrogen-ion concentration approaches zero. As a consequence, when the potential decreases, a value is reached below which direct reduction of water is possible, $H_2O + e \rightarrow 1/2\ H_2 + OH^-$. This reaction is accompanied by further increases in current density as the potential is decreased. The direct reduction of water becomes the dominant reaction at higher potentials as the pH is increased; the data imply that this is the main reduction reaction in deaerated water. The data also indicate, by extrapolation to potentials near –100 mV (SHE), that is, E′ = –59 to –118 mV (SHE) at pH = 1–2, that $i_o$ for the direct reduction of water in acid solution is on the order of $10^{-3}$ mA/m$^2$ (Ref 5).

For reasons stated previously, it is considerably more difficult to construct illustrative polarization curves for the reduction of dissolved oxygen. Reasonable estimates of the exchange current densities, Tafel slopes, and diffusion rates have been used to construct the curves of Fig. 3.17 (Ref 3, 6). These curves, identified by letters, are described as follows:

- *Curves A, A′, A″ and B, B′, and B″:* Conditions for the estimated solid curve, A, A′, A″: Platinum electrode, pH = 0.56 (1N $H_2SO_4$), $P_{O_2}$ = 0.2 atm (air). This curve is representative of reduction reactions in sulfuric acid saturated with air. The equilibrium potential is $E'_{O_2,H^+} = 1229 - 59\text{ pH} + 15 \log P_{O_2} = 1184$ mV (SHE), and the exchange current density is $10^{-2}$ mA/m$^2$. Because of the small solubility of oxygen in water (about 10 ppm), diffusion of oxygen to the interface becomes current limiting at about $10^3$ mA/m$^2$. Diffusion controls the current between +500 and –35 mV (SHE). When the latter potential is reached, hydrogen can be evolved, and with a platinum electrode exhibiting $i_{o,H_2}$ on Pt = $10^4$ mA/m$^2$, a rapid increase in current along section A′ is observed. Additional decrease in potential results in charge-transfer polarization of the hydrogen reaction until diffusion control results in the region of limiting current density along A″. The dashed curve identified as B, B′, B″

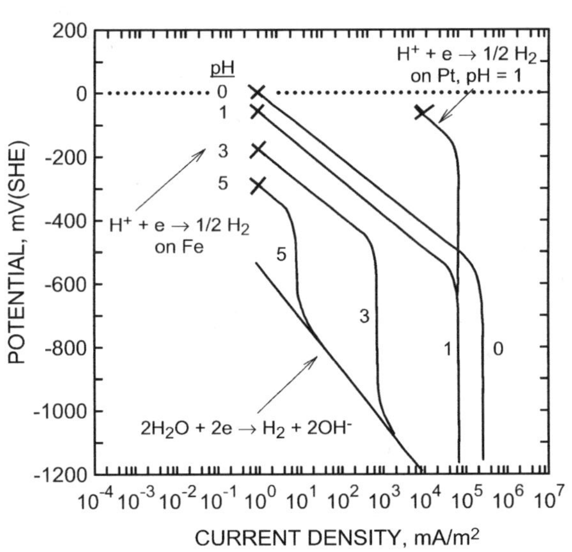


**Fig. 3.16** Cathodic polarization of the hydrogen reduction reaction on iron showing the effect of pH. Curve for platinum shows influence of a metal with much higher exchange current density on the position of the hydrogen reduction curve. Source: Ref 5

represents the experimental measurements for a platinum electrode in 4% NaCl at pH = 1.1. Although these conditions differ slightly from those for the calculated curves, the agreement with the estimated curve (A, A′, A″) is reasonable. At lower potentials, –300 to –1000 mV (SHE), the experimental current density is higher than estimated because of turbulence created at the interface by hydrogen evolution, thus bringing a greater concentration of hydrogen ions to the interface than would occur under stagnant conditions.

- *Curves C, C′, C″ and D, D′, D″:* Conditions for the estimated solid curve, C, C′, C″: Platinum electrode, pH = 7, $P_{O_2} = 0.2$ atm (air). This curve is representative of the reduction reactions in water (pH = 7) saturated with air. The higher pH reduces the equilibrium potential to +800 mV (SHE), and $i_{0,O_2}$ on Pt is estimated to be $4 \times 10^{-5}$ mA/m². On decreasing the potential, charge-transfer polarization occurs along C, current-limiting diffusion polarization along C′, and reduction of water along C″. The dashed experimental curve, D, D′, D″, agrees well with the estimated solid curve.
- *Curve E, E′, E″:* Conditions: Platinum electrode, pH = 7, $P_{O_2} = 10^{-4}$ atm. This curve is representative of partially deaerated water. The partial pressure of oxygen has been reduced from 0.2 to $10^{-4}$ atm (10 ppm to about 5 ppb). Charge-transfer polarization occurs along E, oxygen diffusion limits the current density along E′, and direct reduction of water occurs along E″. The significance of this curve is

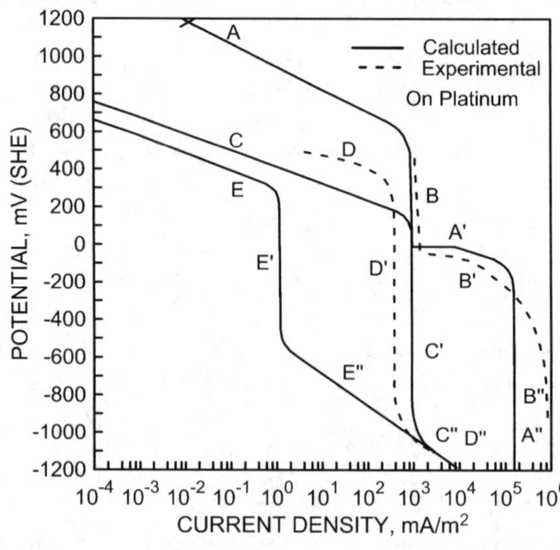


**Fig. 3.17** Theoretical and experimental polarization curves for reduction of oxygen ($O_2 + 4H^+ + 4e \rightarrow 2H_2O$), hydrogen ion ($2H^+ + 2e \rightarrow H_2$), and water ($2H_2O + 2e \rightarrow H_2 + OH^-$) on platinum. Curve A, A′, A″: Theoretical curve for pH = 0.56, $P_{O_2} = 0.2$ atm; curve B, B′, B″: Experimental curve for pH = 1.1, $P_{O_2} = 0.2$ atm; curve C, C′, C″: Theoretical curve for pH = 7, $P_{O_2} = 0.2$ atm; curve D, D′, D″: Experimental curve for pH = 7, $P_{O_2} = 0.2$ atm; curve E, E′, E″: Theoretical curve for pH = 7, $P_{O_2} = 10^{-4}$ atm

that the limiting current density has been decreased by a factor of 1000, from $10^3$ to 1 mA/m$^2$.

The polarization curves in Fig. 3.17 were illustrative of the oxygen, hydrogen-ion, and water-reduction reactions on platinum. In general, platinum exhibits the highest values of exchange current densities, $i_o$, for these reactions of any of the metals. The lower values of exchange current density, particularly in the case of the oxygen reaction, may be due to the presence of oxide films, which are present on most metals. The reactions then occur at the oxide/solution interface rather than at the metal surface. The calculated effect of reducing the exchange current density for the oxygen reaction in an environment of pH = 0.56 and $P_{O_2}$ = 0.2 atm is illustrated in Fig. 3.18. The Tafel regions when the exchange current density has values of $10^{-2}$, $10^{-3}$, $10^{-5}$, and $10^{-7}$ mA/m$^2$ are represented by the upper four curves. These curves merge into a common constant limiting diffusion current of $10^3$ mA/m$^2$. At this current density, diffusion of dissolved oxygen to the interface is the limiting kinetic factor. The current density is constant over a range of potentials and depends only on the oxygen concentration, here corresponding to that established by $P_{O_2}$ = 0.2 atm or about 10 ppm dissolved oxygen. This limiting current density is independent of the exchange current density. At potentials below −33 mV (SHE), hydrogen can be produced

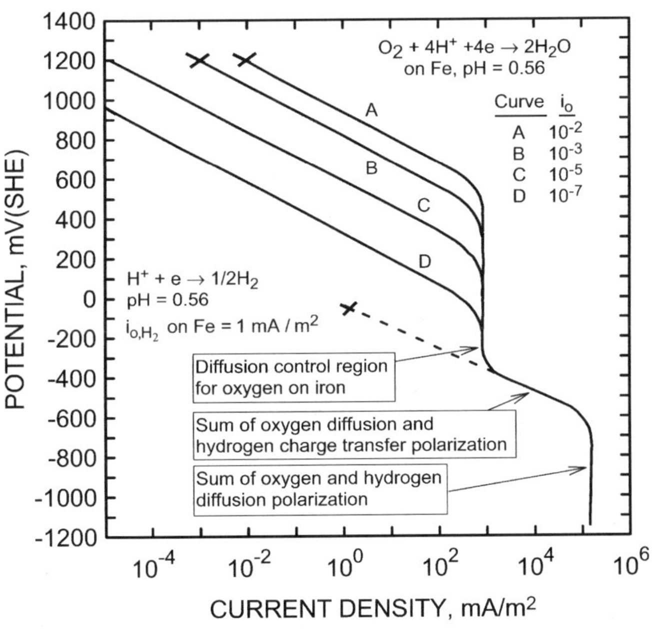


**Fig. 3.18** Illustration of the effect of exchange current density on the polarization curve for oxygen reduction in aerated environments of pH = 0.56 and $P_{O_2}$ = 0.2 atm. Curves converge to the same diffusion limit and are identical when the hydrogen ion reduction is the dominant reaction.

by the reduction of hydrogen ions in this environment of pH = 0.56 (1 N). The Tafel region of the hydrogen ion polarization is shown as the dashed line starting at the exchange current density of 1 mA/m$^2$. Below about –400 mV (SHE) the hydrogen reduction dominates the current density, and the total polarization curve deviates from that of oxygen diffusion control to hydrogen reduction under Tafel control to, finally, hydrogen diffusion control below –800 mV (SHE). It is emphasized that these curves for oxygen reduction cannot generally be measured experimentally at the high potentials on metals such as iron since anodic dissolution of the metal will contribute to the measured current density. There are practical significances to the fact that the kinetics of the oxygen-reduction reaction are slow in the Tafel region (very small $i_o$) and that diffusion control occurs at relatively low current densities due to the small solubility of oxygen. In particular, corrosion processes that are supported by oxygen reduction in these potential ranges occur at rates less than those that would otherwise occur. The corrosion rates are further decreased if deposits form on the surface through which oxygen must diffuse to reach the metal surface. These deposits include thick corrosion product films, settling or adherent inert deposits, or deposits resulting from microbiological activity.

The reduction of ferric iron ions according to the reaction $Fe^{3+} + e \rightarrow Fe^{2+}$ provides a strong cathodic reaction, which may cause the corrosion of a large number of metals and alloys. The reaction is of significance in both industrial environments and laboratory testing environments. The influence results from the relatively high half-cell potential of the reaction, the kinetics being rapid near the half-cell potential due to the relatively large exchange current density, and the high limiting current density under diffusion control (Ref 7). The standard half-cell potential is +770 mV (SHE), but the actual potential is usually higher since the $Fe^{3+}/Fe^{2+}$ concentration ratio is generally much greater than unity, making the concentration-dependent term in the Nernst equation a positive quantity. These characteristics are illustrated by the cathodic polarization curves in Fig. 3.19 for reduction on platinum at concentrations of 100 and 10,000 ppm $Fe^{3+}$. The curves were determined under nitrogen deaerated conditions starting at the open-circuit potential and scanning in the negative direction. Stagnant conditions were maintained in the 100 ppm solution during initial polarization down to +400 mV (SHE). Diffusion control dominates in the range 600 to 400 mV (SHE). The limiting diffusion current density immediately increases on agitation by direct sparging of the nitrogen into the solution, the increased interface velocity of the solution decreasing the diffusion boundary thickness. The current density increases again near –100 mV (SHE) due to hydrogen ion reduction, the hydrogen ions resulting from the hydrolysis of $Fe^{3+}$ and $Fe^{2+}$ ions to produce relatively low pH solutions. In the 10,000 ppm nitrogen-sparged solution, the limiting diffusion current density is greater by a factor of about 100 as would be predicted from Eq 3.80. An increase in current density due to hydrogen ion reduction is not observed since at this higher concentration, the ferric ion reduction dominates over hydrogen ion reduction.

The influence of the substrate on which the $Fe^{3+}$ reduction is occurring is illustrated by the curves in Fig. 3.20. Cathodic polarization

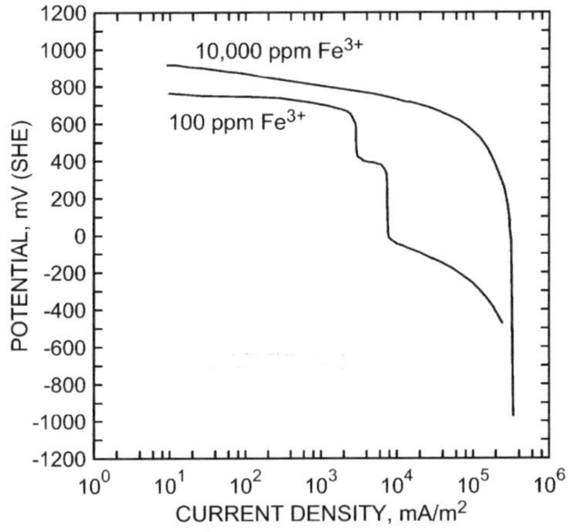


**Fig. 3.19** Cathodic polarization curves for 100 and 10,000 ppm $Fe^{3+}$ (as $FeCl_3$) on platinum in nitrogen-deaerated solution. The increase in current density at 400 mV (SHE) is due to a velocity effect in introducing nitrogen sparging into the solution. The limiting current density is increased by a factor of about 100 on increasing the concentration from 100 to 10,000 ppm. The increase in current density near –100 mV (SHE) is due to hydrogen ion reduction resulting from a decrease in pH due to $Fe^{3+}$ hydrolysis.

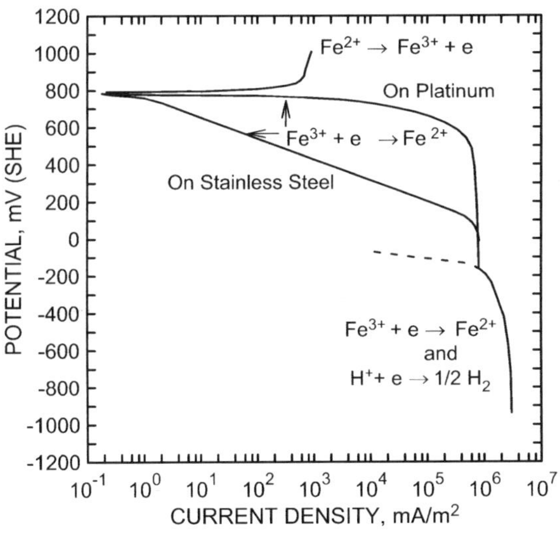


**Fig. 3.20** Polarization curves for $Fe^{3+}$ reduction ($Fe^{3+} + e \rightarrow Fe^{2+}$) on platinum and on type 316 stainless steel, with $a_{Fe^{3+}} = 1$ and $a_{Fe^{2+}} = 0.1$ in chloride solution. The exchange current density is lower on the passive film of the stainless steel. The inflection in the curve near –200 mV (SHE) results from contribution to the current density due to hydrogen ion reduction resulting from the hydrolysis of the $Fe^{3+}$ and $Fe^{2+}$ ions.

curves were determined using platinum and type 316 stainless steel substrates. The chloride solution in this case was 1.0 M in $Fe^{3+}$ and 0.1 M in $Fe^{2+}$ ions in which the equilibrium half-cell potential for the reaction, $Fe^{3+} + e = Fe^{2+}$, is +800 mV (SHE). That the open circuit potential, the potential prior to starting the downscan, is approximately this value indicates that the exchange current density for the reaction is relatively large. The continuous curvature of the polarization curve during the initial downscan precludes detection of a linear Tafel region that could be extrapolated back to the equilibrium potential to give an exchange current density. An approximate value for the exchange current density is obtained by assuming a Tafel slope of 100 mV per log decade, placing a line tangent to the experimental curve with this slope and extrapolating back to the open circuit potential, 800 mV (SHE). An exchange current density of approximately $10^4$ mA/m$^2$ is obtained for the $Fe^{3+}$ reduction on platinum. Extrapolation of the linear portion of the polarization curve for $Fe^{3+}$ on type 316 stainless steel to an open circuit potential indicates that the exchange current density is about 1 mA/m$^2$. Thus, the kinetics of the $Fe^{3+}$ reduction is about $10^4$ greater on platinum than on stainless steel. However, the position of the polarization curve becomes independent of the substrate at potentials below 100 mV (SHE) since diffusion in the solution becomes the controlling factor independent of the substrate. Hydrolysis of $Fe^{3+}$ and $Fe^{2+}$ ions occurs, resulting in sufficient hydrogen ion concentration to allow the reduction of hydrogen ions to contribute to the current density below about –200 mV (SHE).

If the potential scan is positive to the open-circuit potential, the anodic branch of the polarization corresponding to $Fe^{2+} \rightarrow Fe^{3+} + e$ is measured. A short section of this branch is shown in Fig. 3.20. It is evident that the polarization quickly reaches diffusion control.

It is shown in the next chapter that nitrites can be used as passivating inhibitors for corrosion of iron in near-neutral solutions. Since the basis for accomplishing this is related to the polarization characteristics of the reduction of the nitrite ion, brief consideration is given here to the reaction and to the form of the experimentally determined polarization curve for this ion. The curve is shown in Fig. 3.21. Although several reactions have been proposed for the reduction of this nitrite ion, the following is considered here:

$$NO_2^- + 8H^+ + 6e \rightarrow NH_4^+ + 2H_2O \qquad \text{(Eq 3.91)}$$

The curves in Fig. 3.21 apply to a platinum substrate in an environment of pH = 7, $a_{NO_2^-} = 0.01$ and $a_{NH_4^+} = 10^{-5}$. The equilibrium potential calculated from the Nernst equation is 250 mV (SHE). The reduction branch of the curve shows a transition from Tafel control to diffusion control with a limiting diffusion current density of $10^3$ mA/m$^2$, followed at lower potentials by the reduction of water. An anodic branch

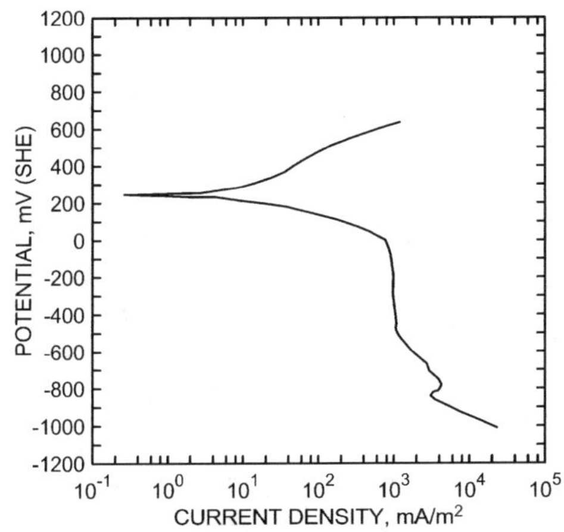


**Fig. 3.21** Anodic and cathodic polarization curves for nitrite ion on platinum. Assumed reduction reaction is $a_{NO_2^-} + 8H^+ + 6e \rightarrow NH_4^+ + 2H_2O$. Equilibrium half-cell potential corresponds to $a_{NO_2^-} = 0.1$, $a_{NH_4^+} = 10^{-5}$, and pH = 7. Limiting current density is $10^3$ mA/m$^2$.

starting at the open-circuit potential is also shown but is not involved in the analysis of the inhibiting action of the nitrite ion.

# Chapter 3 Review Questions

1. Define E, E′, $i_o$, $\alpha$, $\beta_{ox}$, $\beta_{red}$, $i_{ox}$, $i_{red}$, $i_{ex,ox}$, $i_{ex,red}$, $i_{D,ox}$, and $i_{D,red}$.
2. The following problem is designed to provide understanding of Tafel plots for individual half-cell reactions and the form of experimental polarization curves to be expected based on the theory. Assume that for a given metal, M,

   Area: $A_M = 50$ m$^2$
   Equilibrium half-cell potential: $E'_M = -500$ mV (SHE)
   $i_{o,M} = 1$ mA/m$^2$
   $\beta_{ox,M} = 80$ mV/log decade
   $\beta_{red,M} = 60$ mV/log decade

   (Recall that the equations for the polarization involve ratios of currents or current densities, and therefore, the expressions are of the same form since the area factor cancels. Obviously, the numerical scale against which the plots are made will depend on the need to plot in terms of current or current density.)

   a. On a copy of the 7-cycle semilog paper provided (Fig. 3.22), use coordinate ranges of –800 to –200 mV (SHE), and $10^{-1}$ to $10^{+6}$ mA. Plot the anodic Tafel line ($E_M$ versus log $i_{ox,M}$) using Eq 3.47.

b. Plot the cathodic Tafel line ($E_M$ versus log $i_{red,M}$) using Eq 3.48.

c. Plot the polarization curves that should result from experimental measurements of the polarized potential, $E_M$, versus log $|I_{ex}|$.

Note that experimentally, $E_M$ is set and the resulting $I_{ex}$ measured for potentiostatic polarization, and $I_{ex}$ is set and $E_M$ measured in galvanostatic polarization. In either case, the external current must be the difference between the oxidation and reduction components over the metal surface, $I_{ex,M} = I_{ox,M} - I_{red,M}$. Therefore, curves can be derived having the form of experimental curves by plotting points representing the difference between the Tafel curves for progressively changed values of $E_M$. The resulting Tafel and derived experimental curves should be similar to Fig. 3.11.)

3. From the following data for the polarization of iron, make a reasonable plot of the anodic polarization curve over the current density range from $i_{o,Fe}$ to $i_{ox,Fe} = 10^{+4}$ mA/m$^2$.

$$i_{o,Fe} = 10^{-1} \text{ mA/m}^2$$
$$\beta = +50 \text{ mV}$$
$$a_{Fe^{2+}} = 10^{-6}$$

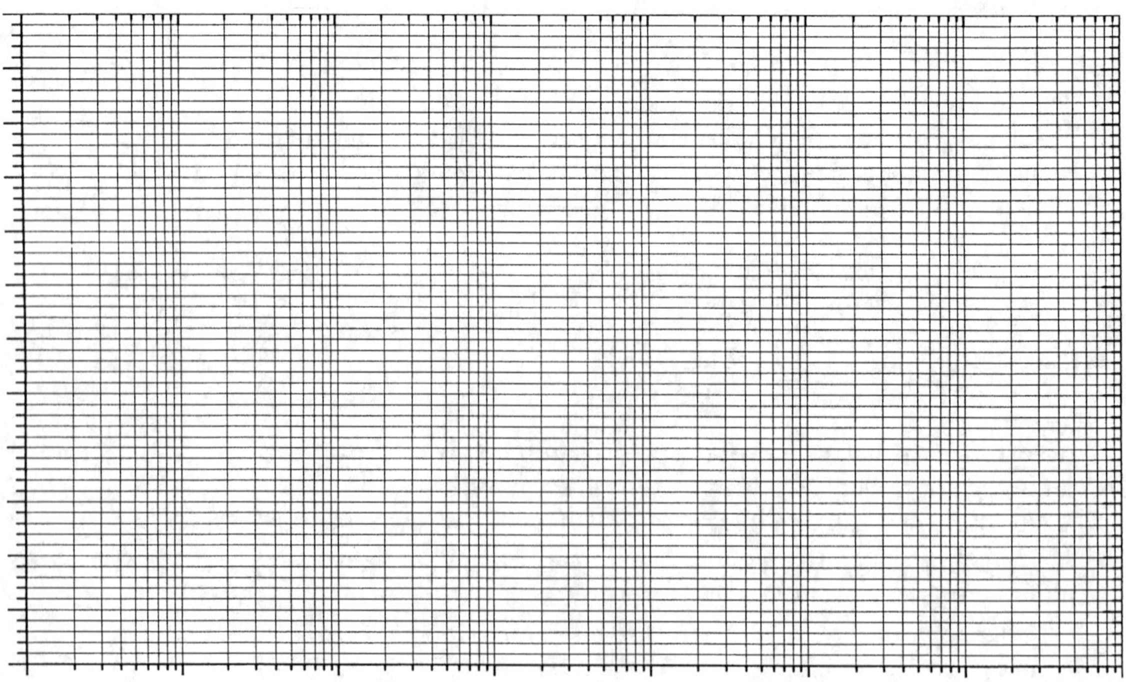

**Fig. 3.22** 7-cycle semilog graph paper

4. From the following data for the polarization of the hydrogen evolution reaction on iron at a pH = 4, plot the cathodic polarization curve from $i_{o,H_2}$ on Fe to $i_{D,red,H_2}$:

$$i_{o,H_2} \text{ on Fe} = 10 \text{ mA/m}^2$$
$$\beta_{red,H_2} \text{ on Fe} = 100 \text{ mV}$$
$$i_{D,red,H_2} = 10^{+4} \text{ mA/m}^2$$

5. Plot the cathodic polarization curve for the hydrogen reaction on copper using the data in problem 4 but with a change in the value of the exchange current density to $i_{o,H_2}$ on Cu = 1 mA/m$^2$. Why should the polarization curves for hydrogen evolution on copper and iron terminate at the same $i_{D,red}$ value?

## References

1. J.Z. Tafel, *Phys. Chem.*, Vol 50, 1905, p 641
2. J.O. Bockris and A.K.N. Reddy, *Modern Electrochemistry*, Vol 2, Plenum Press, 1973, p 632
3. K.J. Vetter, *Electrochemical Kinetics,* Academic Press, 1967, p 104–395
4. J.M. West, *Electrodeposition and Corrosion Processes,* D. Van Nostrand Co., New York, 1965, p 27–43
5. M. Stern, The Electrochemical Behavior, Including Hydrogen Overvoltage, of Iron in Acid Environments, *J. Electrochem. Soc.*, Vol 102, 1955, p 609–616
6. J.P. Hoare, *The Electrochemistry of Oxygen*, John Wiley & Sons, 1968, p 117
7. A.C. Makrides, Kinetics of Redox Reactions on Passive Electrodes, *J. Electrochem. Soc.,* Vol 111, 1964, p 392–399

CHAPTER **4**

# Kinetics of Coupled Half-Cell Reactions

If two or more electrochemical half-cell reactions can occur simultaneously at a metal surface, the metal acts as a mixed electrode and exhibits a potential relative to a reference electrode that is a function of the interaction of the several electrochemical reactions. If the metal can be considered inert, the interaction will be between species in the solution that can be oxidized by other species, which, in turn, will be reduced. For example, ferrous ions can be oxidized to ferric ions by dissolved oxygen and the oxygen reduced to water, the two processes occurring at different positions on the inert metal surface with electron transfer through the metal. If the metal is reactive, oxidation (corrosion) to convert metal to ions or reduction of ions in solution to the neutral metal introduces additional electrochemical reactions that contribute to the mixed electrode.

The current model of the mixed electrode is one of uniform transport of cathodic species to the metal surface and anodic species from the surface with no attempt to define sites at which the anodic and cathodic reactions occur (Ref 1). The two reactions are assumed to occur over a common area that is assigned to each reaction when expressing a current density. In contrast, the surface may be modeled as having distinct areas at which only the anodic or the cathodic reaction is occurring. In this case, distinct local electrochemical cells exist with local currents flowing between them. Practically, there is a continuum of models extending from the mixed electrode surface first described to the surface consisting of macroscopic local cells, each associated with a single electrochemical reaction. Even the surfaces of pure metals are nonuniform at the microscopic level consisting of grains of different orientation and with surface defects such as grain boundaries, emerging dislocations, and steps in the crystal lattice. The surface is a microdistribution of anodic and cathodic sites. A second level of nonuniformity exists for multiphase and nonhomogeneous alloys where different phases or nonuniform compositions within a single phase provide preferred anodic or cathodic sites. And finally, there is the classic case of iron rivets in essentially inert copper leading to the iron being almost exclusively anodic and corroding, the cathodic reaction being supported on the copper surface. Even in this case, both the iron and copper are, in themselves, mixed electrodes as initially defined but form a macroscopic mixed electrode of definable sites for net electrochemical reaction. The reality of these differences becomes apparent when a reference electrode is used to measure a metal potential, and the values are found to depend on the location of the electrode relative to the surface being measured.

A particularly simple illustration is the case of iron in a deaerated acid in which the corrosion (oxidation) of the iron by the reduction of hydrogen ions to hydrogen gas establishes a mixed electrode. The potential of the resulting electrode must lie between that of the equilibrium potential for iron and the equilibrium potential for the hydrogen ion reaction. The potential that is measured, however, will depend both on the kinetics of the individual reactions and on the position of a reference electrode relative to the sites on the metal surface at which the oxidation and reduction reactions are occurring. In the limiting condition of these sites separated by atomic dimensions, a single mixed electrode potential is measured independent of position in the solution, and the value will be a function of the electrode reaction kinetics. If the oxidation (anodic) and reduction (cathodic) sites are separated by dimensions large relative to a reference electrode, the mixed potential measured by the reference electrode will depend upon position. This condition allows location of anodic and cathodic sites on the metal surface and, therefore, measurement of the distribution of corrosion. The kinetics of single electrode reactions are discussed in Chapter 3 in which it is demonstrated that the kinetics are governed by the exchange current density and Tafel slope in the region of charge-transfer polarization. In addition, diffusion processes may become important and even control the kinetics. The present chapter is concerned with the behavior of mixed electrodes and, in particular, how these electrodes relate to corrosion.

The conventional approach to corrosion is to start directly with the concept of a mixed electrode of indistinguishable distribution of sites for the anodic and cathodic reactions. The approach taken in this chapter is to first examine the behavior of distinguishable anodic and cathodic sites. This is the classical case of galvanic couples of joined dissimilar metals in contact with a common solution. In this case, local movement of a reference electrode through the solution can map the current paths between anodic and cathodic sites and can thereby locate their positions. Furthermore, the effects of size and distribution of the sites can be examined as well as the influence of the specific resistivity of the solution. Finally, in the limit of these sites becoming sufficiently small that they are indistinguishable relative to the scale of examination, the analysis of the corrosion phenomena is essentially the same as results from the micro-mixed electrode theory.

Before developing the concept of the mixed electrode in greater detail, it is important to establish an understanding of the relationship between the potential difference across the metal/solution interface and the potential difference within the solution.

## Relationship between Interface Potentials and Solution Potentials

In Chapter 2 (in the section "Interface Potential Difference and Half-Cell Potential"), the equilibrium half-cell potential for the metal reaction, $E'_M$, was defined relative to potentials $\phi$ as follows:

$$E'_M = (\phi'_{M^o} - \phi'_{M^+}) - (\phi'_{H_2} - \phi'_{H^+}) \qquad \text{(Eq 4.1)}$$

$$E'_M = (\phi'_{M^o} - \phi'_{M^+}) - \Delta\phi_{SHE} \qquad \text{(Eq 4.2)}$$

where primes indicate values at equilibrium, $\phi'_{M^o}$ is the potential in the metal, $\phi'_{M^+}$ is the potential in the solution near the metal surface, and $\phi'_{H_2}$ and $\phi'_{H^+}$ have similar meanings relative to the standard hydrogen electrode (SHE). In Chapter 3 (in the section "Charge-Transfer Polarization"), the definition was written in general terms to encompass nonequilibrium conditions:

$$E_M = (\phi_{M^o} - \phi_{M^{m+}}) - \Delta\phi_{SHE} \qquad \text{(Eq 4.3)}$$

In these prior discussions, only the metal reaction was under consideration. The equivalent general definition for species in solution ($X^{X+}$ and X) capable of undergoing reduction or oxidation at the metal surface is:

$$E_X = (\phi_{X^o} - \phi_{X^{x+}}) - \Delta\phi_{SHE} \qquad \text{(Eq 4.4)}$$

Thus, the E values (relative interface potential differences, or interface potentials) represent differences in potentials across the metal/solution interface minus the potential difference across the standard hydrogen reference electrode interface. The E values are physically measured by attaching one lead of an electrometer to the metal, the other lead to a reference electrode in the solution and very close to the metal surface (a point discussed further in Chapter 6). If the positive electrometer lead is connected to the metal, the sign of the electrometer read-out will provide the correct sign for E. In practical measurements, the SHE is generally not employed. Rather, for convenience, another reference electrode such as the saturated calomel electrode (SCE) or the saturated Ag/AgCl electrode might be employed. When this is done, the measured potential relative to a given reference electrode is $E_{meas,ref}$, which is related to E by the expression (see the section "Interpretation of Charge-Transfer Polarization from Experiment" in Chapter 3):

$$E = E_{meas,ref} + E_{ref} \quad \text{(Eq 4.5)}$$

where $E_{ref}$ is the potential of the reference electrode relative to the SHE (Table 2.2 in Chapter 2 provides selected $E_{ref}$ values).

Under conditions of steady-state corrosion, during which net oxidation is occurring at a given anodic site ($M \rightarrow M^{m+} + me$) and net reduction at a given cathodic site ($X^{X+} + xe \rightarrow X$), the potentials at the anodic site and cathodic site, respectively, are given by:

$$E''_M = (\phi''_{M^o} - \phi''_{M^{m+}}) - \Delta\phi_{SHE} \quad \text{(Eq 4.6)}$$

and

$$E''_X = (\phi''_{X^o} - \phi''_{X^{x+}}) - \Delta\phi_{SHE} \quad \text{(Eq 4.7)}$$

where the double primes indicate the steady-state corrosion condition, $\phi''_{M^o}$ and $\phi''_{X^o}$ represent the potentials of the metal at the anodic and cathodic sites, respectively, and $\phi''_{M^{m+}}$ and $\phi''_{X^{x+}}$ represent the potentials in the solution at the anodic and cathodic sites, respectively. In order to more clearly associate the potentials $\phi$ in Eq 4.6 and 4.7 with either the metal or solution, and either the anodic or cathodic sites, the following changes in designations will be introduced: $\phi_{M,a} = \phi''_{M^o}$, $\phi_{S,a} = \phi''_{M^{m+}}$, $\phi_{M,c} = \phi''_{X^o}$, and $\phi_{S,c} = \phi''_{X^{x+}}$, where the subscripts M and S refer to the metal and solution, and the subscripts a and c refer to the anodic and cathodic sites. With these designations, Eq 4.6 and 4.7 become:

$$E''_M = (\phi_{M,a} - \phi_{S,a}) - \Delta\phi_{SHE} \quad \text{(Eq 4.8)}$$

and

$$E''_X = (\phi_{M,c} - \phi_{S,c}) - \Delta\phi_{SHE} \quad \text{(Eq 4.9)}$$

With reference to Fig. 4.1, since the corrosion process is taking place, $E''_X$ at the cathodic site has to be greater than $E''_M$ at the anodic site such that conventional current ($I_{corr}$) flows *in the metal* from the higher potential site (cathode) to the lower potential site (anode); electrons flow in the opposite direction. The driving potential difference responsible for the corrosion process is $(E''_X - E''_M)$, a positive quantity. From Eq 4.8 and 4.9:

$$(E''_X - E''_M) = (\phi_{S,a} - \phi_{S,c}) - (\phi_{M,a} - \phi_{M,c}) \quad \text{(Eq 4.10)}$$

However, since the metal is an excellent electrical conductor, differences in potential within the metal are generally negligible (i.e., $(\phi_{M,a} - \phi_{M,c}) \approx 0$). Therefore:

$$(E''_X - E''_M) = (\phi_{S,a} - \phi_{S,c}) \quad \text{(Eq 4.11)}$$

In Eq 4.11, $(E''_X - E''_M)$ is positive since $E''_X$ at the cathodic site is greater than $E''_M$ at the anodic site. Thus, the potential in the solution at the anodic site, $\phi_{S,a}$, is greater than the potential in the solution at the cathodic site, $\phi_{S,c}$, which is consistent with the overall electrochemical corrosion circuit. It follows that the driving potential difference for conventional current flow ($I_{corr}$) *in the solution* is:

$$\Delta\phi_S = (\phi_{S,a} - \phi_{S,c}) \quad \text{(Eq 4.12)}$$

with the current flowing from the higher potential site (anode) to the lower potential site (cathode). Within the solution, the potential will de-

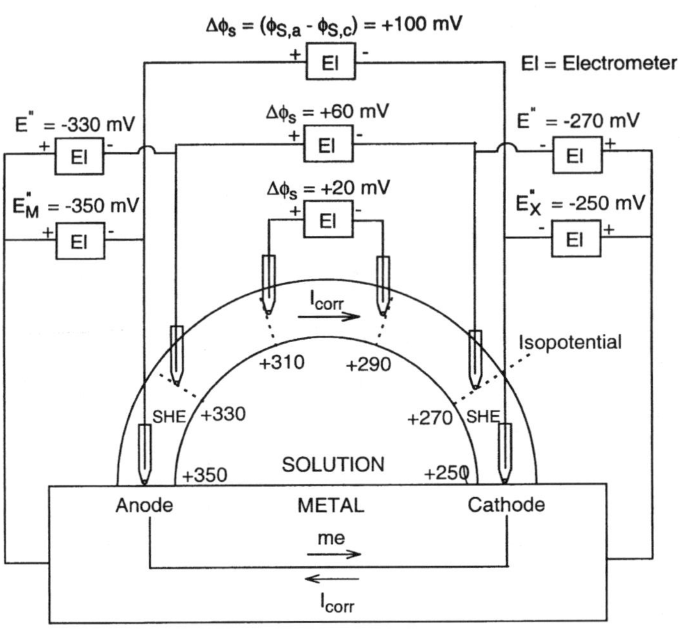


**Fig. 4.1** Schematic representation of measurements of potentials along a path from anode to cathode area on a corroding surface

crease continuously from $\phi_{S,a}$ at the anodic site to $\phi_{S,c}$ at the cathodic site.

It is only possible to physically measure the quantities $E''_M$, $E''_X$, and $\Delta\phi_S$, where the $\Delta\phi_S$ measurement is between any two points in the solution. With reference to Eq 4.8 and 4.9 for $E''_M$ and $E''_X$, the quantities $\phi_{M,a} \approx \phi_{M,c}$ and $\Delta\phi_{SHE}$ are constants, but unknown constants. In Eq 4.8 and 4.9, let the constant quantities $(\phi_{M,a} - \Delta\phi_{SHE})$ and $(\phi_{M,c} - \Delta\phi_{SHE})$ be represented by k, where k is another unknown constant:

$$(\phi_{M,a} - \Delta\phi_{SHE}) = (\phi_{M,c} - \Delta\phi_{SHE}) = k \quad \text{(Eq 4.13)}$$

Then, upon rearrangement:

$$\phi_{S,a} = (k - E''_M) \quad \text{(Eq 4.14)}$$

$$\phi_{S,c} = (k - E''_X) \quad \text{(Eq 4.15)}$$

and the potential difference in the solution (Eq 4.12) becomes:

$$\Delta\phi_S = (\phi_{S,a} - \phi_{S,c}) = (k - E''_M) - (k - E''_X) \quad \text{(Eq 4.16)}$$

Since it is apparent from Eq 4.16 that the unknown constant k, regardless of its value, will always cancel, it is convenient to define k as zero. Then, from Eq 4.14 and 4.15:

$$\phi_{S,a} = -E''_M \qquad \text{(anode)} \quad \text{(Eq 4.17)}$$

$$\phi_{S,c} = -E''_X \qquad \text{(cathode)} \quad \text{(Eq 4.18)}$$

or, in general:

$$\phi_S = -E'' \quad \text{(Eq 4.19)}$$

In order to illustrate the above principles, with reference to Fig. 4.1, assume that $E''_M$ (anode) = −350 mV(SHE) and $E''_X$ (cathode) = −250 mV (SHE). Since $(E''_X - E''_M)$ is a positive quantity (+100 mV), corrosion will occur. Furthermore, $\phi_{S,a}$ (anode) = +350 mV, and $\phi_{S,c}$ (cathode) = +250 mV. Under these conditions, with the use of a SHE reference electrode and assuming a semicircular current path in the solution, experimental measurements with an electrometer—with the positive (high, red) and negative (low, black, common) leads connected as shown—will indicate the potential values shown in Fig. 4.1. In the solution, the potential will vary from +350 mV at the anode to +250 mV at the cathode. In Fig. 4.1, cross sections of constant-potential (isopotential) surfaces are schematically represented as dotted lines at 20 mV increments.

## A Simple Model of the Galvanically Coupled Electrode

It is implied in the introduction to this chapter that the anodic and cathodic sites involved may be very small and evenly distributed or relatively large and widely distributed. Consider initially the presence of an anodic site undergoing corrosion while surrounded by a large area supporting a cathodic reaction. An example would be a hot-rolled steel plate covered with black oxide (magnetite) but from which a small strip of the oxide has been removed exposing bare steel. In aerated near-neutral environments, the reduction of dissolved oxygen is usually the major cathodic reaction, and the oxide is a sufficient electron conductor to support this reaction on its surface. The oxide surface thus supports the dissolution of the iron at the unprotected site by accepting electrons from the anodic reaction. Oxygen is also available at the anodic site and it contributes to the corrosion locally, but if the cathode/anode area ratio is large, the rate of corrosion will be determined largely by the oxygen reduction on the oxide. Additional examples would include the dispersion of second-phase particles in an alloy in which the matrix phase preferentially supports a cathodic reaction, the anodic dissolution of grain-boundary areas relative to exposed grains, and the selective attack at scratches on a metal surface. Extreme, but frequently very serious, cases involve connections of small areas of an active metal (iron) to large areas of a relatively inactive metal (copper). Actually, in all of these cases, both the anodic and cathodic sites will be mixed electrodes on a microscale. This micro local mixed electrode behavior is not considered in what immediately follows; rather, single half-cell reactions are assumed to occur at the individual sites.

As a simple model to illustrate the above variables, consider a surface as shown in Fig. 4.2 consisting of alternate anodic and cathodic strips (e.g., uniform scratches through the oxide coating of a hot-rolled steel

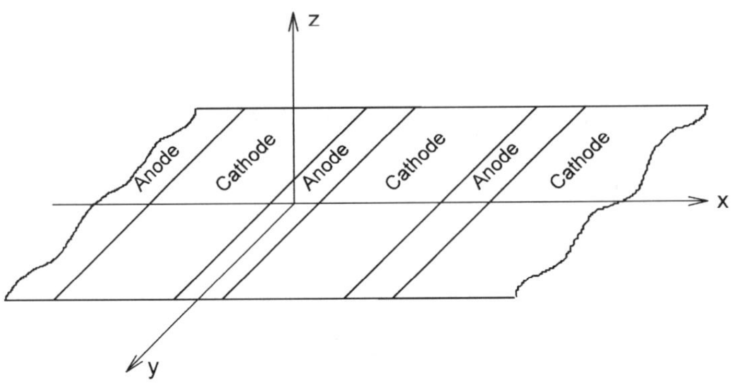


**Fig. 4.2** Array of anodic and cathodic reaction surfaces for mathematical modeling of potentials and currents in an electrolyte

plate). For reference, the origin of a set of coordinate axes is placed in the center of the anodic strip with the z-axis extending vertically into the solution. The y-axis is parallel to the center of the anodic strip, and the x-axis is perpendicular to the strips in the surface. For this simple physical model and with simplifying assumptions, mathematical expressions can be established allowing location of constant potential (isopotential) surfaces in the solution and description of the flow of current in the corrosive environment above the metal surface (Ref 2). The parameters of the model may be divided into those governing the electrochemical behavior and those governing the current distribution of the metal/environment system. The electrochemical parameters are the difference in thermodynamic equilibrium potentials ($E_{cell} = E'_X - E'_M$ calculated by application of the Nernst equation) and the polarization behaviors of the anodic and cathodic reactions. The current distribution parameters are the relative geometries of the anodic and cathodic areas, the specific resistivity of the solution and any other resistances to current flow such as those existing at interfaces and within the metal between anode and cathode areas.

Figures 4.3(a) and (b) are sections in the zx-plane showing the distribution of potential ($\phi$) in the solution as cross sections of imaginary surfaces in the solution of equal potential (isopotentials) and the distribution of current as current channels with cross sections defined by traces of the surfaces ...(n – 1), n, (n + 1)... perpendicular to the isopotentials. These traces are located such that each current channel carries the same total current. Figure 4.3(a) applies to an environment of higher resistivity (e.g., water with specific resistivity of 1000 ohm-cm) and Fig. 4.3(b) to an environment of lower resistivity (e.g., salt brine, 50 ohm-cm). The figures are representative of anodic and cathodic reactions, which, if uncoupled, would have equilibrium half-cell potentials of $E'_M = -1000$ mV and $E'_X = 0$ mV and would, therefore, produce a thermodynamic driving force of $E_{cell} = E'_X - E'_M = +1000$ mV. This positive $E_{cell}$ indicates that corrosion will occur when the reactions are coupled. For the example of Fig. 4.3(a), the high solution resistivity allows the potential $E''_M$ at the anode to approach its equilibrium value ($E'_M = -1000$ mV) and, therefore, allows the potential in the solution at the anode interface, $\phi_{S,a}$, to approach +1000 mV (recall that $\phi_{S,a} = -E''_M$). The first isopotential above the anode, 900 mV, approaches this value. The solution isopotentials are observed to decrease progressively and approach 0 mV at the cathode reaction site.

The figures span the distance from the center of an anodic strip (0.5 cm wide) to the center of an adjacent cathodic strip 1.5 cm wide (i.e., the center-to-center distance for the strips is 1.0 cm). It is assumed that the anodic and cathodic reactions are confined to the respective areas, as stated above. Current flows in the solution as positive ions from the anodic area where the reaction, $M \rightarrow M^{m+} + me$, occurs to the cathodic

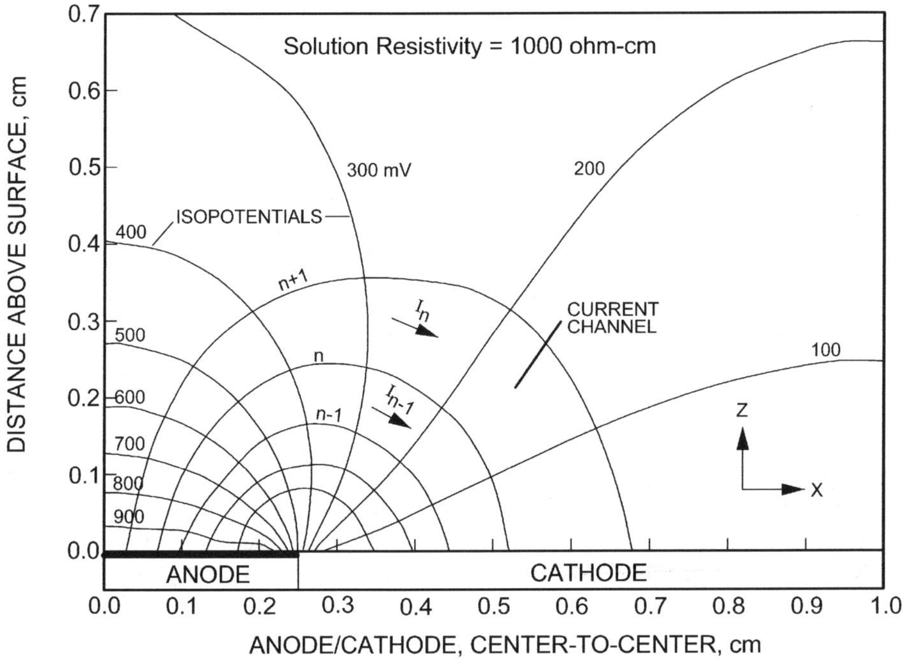


**Fig. 4.3(a)** Potential and current distribution in electrolyte between anode and cathode. Solution-specific resistivity is 1000 ohm-cm. Current channels between boundaries (..., n – 1 and n, and n and n + 1, ...) conduct the same current (..., $I_{n-1} = I_n$, ...). In this example, $I_n$ = 100 μA per cm in the y-direction.

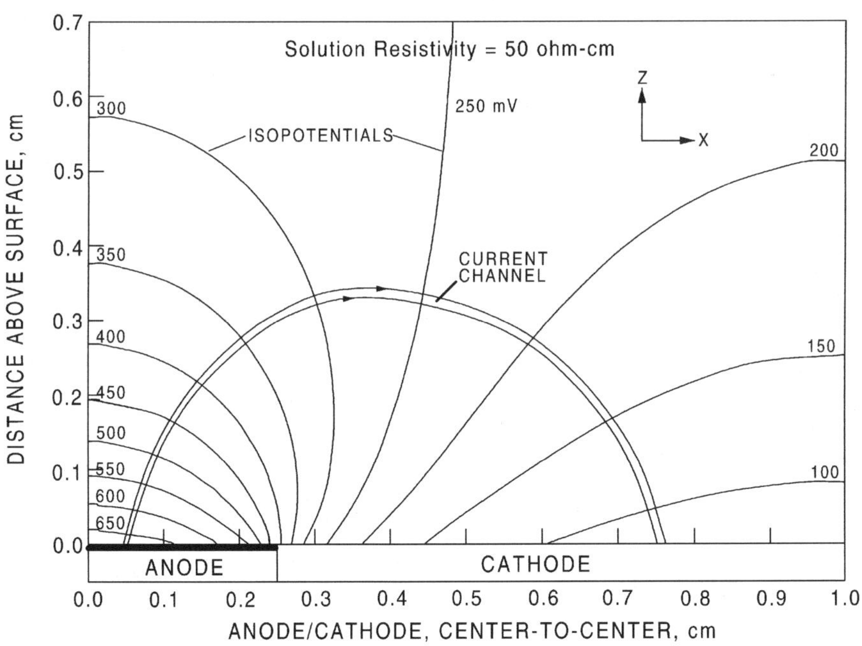


**Fig. 4.3(b)** Potential and current distribution in electrolyte with specific resistivity of 50 ohm-cm. Only one current channel is shown. These become progressively more narrow as the anode/cathode junction is approached. Current channels conduct the same current as in Fig. 4.3(a).

area where the cathodic reaction, $X^{x+} + xe \rightarrow X$, occurs; negative ions contribute to the current by flowing in the opposite direction. The current results from the potential gradient established in the solution (from $\phi_{S,a}$ at a given anodic site to $\phi_{S,c}$ at the corresponding cathodic site) as a consequence of the polarized half-cell potentials between the metal and the solution at a given anodic site ($E''_M$) and a corresponding cathodic site ($E''_X$). These are polarized interface potentials ($E''$) because a current is passing, the interface potential being related to the local current density by the polarization curve for the particular reaction.

Another governing relationship, however, is Ohm's law, which leads to a dependency of the corrosion current on both the polarization characteristics of the anodic and cathodic reactions and on the total electrical resistance of the system, $R_{total}$. $R_{total}$ includes the resistance in the metal between anodic and cathodic areas, $R_M$; a metal junction resistance if different metals are associated with the two areas, $R_{ac}$; any anode- or cathode-solution interface resistance, $R_{ai}$ and $R_{ci}$; and the solution resistance, $R_S$. The latter depends on the specific resistivity or conductivity of the solution and the geometry of the anode-solution-cathode system.

Since a major variable governing corrosion is frequently the solution resistivity, emphasis is placed on analyzing qualitatively how this can be an important factor. The flux of current from anode to cathode will follow approximately semicircular channels, perpendicular to the isopotential surfaces, for the simple geometry shown in Fig. 4.3(a) and (b). The current-channel boundary surfaces have been drawn so as to define channels of fluid extending from the anode to the cathode with a

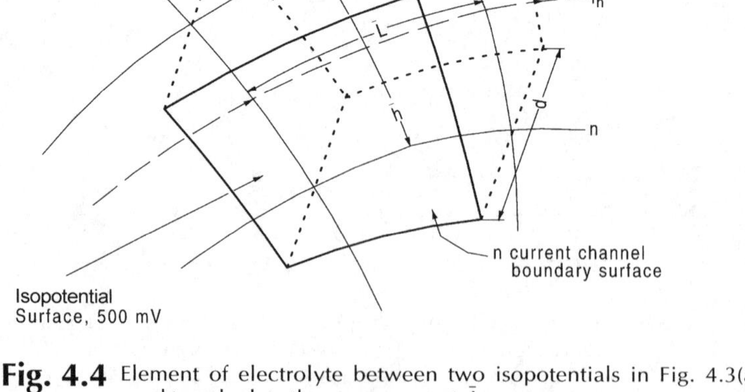


**Fig. 4.4** Element of electrolyte between two isopotentials in Fig. 4.3(a) used to calculate the mean current, $\bar{I}_n$

spacing such that each channel conducts the same amount of current, 100 μA per cm in the y direction. For purposes of calculation, an element of the solution is defined for analysis (Ref 3). An element defined by the 500 and 400 mV isopotential surfaces and the current channel boundaries n and n + 1 in Fig. 4.3(a) is shown in Fig. 4.4. The element (and the channel) is assigned the constant depth, d, in the y-direction. The mean height of the element is $\bar{h}$, and the mean distance between isopotentials is $\bar{L}$. The mean current, $\bar{I}_n$, flowing through the element, and therefore the channel, is given by $\bar{I}_n = \Delta\phi_S/R$, where $\Delta\phi_S = \phi_a - \phi_c$ (with $\phi_a$ and $\phi_c$ corresponding to the isopotentials closer to the anodic site and cathodic site, respectively), and R is the resistance of the element. The resistance is calculated from the specific resistivity of the solution ($\rho$) and the element dimensions, $R = \rho\bar{L}/\bar{A} = (\rho\bar{L})/(\bar{h}d)$, where $\bar{A}$ is the mean area of the channel. It is useful to assign d = 1 cm. The mean current is then $\bar{I}_n = (\Delta\phi_S/\rho)/(\bar{L}/\bar{h})$. The isopotentials and current-channel boundary lines have been drawn in Fig. 4.3(a) with $\bar{h} \approx \bar{L}$. Hence, the mean current through each channel is $\bar{I}_n = 0.1/1000$ A or 100 μA. If this current is divided by the area intercepted by the channel at the anode surface, the current density, which is proportional to the corrosion rate, is obtained. It is evident from Fig. 4.3(a) that $\bar{h}$, and therefore $\bar{A}$, increases with distance from the anode/cathode junction, and hence, the corrosion rate decreases with this distance.

The effect of the specific resistivity of the environment is shown by the isopotentials and current distribution in Fig. 4.3(b) compared with those in Fig. 4.3(a). The current channels in Fig. 4.3(b) have been constructed to carry the same mean current, $\bar{I}_n = 100$ μA, as in Fig. 4.3(a). Since the current-channel boundary lines are so close together in Fig. 4.3(b), only one representative channel is shown. Thus, the effect of changing the resistivity from 1000 to 50 ohm-cm is to decrease the mean area of a channel and hence increase the current density at the interface. Also, the current is more uniformly distributed over the anode for the low-resistivity environment, and the total amount of corrosion is larger. These differences are shown by the corrosion penetration profiles in Fig. 4.5(a) and (b). In the higher-resistivity environment, the penetration is very small at the center of the anode but increases significantly at the anode/cathode junction. In contrast, the low-resistivity environment results in much larger penetration. The profiles of the corroding anode interface are similar for the two environments, but the ratio of penetration at the interface to that at the center of the anode is about 16 to 1 in the high-resistivity environment but only 1.7 to 1 for the low-resistivity environment. Thus, the corrosion is more uniform across the anode area in the low-resistivity environment as can be concluded from comparison of the distribution of corrosion current along the metal/environment interface in Fig. 4.3(a) and (b).

The distribution of potential in the solution along the solution/metal interface is shown in Fig. 4.6. If the anode and cathode areas are not connected, they will exhibit their thermodynamic or "open circuit" potentials, with the potentials in the solution at the anode and cathode being equal to +1000 mV and 0 mV, respectively. When the anode and cathode areas are in contact, current will pass causing polarization of the interface reactions. With a solution-specific resistivity of 1000 ohm-cm, the solution potential at the center of the anode is decreased

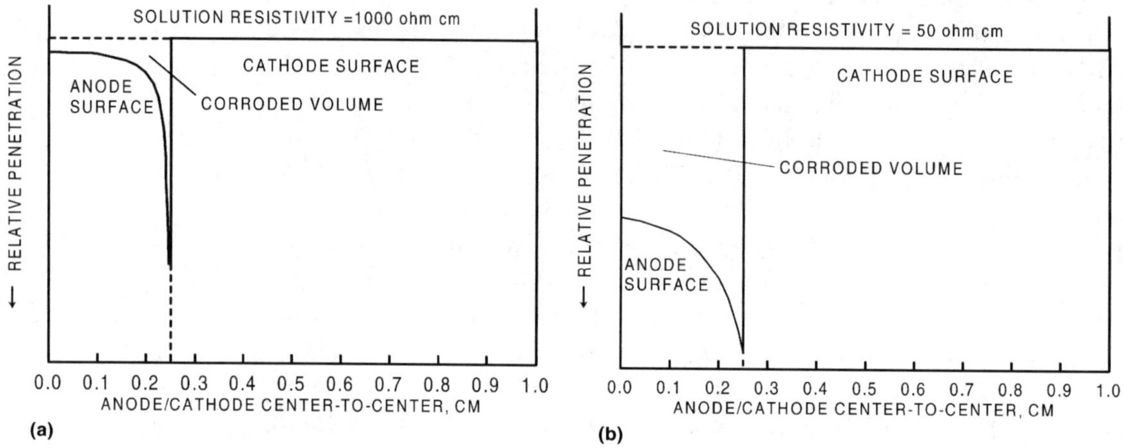


**Fig. 4.5** Corrosion penetration profiles. (a) Corresponding to potential and current distribution of Fig. 4.3(a). (b) Corresponding to potential and current distribution of Fig. 4.3(b)

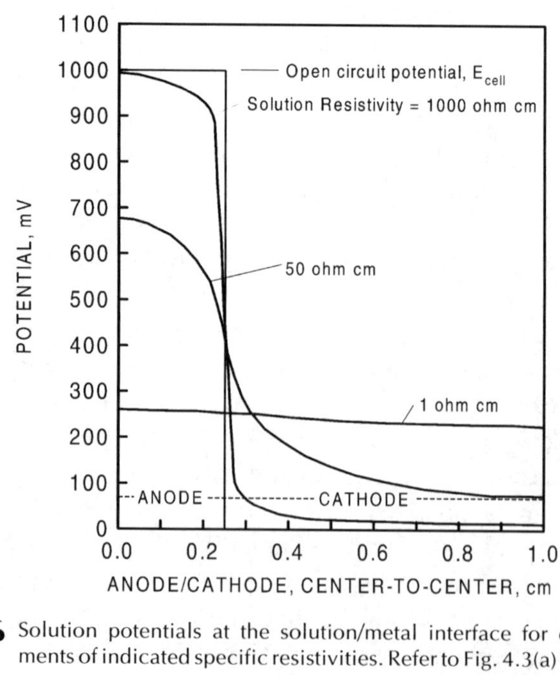


**Fig. 4.6** Solution potentials at the solution/metal interface for environments of indicated specific resistivities. Refer to Fig. 4.3(a) and (b).

slightly and that at the center of the cathode increased slightly. The solution potentials at the solution/metal interface change relatively small amounts until within about 0.04 cm of the anode/cathode junction, where the potential changes rapidly. With a specific resistivity of 50 ohm-cm, the polarization decreases the solution potential at the center of the anode to 680 mV and raises the solution potential at the center of the cathode to 75 mV. The potential change across the junction is spread more than shown for the high-resistivity environment. An additional curve is shown in Fig. 4.6 for an environment of about 1.0 ohm-cm resistivity; it is seen for this case that the potential profile is almost flat at 250 mV.

If two reference electrodes connected through an electrometer are employed, as indicated in Fig. 4.1, the differences in solution potential, $\Delta\phi_S$, between any two points in the solution can be measured. Carrying this measurement technique a step further, with reference to the solution-potential distribution in Fig. 4.3(a) for the highest solution resistivity (1000 ohm-cm), if the first reference electrode is placed and maintained near the surface at the center of the cathode (1.0 cm), and the second reference electrode (connected to the positive electrometer lead) is placed near the surface at the center of the anode (0.0 cm), the reading will approach (but not quite equal) +1000 mV; that is, the reading will be approximately $\Delta\phi_S = (990 - 10) = +980$ mV, as indicated by the potential difference in Fig. 4.6. If the same measurement is conducted for the lower-resistivity solution (50 ohm-cm) shown in Fig. 4.3(b), the reading would be $\Delta\phi_S = (680 - 75) = +605$ mV, in accordance with Fig. 4.6. Finally, if the same measurement is conducted for the lowest resistivity solution in Fig. 4.6 (1.0 ohm-cm), the reading would be $\Delta\phi_S = (260 - 225) = +35$ mV. If the second reference electrode were maintained very close to the metal surface and scanned parallel to the stationary first reference electrode (at the center of the cathode, 1.0 cm), the $\Delta\phi_S$ reading would vary from +980 to 0 mV for the 1000 ohm-cm solution, from +605 to 0 mV for the 50 ohm-cm solution, and from +35 to 0 mV for the 1.0 ohm-cm solution, all in accordance with the solution-potential distributions at the metal surface shown in Fig. 4.6. Thus, such a scanning technique, with two reference electrodes connected through an electrometer, can identify anodic and cathodic sites at the metal surface, the highest (most positive) $\Delta\phi_S$ value corresponding to the center of the anodic site and the lowest $\Delta\phi_S$ value corresponding to the center of the cathodic site. If the specific resistivity of the solution results in the potential distribution of Fig. 4.3(a) (high resistivity), the anodic and cathodic areas can be easily located. For the lower-resistivity solution corresponding to Fig. 4.3(b), the change in solution potential is considerably less. For even lower-resistivity solutions, the changes in solution potential may be too small to allow practical detection of the two areas.

An alternate measurement technique may be employed to determine the interface potentials, E″, during the steady-state corrosion process. With reference to Fig. 4.1, if a single reference electrode is employed, connected through an electrometer to the metal (with the positive electrometer lead connected to the metal), the readings will correspond to E″ values. It should be recalled that $E'' = -\phi_S$. Thus, with reference to Fig. 4.6 for the 1000 ohm-cm solution, if the single reference electrode is placed very close to the metal surface at the center of the anode (0.0 cm), the electrometer reading will be $E''_M = -\phi_{S,a} = -990$ mV (SHE), and at the center of the cathode the reading will be $E''_X = -\phi_{S,c} = -10$ mV (SHE). For the 50 ohm-cm and 1.0 ohm-cm solutions, the readings will be $E''_M = -680$ mV (SHE) and $E''_X = -75$ mV (SHE), and $E''_M = -260$ mV (SHE) and $E''_X = -225$ mV (SHE), respectively. If the reference electrode is scanned parallel to the surface, the E″ values will vary from a minimum at the center of the anode to a maximum at the center of the cathode, with the E″ values corresponding to the negatives of the solution potentials shown in Fig. 4.6.

It should be noted from Fig. 4.3(a) and (b) that if scans are made to determine either $\Delta\phi_S$ or $E'' = -\phi_S$ parallel to the surface at increasing distances away from the surface, the potential variations become progressively less and more uniform as the solution resistivity decreases. For example, in Fig. 4.3(a) (higher-resistivity solution), at 0.3 cm above the surface, the interface potential at the center of the anodic area is $E'' \approx -480$ mV (SHE) and at the center of the cathodic area $E'' \approx -110$ mV (SHE), a difference of 370 mV. From Fig. 4.3(b) (lower-resistivity solution), the values are $E'' \approx -385$ and $E'' \approx -160$ mV (SHE), respectively, a difference of only 225 mV. It can be shown that as the distance into the environment increases to large values relative to the sizes of the anodic and cathodic areas, a single interface potential is measured having a value that approaches $[E'_M + (1 - f_a)(E'_X - E'_M)]$ where $f_a$ is the fraction of the surface that is anodic, and $E'_M$ and $E'_X$ are the equilibrium half-cell potentials for the $M = M^{m+} + me$ and $X = X^{x+} + xe$ reactions, respectively. For Fig. 4.3(a) and(b), this single value would be $E'' \approx -250$ mV (SHE). That is, at large distances a reference electrode indicates a single "mixed" potential, although the corroding surface is a distribution of local anodes and cathodes. From a practical standpoint, as the distribution of cathodes and anodes becomes microscopic in scale, a single electrode potential is measured independent of position. It is effectively a mixed electrode potential called the corrosion potential, $E_{corr}$.

The concept of a single $E_{corr}$ (measured in most instances where a surface is corroding uniformly on a macroscopic scale) can be emphasized by again referring to Fig. 4.3(a) and (b). The latter figure showed potential and current distributions for an environment having a specific resistivity $\frac{1}{20}$ that of Fig. 4.3(a). It can be shown that the distributions in Fig.

4.3(b) also would apply if the resistivity remained high (the same as for Fig. 4.3a), but the absolute sizes of the anodic and cathodic regions were decreased by a factor $\frac{1}{20}$ (i.e., the surface was a distribution of anodes 0.025 cm wide and cathodes 0.075 cm wide). This similarity of current and potential distribution is due to the fact that as the average distance between anodes and cathodes decreases, the average resistance between the two decreases, leading to larger current densities, which in turn causes the amount of interface reaction polarization to increase even though the specific resistivity is the same.

The processes in real corroding systems are obviously more complicated than represented by this model. Useful quantitative calculation of the distribution of current density, and hence corrosion rate along the surface, based on the polarization curves for the anodic and cathodic reactions and on the geometry of the anodic and cathodic sites is very complex. In principle, computer-based techniques can be used if exact polarization curves and the geometry of the anodic and cathodic areas are available. For most industrially important situations, this information is not available. Also, time-dependent factors, such as film formation, make quantitative calculations of long-time corrosion rates very uncertain. The theory underlying these calculations, however, has been useful in interpreting observations in research and in industrial situations.

## A Physical Representation of the Electrochemical Behavior of Mixed Electrodes

In the following discussion, a physical interpretation is given to the corrosion process leading to a graphical understanding of the interrelationships among the distribution of corrosion current density, measured potentials, and the polarization characteristics of the anodic and cathodic reactions. These relationships are developed initially with reference to defined local anodic and cathodic areas represented by Fig. 4.7 based on Fig. 4.3(a). Analysis of this model leads to the limiting case of uniform corrosion (very small anodic and cathodic areas) and the estimation of values of macroscopically uniform corrosion rates. As previously discussed, the lines, ...n, (n + 1), ..., define channels passing equal current, these channels having a solution resistance, $R_S$, which increases with distance from the anode/cathode interface. For a complete circuit that includes the metal/solution interfaces and the metal, the anode interface resistance, $R_{ai}$, and the cathode interface resistance, $R_{ci}$, may be significant. And, if the anode and cathode areas are associated with different materials, a resistance, $R_{ac}$, between them must be considered. In the following example, assume that the anode and cathode areas are known and that the interface and metal-path resistances are small compared with $R_S$. The driving potential difference for the local nth current channel is $(\phi_{S,a} - \phi_{S,c})_n = (E''_X - E''_M)_n$ where $E''_X$ and $E''_M$ are the polarized interface potentials at the cathodic and anodic surfaces, respectively, for the nth channel during steady-state corrosion. The current entering the solution at the anodic interface is $(I_{net,ox} = I_{ox,M} - I_{red,M})_n$ (refer to the section "Charge-Transfer Polarization" in Chapter 3). The net current at the cathodic interface is $(I_{net,red} = I_{ox,X} - I_{red,X})_n$. In the following example, the contributions of $I_{red,M}$ and $I_{ox,X}$ are considered to be negligible (a generally valid assumption when $E_{corr}$ is removed from $E'_M$ and $E'_X$ by more than 50 mV). Therefore, $I_{net,ox} \approx (I_{ox,M})_n$ at the anodic interface and $I_{net,red} \approx -(I_{red,X})_n$ at the cathodic interface. Under the freely corroding conditions of Fig. 4.7, the corrosion current must equal both the anodic and the cathodic currents, $(I_{corr} = I_{ox,M} = I_{red,X})_n$. In addition, Ohm's law must be satisfied for each path:

$$(I_{corr})_n = \frac{(E''_X - E''_M)_n}{(R_{total})_n} = \frac{(E''_X - E''_M)_n}{(R_S)_n} \quad \text{(Eq 4.20)}$$

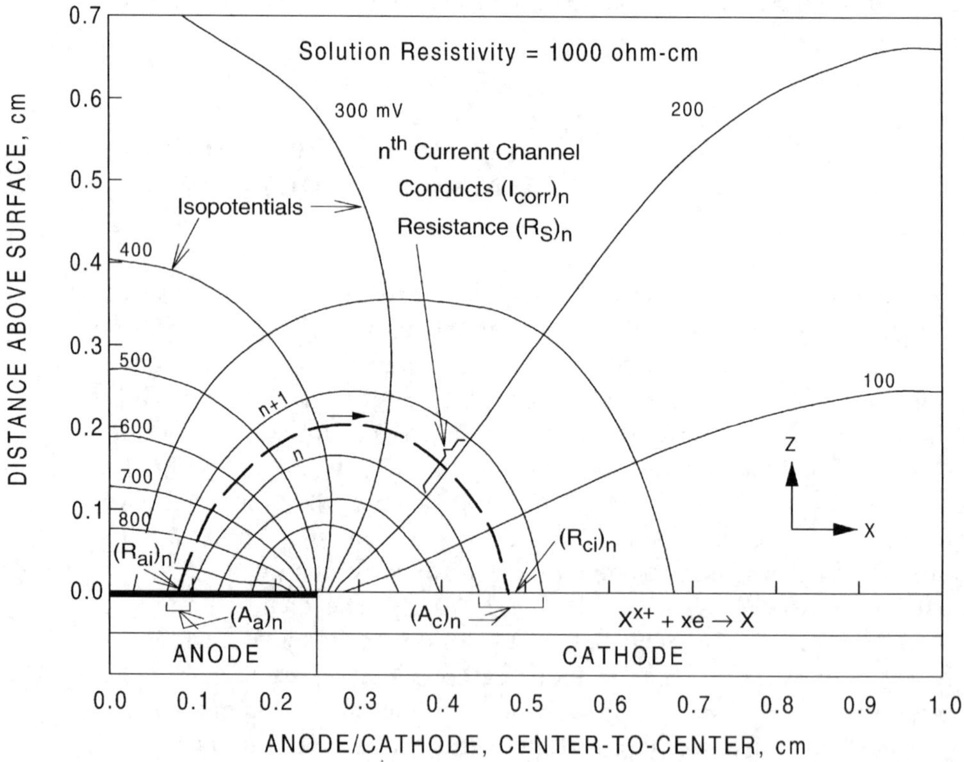


**Fig. 4.7** Potential and current distribution in an environment of specific resistivity, 1000 ohm-cm. Parameters relating to one (nth) current channel

The potentials at the cathodic and anodic sites are functions of the current density. From Chapter 3, under charge-transfer polarization conditions, Tafel equations of the forms of Eq 3.47 and 3.48 lead to:

$$E_X = E'_X - \beta_{red,X} \log \frac{I_{red,X} / A_c}{i_{o,X}} \quad \text{(Eq 4.21)}$$

and

$$E_M = E'_M + \beta_{ox,M} \log \frac{I_{ox,M} / A_a}{i_{o,M}} \quad \text{(Eq 4.22)}$$

Under freely corroding conditions, when $I_{corr} = I_{ox,M} = I_{red,X}$, Eq 4.21 and 4.22 for the cathodic and anodic reactions become:

$$E''_X = E'_X - \beta_{red,X} \log \frac{I_{corr} / A_c}{i_{o,X}} \quad \text{(Eq 4.23)}$$

and

$$E''_M = E'_M + \beta_{ox,M} \log \frac{I_{corr} / A_a}{i_{o,M}} \quad \text{(Eq 4.24)}$$

Therefore, the Ohm's law expression (Eq 4.20) for the nth current channel can be written as:

$$(I_{corr})_n = \frac{\left[E'_X - \beta_{red,X} \log \frac{I_{corr} / A_c}{i_{o,X}}\right]_n - \left[E'_M + \beta_{ox,M} \log \frac{I_{corr} / A_a}{i_{o,M}}\right]_n}{(R_S)_n} \quad \text{(Eq 4.25)}$$

Equation 4.25 can be interpreted in relationship to the conventional plotting of linear or Tafel polarization behavior of the anodic and cathodic reactions. For this purpose, the individual anodic and cathodic curves are plotted as functions of the total current rather than current density. For any channel (e.g., the nth channel), the oxidation current at the anode is $(I_{ox,M} = i_{ox,M} A_a)_n$ where $A_a$ is the area of the nth channel at the anode/solution interface. Similarly for the cathode interface, the reduction current is $(I_{red,X} = i_{red,X} A_c)_n$. The polarization curves are plotted using exchange currents, $I_o$, obtained by multiplying the exchange current densities by the respective areas, and the Tafel slopes. The polarization curves have the relative forms illustrated in Fig. 4.8. Any vertical separation between the two curves is a potential difference driving the current from the anodic to the cathodic surface in the channel. This difference in potential must be such that Eq 4.20 (and Eq 4.25) is satisfied. The difference is determined graphically by determining $(E''_X - E''_M)_n$ at selected values of the current until a potential difference is found such that when divided by $(R_S)_n$, the resulting current has the same value as given along the log I axis. This current will be $(I_{corr})_n$

for the nth channel and on division by $(A_a)_n$ will give the corrosion current density, $(i_{corr})_n$, at this location on the anode interface. The local corrosion rate can be calculated from this corrosion current density. This interpretation of Eq 4.20 (and Eq 4.25) applies to each of the channels and accounts for the corrosion depth profiles of Fig. 4.5(a) and (b).

As the solution resistance decreases, this analysis indicates that the conditions satisfying Eq 4.20 move toward the intersection of the two polarization curves in Fig. 4.8. A decrease in resistance between the anodes and cathodes results when the specific resistivity of the solution is decreased and will occur even for higher-resistivity environments if the anodic and cathodic areas are very small and separated by small distances. Under these conditions, corrosion will appear to be uniform on a macroscopic scale, and movement of a reference electrode in the solution will measure a single $E_{corr}$ independent of position with a value approaching the potential at which the anodic and cathodic polarization curves intersect in Fig. 4.8. To appreciate how small this driving potential difference may be, consider an anodic area of 1 $cm^2$ ($10^{-4}$ $m^2$) in a large cathodic area exposed to a relatively low resistivity environment such that $R_S = 10$ ohms and that the conditions are such as to cause the practically small current of $10^{-2}$ mA. The anodic current density is then 100 $mA/m^2$, which for iron would be a corrosion penetration rate of about 125 μm/year (5 mpy). The driving potential supporting this corrosion would have the very small value of $(10^{-2}$ mA)(10 ohm) = 0.1 mV, a difference so small that it cannot be represented graphically in Fig. 4.8.

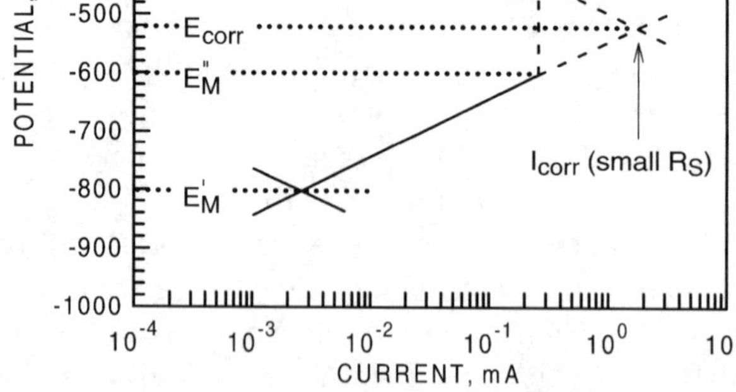


**Fig. 4.8** Tafel polarization curves for anodic and cathodic reactions as related to the nth current channel in Fig. 4.7, illustrating the dependence of the corrosion current, $I_{corr}$, on the solution resistance, $R_S$

If either or both of the curves representing single half-cell polarization behavior deviate from linearity due to diffusion control, the intersection will occur at smaller values of corrosion current, resulting in smaller corrosion rates. This effect is illustrated in Fig. 4.9, where in all cases $I_{corr}$ with diffusion control is less than $I_{corr}$ without diffusion control. It should be noted that the corrosion potential, $E_{corr}$, may increase or decrease when the corrosion is under diffusion control as compared with that which would be observed in the absence of such control. The influence of fluid velocity is represented by displacement of the diffusion control (curved) sections of each curve to larger values of current in accordance with the discussion in Chapter 3 relating velocity to the thickness of the diffusion boundary layer. It should be clear, as shown in Fig. 4.10, that an increase in fluid velocity will increase the corrosion rate until the velocity is sufficiently large to move the diffusion control range to current densities greater than the intersection of the linear or Tafel section of the anodic polarization curve with the polarization curve for the cathodic reaction. Thus, at sufficiently large velocities, the corrosion rate becomes constant (i.e., independent of velocity).

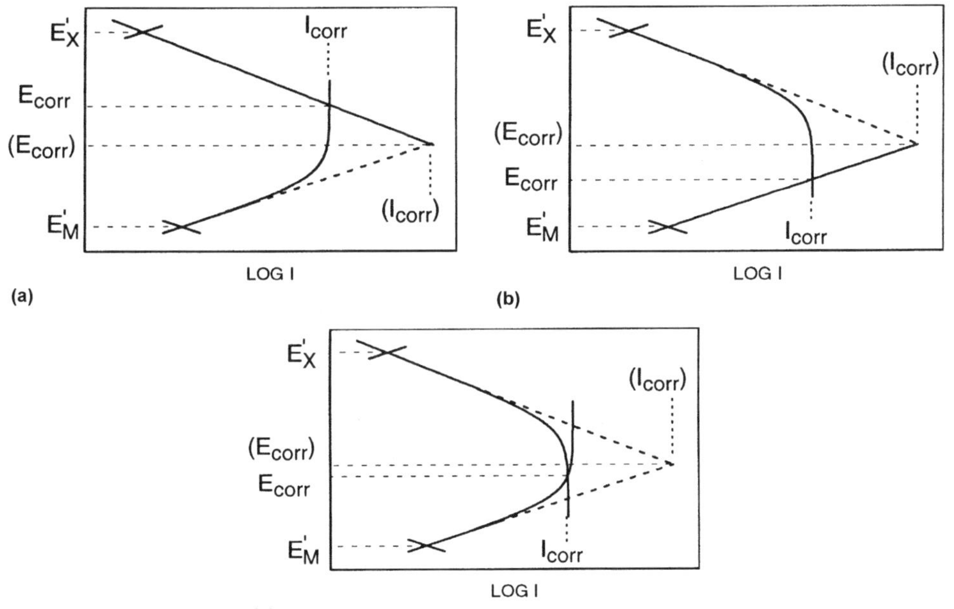


**Fig. 4.9** Influence of relative positions and shapes of anodic and cathodic polarization curves on the corrosion current, $I_{corr}$. (a) Anodic diffusion control. (b) Cathodic diffusion control. (c) Anodic and cathodic diffusion control. $E_{corr}$ and $I_{corr}$ refer to corrosion under diffusion control. ($E_{corr}$) and ($I_{corr}$) refer to corrosion without diffusion control.

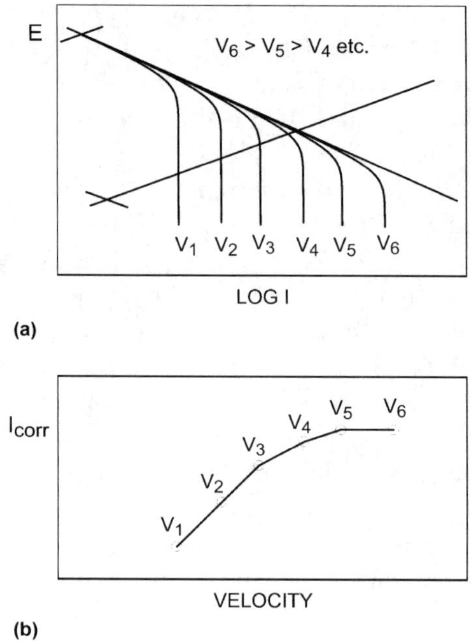


**Fig. 4.10** (a) Schematic representation of the influence of fluid velocity on the corrosion current as the intersection of a velocity-dependent cathodic polarization curve with the anodic polarization curve. (b) The resulting dependence of the corrosion current on fluid velocity

## Interpretation of $E_{corr}$

A reference electrode scanned along the metal surface will measure the series of $(E''_X)_n$ and $(E''_M)_n$ interface potentials. From these values, solution potentials ($\phi_S$) at the metal/solution interface may be calculated ($\phi_S = -E''$) and presented as in Fig. 4.6. When the anodic and cathodic sites are microscopic relative to the size and position of the reference electrode, identity of the anodic and cathodic sites on a macroscale is lost, and a single mixed or corrosion potential, $E_{corr}$, is measured as discussed previously. There is essentially a uniform flux of metal ions from the surface, and cathodic reactants to the surface, which constitute anodic and cathodic currents. Since the relative areas to which these currents apply usually are not known, the total area is taken as the effective area for each reaction. It is these currents, however, that mutually polarize the anodic reaction potential from $E'_M$ up to $E_{corr}$ and the cathodic reaction potential from $E'_X$ down to $E_{corr}$.

# Faraday's Law

Faraday's law is the connecting relationship between the corrosion current density, $i_{corr} = I_{corr}/A_a$, and other expressions of "corrosion rate," such as corrosion intensity (CI), in units of mass-loss per unit area per unit time, and corrosion penetration rate (CPR) in units of loss-in-dimension perpendicular to the corroding surface per unit time.

To retain emphasis on corrosion processes, Faraday's law will be derived with reference to the generalized metal oxidation reaction, $M \rightarrow M^{m+} + me$. In Fig. 4.11, an anodic area, $A_a$, is shown over which $I_{net,ox} = I_{ox,M} - I_{red,M} = I_{corr} \approx I_{ox,M}$. The current flows to the surface counter to the electrons and enters the solution as positive ions (cations), $M^{m+}$. Since metal is lost to the solution, corrosion occurs over areas where internal current flows to the metal surface or, conversely, where current is flowing from the surface in the aqueous environment, a useful general rule in the analysis of corroding systems.

Consider that the corrosion current, $I_{corr}$, is expressed in amperes (A) or coulombs (C) per second (s). The unit of positive electricity (equivalent to the magnitude of the charge on the electron but with opposite sign) has a charge of $1.60 \times 10^{-19}$ coulombs and will be designated $e^+$. Each ion formed by detachment from the surface contributes $me^+$ coulombs to the current. W grams (g) of metal entering the solution in t seconds contributes W/Mt moles per second, where M (g/mol) is the atomic mass. Multiplying by Avogadro's number, $N_o$, gives $(W/Mt)N_o$ ions per second. The product of the ions per second and the charge per ion gives the current; thus:

$$I_{corr}\ (\text{C/s or A}) = (WN_o/Mt)(me^+) \qquad \text{(Eq 4.26)}$$

or

$$I_{corr} = (Wm/M)(N_o e^+)(1/t) \qquad \text{(Eq 4.27)}$$

$$I_{corr} = M'F/t \qquad \text{(Eq 4.28)}$$

where $M' = Wm/M$ is the number of electrochemical equivalents (ee) entering the solution (recall that one ee is the number of moles of material that will produce one mole or Avogadro's number of electrons; that is, 1 ee = 1/m mol of metal), and F is Faraday's constant (the charge of 1 mol of electrons, F = 96,485 C/mol of electrons = 96,485 C/ee = 96,485 J/volt-ee = 96.485 kJ/volt-ee = 23,060 calories/volt-ee).

If Eq 4.27 is solved for W/t and then divided by the anode area, $A_a(cm^2)$, an expression for the corrosion intensity (CI) is obtained:

$$CI\ (g/cm^2 \cdot s) = \frac{M(I_{corr}/A_a)}{m(N_o e^+)} \qquad \text{(Eq 4.29)}$$

$$CI\ (g/cm^2 \cdot s) = \frac{M i_{corr}}{mF} \quad \text{(Eq 4.30)}$$

where $i_{corr}$ is the corrosion current density in A/cm$^2$.

If Eq 4.30 is divided by the density of the material, $\rho$ (g/cm$^3$), an expression for the corrosion penetration rate (CPR) is determined:

$$CPR\ (cm/s) = \frac{M i_{corr}}{mF\rho} \quad \text{(Eq 4.31)}$$

The expressions for CI and CPR, Eq 4.30 and 4.31, can be easily converted to more convenient and traditional sets of units. For example, the CI in units of grams per m$^2$ per year (g/m$^2$ · y) is given by:

$$CI\ (g/m^2 \cdot y) = 0.327 \frac{M i_{corr}}{m} \quad \text{(Eq 4.32)}$$

where, in this expression, $i_{corr}$ is in mA/m$^2$. The CPR in μm/year is given by:

$$CPR\ (\mu m/year) = 0.327 \frac{M i_{corr}}{m\rho} \quad \text{(Eq 4.33)}$$

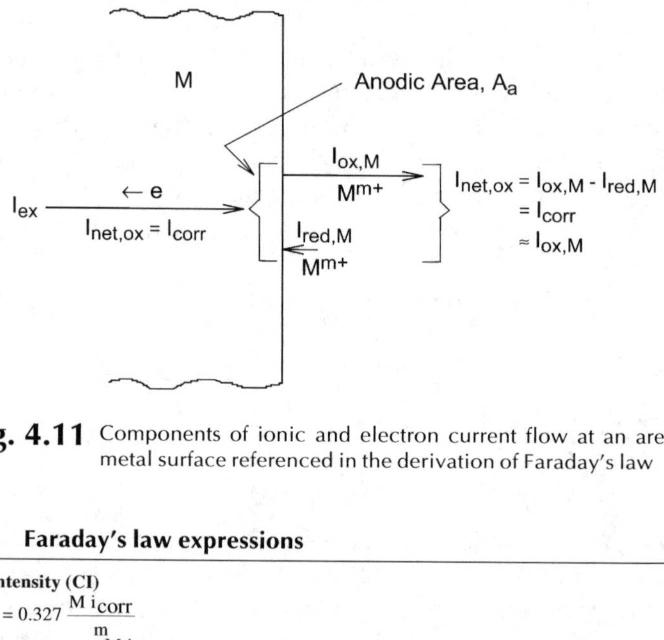


**Fig. 4.11** Components of ionic and electron current flow at an area of metal surface referenced in the derivation of Faraday's law

**Table 4.1 Faraday's law expressions**

**Corrosion Intensity (CI)**

$CI\ (g/m^2 \cdot y) = 0.327 \frac{M i_{corr}}{m}$

$CI\ (m\,g/cm^2 \cdot y) = 0.0327 \frac{M i_{corr}}{m}$

**Corrosion Penetration Rate (CPR)**

$CPR\ (\mu m/y) = 0.327 \frac{M i_{corr}}{m\rho}$

$CPR\ (mm/y) = 0.327 \times 10^{-3} \frac{M i_{corr}}{m\rho}$

$CPR\ (mpy) = 0.0129 \frac{M i_{corr}}{m\rho}$

Note: M, g/mol; m, oxidation state or valence; $\rho$, g/cm$^3$; $i_{corr}$, mA/m$^2$; y, year; and mpy = mils (0.001 in.) per year where $i_{corr}$ is in $mA/m^2$. Other expressions for CI and CPR in various sets of units are given in Table 4.1.

In the previous discussion, Faraday's law was derived on the basis that the net metal oxidation current, $I_{net,ox}$, was equal to the corrosion current, $I_{corr}$, at the corrosion potential, $E_{corr}$. Although this is by far the most common way in which Faraday's law is applied in the analysis of corrosion, it should be noted that the law is quite general in terms of relating currents to electrochemical reaction rates. For example, in Eq 4.30 and 4.31, if $i_{corr}$ is replaced with $i_{net,ox}$ (or $i_{ox,M}$ if $i_{red,M}$ is negligible), the equations allow calculations of CI and CPR at any potential. Alternately, the net reduction rate at any potential (including $E_{corr}$) can be obtained from Eq 4.27 upon replacement of $I_{corr}$ with $I_{net,red}$.

## Effects of Cathode-to-Anode Area Ratio

The cathode-to-anode area ratio is frequently a critical factor in corrosion. (This is true when well-defined cathodes and anodes exist. With mixed electrode behavior, where cathodic and anodic reactions occur simultaneously, separate areas are not readily distinguishable, and $A_a$ is assumed equal to $A_c$.) Discussion of the influence of this ratio will be restricted to the case of a small total-corrosion-circuit resistance leading to the anodic and cathodic reactions occurring at essentially the same potential, $E_{corr}$, as described previously. In Fig. 4.12, three different values of corrosion current, $I_{corr}$, and corrosion potential, $E_{corr}$, are shown for three cathode areas relative to a fixed anode area of 1 $cm^2$. For these cases, a reference electrode placed anywhere in the solution

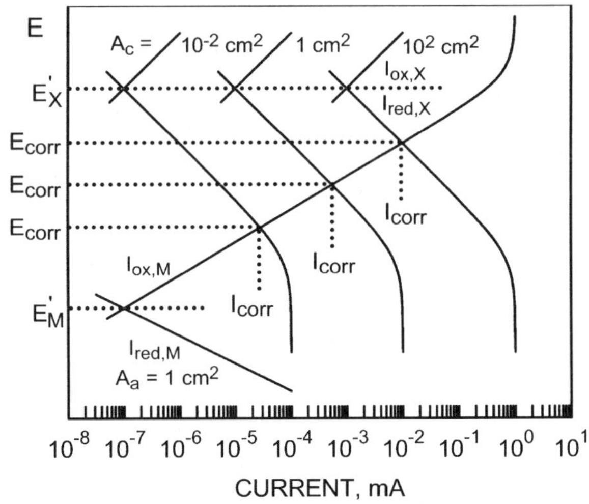


**Fig. 4.12** Schematic representation of the effect on $I_{corr}$ of different cathodic areas, $A_c$, and a constant anodic area, $A_a$

will result in an electrometer reading, $E_{M,ref}$, from which the corrosion potential is determined on the standard hydrogen electrode scale (i.e., $E_{corr} = E_{M,ref} + E_{ref}$) (see the section "Interpretation of Charge-Transfer Polarization from Experiment" in Chapter 3). As the ratio $A_c/A_a$ increases, the corrosion current increases. The important consequence is that the corrosion current density, $i_{corr}$, also increases (i.e., $I_{corr}$ is larger for the same $A_a$). Hence, from Faraday's law, the corrosion penetration rate increases by an amount proportional to the increase in the cathode-to-anode area ratio, $A_c/A_a$. Thus, from the requirement at $E_{corr}$ that $I_{ox,M} = I_{red,X}$ (assuming $I_{red,M}$ and $I_{ox,X}$ to be negligible):

$$i_{ox,M} A_a = i_{red,X} A_c \qquad \text{(Eq 4.34)}$$

from which:

$$i_{corr} = i_{ox,M} = \frac{A_c}{A_a} i_{red,X} \qquad \text{(Eq 4.35)}$$

## Interpretation of Experimental Polarization Curves for Mixed Electrodes (Ref 4–6)

The earlier sections of this chapter discuss the mixed electrode as the interaction of anodic and cathodic reactions at respective anodic and cathodic sites on a metal surface. The mixed electrode is described in terms of the effects of the sizes and distributions of the anodic and cathodic sites on the potential measured as a function of the position of a reference electrode in the adjacent electrolyte and on the distribution of corrosion rates over the surface. For a metal with fine dispersions of anodic and cathodic reactions occurring under Tafel polarization behavior, it is shown (Fig. 4.8) that a single mixed electrode potential, $E_{corr}$, would be measured by a reference electrode at any position in the electrolyte. The counterpart of this mixed electrode potential is the equilibrium potential, $E'_M$ (or $E'_X$), associated with a single half-cell reaction such as Cu in contact with $Cu^{2+}$ ions under deaerated conditions. The forms of the anodic and cathodic branches of the experimental polarization curves for a single half-cell reaction under charge-transfer control are shown in Fig. 3.11. It is emphasized that the observed experimental curves are curved near $i_o$ and become asymptotic to $E'_M$ at very low values of the external current. In this section, the experimental polarization of mixed electrodes is interpreted in terms of the polarization parameters of the individual anodic and cathodic reactions establishing the mixed electrode. The interpretation then leads to determination of the corrosion potential, $E_{corr}$, and to determination of the corrosion current density, $i_{corr}$, from which the corrosion rate can be calculated.

In review, consider a mixed electrode at which one net reaction is the transfer of metal to the solution as metal ions, and the other net reaction is the reduction of chemical species in the solution such as $H^+$, $O_2$, $Fe^{3+}$, or $NO_2^-$ on the metal surface. For purposes of the present discussion, no attempt is made to define the individual sites for the anodic (net oxidation) and cathodic (net reduction) reactions. They may be homogeneously distributed, resulting in uniform corrosion, or segregated, resulting in localized corrosion. In the latter case, the cathode-to-anode area ratio is of practical importance in determining the rate of penetration at anodic areas.

The half-cell reactions are again represented as:

$$X \leftrightarrow X^{X+} + xe \quad \text{(Eq 4.36)}$$

and

$$M \leftrightarrow M^{m+} + me \quad \text{(Eq 4.37)}$$

with the equilibrium potential of the X reaction being greater than that for the metal reaction, $E'_X > E'_M$. Hence, at a mixed potential between $E'_X$ and $E'_M$, reaction 4.36 will undergo net reduction, $X^{X+} + xe \rightarrow X$, and reaction 4.37 will undergo net oxidation $M \rightarrow M^{m+} + me$. Schematic oxidation and reduction curves for each half-cell reaction under charge-transfer conditions are shown in Fig. 4.13 (i.e., E versus log $I_{ox,X}$, log $I_{red,X}$, log $I_{ox,M}$, and log $I_{red,M}$). Note that the curves for the individual reactions are based on knowing the respective values for E′, $I_o = A\, i_o$, and β. Under charge-transfer conditions, each oxidation or reduction current is related to the potential through the appropriate Tafel equation (Chapter 3, Eq 3.47 and 3.48). For the oxidation component of the metal reaction:

$$E = E'_M + \beta_{ox,M} \log \frac{I_{ox,M}}{I_{o,M}} \quad \text{(Eq 4.38)}$$

or

$$I_{ox,M} = I_{o,M}\, e^{2.3(E-E'_M)/\beta_{ox,M}} \quad \text{(Eq 4.39)}$$

For the reduction component of the metal reaction:

$$E = E'_M - \beta_{red,M} \log \frac{I_{red,M}}{I_{o,M}} \quad \text{(Eq 4.40)}$$

or

$$I_{red,M} = I_{o,M}\, e^{-2.3(E-E'_M)/\beta_{red,M}} \quad \text{(Eq 4.41)}$$

For the oxidation component of the X reaction:

$$E = E'_X + \beta_{ox,X} \log \frac{I_{ox,X}}{I_{o,X}} \qquad \text{(Eq 4.42)}$$

or

$$I_{ox,X} = I_{o,X}\, e^{2.3(E-E'_X)/\beta_{ox,X}} \qquad \text{(Eq 4.43)}$$

For the reduction component of the X reaction:

$$E = E'_X - \beta_{red,X} \log \frac{I_{red,X}}{I_{o,X}} \qquad \text{(Eq 4.44)}$$

or

$$I_{red,X} = I_{o,X}\, e^{-2.3(E-E'_X)/\beta_{red,X}} \qquad \text{(Eq 4.45)}$$

For the isolated corroding surface (i.e., no external current), the total rate of oxidation must equal the total rate of reduction. This condition, in terms of currents, is expressed by:

$$\Sigma I_{ox} = \Sigma I_{red} \qquad \text{(Eq 4.46)}$$

where the sums are taken for all species involved in the reactions. For the species X, M, $X^{X+}$, and $M^{m+}$:

$$I_{ox,X} + I_{ox,M} = I_{red,X} + I_{red,M} \qquad \text{(Eq 4.47)}$$

The sums of the currents resulting from the oxidation and from the reduction reactions are also shown in Fig. 4.13 as a function of potential. The steady-state corrosion condition of $\Sigma I_{ox} = \Sigma I_{red}$ corresponds to the intersection of the $\Sigma I_{ox}$ and $\Sigma I_{red}$ lines, which identifies the corrosion potential, $E_{corr}$. The solution ohmic resistance is assumed to be very small for the present interpretation.

It is noted that in the example of Fig. 4.13, the $I_{ox,X}$ and $I_{ox,M}$ curves are close (within a factor of 10), and the $I_{red,M}$ and $I_{red,X}$ curves are similarly close. These conditions result in $\Sigma I_{ox}$ being observably greater than $I_{ox,M}$ and $\Sigma I_{red}$ being observably greater than $I_{red,X}$. In the example, the conditions also result in $E_{corr}$ being within 50 mV of both equilibrium potentials, $E'_M$ and $E'_X$. These conditions were selected to clearly illustrate the principles under discussion. Generally, however, these relative positions are not typical of corroding systems. Rather, $E_{corr}$ differs from both $E'_M$ and $E'_X$ by more than approximately 50 mV, which is the result of $I_{ox,X}$ and $I_{ox,M}$, and $I_{red,M}$ and $I_{red,X}$, differing by factors greater than 10. As a consequence, typically, $\Sigma I_{ox} \approx I_{ox,M}$ and $\Sigma I_{red} \approx I_{red,X}$.

A schematic representation of two electrochemical reactions establishing a mixed electrode at a metal surface is shown in Fig. 4.14(a). Each reaction will have an oxidation component and a reduction component as discussed in Chapter 3. These component currents cannot be directly measured because they are internal to the metal and surrounding electrolyte. Figures 4.14(b) and (c) represent wires through which an external current can be passed to or from the metal. It is important to note that only the external current can be measured. It is defined as the difference between the total-oxidation and total-reduction currents at the metal surface, or:

$$I_{ex} = \Sigma I_{ox} - \Sigma I_{red} = (I_{ox,M} + I_{ox,X}) - (I_{red,X} + I_{red,M}) \qquad \text{(Eq 4.48)}$$

Thus, when $\Sigma I_{ox} > \Sigma I_{red}$, $I_{ex}$ will be positive and identified as $I_{ex,ox}$ (i.e., net oxidation occurs at the electrode surface and produces an external anodic current). Conversely, when $\Sigma I_{red} > \Sigma I_{ox}$, $I_{ex}$ will be negative and

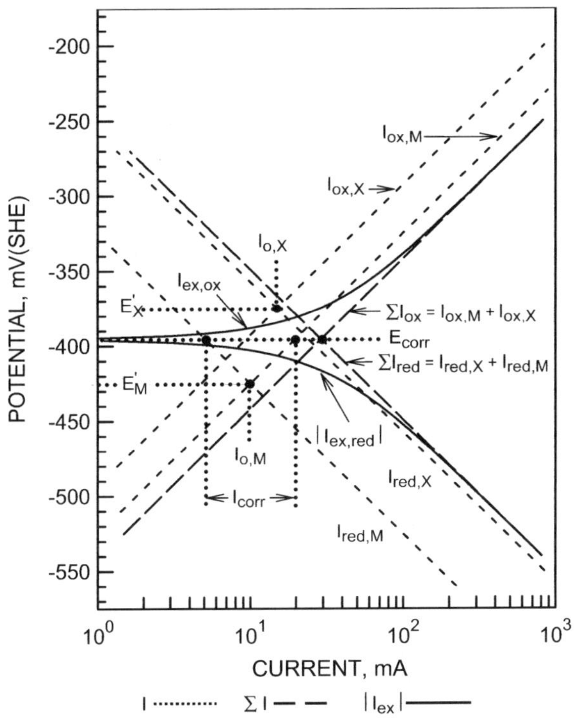


**Fig. 4.13** Relationship of the mixed-electrode cathodic and anodic polarization curves (solid lines) to the oxidation and reduction components (dashed lines) of the individual anodic and cathodic reactions

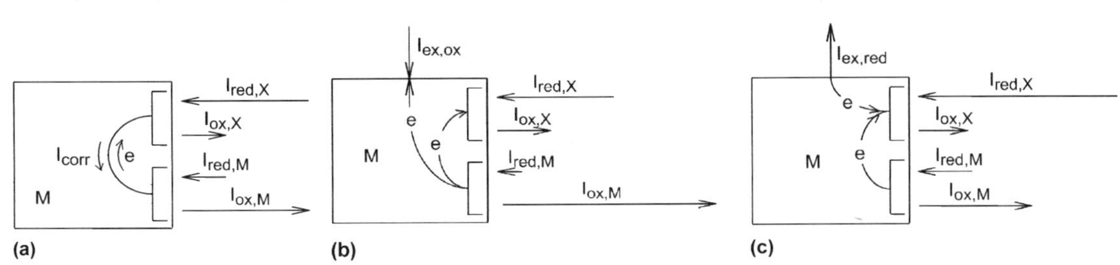


**Fig. 4.14** Representation of a mixed electrode with anodic reactant, M, and cathodic reactant, X. (a) Freely corroding condition. (b) Net external oxidation current. (c) Net external reduction current

identified as $I_{ex,red}$ (i.e., net reduction occurs at the electrode surface and produces an external cathodic current). If the arrows in Fig. 4.14(b) and (c) represent the magnitude of the currents, then in Fig. 4.14(b), net oxidation is occurring and the external current is $I_{ex,ox} > 0$; in Fig. 4.14(c), net reduction is occurring and the external current is $I_{ex,red} < 0$.

It is emphasized, however, that unless $E < E'_M$, the metal reaction will always undergo net oxidation, and therefore, the corrosion rate, expressed as a current, will be:

$$I'_{corr} = I_{ox,M} - I_{red,M} \quad \text{(Eq 4.49)}$$

where $I'_{corr}$ denotes the corrosion current at any potential, E. This is to distinguish the general case of arbitrary E from the specific use of $I_{corr}$ to designate the corrosion current at the corrosion potential, $E_{corr}$, where $I_{ex} = 0$. For this condition, setting $I_{ex} = 0$ in Eq 4.48 results in the following important relationship:

$$(I_{ox,M} - I_{red,M}) = (I_{red,X} - I_{ox,X}) = I_{corr} \quad \text{(Eq 4.50)}$$

The $I_{corr}$ shown in Fig. 4.14(a) is consistent with this relationship.

An analytical expression for the external current is obtained on substitution of Eq 4.39, 4.41, 4.43, and 4.45 into Eq 4.48:

$$I_{ex} = I_{o,M}\, e^{2.3(E-E'_M)/\beta_{ox,M}} + I_{o,X}\, e^{2.3(E-E'_X)/\beta_{ox,X}} - I_{o,X} e^{-2.3(E-E'_X)/\beta_{red,X}} - I_{o,M}\, e^{-2.3(E-E'_M)/\beta_{red,M}} \quad \text{(Eq 4.51)}$$

With reference to Fig. 4.13, Eq 4.51 is the sum of the values of the currents of the oxidation Tafel curves minus the sum of the values of the currents of the reduction Tafel curves (i.e., $I_{ex} = \Sigma I_{ox} - \Sigma I_{red}$) at any value of E. Since $I_{ex}$ changes from a negative to a positive quantity on increasing E from $E < E_{corr}$ to $E > E_{corr}$ (a discussion follows Eq 4.48), the equation is plotted as log $|I_{ex,red}|$ versus E for $E < E_{corr}$ (the lower solid curve in Fig. 4.13, net reduction) and as log $I_{ex,ox}$ versus E for $E > E_{corr}$ (the upper solid curve, net oxidation). Both curves approach very low values of current as $E \rightarrow E_{corr}$. The log $I_{ex,ox}$ curve becomes asymptotic to the log $\Sigma I_{ox}$ curve for $E >> E_{corr}$, and the log $|I_{ex,red}|$ curve becomes asymptotic to the log $\Sigma I_{red}$ curve for $E << E_{corr}$.

These limiting conditions (i.e., when $E = E_{corr}$, $E >> E_{corr}$ and $E << E_{corr}$) are analyzed as follows. Since $E_{corr}$ is the free or open-circuit corrosion potential, $I_{ex}$ must equal zero at this potential and, therefore, the curves of log $I_{ex,ox}$ and log $|I_{ex,red}|$ must approach very low values when plotted on logarithmic coordinates as observed in Fig. 4.13. At large positive deviations from $E_{corr}$, reference to Fig. 4.13 shows that $I_{red,M}$ and $I_{red,X}$ become negligible, which allows Eq 4.48 to be written as:

$$I_{ex,ox} = I_{ox,M} + I_{ox,X} = \Sigma I_{ox} \qquad (E >> E_{corr}) \quad \text{(Eq 4.52)}$$

Therefore, the log $I_{ex,ox}$ solid curve becomes asymptotic to the log $\Sigma I_{ox}$ curve as occurs in Fig. 4.13. Conversely, at large negative deviations from $E_{corr}$, $I_{ox,M}$ and $I_{ox,X}$ become negligible, which allows Eq 4.48 to be written as:

$$I_{ex,red} = -(I_{red,X} + I_{red,M}) = -\Sigma I_{red} \qquad (E << E_{corr}) \qquad \text{(Eq 4.53)}$$

or

$$|I_{ex,red}| = \Sigma I_{red} \qquad \text{(Eq 4.54)}$$

In this case, the log $|I_{ex,red}|$ solid curve becomes asymptotic to the log $\Sigma I_{red}$ curve as occurs in Fig. 4.13.

The above analysis of a mixed electrode in terms of the current components is usually simplified under several common, and often very accurate, assumptions. With reference to Fig. 4.13, if the intersection of the $\Sigma I_{ox}$ and the $\Sigma I_{red}$ lines occurs at a potential, $E_{corr}$, that deviates by more than approximately 50 mV from both equilibrium potentials, $E'_X$ and $E'_M$, the contributions of $I_{o,X}$ and $I_{red,M}$ become insignificant, and the analysis of the corrosion is based on the intersection of the $I_{red,X}$ and $I_{ox,M}$ lines. These individual Tafel lines are plotted (dashed lines) in Fig. 4.15. $E_{corr}$ and $I_{corr}$ are identified, again assuming that $R_{total}$ is very small.

Under these assumptions and at $E < E_{corr}$, the external cathodic (net reduction) current is, from Eq 4.48:

$$I_{ex,red} = (I_{ox,M} - I_{red,X}) < 0 \qquad \text{(Eq 4.55)}$$

At $E > E_{corr}$, the external anodic (net oxidation) current is:

$$I_{ex,ox} = (I_{ox,M} - I_{red,X}) > 0 \qquad \text{(Eq 4.56)}$$

Substituting the appropriate Tafel relationships into Eq 4.55 and 4.56 gives:

$$I_{ex,red} = I_{o,M}\, e^{2.3(E-E'_M)/\beta_{ox,M}} - I_{o,X}\, e^{-2.3(E-E'_X)/\beta_{red,X}} \qquad \text{(Eq 4.57)}$$

and

$$I_{ex,ox} = I_{o,M}\, e^{2.3(E-E'_M)/\beta_{ox,M}} - I_{o,X}\, e^{-2.3(E-E'_X)/\beta_{red,X}} \qquad \text{(Eq 4.58)}$$

Equations 4.57 and 4.58 are plotted in terms of log $|I_{ex,red}|$ and log $I_{ex,ox}$ as the lower and upper solid curves, respectively, in Fig. 4.15. A major significance of these equations is that they are expressions involving E and $I_{ex}$, both of which are experimentally measurable, with the parameters $I_{o,M}$, $E'_M$, $\beta_{ox,M}$, $I_{o,X}$, $E'_X$, and $\beta_{red,X}$ characterizing the anodic and cathodic reactions. Therefore, if the parameters are known, the equations can be used to compare experimental values of E and $I_{ex}$ with those calculated and thereby provide information on the validity of the theory on which the analysis has been based. Alternatively, assuming that the theory is correct, the equations can be converted to forms in which experimental values of E and $I_{ex}$ allow determination of the parameters and, from their values, insight on the nature of the reactions. In addition, an experimental value for $I_{corr}$, and hence corrosion rate, is determined, providing that $R_{total}$ is very small. To understand the basis for determining the parameters in Eq 4.57 and 4.58 from experimental data, it is helpful to convert the forms of these equations to ones in which E is expressed as a function of $I_{ex,red}$ and $I_{ex,ox}$.

The net cathodic polarization curve, $I_{ex,red}$, is analyzed as follows. The Tafel equation for the reduction of cathodic species, X (Eq 4.44), is:

$$E = E'_X - \beta_{red,X} \log \frac{I_{red,X}}{I_{o,X}} \quad \text{(Eq 4.59)}$$

Rearranging Eq 4.55 gives, for $I_{red,X}$:

$$I_{red,X} = I_{ox,M} - I_{ex,red} \quad \text{(Eq 4.60)}$$

Substituting Eq 4.60 into Eq 4.59 gives:

$$E = E'_X - \beta_{red,X} \log \frac{I_{ox,M} - I_{ex,red}}{I_{o,X}} \quad \text{(Eq 4.61)}$$

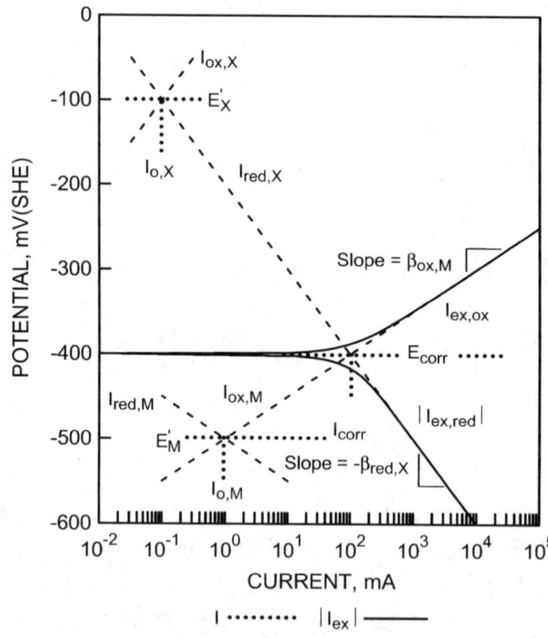


**Fig. 4.15** Mixed-electrode cathodic and anodic polarization curves (solid lines) based on the reduction component of the cathodic reaction and the oxidation component of the anodic reaction (compare with Fig. 4.13)

At potentials sufficiently negative to $E_{corr}$ (normally about 50 mV), $I_{ox,M}$ becomes negligible; consequently, from Eq 4.60, $I_{red,X} = -I_{ex,red}$. Therefore, Eq 4.59 and 4.61 become equivalent, or:

$$E = E'_X - \beta_{red,X} \log \frac{-I_{ex,red}}{I_{o,X}} = E'_X - \beta_{red,X} \log \frac{I_{red,X}}{I_{o,X}} \qquad \text{(Eq 4.62)}$$

which shows that the Tafel equation involving the $I_{red,X}$ current also can be written, under the condition of $E << E_{corr}$, in terms of $I_{ex,red}$, which is experimentally measurable. In this limit, the external cathodic current, $(-I_{ex,red})$ or $|I_{ex,red}|$, plotted as $\log |I_{ex,red}|$ in Fig. 4.15, becomes linear and establishes a section of the single $I_{red,X}$ Tafel line. Three basically important quantities are obtained having established this Tafel line. First, it can be extrapolated to the equilibrium potential for the X reaction, $E'_X$, at which the current is an experimental value for the exchange current $I_{o,X}$. Second, the slope of this line represents an experimental value for $-\beta_{red,X}$. Third, it follows from earlier discussion that at the steady-state corrosion potential, $E_{corr}$, $I_{ex,red} = I_{ex,ox} = 0$. Therefore, from Eq 4.55 or 4.56:

$$I_{ox,M} = I_{red,X} = I_{corr} \qquad \text{(Eq 4.63)}$$

From $I_{corr}$, the total amount of corrosion can be calculated from Faraday's law, and by dividing $I_{corr}$ by the corroding area, the corrosion current density and hence the corrosion intensity or corrosion penetration rate is determined. Thus, the intersection of the extrapolated Tafel line with $E = E_{corr}$ gives an experimentally determined value for $I_{corr}$.

A similar analysis for an external anodic (net oxidation) current, $I_{ex,ox}$, leads to the following Tafel-based equations under conditions that the E is sufficiently greater than $E_{corr}$ to make $I_{red,X}$ negligible:

$$E = E'_M + \beta_{ox,M} \log \frac{I_{ex,ox}}{I_{o,M}} = E'_M + \beta_{ox,M} \log \frac{I_{ox,M}}{I_{o,M}} \qquad \text{(Eq 4.64)}$$

Consequently, under these conditions and with reference to Fig. 4.15, the solid curve for the external anodic current, $I_{ex,ox}$, becomes linear and establishes the individual $I_{ox,M}$ Tafel line. In this case, extrapolation of the linear section to $E'_M$ gives an experimentally determined value for $I_{o,M}$, the slope is $\beta_{ox,M}$, and the intersection of the extrapolated line with $E = E_{corr}$ gives the same experimental value for $I_{corr}$.

Two additional comments are in order with respect to Fig. 4.15. First, it should be noted that the extrapolated $I_{red,X}$ and $I_{ox,M}$ Tafel lines cross at $E_{corr}$ and $I_{corr}$ and, therefore, can be used to establish values for these quantities. Second, an interpretation of $E_{corr}$ and $I_{corr}$ is that the coupling of the anodic and cathodic reactions on the surface has resulted in currents that have polarized the anodic reaction from $E'_M$ up to $E_{corr}$ and the cathodic reaction from $E'_X$ down to $E_{corr}$. This current is $I_{corr} = I_{ox,M} = I_{red,X}$.

The concepts associated with an analysis of Fig. 4.15 are reemphasized by examining the information derivable from *experimental* polarization curves (i.e., E versus log $|I_{ex}|$ curves). In general, the following are available from measurements or calculations: $E'_X$, $E'_M$, the cathodic polarization curve, the anodic polarization curve, and $E_{corr}$ from asymptotic values of the polarization curves as $I_{ex,red} \rightarrow 0$ and $I_{ex,ox} \rightarrow 0$. The slopes of the linear segments of the experimental polarization curves permit estimates of the Tafel constants, $\beta_{red,X}$ and $\beta_{ox,M}$. Extrapolations of the linear portions of the polarization curves through their intersections with ordinate values of $E_{corr}$, $E'_X$, and $E'_M$, respectively, permit estimations of $I_{corr}$, $I_{o,X}$, and $I_{o,M}$. Unfortunately, well-defined linear portions of experimental polarization curves are not always observed, and the method has limitations. This is particularly the case when diffusion or solid corrosion products introduce controlling reaction rate mechanisms. An alternative method of analysis that uses mathematical modeling methods to obtain values for the parameters characterizing the anodic and cathodic reactions is presented in Chapter 6.

The foregoing discussion developed individual expressions for the external cathodic and anodic currents, $I_{ex,red}$ and $I_{ex,ox}$. Although this approach was instructive, it was not necessary mathematically. Note that the external current, whether reduction or oxidation, was consistently defined as the sum of the individual oxidation currents minus the sum of individual reduction currents (Eq 4.48). In general then, the external current is defined as:

$$I_{ex} = \Sigma I_{ox} - \Sigma I_{red} \quad \text{(Eq 4.65)}$$

or when the half-cell reactions, $X \leftrightarrow X^{x+} + xe$ and $M \leftrightarrow M^{m+} + me$, are involved:

$$I_{ex} = (I_{ox,M} + I_{ox,X}) - (I_{red,X} + I_{red,M}) \quad \text{(Eq 4.66)}$$

At potential ranges where $I_{ex} < 0$, that is, when $E < E_{corr}$, the external current is cathodic (net reduction), and at potential ranges where $I_{ex} > 0$, that is, when $E > E_{corr}$, the external current is anodic (net oxidation). Thus, the sign of $I_{ex}$ is sufficient to identify whether it is an external cathodic or anodic current. An expression for the external current is obtained on substitution of the individual Tafel relationships in Eq 4.66:

$$\begin{aligned} I_{ex} &= I_{o,M}\, e^{2.3(E-E'_M)/\beta_{ox,M}} + I_{o,X}\, e^{2.3(E-E'_X)/\beta_{ox,X}} \\ &\quad - I_{o,X}\, e^{-2.3(E-E'_X)/\beta_{red,X}} - I_{o,M} e^{-2.3(E-E'_M)/\beta_{red,M}} \end{aligned} \quad \text{(Eq 4.67)}$$

This is the same relationship as Eq 4.51. In most metal/environment conditions, $I_{corr}$ will be on the order of $10 \times I_o$, or greater, for both half-cells, under which conditions $I_{ox,X}$ and $I_{red,M}$ are negligible. Equation 4.66 then reduces to:

$$I_{ex} = I_{ox,M} - I_{red,X} \qquad \text{(Eq 4.68)}$$

and Eq 4.67 reduces to:

$$I_{ex} = I_{o,M}\, e^{2.3(E - E'_M)/\beta_{ox,M}} - I_{o,X}\, e^{-2.3(E - E'_X)/\beta_{red,X}} \qquad \text{(Eq 4.69)}$$

Equation 4.69 is now used to establish an important relationship between $I_{ex}$, $E_{corr}$, and $I_{corr}$. Under the specific case of free corrosion, $E = E_{corr}$, and with $I_{ex} = 0$, $I_{ox,M} = I_{red,X} = I_{corr}$ (see also Eq 4.63). Using these conditions:

$$I_{corr} = I_{o,M}\, e^{2.3(E_{corr} - E'_M)/\beta_{ox,M}} = I_{o,X}\, e^{-2.3(E_{corr} - E'_X)/\beta_{red,X}} \qquad \text{(Eq 4.70)}$$

Division of Eq 4.70 into Eq 4.69 results in the desired relationship that will be used in Chapter 6 devoted to electrochemical measurement techniques:

$$I_{ex} = I_{corr}\left(e^{2.3(E-E_{corr})/\beta_{ox,M}} - e^{-2.3(E-E_{corr})/\beta_{red,X}}\right) \qquad \text{(Eq 4.71)}$$

## Summary of the Form and Source of Polarization Curves

Much of the previous discussion is directed toward the experimental determination of polarization curves from which parameters characterizing half-cell reactions are derived. These parameters are the exchange current density, $i_o$; the Tafel slope, $\beta$; and the limiting current density for diffusion polarization, $i_D$. It should be appreciated that E and $I_{ex}$ are the experimentally measurable variables used in evaluating these parameters. The equivalent of $I_{ex}$ in any *section* of a corroding system is a current to or from the section originating in corrosion currents generated by coupling to other sections of the system, stray electrical currents generated by electrical equipment used in the vicinity of the system, or external sources designed to control corrosion. Currents established under the latter conditions are referred to as cathodic or anodic protection currents and are discussed later.

Knowledge of the parameters of the individual electrode reactions permits writing expressions for the individual oxidation or reduction curves (see the section "Complete Polarization Curves for a Single Half-Cell Reaction" in Chapter 3). Thus, the expression for the cathodic-reactant reduction reaction:

$$X^{x+} + xe \rightarrow X \qquad (\text{e.g., } H^+ + e \rightarrow \tfrac{1}{2}H_2) \qquad \text{(Eq 4.72)}$$

at 25 °C is:

$$E = E'_X + \eta_{CT,red,X} + \eta_{D,red,X} \qquad \text{(Eq 4.73)}$$

$$E = E^o_X + \frac{59}{x}\log a_{X^{x+}} - \beta_{red,X}\log\frac{I_{red,X}}{I_{o,X}} - \frac{59}{x}\log\frac{I_{D,red,X}}{I_{D,red,X} - I_{red,X}} \qquad \text{(Eq 4.74)}$$

The expression for the metal oxidation reaction:

$$M \rightarrow M^{m+} + me \qquad (\text{e.g., } Fe \rightarrow Fe^{2+} + 2e) \qquad \text{(Eq 4.75)}$$

at 25 °C is:

$$E = E'_M + \eta_{CT,ox,M} + \eta_{D,ox,M} \qquad \text{(Eq 4.76)}$$

$$E = E^o_M + \frac{59}{m}\log a_{M^{m+}} + \beta_{ox,M}\log\frac{I_{ox,M}}{I_{o,M}} + \frac{59}{m}\log\frac{I_{D,ox,M}}{I_{D,ox,M} - I_{ox,M}} \qquad \text{(Eq 4.77)}$$

It should be recalled that all currents in Eq 4.74 and 4.77 are positive quantities.

## Estimation of $E_{corr}$ and $I_{corr}$ for Iron as a Function of pH

Very careful measurements of the anodic polarization of iron by Kelly (Ref 7) resulted in the proposal of five kinetics steps, the sum of which result in the simple oxidation reaction, $Fe \rightarrow Fe^{2+} + 2e$. The proposed steps are:

$$Fe + H_2O \rightarrow Fe(H_2O)_{ads} \qquad \text{(Eq 4.78)}$$

$$Fe(H_2O)_{ads} \rightarrow Fe(OH^-)_{ads} + H^+ \qquad \text{(Eq 4.79)}$$

$$Fe(OH^-)_{ads} \rightarrow (FeOH)_{ads} + e \qquad \text{(Eq 4.80)}$$

$$(FeOH)_{ads} \rightarrow (FeOH)^+ + e \qquad (\text{rate determining}) \qquad \text{(Eq 4.81)}$$

$$(FeOH)^+ + H^+ \rightarrow Fe^{2+} + H_2O \qquad \text{(Eq 4.82)}$$

The sum of Eq 4.78 to 4.82 is:

$$Fe \rightarrow Fe^{2+} + 2e \quad \text{(Eq 4.83)}$$

The proposed rate-determining step is noted in the above sequence of reaction steps. It also should be noted that one of the steps involves the hydrogen ion, and therefore, the kinetics of the dissolution of iron becomes a function of the pH, although the overall reaction and the equilibrium potential of iron is independent of pH. Kelly's results have been used to approximate polarization curves for the oxidation of iron (A = 1 $m^2$) at pH values of 1, 3, and 5 in Fig. 4.16.

Approximate polarization curves for the hydrogen-reduction reaction, $H^+ + e \rightarrow \frac{1}{2}H_2$, were shown in Fig. 3.16 for pH values of 1, 3, and 5. These curves also are shown in Fig. 4.16, where the abscissa is in terms of current, I, rather than current density, i (Ref 8).

The polarization curves in Fig. 4.16 permit an estimate of $I_{corr}$ as the intersection of pairs of oxidation and reduction curves corresponding to the condition that $I_{ox} = I_{red} = I_{corr}$. Actually, in this case, the corrosion is uniform, and the anodic and cathodic reactions are assumed to occur uniformly over the surface. Under this assumption, unit area is taken for analysis (A = 1 $m^2$), and either E versus log i or E versus log I curves can be used in the analysis. However, the use of E versus log i curves obscures the fundamental basis on which corrosion rates are estimated

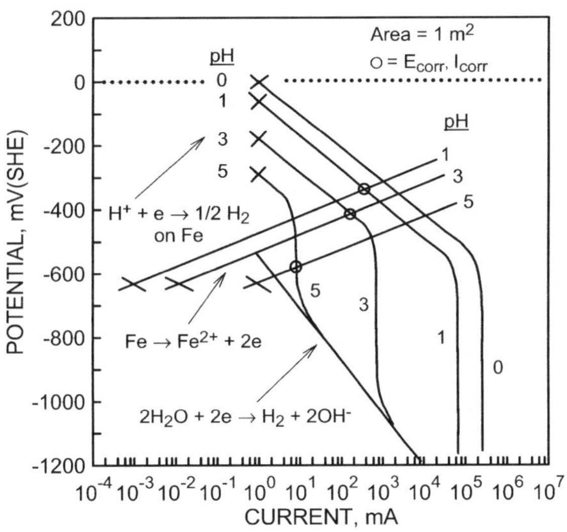


**Fig. 4.16** Estimation of $E_{corr}$ and $I_{corr}$ for iron at the indicated values of pH. Curves for hydrogen-ion reduction are based on experimental values of the polarization parameters governing the polarization curves. The anodic polarization curves for iron show a dependence on pH due to the influence of hydrogen ion concentration on the kinetic steps in the iron oxidation. Based on Ref 7 and 8

(i.e., from superposition of oxidation and reduction polarization behavior with the criterion that $I_{ox} = I_{red} = I_{corr}$).

The intersection of pairs of curves corresponding to the same pH gives $E_{corr}$ and $I_{corr}$ for the particular environment. For the present example, the results are:

| pH | $E_{corr}$, mV (SHE) | $I_{corr}$, mA |
|---|---|---|
| 1 | −330 | 400 |
| 3 | −410 | 180 |
| 5 | −580 | 6 |

As just substantiated, these numbers apply to unit area (1 $m^2$), and therefore, the values in the right-hand column may be taken as corrosion current densities. The corrosion penetration rate (CPR) can then be calculated from Faraday's law. For iron, CPR (μm/year) = 1.16 $i_{corr}$, where $i_{corr}$ is the corrosion current density in $mA/m^2$.

## Interpretation of Inhibitor Effects in Terms of Polarization Behavior

Soluble species other than corrosion-product ions, and species involved in cathodic reactions supporting corrosion, can have major effects on both the anodic and cathodic reactions involved in the corrosion process. These species may be either ionic or nonionic, the latter generally being organic and frequently having a polar molecular structure. These species can influence the kinetic mechanism of anodic dissolution, or the supporting cathodic reactions, or both. The influence is reflected in changes in the values of the exchange current density, $i_o$, and the Tafel slope, β; other aspects of the polarization curve may be altered if the additional species either enhance or decrease the tendency for corrosion products to form protective films. Species decreasing $i_o$ and/or increasing β are called inhibitors. If inhibitors act through adsorption to the surface, they may do so through an effect on $i_o$ or β, or their effect may be to decrease the surface area available to either the anodic or cathodic reaction. Examples of effects of inhibitors in decreasing corrosion are shown in Fig. 4.17. Figure 4.17(a) shows the effect of an inhibitor influencing the cathodic reaction; Fig. 4.17(b) shows the corresponding response to an anodic inhibitor, and Fig. 4.17(c) shows the response when both reactions are influenced. The effect of the inhibitor is shown in each case and the effect on $I_{corr}$ is indicated. It is significant to note that for an anodic inhibitor, if a decrease in $i_o$ is due to inhibitor adsorption effectively decreasing the area, then if the anodic area is not completely covered, the cathode/anode area ratio will be increased to a

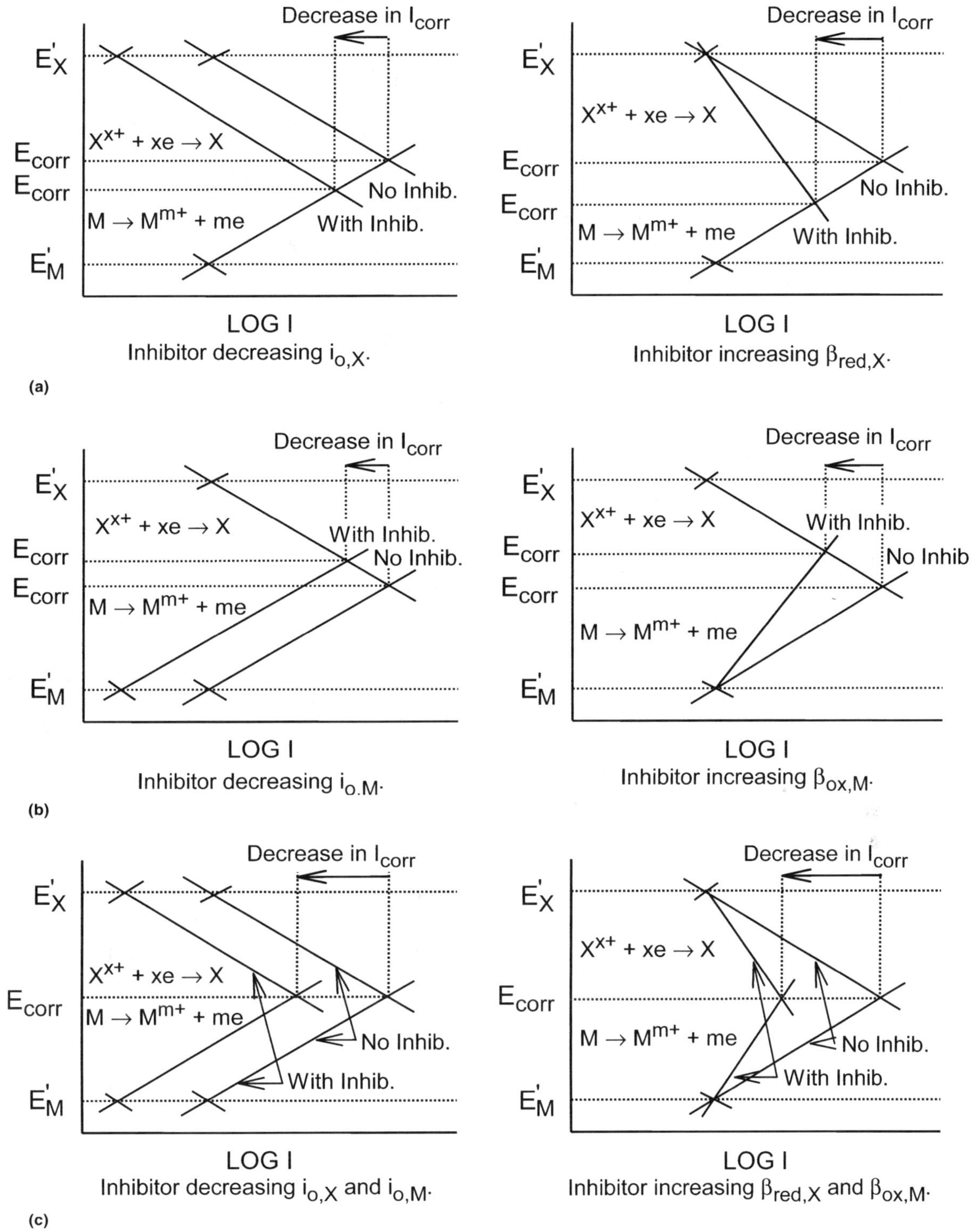


**Fig. 4.17** Schematic examples of the effects of changes in the relative positions of anodic and cathodic polarization curves due to inhibitors, with the resultant $E_{corr}$ and $I_{corr}$ values. (a) Effects of cathodic inhibitor. Note that $I_{corr}$ is decreased and $E_{corr}$ is decreased. (b) Effects of anodic inhibitor. Note that $I_{corr}$ is decreased and $E_{corr}$ is increased. (c) Effects of cathodic and anodic inhibitor

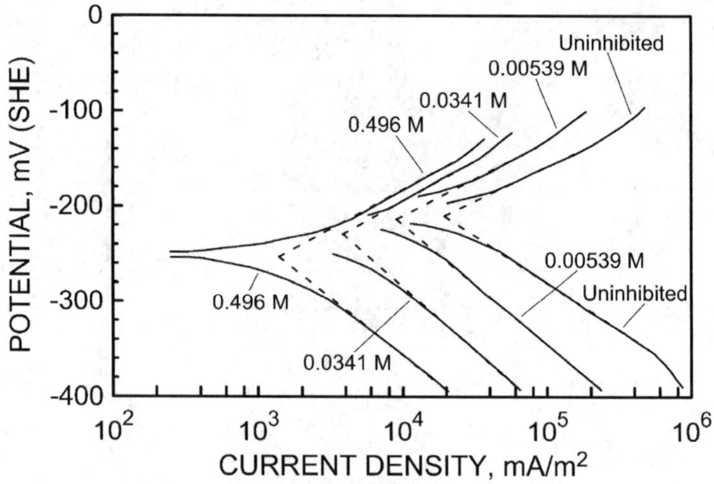


**Fig. 4.18** Polarization curves for iron in deaerated 6 N HCl with $NH_2$-$(CH_2)_3$-$NH_2$ inhibitor (molar concentrations are indicated). Redrawn from Ref 9

high value, and severe pitting will occur at exposed anodes. For this reason, anodic inhibitors must be used with caution, and cathodic inhibitors are generally preferred.

An example of the effect of increasing concentrations of a diamine-type organic inhibitor ($NH_2$-$(CH_2)_3$-$NH_2$) on the corrosion of iron in 6 N HCl is shown in Fig. 4.18 (Ref 9). Under uninhibited conditions, $E_{corr} \approx -210$ mV (SHE) and $i_{corr} \approx 20{,}000$ mA/m$^2$. The effect of increasing inhibitor concentration is to decrease both $E_{corr}$ and $i_{corr}$, the latter being reduced by a factor of about ten at the largest inhibitor concentration shown. Since the Tafel slopes remain essentially the same, and $E_{corr}$ is changed a relatively small amount, it is concluded that the major influence of the inhibitor is to decrease the exchange current densities of both the anodic and cathodic reactions. A mechanism for this effect is adsorption of the inhibitor to the metal/solution interface, thereby decreasing the metal ion transfer rate between the metal and the environment.

## Galvanic Coupling (Ref 10, 11)

When two metals or alloys are joined such that electron transfer can occur between them and they are placed in an electrolyte, the electrochemical system so produced is called a galvanic couple. Coupling causes the corrosion potentials and corrosion current densities to change, frequently significantly, from the values for the two metals in the uncoupled condition. The magnitude of the shift in these values depends on the electrode kinetics parameters, $i_o$ and $\beta$, of the cathodic and anodic reactions and the relative magnitude of the areas of the two metals. The effect also depends on the resistance of the electrochemical circuit including the resistance that exists at the junction between the two metals. Four cases are described, three assuming Tafel behavior for all reactions and one showing the analysis when diffusion is a controlling factor. The areas of the two metals are assumed not to change for the four cases.

## *Case I: Galvanically Coupled Metals with Similar Electrochemical Parameters*

The polarization behavior of two metals, A and B, along with the polarization curve for the hydrogen evolution reaction on each metal is shown in Fig. 4.19. Metal A has an equilibrium half-cell potential slightly more positive than B; otherwise the behaviors are similar, the slopes of the curves being approximately the same and the exchange current densities not differing by more than a factor of 10. Consider first the corrosion behavior of the individual metals (i.e., when they are not in electrical contact). Metal A corrodes with the conditions at the point identified by $E_{corr,A}$ and $I_{corr,A}$; similarly, the conditions for metal B are $E_{corr,B}$ and $I_{corr,B}$. The corrosion of each metal is due to the cathodic hydrogen-ion-reduction reaction. The four polarization curves have positions such that the corrosion current for each separate metal is approximately the same. It should be noted that in this analysis (also in the ones that follow), the oxidation curves for the metals and the reduction curves for the hydrogen reaction are the only ones considered. It will be recalled that this is a valid approximation if the corrosion potentials are reasonably different (≈50 mV) from the equilibrium half-cell potentials.

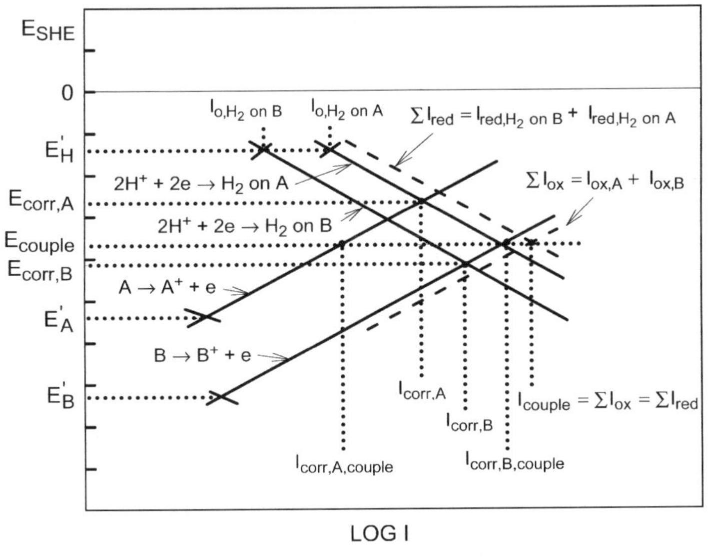


**Fig. 4.19** Schematic representation of polarization curves for the analysis of galvanic coupling when the coupled metals have similar electrochemical parameters. Tafel polarization is represented.

When the metals are coupled, conservation of charge requires that the total oxidation current must equal the total reduction current, $\Sigma I_{ox} = \Sigma I_{red}$. Thus, the two oxidation and the two reduction curves must

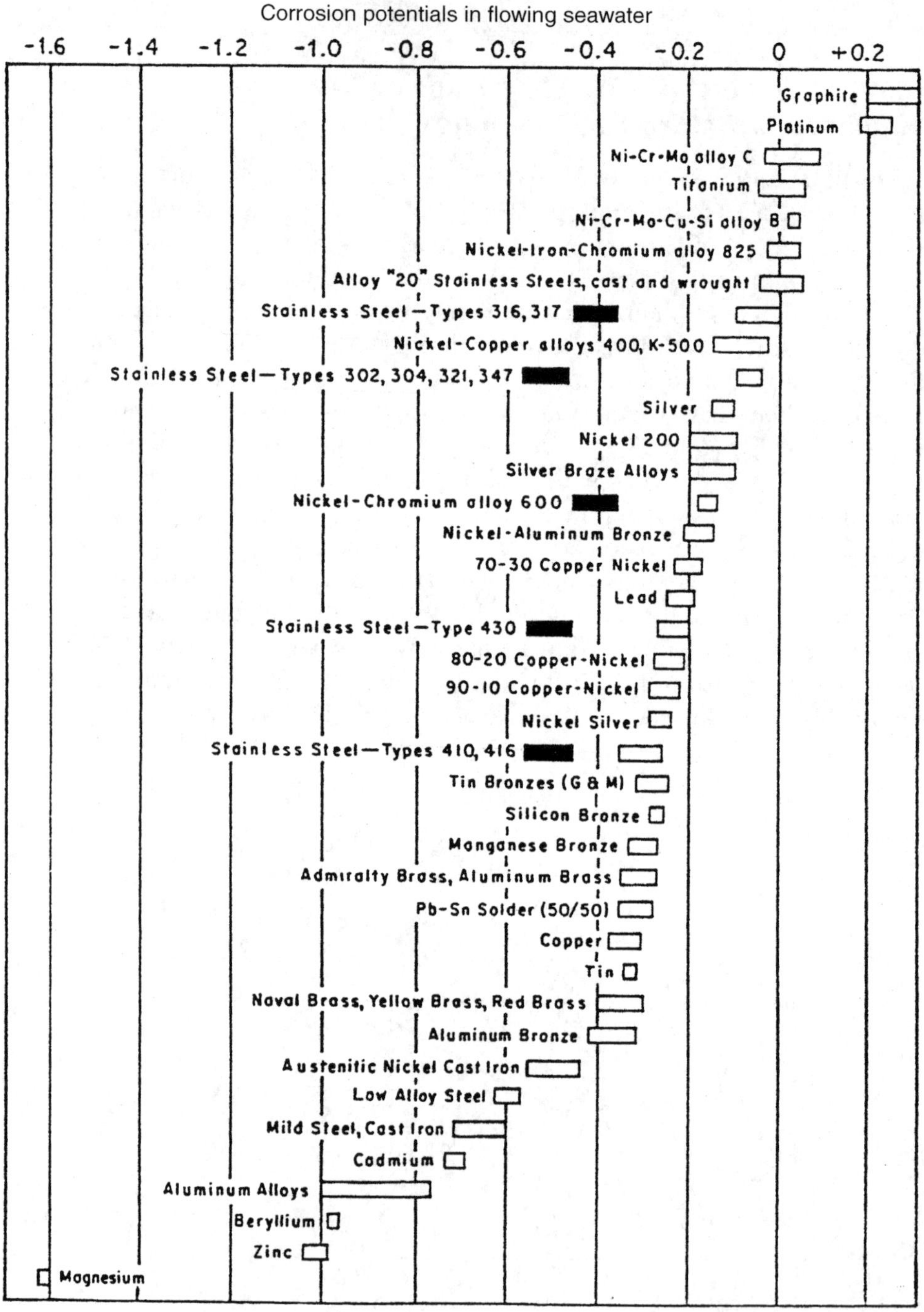


**Fig. 4.20** Galvanic series of various metals in flowing seawater at 2.4 to 4.0 m/s at 5 to 30 °C (volts vs. saturated calomel reference electrode). Note: Dark boxes indicate active behavior of active-passive alloys. Source: Ref 12 and 13

be added in terms of currents at any potential. These sums are given by the dashed lines identified as $\Sigma I_{ox}$ and $\Sigma I_{red}$. The steady-state condition for a low-resistance circuit is given by the intersection of these two dashed lines (i.e., by the point identified as $E_{couple}$, where $I_{couple} = \Sigma I_{ox} = \Sigma I_{red}$). It is important to appreciate that this intersection establishes the electrical potential of the metals on the hydrogen scale and each will have the value $E_{couple}$ if the circuit resistance is low. The behavior of the individual metals when coupled is then determined by the magnitude of the currents on each when at a potential corresponding to $E_{couple}$. As a consequence, metal A corrodes at the rate $I_{corr,A,couple}$ and metal B corrodes at the rate $I_{corr,B,couple}$. The effect of the coupling is thus to decrease the corrosion rate of A from $I_{corr,A}$ to $I_{corr,A,couple}$ and to increase the corrosion rate of B from $I_{corr,B}$ to $I_{corr,B,couple}$. It should be evident that the magnitude of these changes of corrosion rate will depend upon the particular metals that are coupled and the values of the parameters establishing the positions of the polarization curves. The pH of the environment and the metal-ion concentrations are also variables.

Several results of this analysis should be noted. First, the coupled corrosion potential, $E_{couple}$, is located *between* the uncoupled corrosion potentials, $E_{corr,A}$ and $E_{corr,B}$. Next, the metal with the more negative uncoupled corrosion potential ($E_{corr,B}$) experiences an increase in corrosion rate in the galvanic couple, whereas the metal with the more positive uncoupled corrosion potential ($E_{corr,A}$) experiences a decrease in corrosion rate in the galvanic couple. Within the couple, the former metal is called the *anode,* and the latter metal, the *cathode.* Another ramification of this analysis that should be appreciated is as follows. Certainly if Tafel behavior is exhibited (exceptions may arise when active-passive or diffusion-control behavior is involved), as the uncoupled corrosion potentials ($E_{corr,A}$ and $E_{corr,B}$) become more widely separated, the corrosion rate of the anode in the couple progressively increases relative to its uncoupled value, and the corrosion rate of the cathode in the couple progressively decreases relative to its uncoupled value. Thus, to minimize galvanic effects, one would select metals or alloys with similar uncoupled corrosion potentials. To provide qualitative guidelines on selection of metals or alloys that must be coupled, "galvanic series" have been experimentally determined (i.e., rankings of materials based on their uncoupled corrosion potentials). An example is shown in Fig. 4.20 (Ref 12, 13).

## *Case II: Galvanic Coupling of a Metal to a Significantly More Noble Metal*

This case is illustrated in Fig. 4.21. The only change from Case 1 is the position of the oxidation curve for metal A, which is now placed sufficiently positive that its equilibrium half-cell potential is above that for

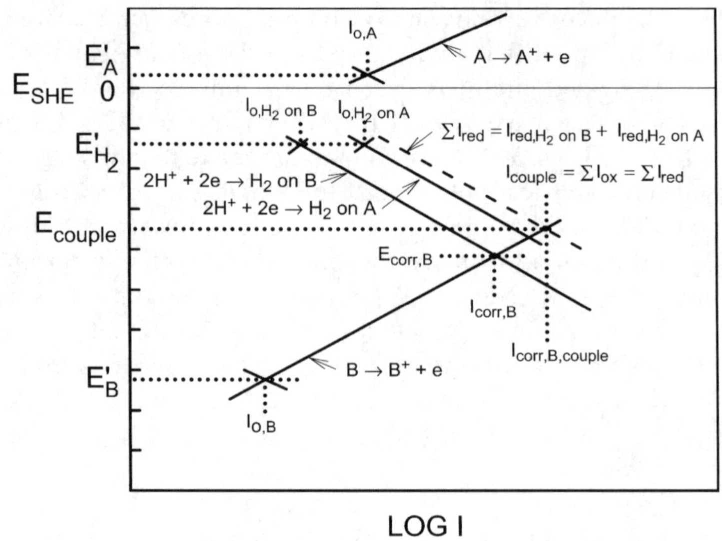


**Fig. 4.21** Schematic representation of polarization curves for the analysis of galvanic coupling when one metal is significantly more noble. Tafel polarization is represented.

the hydrogen reaction. Hence, A does not corrode; it acts as a noble metal; however, when coupled to B it can provide a surface on which hydrogen is evolved. Thus, the two hydrogen reduction curves are added to give $\Sigma I_{red}$ (dashed line), which intersects the oxidation curve for B at the point identified by $I_{corr,B,couple}$. The coupling has increased the corrosion rate from $I_{corr,B}$ to $I_{corr,B,couple}$, and the corrosion potential has increased from $E_{corr,B}$ to $E_{couple}$. It should be noted that the controlling factor in establishing the effect of metal A on the corrosion rate of metal B is not the nobility of A, that is, how positive $E'_A$ is, but rather how effective the surface of A is in evolving hydrogen ($i_{o,H_2 \text{ on } A}$ and $\beta_{red,H_2 \text{ on } A}$). More generally, the coupled cathodic surface needs only to be an electron conductor to allow access of electrons from the anodic reaction to the cathodic reactant. As a consequence, an oxide-coated surface, such as the black oxide on hot-rolled steel, can function as a cathodic surface and therefore act as part of a couple with any region of the steel from which the oxide has been removed to expose the underlying base metal that corrodes as an anodic area.

## *Cases III and IV: Galvanically Coupled Metals: One Metal Significantly Active*

Two cases are considered, one with hydrogen-ion reduction supporting the corrosion (Case III) and the other representative of aerated conditions in which the reduction of oxygen is the governing cathodic reaction (Case IV). The first example, Case III, is shown in Fig. 4.22 in which the position of the oxidation curve for B is sufficiently negative to A that the condition, $\Sigma I_{red} = \Sigma I_{ox}$, causes $E_{couple}$ to have the same value as the equilibrium potential of metal A. Thus, A no longer corrodes; however, the corrosion rate of B has increased from $I_{corr,B}$ to $I_{corr,B,couple}$. Note that A does not corrode because at $E_{couple} = E'_A$, the oxidation and reduction curves for A cross, namely, $I_{ox,A} = I_{red,A} = I_{o,A}$.

In essentially neutral environments (pH = 7), in contact with air, the controlling reaction is the reduction of dissolved oxygen. For Case IV, the effects of galvanic coupling under conditions of oxygen diffusion control are analyzed by reference to Fig. 4.23. Again, metal B is repre-

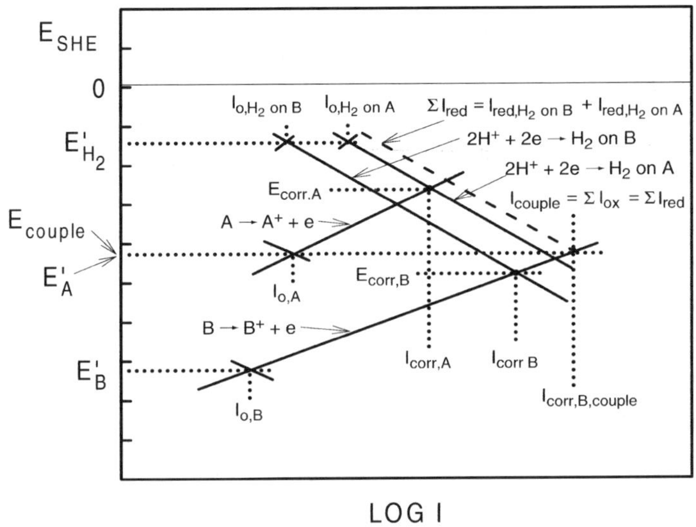


**Fig. 4.22** Schematic representation of polarization curves for the analysis of galvanic coupling when one metal is significantly more active. Tafel polarization is represented.

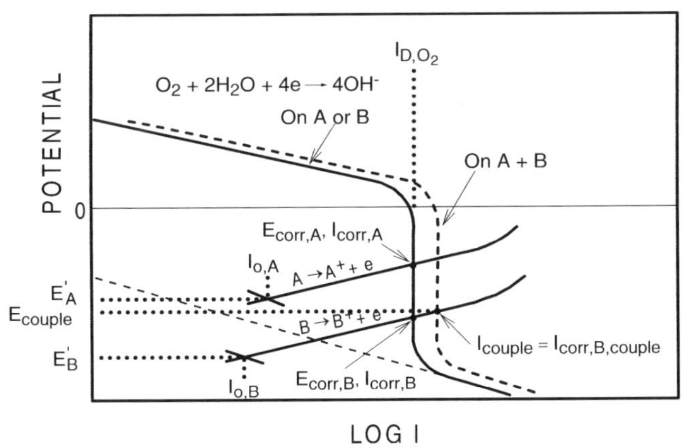


**Fig. 4.23** Schematic representation of polarization curves for the analysis of galvanic coupling when diffusion control of the oxygen reduction reaction is the dominant factor governing the corrosion rate

sented as having the more active equilibrium half-cell potential relative to metal A. $E_{corr}$ and $I_{corr}$ for the uncoupled individual metals are identified by the intersection of the respective metal oxidation and oxygen-reduction curves. Since the corrosion of both metals is under oxygen diffusion control, $I_{corr}$, and hence the corrosion rate (in terms of corrosion current density), is the same for each metal. When B is coupled to an equal area of A, the total surface supporting the oxygen reaction is doubled; this total oxygen reduction is given by the dashed curve. Since the oxidation curve for A is negligible relative to the oxidation curve for B, metal B provides the total anodic current. Therefore, the intersection of the oxidation curve for B and the total oxygen-reduction curve establishes $E_{couple}$. Since $E_{couple} < E'_A$, metal A does not corrode; but the corrosion of B is doubled, increasing from $I_{corr,B}$ to $I_{corr,B,couple}$.

In both of the cases just considered, for the purpose of protecting A from corrosion, B is referred to as a *sacrificial anode.*

# Cathodic Protection (Ref 14)

Cathodic protection is the process whereby the corrosion rate of a metal is decreased or stopped by decreasing the potential of the metal from $E_{corr}$ to some lower value and in the limit to $E'_M$, the thermodynamic equilibrium half-cell potential. At this potential, $i_{ox,M} = i_{red,M} = i_{o,M}$, and net transfer of metal ions to the solution no longer occurs. This is the criterion for complete cathodic protection (i.e., $E = E'_M$).

Cathodic protection is generally accomplished by one of the following two methods.

## *Cathodic Protection by Sacrificial Anodes*

Cases III and IV, discussed in the previous section on galvanic coupling, illustrate the principle of cathodic protection using sacrificial anodes. Specific examples are the coupling of zinc or magnesium to iron. In the examples analyzed with reference to Fig. 4.22 and 4.23, the polarization curves for the reactions involved were such that $E_{couple}$ was reduced to or below $E'_A$. The criterion for cathodic protection was thus met. It is emphasized that $E_{couple}$ depends not only on the electrochemical parameters of the system ($E'$, $i_o$, and $\beta$ for each reaction) but also on the relative sizes and shapes of the anodic and cathodic areas, the relative distance between these areas, the resistivity of the environment, the metallic path resistance between anodes and cathodes, and the fluid velocity. In Fig. 4.22 and 4.23, the electrical resistance of the circuit, including metal and solution paths, was assumed to be negligibly small. This allowed establishing $E_{couple}$ in terms of the intersection of the

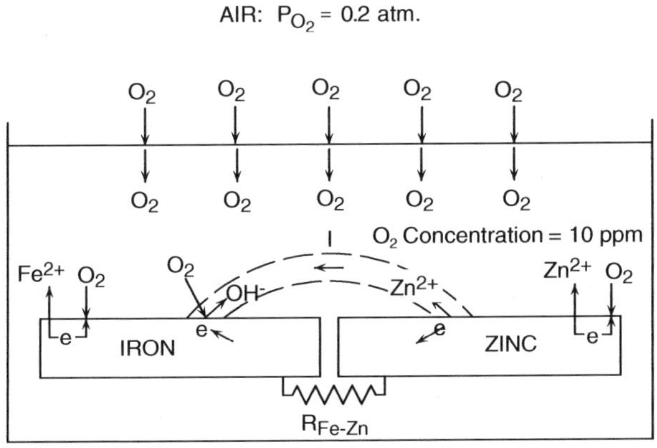


**Fig. 4.24** Representation of variables involved in galvanic interaction between iron and zinc

curves representing $\Sigma I_{ox}$ and $\Sigma I_{red}$. If the area of metal B is decreased, the curves for both reactions associated with B (anodic dissolution of B and cathodic reduction on B) will move to the left or to lower values of current proportional to the decrease in area. As a result, $E_{couple}$ will increase, and if $E_{couple} > E'_A$, metal A is no longer completely protected. Obviously, the values of $I_{o,B}$, $\beta_{ox,B}$, $I_{o,H_2 \text{ on B}}$, and $\beta_{red,H_2 \text{ on B}}$ for a given B to A area ratio will govern the value of $E_{couple}$ and, therefore, whether cathodic protection will be accomplished.

The influences of the geometry and spacing of metals A and B, and the circuit resistance, on the ability of B to cathodically protect A are complicated. The factors involved will be examined by reference to the particularly simple arrangement in Fig. 4.24, which uses iron and zinc as representative metals in an aerated environment. Plates of iron and zinc are joined by a variable resistance connection that can be varied from $R_{Fe\text{-}Zn} = 0$ to $R_{Fe\text{-}Zn} = \infty$, the latter corresponding to two individual uncoupled pieces of metal. When the two metals are directly joined, there will be negligible resistance between them and the metals will be at essentially the same potential. A flux of current, however, will pass in the aqueous environment from the zinc to the iron as shown. The current density at the surface will diminish with increasing distance from the iron/zinc junction because of the progressively larger resistance of fluid elements (illustrated by dashed lines) away from the junction between the metals. Near the junction, the current path length, and hence the solution resistance, will approach zero, and the potential at the junction will approach $E_{couple,Fe\text{-}Zn}$. Progressively away from the junction, the length of fluid elements increases, and the current density decreases. When the current density is decreased to a value that no longer results in an iron surface-potential of $E'_{Fe}$, the required potential for total protection, corrosion of the iron will occur. As a consequence, in the arrangement of Fig. 4.24, the zinc will protect the iron to a certain distance from the metal junction beyond which corrosion will be observed. The larger the specific resistivity of the environment, the shorter the distance over which the iron will be protected. A direct consequence is that sacrificial zinc anodes can be spaced farther apart on structures in seawater (specific resistivity $\approx$ 10 ohm-cm) than in fresh water (specific resistivity $\approx$ 5000 ohm-cm).

## *Cathodic Protection by Impressed Current*

Cathodic protection also can be accomplished by lowering the electrode potential to $E'_M$, the equilibrium potential for the metal to be protected, by an external power source. The circuit used to accomplish this is the same as shown in Fig. 2.12. With slight modification, it is again shown in Fig. 4.25 in which the metal to be protected is iron and the cathodic reaction supporting corrosion is either hydrogen-ion reduction, oxygen reduction, or both.

Interpretation of cathodic protection of iron in an environment of pH = 1 may be made by reference to Fig. 4.26. Without an external current, steady-state corrosion occurs under the conditions, $E_{corr}$ and $i_{corr}$. If electrons are supplied to the metal, the potential will decrease, and at any arbitrary reduction of potential (e.g., $E_1$), a current balance requires that $I_{ex} = I_{ox,M} - I_{red,X}$, or $i_{ex}A = i_{ox,M}A - i_{red,X}A$ for a given area A (assuming that $A_c = A_a = A$), or $i_{ex} = i_{ox,M} - i_{red,X}$. This external current density is represented in Fig. 4.26 as the span between the respective polarization curves at $E_1$. It is evident that for corrosion to be stopped, E must be reduced to $E'_{Fe}$, and to maintain this protection, the external

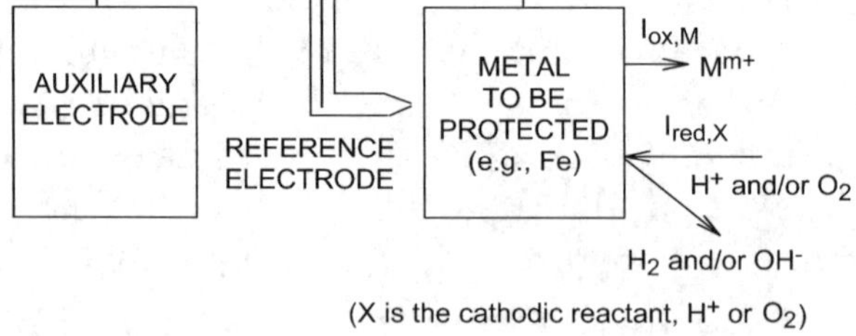


**Fig. 4.25** Components used to impose and monitor conditions providing cathodic protection by an impressed external current. Note: Power supply may be either a galvanostat or a potentiostat. In the latter, the electrometer provides feedback to the potentiostat to control to constant potential. Electrometer provides check to show that the metal is at the protection potential.

current density will be $i_{ex,complete\ protection}$. But since there is no longer any net current associated with the iron, the entire external current ($I_{ex,\ complete\ protection} = i_{ex,\ complete\ protection}$ A) will be consumed in evolving hydrogen. This current will represent an operating cost to maintain protection. Also, the hydrogen evolved may be sufficient to cause an explosion hazard, and caution should be invoked to avoid an accident.

In aerated neutral environments, corrosion will be supported by the cathodic oxygen reaction and will normally occur under oxygen-diffu-

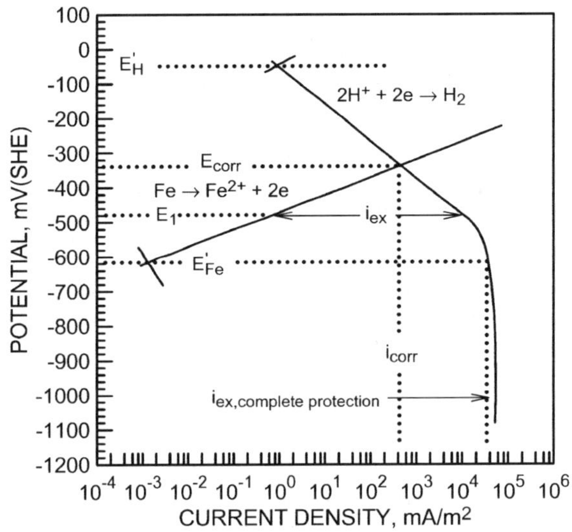


**Fig. 4.26** Schematic polarization curves used in the analysis of cathodic protection by an impressed external current. Cathodic reaction is under Tafel control.

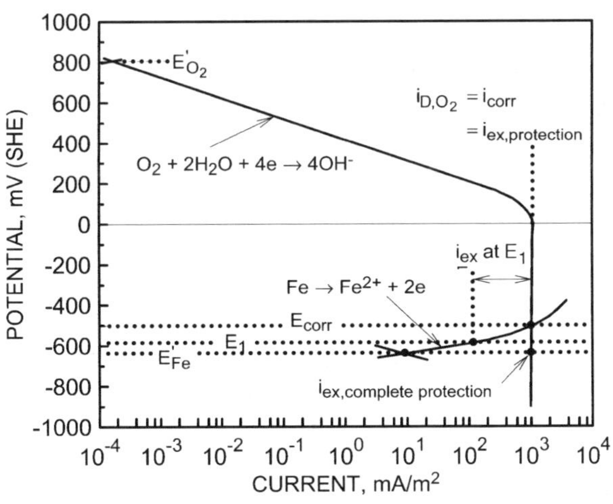


**Fig. 4.27** Schematic polarization curves used in the analysis of cathodic protection by an impressed external current. Cathodic reaction is under diffusion control.

sion control. In the case of iron, the corrosion product may be a loose-to-adherent oxide scale that can further control access of oxygen to the surface. Representative schematic polarization curves for iron and oxygen are shown in Fig. 4.27 in which the corrosion current density is equal to the limiting diffusion current density for oxygen reduction. Partial cathodic protection is represented by decreasing the iron potential to $E_1$; full protection will occur when the external power source depresses the potential to the equilibrium value of $E'_{Fe}$. The external current density required to maintain this potential is also shown. Under complete protection, the external current, $I_{ex,complete\ protection} = (i_{ex,complete\ protection})(A)$, is supplying the current for the diffusion-controlled oxygen-reduction reaction, $I_{D,O_2} = (i_{D,O_2})(A)$. As with the previous example, this current represents a cost of protection.

### *Cathodic Protection: Hydrogen Embrittlement*

For metals susceptible to hydrogen embrittlement, there can be an adverse effect of applying cathodic protection to control corrosion. Cathodic protection, by its definition, involves lowering the potential of the corroding system below $E_{corr}$. Particularly for metals corroding due to cathodic reduction of hydrogen ions or water, this lowering of the potential results in increased rates of formation of hydrogen. The reduction at the surface is to atomic hydrogen, which is then either converted to hydrogen gas and escapes or diffuses into the metal. The greater the rate of formation of atomic hydrogen is (i.e., the lower the potential), the greater the transport of hydrogen into the metal will be. By mechanisms discussed in detail in Chapter 7, atomic hydrogen trapped in the metal can result in severe embrittlement. Metals differ significantly in their susceptibility to hydrogen embrittlement as do the alloys of a base metal. Variables for a given material include temperature, time, surface condition, and species in the environment that influence the interface mechanisms controlling the transport of hydrogen atoms into the metal. Thermal and mechanical treatment of the metal also may be significant. Alloy steels heat treated to high strengths are particularly susceptible to hydrogen embrittlement and hence can be significantly embrittled by cathodic protection.

## Example Calculations of Corrosion Potentials, Corrosion Currents, and Corrosion Rates for Aerated and Deaerated Environments, and the Effects of Galvanic Coupling

The objective of this example is to illustrate the use of data characterizing the requisite half-cell reactions to estimate corrosion rates. In this example, the corrosion behaviors of metals A and B, and A-B couples, in aerated and deaerated environments at pH = 4.5 are examined. A and B qualitatively approximate iron and nickel, respectively, without consideration of the effects of corrosion-product films on the linearity of the polarization curves. Reasonable values for the kinetic parameters ($i_o$, $\beta$, $i_D$) are used in theoretical expressions to plot idealized polarization curves for the electrochemical reactions involved. The resulting curves are used to estimate, quantitatively within the validity of the assumptions, corrosion potentials, currents, and rates. The four polarization curves plotted in Fig. 4.28 correspond to the following conditions:

- *Curve A:* Anodic reaction for A, $A \rightarrow A^{2+} + 2e$, Area = 10 $cm^2$ ($10^{-3}$ $m^2$), $a_{A^{2+}} = 10^{-7}$, $i_{o,A^{2+}} = 1$ $mA/m^2$, $\beta_{ox,A} = 70$ mV
- *Curve B:* Anodic reaction for B, $B \rightarrow B^{2+} + 2e$, Area = 10 $cm^2$ ($10^{-3}$ $m^2$), $a_{B^{2+}} = 10^{-6}$, $i_{o,B^{2+}} = 0.5$ $mA/m^2$, $\beta_{ox,B} = 50$ mV
- *Curve C:* Cathodic reaction of hydrogen on A or B, $H^+ + e \rightarrow \frac{1}{2}H_2$, Area = 10 $cm^2$ ($10^{-3}$ $m^2$), pH = 4.5, $i_{o,H^+} = 0.4$ $mA/m^2$, $i_{D,H^+} = 10^{+2}$ $mA/m^2$, $\beta_{red,H} = 100$ mV
- *Curve D:* Cathodic reaction of hydrogen on A or B, $\frac{1}{4}O_2 + H^+ + e \rightarrow \frac{1}{2}H_2O$, Area = 10 $cm^2$ ($10^{-3}$ $m^2$), pH = 4.5, $i_{o,O_2} = 0.3$ $mA/m^2$, $i_{D,O_2} = 10^{+3}$ $mA/m^2$, $\beta_{red,O_2} = 80$ mV, $P_{O_2} = 0.2$ atm

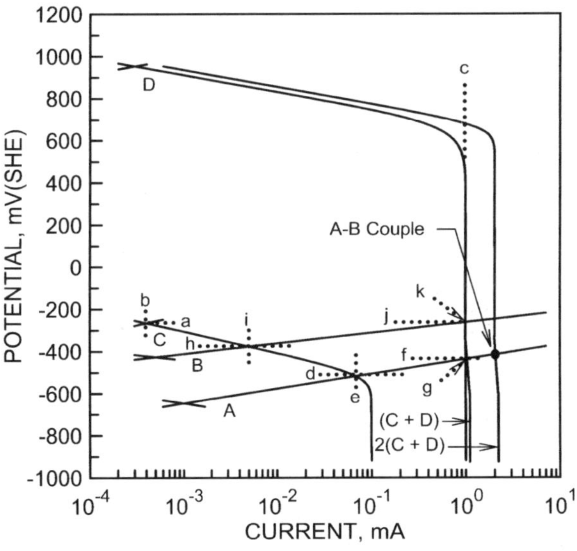


**Fig. 4.28** Idealized anodic polarization curves for metals A and B and for hydrogen and oxygen reduction. An explanation for the use of these curves for estimating the corrosion potentials, currents and rates for aerated and deaerated environments and for galvanic coupling can be found in the text.

- *Curve C + D:* Sum of the hydrogen and oxygen polarization curves on A or B
- *Curve 2(C + D):* Sum of the hydrogen and oxygen polarization curves on A and B coupled

A convenient approach for plotting the Tafel or charge-transfer region of a given curve is as follows. The equilibrium potential, E′, is calculated with the data provided by the Nernst equation. Then, the $(E', I_o)$ point is located. Next, a point is located at $(E' \pm \beta, 10\,I_o)$, where the positive sign refers to an oxidation curve, the negative sign to a reduction curve. A straight line on the semilog plot is then drawn through the two points. Justification for this convenient approach is, of course, based on the Tafel equation, $E = E' \pm \beta \log I/I_o$: when $I = I_o$, $E = E'$; and when $I = 10I_o$, $E = E' \pm \beta$.

Corrosion behaviors based on the polarization curves in Fig. 4.28 are analyzed as follows:

- Indicated on the curves are points identified by the letters a to k. These points correspond to the following quantities that are representative of calculations that can be made from the preceding data and from the curves. Positions on the curves such as $E'_{B,B^{2+}}$, $E'_{A,A^{2+}}$, and $E'_{O_2,H^+,H_2O}$ may be calculated from the data.

  a. $E'_{H_2,H^+} = -59\ pH = -59\ (4.5) = -266$ mV (SHE)
  b. $I_{o,H^+} = A_c\,(i_{o,H^+}) = 10^{-3}\ (0.4) = 4 \times 10^{-4}$ mA
  c. $I_{D,O_2} = A_c\,(i_{D,O_2}) = 10^{-3}\ (10^{+3}) = 1.0$ mA
  d. $E_{corr,A} = -530$ mV (SHE) for A in the *deaerated* solution
  e. $I_{corr,A} = 8.5 \times 10^{-2}$ mA for A in the *deaerated* solution
  f. $E_{corr,A} = -450$ mV (SHE) for A in the *aerated* solution
  g. $I_{corr,A} = 1.05$ mA for A in the *aerated* solution
  h. $E_{corr,B} = -380$ mV (SHE) for B in the *deaerated* solution
  i. $I_{corr,B} = 6.5 \times 10^{-3}$ mA for B in the *deaerated* solution
  j. $E_{corr,B} = -280$ mV (SHE) for B in the *aerated* solution
  k. $I_{corr,B} = 1.0$ mA for B in the *aerated* solution

- The corrosion penetration rate for B in the aerated solution in μm/year is calculated as follows (Table 4.1):

  a. From Faraday's law: CPR (μm/year) = 0.327 (M/mρ) $i_{corr}$, where M = atomic mass (g/mol), m = ion valence, ρ = density (g/cm$^3$), and $i_{corr}$ = corrosion current density (mA/m$^2$)
  b. For B (assuming Ni): M = 58.71, m = 2, ρ = 8.9 g/cm$^3$, $i_{corr} = (1.0/10^{-3})$ mA/m$^2$ = $10^3$ mA/m$^2$, and CPR = 1,080 μm/year

- It is important to compare the rates of corrosion of A and B in the aerated and then the deaerated solution. *In the aerated solution,* A is corroding at the point (f,g) and B at the point (j,k). The corrosion rates in terms of the corrosion currents are approximately the same because each is corroding under oxygen diffusion control. *In the deaerated solution,* corrosion is supported by the hydrogen reaction only, and for A this occurs at point (d,e) and for B at point (h,i). A is corroding about 15 times faster because of the relative position of these two metals electrochemically.

- Although one of the conditions in this example is stated to be dearation, actually it is impossible to remove all oxygen from the system. It is useful to consider what level of partial pressures of oxygen would be required to reduce the corrosion due to oxygen down to a specified level. As an example, the following treatment estimates the partial pressure of oxygen allowed if the corrosion due to oxygen is not to exceed 10% of that due to the acidity. The analysis will be restricted to metal A. The corrosion current of A due to the acidity is $8.5 \times 10^{-2}$ mA. For the contribution due to dissolved oxygen to be 10% of this value, the corrosion current associated with oxygen should not exceed $8.5 \times 10^{-3}$ mA, or approximately $10^{-2}$ mA. Since in the potential range for the corrosion of A, the oxygen reaction is under diffusion control (vertical section of curve D), the curve will be shifted proportional to the oxygen concentration (see Eq 3.80, Chapter 3, in which $C_{O_2}$ is made proportional to $P_{O_2}$). Therefore, to shift the oxygen curve from 1.0 mA to $10^{-2}$ mA, $P_{O_2}$ would need to be reduced from 0.2 atm to 0.002 atm.
- There are four corrosion situations in this example (i.e., A and B under aerated and deaerated conditions). Of these four, it is of interest to consider which one or ones would show the greatest change in corrosion rate if the solution velocity were increased. Since the charge-transfer section of a reduction curve is not affected by increase in velocity, intersections of the polarization curves defining corrosion conditions in the diffusion-controlled range for the cathodic reactions are sought. This condition is met for A and B in the aerated conditions. Thus, an increase in velocity would move curves D and (C + D) to higher values of current and, therefore, to higher corrosion currents. A in the deaerated solution would be slightly velocity sensitive, and the corrosion rate of B due to acidity alone should not be velocity sensitive.
- Consider the effect on A of coupling it to the B in the aerated solution. Because the anodic currents for B are so much smaller than for A, the total anodic curve is essentially equal to the anodic curve for A. Both B and A surfaces are now available as cathodic reaction sites with a total area of 20 $cm^2$ ($2 \times 10^{-3}$ $m^2$), and both cathodic reactions occur on these surfaces. The total cathodic curve is now 2(C + D). The new corrosion condition is labeled, A-B Couple,

from which it is observed that relative to point (f,g), $E_{couple}$ is about –430 mV (SHE) and $I_{corr,A,couple}$ is 2.2 mA. The corrosion current has been increased from 1.05 to 2.2 mA, or doubled. B no longer corrodes since the potential of the A-B couple is lower than the equilibrium potential for B.

## Chapter 4 Review Questions

1. Assume the homogeneous corrosion of iron in a deaerated acid solution at pH = 4 (i.e., the anodic and cathodic reactions are occurring uniformly over any unit area). Plot the anodic polarization curve for iron and the cathodic polarization curve for the hydrogen reaction. Estimate $E_{corr}$, $i_{corr}$, and the corrosion penetration rate in μm/year.
Given:

$$i_{o,Fe} = 10^{-1}\ mA/m^2$$

$$\beta_{ox,Fe} = 50\ mV$$

$$a_{Fe^{2+}} = 10^{-6}$$

$$i_{o,H_2\ on\ Fe} = 10\ mA/m^2$$

$$\beta_{red,H_2\ on\ Fe} = 100\ mV$$

$$i_{D,red,H_2} = 10^{+4}\ mA/m^2$$

2. Under the same solution conditions as in problem 1, suppose that 1 $cm^2$ ($10^{-4}$ $m^2$) of iron is to be galvanically coupled to 1 $cm^2$ of copper (i.e., equal areas).
Given:

$$i_{o,Cu} = 10^{-1}\ mA/m^2$$

$$\beta_{ox,Cu} = 50\ mV$$

$$a_{Cu^{2+}} = 10^{-6}$$

$$i_{o,H_2\ on\ Cu} = 1.0\ mA/m^2$$

$$\beta_{red,H_2\ on\ Cu} = 100\ mV$$

   a. Before coupling, estimate $i_{corr,Fe}$ and $i_{corr,Cu}$.
   b. After coupling, estimate $E_{couple}$, $i_{corr,Fe,couple}$, and $i_{corr,Cu,couple}$.

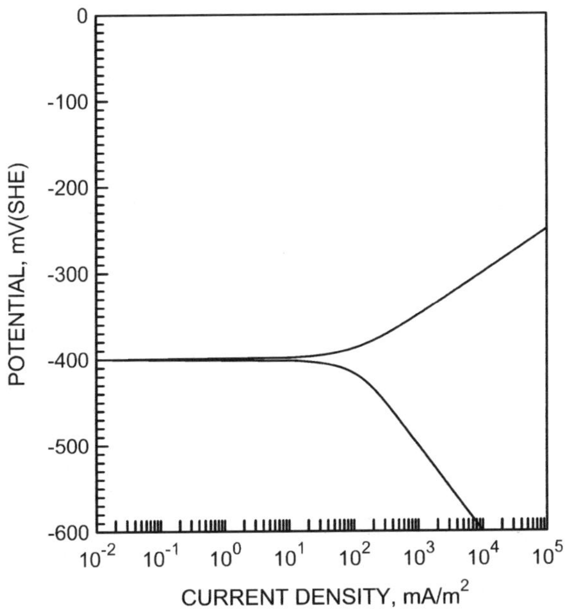


**Fig. 4.29** Representative, experimental polarization curves for metal M in a deaerated acid solution

3. Under the same solution conditions as in problem 2, estimate the corrosion current density for Fe if 1 $cm^2$ ($10^{-4}$ $m^2$) of iron is coupled to 100 $cm^2$ ($10^{-2}$ $m^2$) of Cu. This situation illustrates the effect of using iron (or steel) bolts to hold copper plates together in a corrosive environment.
4. Representative, experimental anodic and cathodic polarization curves for metal M in a deaerated acid solution are given in Fig. 4.29. The equilibrium potentials for the half-cell reactions, $M = M^{3+} + 3e$ and $2H^+ + 2e = H_2$, are –500 mV (SHE) and –100 mV (SHE), respectively. The atomic mass and density for the metal are 42 g/mol and 8.2 g/$cm^3$, respectively. Evaluate the following: (a) $E_{corr}$; (b) $i_{corr}$; (c) $\beta_{ox,M}$; (d) $i_{o,M}$; (e) $\beta_{red,H}$; (f) $i_{o,H}$; (g) solution pH; (h) CPR (μm/year).
5. Suppose metal M in problem 4 is to be cathodically protected by impressed current.
   a. Give the potential required for complete protection from corrosion. Briefly explain your answer.
   b. Determine the resultant external current density corresponding to the potential required for complete protection.

## Answers to Chapter 4 Review Questions

1. $E_{corr} \cong -420$ mV (SHE), $i_{corr} \cong 700$ mA/$m^2$, CPR $\cong$ 800 μm/year

2. (a) $i_{corr,Fe} \cong 700$ mA/m$^2$; $i_{corr,Cu} = 0$; (b) $E_{couple} \cong -420$ mV (SHE); $i_{corr,Fe,couple} \cong 800$ mA/m$^2$; $i_{corr,Cu,couple} = 0$
3. $i_{corr,Fe,couple} \cong 4200$ mA/m$^2$
4. (a) $E_{corr} = -400$ mV (SHE); (b) $i_{corr} \cong 100$ mA/m$^2$; (c) $\beta_{ox,M} \cong 50$ mV; (d) $i_{o,M} \cong 1$ mA/m$^2$; (e) $\beta_{red,H} \cong 100$ mV; (f) $i_{o,H} \cong 0.1$ mA/m$^2$; (g) solution pH = 1.7; (h) CPR $\cong$ 56 μm/year
5. (a) E $\leq -500$ mV (SHE); (b) $\cong$ 1000 mA/m$^2$

## References


1. C. Wagner and Z.Z. Traud, *Electrochem.*, Vol 44, 1938, p 391
2. J.T. Waber, Mathematical Studies of Galvanic Corrosion III. Semi-Infinite Coplanar Electrodes with Equal Constant Polarization Parameters, *J. Electrochem. Soc.,* Vol 102, 1955, p 420–429
3. H.R. Copson, Distribution of Galvanic Corrosion, *J. Electrochem. Soc.,* Vol 84, 1943, p 71–80
4. M. Stern and A.L. Geary, Electrochemical Polarization I. A Theoretical Analysis of the Shape of Polarization Curves, *J. Electrochem. Soc.,* Vol 104, 1957, p 56–63
5. M. Stern and A.L. Geary, Electrochemical Polarization II. Experimental Verification, *J. Electrochem. Soc.,* Vol 104, 1957, p 559-565
6. M. Stern, Electrochemical Polarization III. Further Aspects of the Shape of Polarization Curves, *J. Electrochem. Soc.,* Vol 104, 1957, p 645–655
7. E.J. Kelly, The Active Iron Electrode I. Iron Dissolution and Hydrogen Evolution Reactions in Acidic Sulfate Solutions, *J. Electrochem. Soc.,* Vol 112, 1965, p 124–131
8. M. Stern, The Electrochemical Behavior, Including Hydrogen Overvoltage, of Iron in Acid Environments, *J. Electrochem. Soc.,* Vol 102, 1955, p 609–616
9. N. Hackerman and E. McCafferty, Adsorption and Corrosion with Flexible Organic Diamines, *Proc. Fifth International Congress on Metallic Corrosion,* National Association of Corrosion Engineers, 1974, p 542–548
10. R. Baboian, Predicting Galvanic Corrosion Using Electrochemical Techniques, *Electrochemical Techniques for Corrosion Engineering,* R. Baboian, Ed., National Association of Corrosion Engineers, 1986
11. H.P. Hack, Galvanic, *Corrosion Tests and Standards,* R. Baboian, Ed., ASTM, 1995
12. "Standard Guide for Development and Use of a Galvanic Series for Predicting Galvanic Corrosion Performance," G 82-83 (Re-

approved 1993), *Annual Book of ASTM Standards,* Vol 03.02, ASTM, 1995
13. F.L. LaQue, *Marine Corrosion, Causes and Prevention,* John Wiley & Sons, 1975, p 179
14. J.H. Morgan, *Cathodic Protection,* National Association of Corrosion Engineers, 1987

CHAPTER **5**

# Corrosion of Active-Passive Type Metals and Alloys

## Anodic Polarization Resulting in Passivity

When anodic polarization measurements are extended to progressively higher potentials, several potential versus current-density relationships may result depending upon the electrode material and the aqueous environment. For purposes of the present discussion, it is sufficient to describe three types of curves of the forms of Fig. 5.1(a), (c), and (e), all determined when the potential is continuously scanned from the lowest potential of the curve. Figure 5.1(a) shows the anodic polarization curve for copper in deaerated 1 N $H_2SO_4$. In this case, a progressive increase in the potential results in a curve that rises rapidly and becomes essentially vertical at a limiting current density for diffusion-controlled polarization. At sufficiently high potentials, the current density may increase due to the oxidation of $H_2O$ to $O_2$. If the imposed potential is removed and the "free" electrode potential is measured as a function of time, then the smooth decrease in potential shown in Fig. 5.1(b) is observed. This smooth decrease is due to the diffusion of accumulated copper ions from the interface.

The anodic potentiodynamic polarization curve for zinc in 1 N NaOH is shown in Fig. 5.1(c). In this case, the curve again starts to rise due to diffusion polarization but rather suddenly decreases near –800 mV

(SHE) due to formation of a surface coating of $Zn(OH)_2$, which increases the circuit resistance and hence decreases the current density. The decay of the "free" electrode potential is shown in Fig. 5.1(d). In this case, the potential decreases rapidly since the activity of the $Zn^{2+}$ ions is held to a low value because of the relatively low solubility of $Zn(OH)_2$.

The anodic polarization for iron in 1 N $H_2SO_4$ is shown in Fig. 5.1(e) and the change in potential with time in Fig. 5.1(f) when control of the potential is terminated at the maximum potential. With iron, on increasing the potential, a rapid decrease in current density, associated with oxide film formation, occurs near 450 mV (SHE), which finally becomes essentially constant at a current density several orders of magnitude lower than the maximum observed at a slightly lower potential. Again, at higher potentials, oxygen evolution and conversion of the oxide to soluble hexavalent iron ions results in an increase in current density. In contrast to the potential decay curves of Fig. 5.1(b) and (d), the decay curve for iron includes a plateau, called the Flade potential, at which the "free" electrode potential remains essentially constant for a period of time, associated with oxide film dissolution, at approximately the same potential that had resulted in a decreasing current density (oxide film formation) as the potential was initially increased. The potential finally decreases to the initial corrosion potential.

It is evident from Fig. 5.1 that the shape of the anodic polarization curve depends on the electrode material. Although only two environments were considered, the chemical species that are in solution in contact with the electrode material will have a major effect on the form of the potential versus current-density relationship. Materials exhibiting polarization behavior of the form of Fig. 5.1(e) are said to exhibit *passivity* in the particular environment. The passive behavior is characterized by the critical current density, $i_{crit}$, that must be exceeded on an upscan of potential to initiate formation of the passive film; the passivating potential, $E_{pp}$, at which the current density begins to decrease; and by the magnitude of the current density in the passive condition, $i_p$. The magnitude of the change in current density between $i_{crit}$ and $i_p$ is of major significance since this change indicates the effectiveness of the passive film in reducing the dissolution (corrosion) rate at the anode surface. To be of practical significance, the ratio $i_p/i_{crit}$ should be $10^{-2}$ and preferably smaller; ratios as low as $10^{-6}$ are observed. Values of $i_p$ are frequently of the order of 10 mA/m$^2$ corresponding to corrosion rates of about 25 μm/year (1 mpy, or mil per year).

The theoretical predictions or experimental determinations of the composition, thickness, and structure of films responsible for passivity are difficult. The problem of prediction is complicated by many factors, including knowledge of the composition of the solution at the material interface, knowledge of the effects of potentials differing significantly

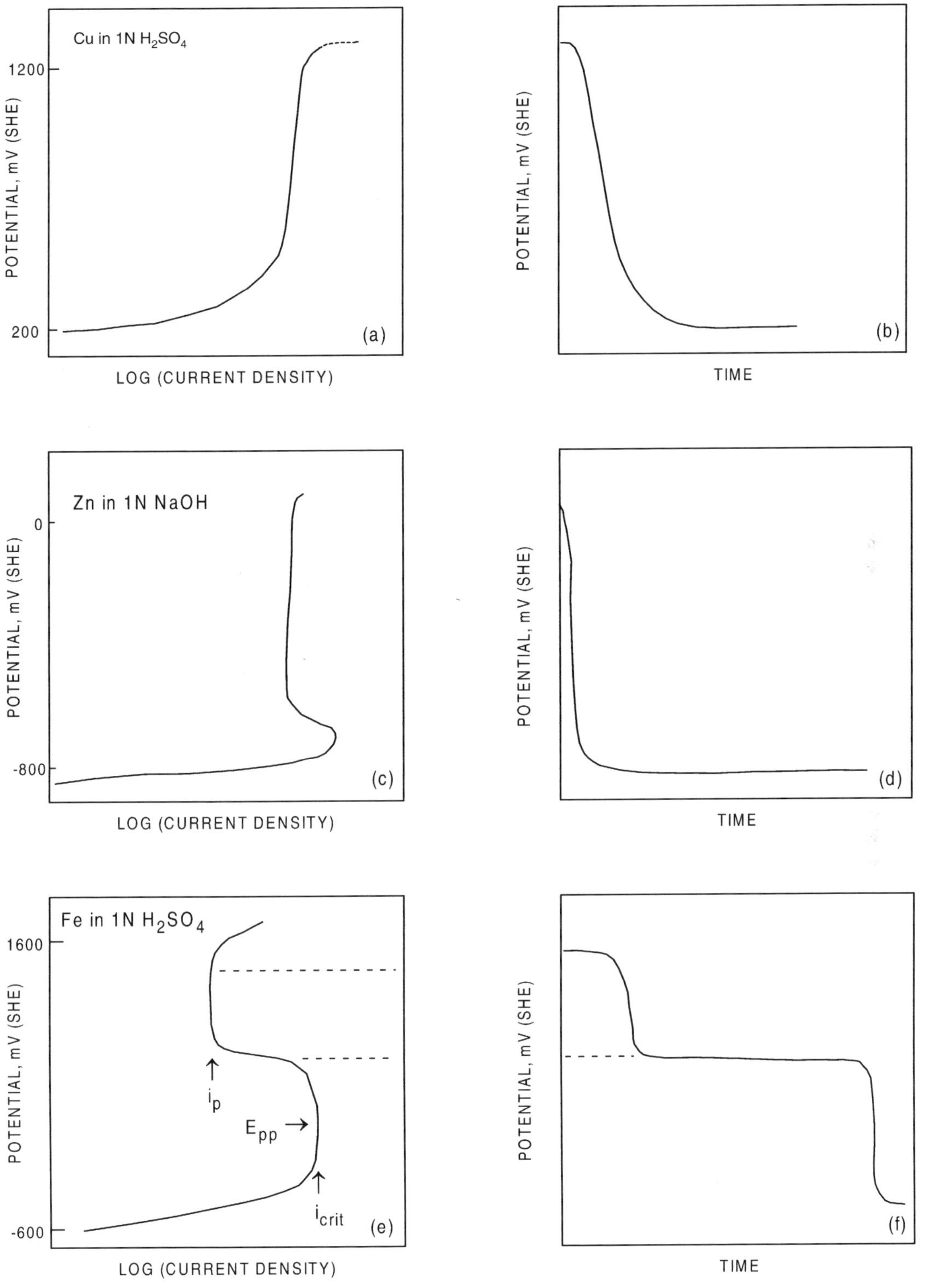


**Fig. 5.1** Schematic representation of several forms of anodic polarization curves and associated potential decay curves following release of potentiostatic control

from equilibrium values on the chemical composition and structure of the film, and knowledge of the effects of kinetic (reaction rate) factors on the composition and structure of the film. The experimental determination of the characteristics of the film is difficult because the film forms on the metal when in contact with a solution, and changes may occur on removal for examination by chemical, optical, x-ray, or electron microscopy techniques. When these surface layers have been examined either directly or after detachment from the underlying metals, some type of oxide structure is usually deduced. Additional information on the structures must be inferred from characteristics of the polarization curves, the potential decay curves, and other electrochemical measurements. A major controversy among investigators concerns whether the initial state of passivation is only a chemisorbed monolayer of oxygen ions or is actually an oxide layer (Ref 1). Passivity can be produced when electrochemical measurements indicate a surface layer of one or, at most, a few atom layers thick, although continued passivation may result in layers tens to hundreds of nanometers thick.

## Significance of the Pourbaix Diagram to Passivity

To illustrate the significance of the Pourbaix diagram to passivity, consider the iron-water system at point A in Fig. 5.2 (Ref 2). At this point, iron at a potential of –620 mV (SHE) is in equilibrium with a so-

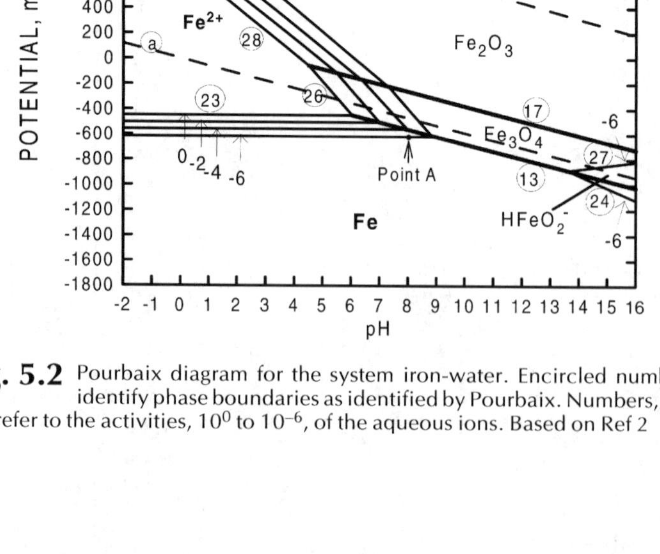


**Fig. 5.2** Pourbaix diagram for the system iron-water. Encircled numbers identify phase boundaries as identified by Pourbaix. Numbers, 0 to –6, refer to the activities, $10^0$ to $10^{-6}$, of the aqueous ions. Based on Ref 2

lution of pH = 8 and $a_{Fe^{2+}} = 10^{-6}$. The iron will remain uncoated and will not corrode, although there will be hydrogen evolution since the potential is below line (a). (It is important to note that the iron must be held potentiostatically at the –620 mV (SHE) for this condition to exist and that electrons for the reaction $H^+ + e \rightarrow \frac{1}{2}H_2$ come from the external current.) Consider two changes of conditions. First, if the potential is increased, iron will tend to go into solution and $Fe_3O_4$ (and/or $Fe(OH)_2$) will form at about –560 mV (SHE) when dissolution of the iron has increased the activity to $a_{Fe^{2+}} = 10^{-4}$. Further increase in potential will cause additional conversion to $Fe_3O_4$, and above about –200 mV (SHE), $Fe_2O_3$ is predicted to form on the surface of the $Fe_3O_4$, the $Fe_2O_3$ then being in contact with the solution. The exact sequence of changes and the protection provided by the oxide films depends on their adherence, their ability to prevent contact of the solution with the underlying metal, and the rate of transport of anions, cations, and electrons through the film. The time of exposure is also a variable.

As a second change of conditions, assume that the pH is increased to 9.0 at the initial –620 mV (SHE). At this pH, $Fe_3O_4$ forms on the iron surface, and an increase in potential would again produce an outer layer of $Fe_2O_3$. It is evident that over the pH range of line 13, and at potentials above this line, iron becomes coated with $Fe_3O_4$ and then $Fe_2O_3$. Pourbaix defines this to be the *region of passivation*, and if the oxides are protective, the *condition of passivity* exists; the iron is said to be in the *passive state* (Ref 3).

Consider next that the iron is in a solution of pH = 4 and $a_{Fe^{2+}} = 10^{-6}$ and that the potential is controlled at –620 mV (SHE). Again, the iron will remain in equilibrium even though $H_2$ is evolved. If the potential is slowly raised, iron will pass into solution to bring the system into equilibrium at each higher potential. For example, at E = –440 mV (SHE), $a_{Fe^{2+}}$ must be unity. If the potential is raised rapidly, only the solution in immediate contact with the iron increases in $a_{Fe^{2+}}$, and depending on this value, $Fe_2O_3$ can form at potentials of about 400 mV (SHE). Thus, rapidly increasing the potential in an acid solution leads to the *possibility* of forming protective or passive films. The Pourbaix diagram, however, is a representation of equilibrium states for the system, and its use to predict behavior under nonequilibrium conditions (such as rapidly increasing the potential of iron into ranges where it cannot exist at equilibrium) is uncertain.

Experimental values of potentials at which passive films are observed to form on iron as a function of pH are plotted in Fig. 5.3 (Ref 2). Lines 23, 26, and 28 for $a_{Fe^{2+}} = 10^{-6}$ and the boundaries of the $Fe/Fe_3O_4$ and $Fe_3O_4/Fe_2O_3$ equilibria according to the Pourbaix diagram are also plotted. It is evident that the experimental data lie near the $Fe_3O_4/Fe_2O_3$ line 17 and its extrapolation, indicating that the formation of $Fe_2O_3$ is necessary to cause passivity.

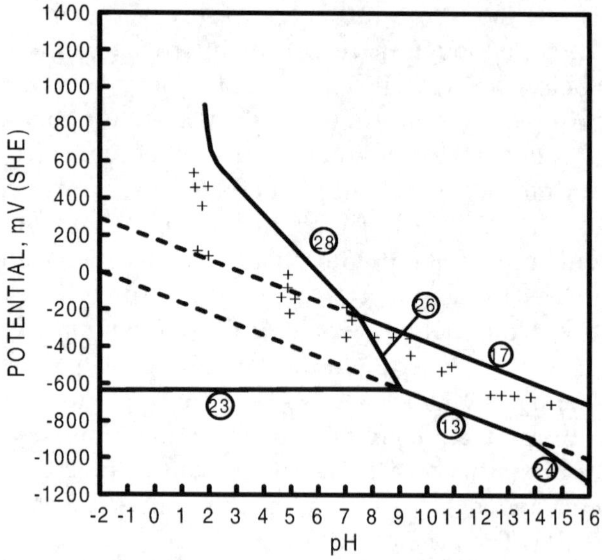


**Fig. 5.3** Experimental values (+ symbols) of the passivating potential, $E_{pp}$, of iron plotted to show the relationship to selected phase boundaries from Fig. 5.2. Dashed lines are extrapolations of lines 13 and 17. Based on Ref 2

## Experimental Observations on the Anodic Polarization of Iron

A representative anodic polarization curve for iron in a buffered environment of pH = 7 is shown in Fig. 5.4. The solid curve is representative of experimental observations; the dashed curve is an extrapolation of the Tafel region to the equilibrium half-cell potential of –620 mV (SHE) and $a_{Fe^{2+}} = 10^{-6}$. This extrapolation allows estimation of an exchange current density of 0.03 mA/m$^2$. The essentially steady minimum current density of the passive state is $i_p$ = 1 mA/m$^2$.

Research on the polarization of iron in a buffered solution of pH = 8.4 and higher has been interpreted to show that a series of electrochemical reactions occur as the polarization potential increases (Ref 4). Reactions 5.1 to 5.5, identified below by letter, are considered to be the dominant reactions in the potential ranges identified by the corresponding letters along the polarization curve in Fig. 5.4:

A:

$$Fe \rightarrow Fe^{2+} + 2e \qquad \text{(Eq 5.1)}$$

B:

$$3Fe + 4H_2O \rightarrow Fe_3O_4 + 8H^+ + 8e \qquad \text{(Eq 5.2)}$$

C:

$$2Fe_3O_4 + H_2O \rightarrow 3Fe_2O_3 + 2H^+ + 2e \qquad \text{(Eq 5.3)}$$

or

$$2Fe^{2+} + 3H_2O \rightarrow Fe_2O_3 + 6H^+ + 2e \qquad \text{(Eq 5.3a)}$$

D:

$$(2 - x)Fe_2O_3 + 3xH_2O \rightarrow 2Fe_x^{6+}Fe_{(2-2x)}^{3+}\square_xO_3 + 6xH^+ + 6xe \qquad \text{(Eq 5.4)}$$

where x is the fraction of the iron lattice sites occupied by $Fe^{6+}$ in the $Fe_2O_3$ crystal structure, and □ represents the vacant iron lattice sites.

E:

$$Fe_2O_3 + 5H_2O \rightarrow 2FeO_4^{=} + 10H^+ + 6e \qquad \text{(Eq 5.5)}$$

The onset of passivity is associated with reaction C, which results in a layer having the sequence of phases shown in Fig. 5.5(a).

Since the $Fe_2O_3$ is in contact with the solution, the surface behaves as an $Fe_2O_3/(Fe^{2+}, H^+)$ electrode with the underlying $Fe_3O_4$ and Fe functioning as electrical conductors to the interface. Reaction D occurs as the potential is increased progressively above $E_{pp}$ and involves formation of a defect oxide (one containing vacant lattice sites) at the outer surface of the $Fe_2O_3$ layer as shown in Fig. 5.5(b). Although the passive layer is electrically conducting, it has not been established whether the low passive current density, $i_p$, is due to the low conductivity by migration of cations and anions through the film, to slow transfer of ions across the interface, or to the low conduction of electrons. In any case,

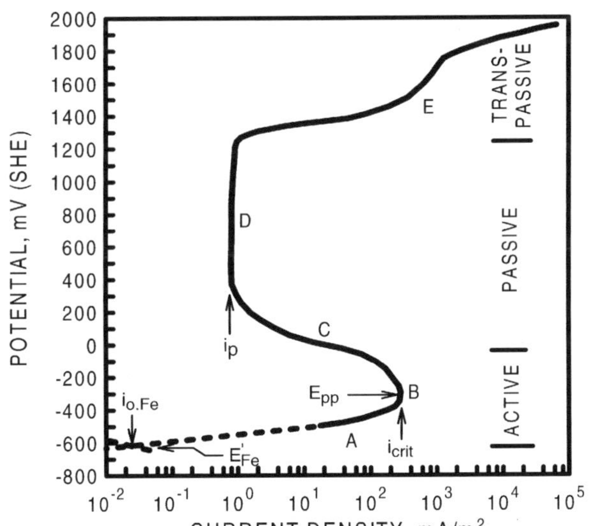


**Fig. 5.4** Representative polarization curve for iron in buffered solution of pH = 7. Dashed curve extends to the half-cell potential of iron with $a_{Fe^{2+}} = 10^{-6}$. Letters along curves relate to reactions (details can be found in the text) that are dominant in the associated potential range.

the film thickens at constant potential with time in the passive potential range reaching a steady-state value in the range of 1 to 5 nm as the potential is increased (Ref 4). This steady-state thickness at a constant current density indicates a steady dissolution by transport of iron ions through the passive film and into the solution. Also note that according to the alternate reaction C, oxide can be formed by reaction with $Fe^{2+}$ in solution. This has been confirmed experimentally by adding $Fe^{2+}$ ions to the environment, which results in a higher current density during the initial stage of film formation (Ref 4). The effect also has been confirmed by the observation that during scanning from the active region, the $Fe^{2+}$ produced in this potential range (as, for example, by decreasing the positive potential scan rate, thereby increasing $Fe^{2+}$) can influence the measured passive current density in the passive potential range.

It will be shown later that the values of $i_{crit}$, $E_{pp}$, and $i_p$, which are the important parameters defining the shape of the active-passive type of polarization curve, are important in understanding the corrosion behavior of the alloy. In particular, low values of $i_{crit}$ enhance the ability to place the alloy in the passive state in many environments. For this reason, the maximum that occurs in the curve at "B" (Fig. 5.4) is frequently referred to as the active peak current density or, in general discussion, as the active peak. It is the limit of the active dissolution current density occurring along the "A" region of the polarization curve.

The above series of reactions indicates that pH should be a major variable affecting the position of the active-passive polarization curve of

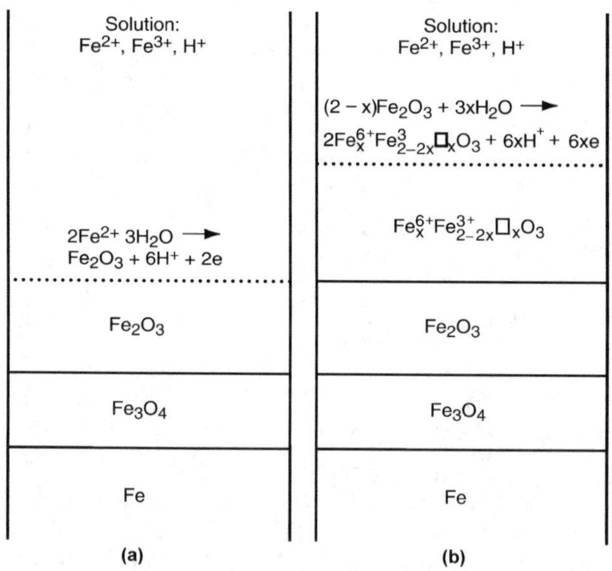


**Fig. 5.5** Proposed sequence of species present from iron substrate to solution of indicated ions. (a) Sequence occurring in the potential range C in Fig. 5.4. (b) Sequence in the potential range D. Solid line, solid/oxide or oxide/oxide interface; dashed line, oxide/solution interface. Based on Ref 4

iron. The effect of even simple changes of pH cannot be interpreted independent of the possible influence of accompanying ions. Generally, environments encountered in service will be even more complex mixtures of ions at any pH. These species may alter the anodic dissolution processes through effects on the kinetics of the interface reactions and by altering the physical and chemical structure of solid corrosion products. Examples of several environmental effects on anodic polarization are discussed subsequently. A representative effect of pH on the anodic polarization of iron is shown in Fig. 5.6. These curves have been obtained from polarization measurements in buffered acid ($H_2SO_4$) solutions of sodium phosphate and phosphoric acid and buffered alkaline (NaOH) solutions of sodium borate and boric acid in a base solution of 0.15 M $Na_3PO_4$ (Ref 5). There are three distinct displacements of the curves with increasing pH: the passivating potential decreases, the critical current density decreases, and the current density in the passive state decreases. The decrease in the passivating potential is consistent with the Pourbaix diagram in that the oxide phases form at progressively lower potentials as the pH increases. Two sets of curves are shown at the higher potentials (i.e., in the transpassive potential range). The solid curves show the current density associated with the dissolution of the iron to $Fe^{3+}$ or $FeO_4^{=}$. Water also can be oxidized to oxygen in this

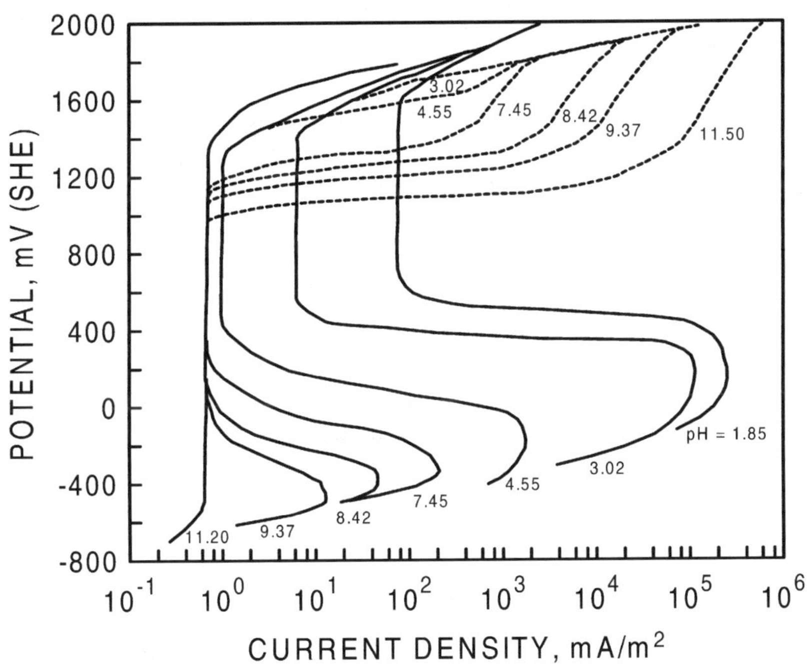


**Fig. 5.6** Anodic polarization curves for iron dissolution (solid curves) and for total current density of iron plus oxygen evolution (dashed curves) after 1 h at steady state in deaerated 0.15 M $Na_3PO_4$ solution. Indicated pH obtained by use of acid and base buffers and additions of $H_2SO_4$ or NaOH. Redrawn from Ref 5

potential range, the onset occurring at lower potentials the higher the pH (line b in Fig. 5.2). The dashed curves in Fig. 5.6 represent the total current density associated with iron dissolution plus oxygen evolution. It is shown subsequently that the decrease in $E_{pp}$ and $i_{crit}$ with increase in pH allows the passive state to be more easily established, which, along with the lower current density in the passive range, provides understanding to the observation that the rate of corrosion of iron is significantly less in alkaline environments. In this respect, it should be noted that at pH = 11.2 iron is already passivated on exposure to the solution and no "active peak" is formed on an increase in potential as occurs in the lower pH environments. It is reemphasized that an anodic polarization curve is sensitive to the anionic species in the environment and to the experimental conditions under which it is determined (e.g., by stepwise holding at each potential for a specified time or by continuously scanning the potential). Figure 5.6 is therefore representative of the general effects of pH on the anodic polarization behavior, but the exact position of the curve will depend on the specific species in the solution and the experimental procedures.

All of the curves in Fig. 5.6 start in the active dissolution potential range and hence do not show the complete polarization curve for the iron extending to the equilibrium half-cell potential as was done in Fig. 5.4. This extension was shown as dashed lines and the equilibrium potential was taken as –620 mV for $a_{Fe^{2+}} = 10^{-6}$. Qualitatively, the basis for estimating how the active regions of the curves in Fig. 5.6 would be extrapolated to the equilibrium potential can be seen by reference to Fig. 4.16. There, the corrosion potential is represented as the intersection of the anodic Tafel curve and the cathodic polarization curve for hydrogen-ion reduction at several pH values. It is pointed out that careful measurements have shown that the anodic Tafel line shifts with pH (Ref 6), this shift being attributed to an effect of the hydrogen ion on the intermediate steps of the iron dissolution.

In mildly alkaline environments (pH = 10 to 13.5), the corrosion rate of iron is very low (<25 μm/year, or 1 mpy) due to the ease with which a protective passive film forms in accordance with the position of the polarization curve for pH = 11.2 in Fig. 5.6. However, the polarization curve moves to higher current densities as the concentration of alkaline species in solution increases. This is illustrated by the set of polarization curves in Fig. 5.7 for iron in boiling solutions of increasing concentration of sodium hydroxide (Ref 7). The increase in the current density in the passive range with increasing concentration indicates an increase in corrosion rate if an environment establishes potentials in the passive range. This is consistent with the Pourbaix diagram for iron (Fig. 5.2), which, at high pH, shows a region in which the stable corrosion product species is soluble hypoferrite ion, $HFeO_2^-$, rather than a solid oxide.

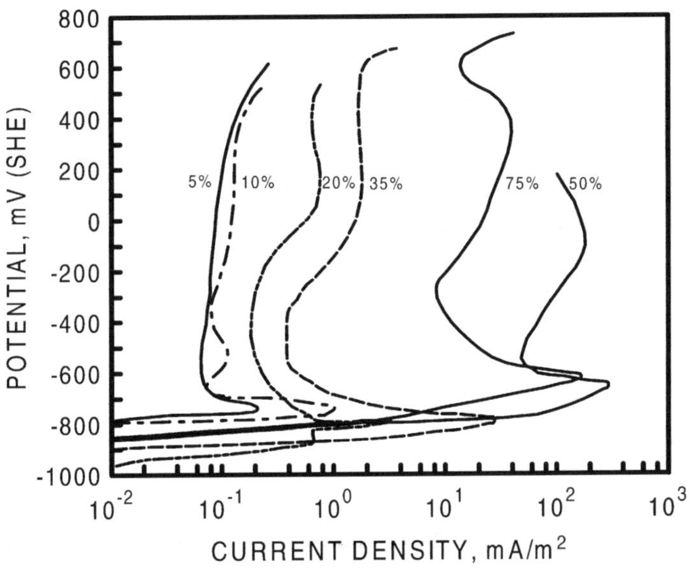


**Fig. 5.7** Polarization curves for mild steel in boiling NaOH solutions of various strengths. (Redrawn from Ref 7)

Iron in concentrated nitric acid exhibits very low rates of corrosion, about 25 μm/year (1 mpy) (Ref 8). This anomalous behavior was observed by Faraday over 150 years ago. In the concentrated acid, the iron immediately forms a thin passive film and remains bright while immersed. The effective noble state imparted by the passive film can be demonstrated by carefully removing a specimen and applying a drop of copper sulfate solution to the surface. Since the copper half-cell potential is more positive than that of iron, normally copper will deposit on an iron surface being reduced by the dissolution of the iron. This is not observed when the copper sulfate solution is first placed on the iron passivated by concentrated nitric acid, indicating that the passivated iron exhibits an electrochemical potential greater than that of $Cu/Cu^{2+}$. When, within a few minutes, the passive film is broken, copper deposits onto the iron substrate and spreads, consuming the passive film. Also, if the nitric acid is diluted and the film broken, a violent reaction occurs as the passive film can no longer be sustained by the diluted acid.

## Relationship of Individual Anodic and Cathodic Polarization Curves to Experimentally Measured Curves

In Chapter 4, analysis of the kinetics of coupled half-cell reactions shows how the corrosion potential and corrosion current density depend on the positions of the anodic and cathodic polarization curves. The anodic polarization curves are generally represented as showing linear or Tafel behavior, and the cathodic curves are shown with both Tafel and diffusion control characteristics. The intersection of these curves allows the corrosion current density and, hence, corrosion rate to be determined. It is shown how the several kinetic parameters governing the positions of the curves determine the intersection and, thus, the corrosion rate.

In a similar manner, the corrosion of an active-passive type alloy is determined by the relative positions of the anodic polarization curve (of the types introduced at the beginning of this chapter) and the polarization curve or curves of cathodic reactants in the aqueous environment. Because of the more complex and varied shapes of the anodic curves of active-passive type alloys, the possible positions of intersections with the several forms of cathodic curves are greater leading to more complex interpretations of the corrosion behaviors. And, since potentiodynamic polarization measurements provide curves representative only of the external or net current densities (i.e., $i_{ex} = i_{net} = \Sigma\, i_{ox}$ [anodic] $- \Sigma\, i_{red}$ [cathodic]) as a function of potential, an understanding of how the positions of the individual anodic and cathodic curves can result in the observed net anodic and cathodic curves is important. This becomes particularly significant when a corrosion behavior is observed and a contribution to an understanding of the factors governing the corrosion is being based on a polarization curve determined experimentally for the alloy/environment combination.

In this section, the relative positions of several schematic anodic and cathodic curves are presented. The sum anodic ($\Sigma\, i_{ox}$) and sum cathodic ($\Sigma\, i_{red}$) curves are shown relative to the individual curves, and then the net curves are shown as representative of what would be observed experimentally. Figure 5.8 shows an anodic polarization curve (M) repre-

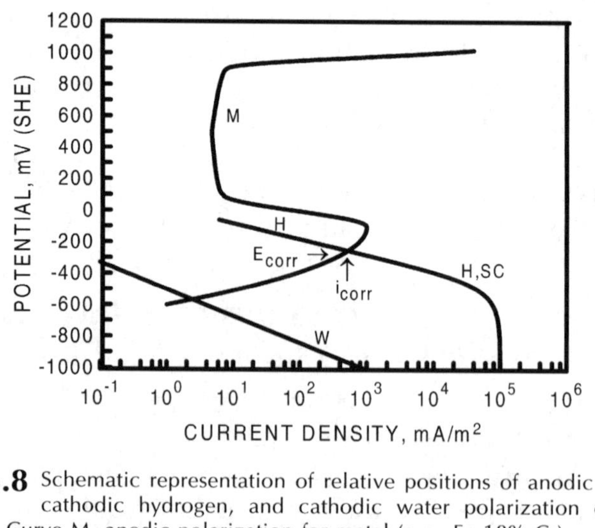


**Fig. 5.8** Schematic representation of relative positions of anodic metal, cathodic hydrogen, and cathodic water polarization curves, pH = 1. Curve M, anodic polarization for metal (e.g., Fe-18% Cr); curve H, cathodic polarization for $H^+$; curve W, cathodic polarization for $H_2O$; curve SC, sum of $H^+$ and $H_2O$ polarization

sentative of an 18% Cr-82% Fe alloy in sulfuric acid at pH = 1. Representative cathodic curves for the hydrogen reaction (H) at this pH and for direct reduction of water (W) also are shown. The sum cathodic curve (SC) at each potential would represent the current density resulting from the two cathodic reactants. In this case, however, the contribution due to water reduction is negligible compared with that for the hydrogen-ion reduction. The environment is assumed to be completely deaerated (sparged with pure nitrogen) so that a curve for the cathodic reduction of oxygen does not appear. The intersection of the anodic and sum cathodic curves occurs in the active section (Fig. 5.4) of the alloy anodic curve and gives values for the corrosion potential, $E_{corr}$, and for the corrosion current density, $i_{corr}$, from which the corrosion rate can be evaluated. The analysis of the corrosion behavior in this case is similar to that of Chapter 4 (refer to Fig. 4.8) since it is restricted to the active Tafel region of the anodic curve.

The individual curves (M, H, and W), the sum cathodic curve (SC), and the net curves (N) are shown in Fig. 5.9. The net curves only are shown in Fig. 5.10. The net curves pass to very low values and become zero at $E_{corr}$, being net cathodic below this potential and net anodic above $E_{corr}$. It is evident from the net curves (Fig. 5.10) that $E_{corr}$ is easily determined but that $i_{corr}$ would be estimated by extrapolation of the Tafel region of the cathodic curve to $E_{corr}$. Also, the portion of the cathodic polarization curve above $E_{corr}$ and the portion of the anodic curve below $E_{corr}$ must be estimated by extrapolation of the experimentally determined portions (Fig. 5.9 and 5.10).

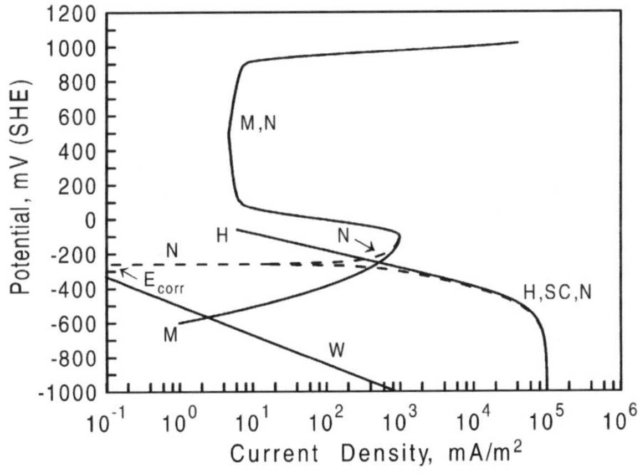


**Fig. 5.9** Schematic representation of relative positions of the net polarization curves to the individual curves for anodic metal polarization and cathodic hydrogen and water polarization, pH = 1. Curve M, anodic polarization for metal (e.g., Fe-18% Cr); curve H, cathodic polarization for $H^+$; curve W, cathodic polarization for $H_2O$; curve SC, sum of cathodic polarization for $H^+$ and $H_2O$; curve N, net curve for anodic and cathodic polarization. Note: Curve N coincides with curve M above –100 mV and with curve H below –350 mV.

In Fig. 5.11, the curve for the cathodic reduction of oxygen (O) has been added corresponding to an aerated environment with $P_{O_2} = 0.2$ atm or 8.5 ppm dissolved oxygen. The position shown is representative of the polarization on a clean surface such as platinum. In general, the position will depend on the particular metal surface supporting the reaction, although the limiting current density of about $10^3$ mA/m$^2$ is representative of aerated solutions. The sum cathodic curve (SC) (i.e., the sum of oxygen, hydrogen-ion, and water reduction), is plotted in Fig. 5.12 and is the effective cathodic curve relating to the corrosion behav-

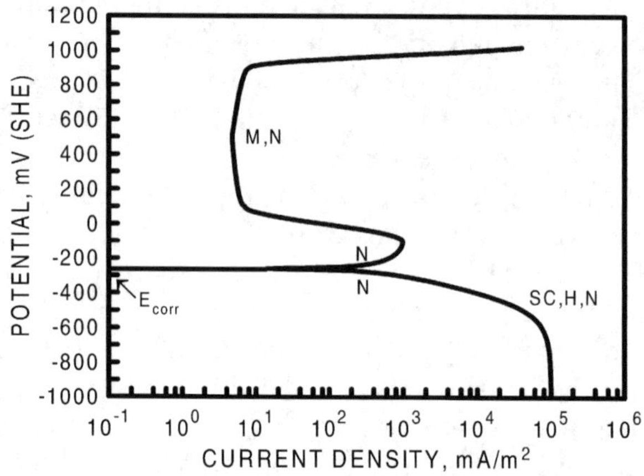


**Fig. 5.10** Schematic representation of the net anodic and cathodic polarization curves, N, for the anodic metal, M, and for the cathodic hydrogen, H, polarization curves. Note that the net curves deviate from curves M and H only near $E_{corr}$. SC is the sum of cathodic polarization for $H^+$ and $H_2O$. pH = 1

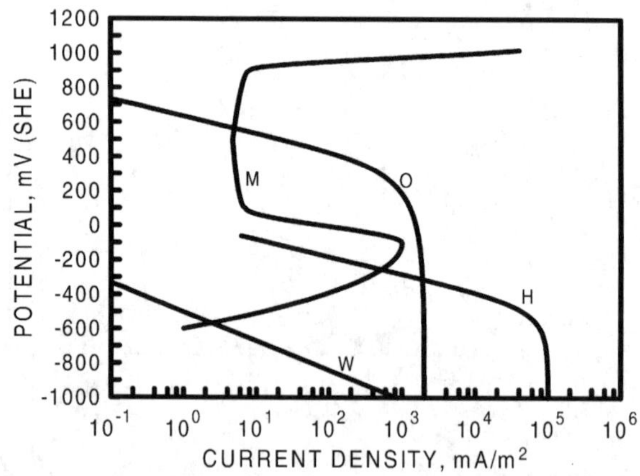


**Fig. 5.11** Schematic representation of relative positions of anodic metal, cathodic oxygen, cathodic hydrogen, and cathodic water reduction polarization curves. pH = 1. $P_{O_2} = 0.2$ atm. Curve M, anodic polarization curve for metal (e.g., Fe-18% Cr); curve H, cathodic polarization curve for $H^+$; curve W, cathodic polarization curve for $H_2O$; curve O, cathodic polarization for $O_2$

ior. The sum cathodic curve and the anodic curve are superimposed in Fig. 5.13 in which the cathodic curve intersects the anodic curve in the passive region of the alloy resulting in a low corrosion rate associated with the protective passive film formed on the metal surface. In this case, the oxygen reaction is completely responsible for establishing $E_{corr}$ and $i_{corr}$ since this potential is greater than the potential at which the hydrogen-ion and water-reduction reactions are thermodynamically possible.

Again, emphasis is placed on the fact that experimental potentiodynamic scans measure only the net current densities (i.e., the difference between the sum anodic and sum cathodic curves). The net curves

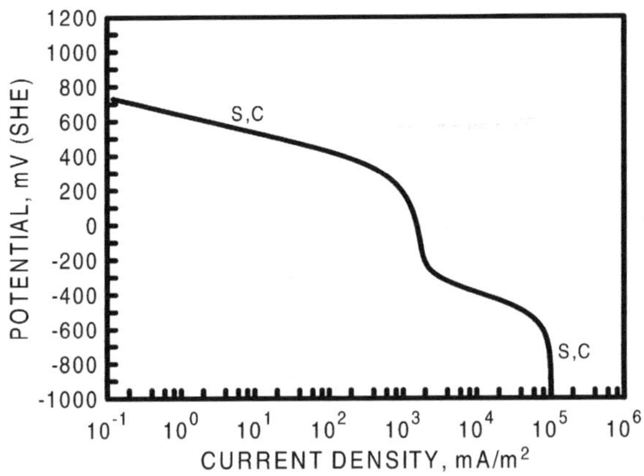


**Fig. 5.12** Sum (SC) of cathodic oxygen, hydrogen, and water polarization curves of Fig. 5.11. Oxygen curve dominates above −300 mV (SHE) and hydrogen curve below −300 mV (SHE). Water reduction makes negligible contribution to the current density. pH = 1. $P_{O_2}$ = 0.2 atm

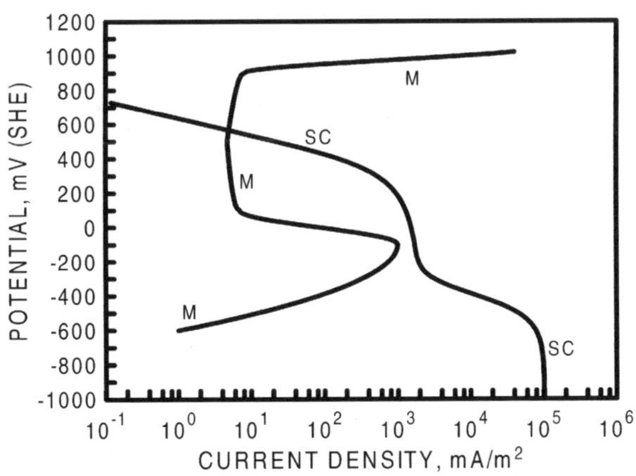


**Fig. 5.13** Relative positions of anodic metal polarization curve, M, and sum cathodic curve, SC, for cathodic oxygen and hydrogen-ion polarization. pH = 1. $P_{O_2}$ = 0.2 atm

(N) are shown relative to the individual curves in Fig. 5.14 and as the curves that would be measured in Fig. 5.15. In this case, the lower portion of the anodic polarization curve is not apparent in the net cathodic curve with the consequence that the shape of the anodic curve in this region cannot be determined under these conditions. A slight decrease in the current density of the cathodic curve near –100 mV (SHE) is due to the underlying active current density peak in the anodic metal curve. If conditions (e.g., decreased oxygen concentration) allowed these curves to become closer, the deviation would be greater, and if they touched, the net curve would become zero, and the experimental curve would be

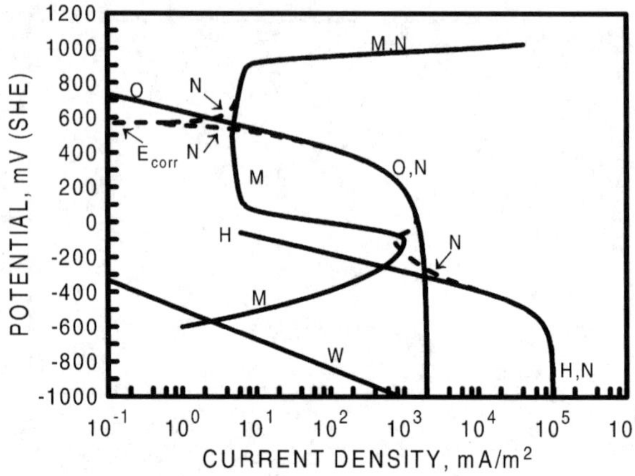


**Fig. 5.14** Net polarization curves, N, associated with the individual anodic and cathodic polarization curves. pH = 1. $P_{O_2}$ = 0.2 atm. Curve M, anodic polarization curve for metal; curve H, cathodic polarization curve for $H^+$; curve W, cathodic polarization curve for $H_2O$; curve O, cathodic polarization curve for $O_2$; curves N, net anodic and cathodic polarization curves.

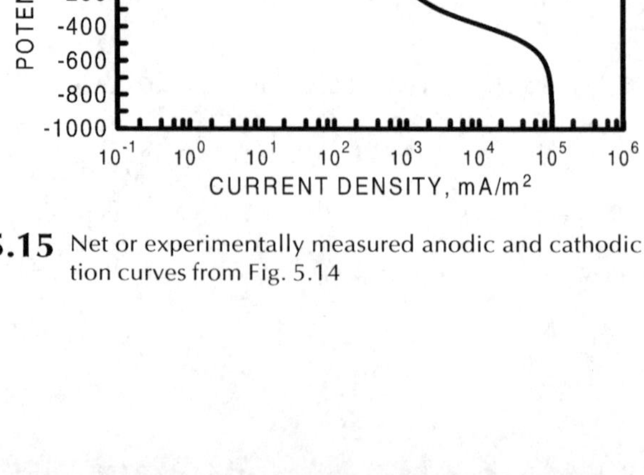


**Fig. 5.15** Net or experimentally measured anodic and cathodic polarization curves from Fig. 5.14

difficult to interpret without some understanding of the synthesis of the net curve from its components.

If the sum cathodic curve for the cathodic reactants passes through the potential region of the anodic (active) current peak, as shown in Fig. 5.16, three intersections occur. The higher-potential intersection is in the passive region of the anodic curve, and the lower-potential intersection is in the active corrosion region. It can be shown that the intermediate intersection is a condition of instability and therefore does not correspond to steady-state corrosion. In fact, only the lower intersection corresponds to the real steady state. If conditions initially exist with a single intersection as in Fig. 5.14 and 5.15, but subsequent changes in environment result in the condition of Fig. 5.16, the passive state may continue to be maintained by the cathodic reaction. However, in time, and in particular if the passive film is damaged, the film will eventually be stripped from the surface, corrosion will shift from passive to active, and the corrosion potential will exhibit a large decrease. This change provides an explanation for the rapid increase in corrosion rate of iron in contact with concentrated nitric acid when the acid is diluted. In other cases, as discussed later in the chapter on localized corrosion (Chapter 7), a slow decrease in the conditions of Fig. 5.15 to those in Fig. 5.16 may occur with pitting, resulting in failure by this mode of corrosion. The net curves are shown along with the sum cathodic and anodic curves in Fig. 5.17. The net curves, representative of experimental measurements, are shown in Fig. 5.18. It is evident that these net curves are still more complex, and their interpretation requires knowledge of how they can be synthesized from the component curves.

For metals such as titanium and chromium, the active peak in the anodic polarization curve may occur below the half-cell potential for the

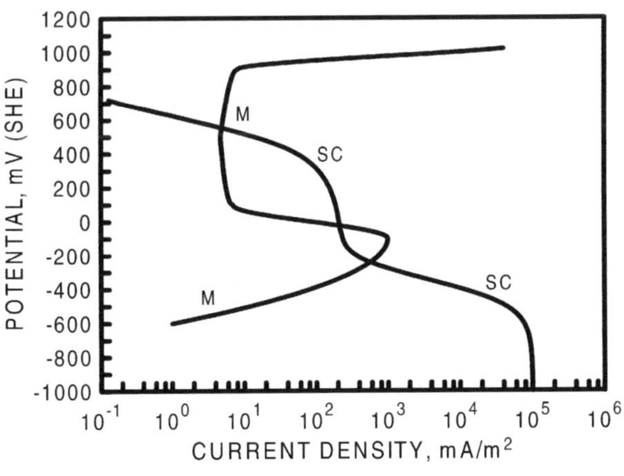


**Fig. 5.16** Effect of reducing dissolved oxygen concentration such that the sum cathodic curve, SC, intersects the anodic polarization curve for the metal, M, at three positions. pH = 1. $P_{O_2} \approx 0.05$ atm

hydrogen reaction. It is then possible for the cathodic polarization curve for the reduction of hydrogen ions in acid solution to intersect the metal anodic curve in both the passive and active potential ranges. In this case, a cathodic "peak" just above the active anodic peak can occur similar to that just described in relation to Fig. 5.17 and 5.18 (where dissolved oxygen was responsible for the cathodic peak). This is shown in Fig. 5.19 for chromium in hydrogen-saturated (deaerated) 1 N $H_2SO_4$ (Ref 9). During the upscan from $E_{corr}$, a single anodic maximum is observed followed by a cathodic "peak" in a potential range where the hydrogen-ion reduction current density exceeds the passive current density of the chromium. At slightly higher potentials, the hydrogen-ion

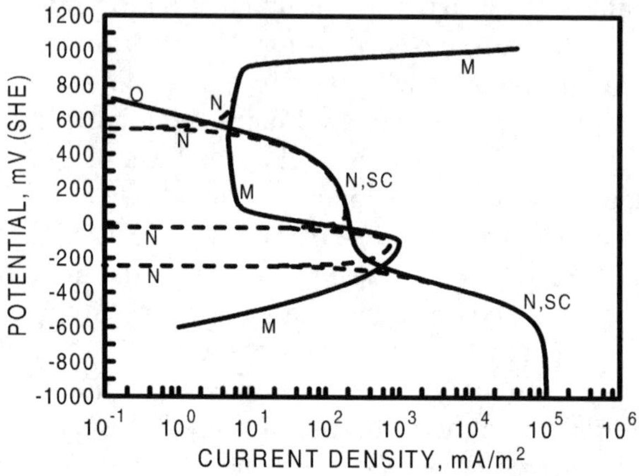


**Fig. 5.17** Net polarization curves, N, resulting from the metal anodic curve, M, and the sum cathodic curve, SC, for the oxygen-reduction and hydrogen-ion-reduction curves. Curves M and SC are from Fig. 5.16. pH = 1. $P_{O_2} \approx 0.05$ atm

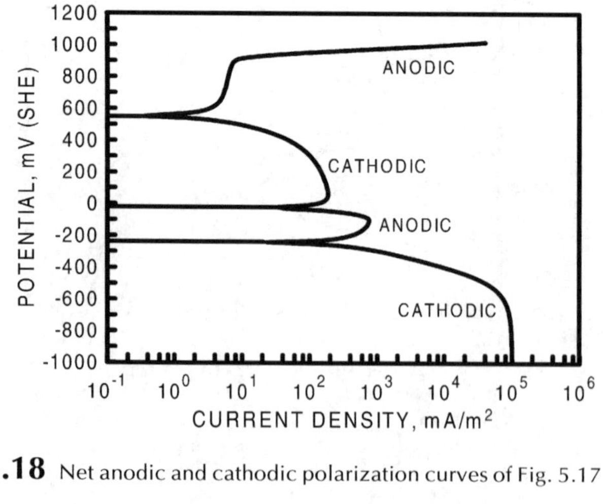


**Fig. 5.18** Net anodic and cathodic polarization curves of Fig. 5.17

reduction no longer occurs, and the anodic curve for the chromium in the passive range is determined.

In summary of this section, two points are emphasized. First, a potentially corrosive environment will establish the types of cathodic reactions that may occur, and also the environment may influence the characteristics of the anodic polarization curve of the exposed metal or alloy. Generally, a metal or alloy is selected such that the environment will place the material in the passive state with a low corrosion rate. Intersection of cathodic and anodic curves in the active potential range is avoided. Corrosion-resistant alloys are designed to meet these criteria. Second, it is important to recognize that polarization curves for various metal/environment conditions that are consulted as guides for materials selection are experimental curves representing the sum of coexisting anodic and cathodic curves. It is frequently necessary to estimate the positions of the component polarization curves responsible for the experimental curve when judging how the corrosion behavior may change with modifications in either the environment or alloy. This cannot be done without knowledge of how combinations of individual cathodic and anodic current densities sum to account for an experimental curve as discussed in this section.

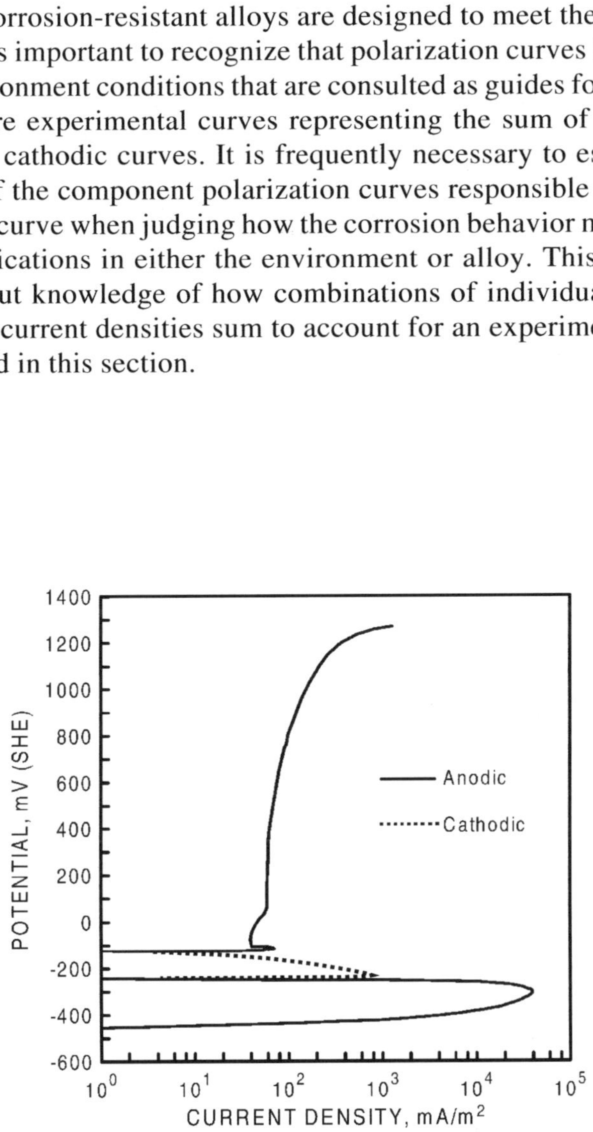


**Fig. 5.19** Potentiostatic polarization curve for pure chromium in hydrogen-saturated (deaerated) 1 N $H_2SO_4$ at 25 °C. Dashed section is a cathodic "peak" where the hydrogen-ion reduction dominates over the passive chromium oxidation. Redrawn from Ref 9

# Anodic Polarization of Several Active-Passive Metals

## *Anodic Polarization of Iron*

Complete or partial anodic polarization curves for iron (Ref 5, 10–12), nickel (Ref 5, 13), chromium (Ref 12, 13), titanium (Ref 14), and molybdenum (Ref 11) are shown in Fig. 5.20. The curves are representative of the metals in 1 N $H_2SO_4$ (pH = 0.56), and since they are experimental curves, they start at the corrosion potential. Hence, a distinct linear Tafel region extending to the equilibrium potential and exchange current density is not always shown. It is emphasized that these curves characterize the behavior in the indicated environment. Their exact position, as determined experimentally, is sensitive to the specific composition of the environment and the experimental technique used, particularly with respect to exclusion of oxygen from the cell and to variables such as the potential scan rate. Although the curves were derived from potentiodynamic polarization measurements, their practical significance relates to the values of the parameters $E_{pp}$ (the passivating potential), $i_{crit}$ (the critical current density), and $i_p$ (the passive current density). It is evident that the curves for iron and nickel are similar, with the latter having a lower $i_{crit}$. The passivating potentials for chromium and titanium are significantly lower, and the current density in the passive state is very low, about 1 mA/m$^2$. Since all of the parameters for titanium characterizing its polarization behavior have these smaller values, titanium is more easily placed into the passive condition than iron, nickel, or chromium.

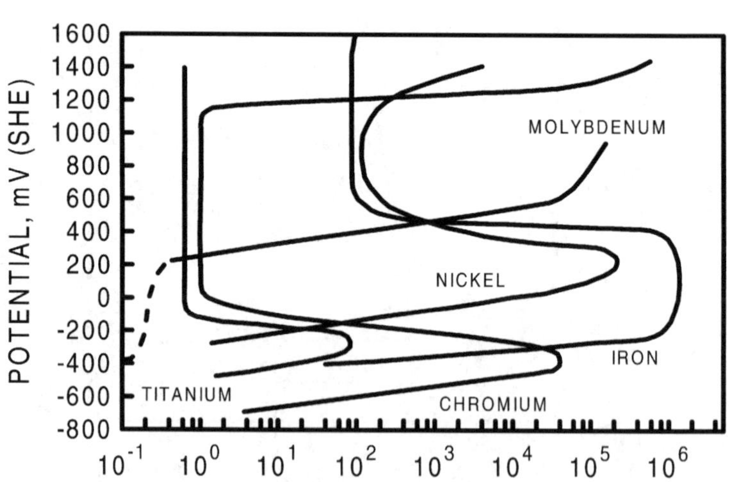


**Fig. 5.20** Representative anodic polarization curves for indicated pure metals in 1 N $H_2SO_4$, pH = 0.56. Linear sections at lower potentials are representative of Tafel behavior. Redrawn from Ref 5, 10–14

Molybdenum exhibits unusual polarization behavior. The initial portion of the curve, shown dashed in Fig. 5.20, is very difficult to determine experimentally because it occurs at very low current densities indicating that the passive state is very rapidly established by traces of dissolved oxygen or by very low concentrations of other cathodic reactants. In fact, many of the published curves show only the transpassive range over which the current density rapidly increases. The implication is that as long as the potential is below 200 mV (SHE), the corrosion rate of molybdenum would be very low and this is observed.

The passive films on these metals are either the conventional oxides or species such as FeOOH, which are related chemically. On nickel the film is related to NiO, on chromium to $Cr_2O_3$, and on titanium to $TiO_2$.

## *Effect of Crystal Lattice Orientation*

Another variable that can influence the shape and position of the anodic polarization curve is the crystal plane and hence atom arrangement that is exposed to the environment. This effect is illustrated in Fig. 5.21, which shows the polarization curves for pure nickel cut to expose (100), (110), and (111) planes to 1 N $H_2SO_4$ (Ref 15). The observation that the current density is crystal orientation dependent indicates that the passive film structure and/or thickness is sensitive to the arrangement of atoms at the surface. Also, the polarization curve of a polycrystalline metal is a complex sum of the curves corresponding with the distribution of orientations of the grains. This effect of crystal orientation is partially responsible for revealing the individual grains at the surface of a metal when etched, particularly for metallographic examination.

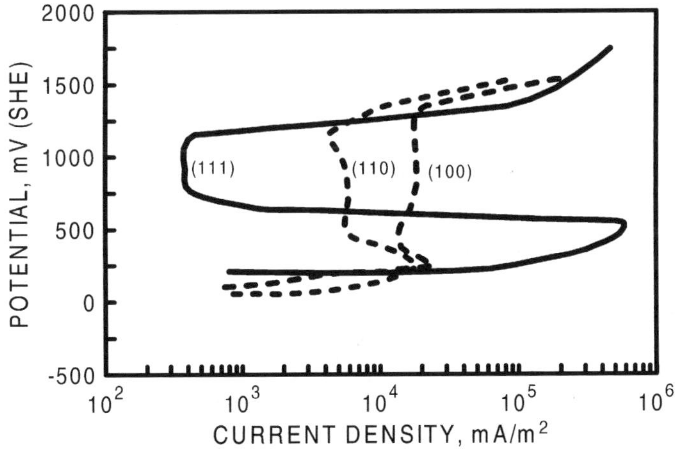


**Fig. 5.21** Anodic polarization curves determined potentiostatically for three low index faces cut from a nickel monocrystal grown parallel to (110), 1 N $H_2SO_4$ at 22–23 °C. Redrawn from Ref 15

## *Anodic Polarization of Aluminum*

The polarization of bare aluminum is essentially impossible to determine because of the rapid formation of an oxide film on contact with air and the persistence of the film in aqueous solutions. This high reactivity relates to the very negative aluminum half-cell potential of –1662 mV (SHE). At pH > 9, the oxide film dissolves, and the bare metal corrodes at progressively greater rates as the pH increases. At pH < 4, the oxide film becomes thermodynamically unstable, but the dissolution rate is usually very small. As a consequence, polarization curves in acid solutions generally represent the polarization behavior on a preexisting passive oxide film. Curves of the form shown in Fig. 5.22 are obtained in 1 N $H_2SO_4$. As discussed in greater detail in Chapter 7, the oxide film in contact with an aqueous environment is complex in physical and chemical structure. The initial air-formed film is $Al_2O_3$, varying from crystalline to amorphous depending on conditions of formation, and contains a distribution of flaws (Ref 16). On contact with water, the film becomes hydrated and changes properties with time which influences the form of the measured polarization curve (Ref 16–18). The oxide film grows by diffusion of aluminum ions from the metal through the film to the oxide/solution interface. The cathodic reactions are reduction of oxygen and hydrogen, with the latter usually predominant. Because of the very small electronic conduction of the passive film, the reduction reactions are essentially inhibited by the passive film (Ref 17). Flaws in the passive film are frequently related to second-phase intermetallic particles at the surface of the substrate aluminum. Since the passive film is less protective over these particles, the rates for both the anodic and cathodic reactions are higher at the flaws, and the observed polarization curve may be associated largely with these localized regions (Ref 19).

The $E_{corr}$ near –600 mV (SHE) in Fig. 5.22 results from the nearly constant current density (passive) anodic curve and a cathodic diffusion-

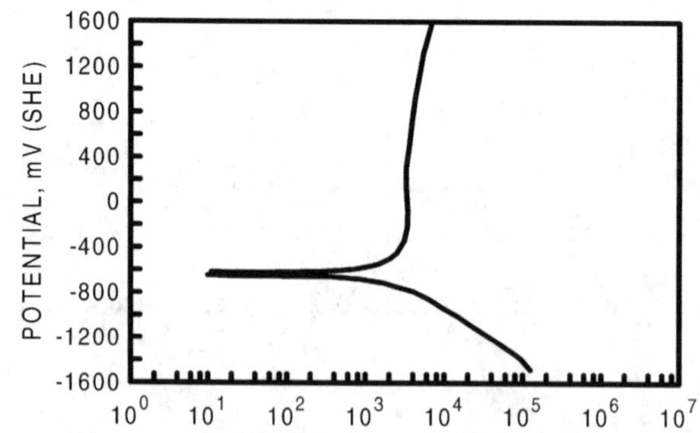


**Fig. 5.22** Polarization curve for aluminum in deaerated 1 N $H_2SO_4$

controlled, hydrogen-reduction curve. There is evidence that dissolved oxygen has a small but indirect effect on $E_{corr}$, changing it to lower potentials rather than higher as is usually observed on increased aeration (Ref 18). It has been proposed that dissolved oxygen influences the structure of the oxide film such that the diffusion rate of hydrogen ions to the metal interface is decreased. Thus, the polarization of the hydrogen reduction reaction is depressed over that observed for the deaerated environment and $E_{corr}$ is lowered.

It should be noted that there is no evidence of a peak or local maximum in the anodic curve related to a transition from the active to passive state. This is a result of the preexisting air-formed oxide film that effectively prepassivates the aluminum. Only under very restricted conditions is it possible to produce a sufficiently active surface to allow measurement of the active-to-passive transition (Ref 18).

## *Anodic Polarization of Copper*

The anodic polarization curve for copper in 1 N $H_2SO_4$ is shown in Fig. 5.23 (Ref 20). In contrast to aluminum, copper is thermodynamically stable in oxygen-free acid solutions, and the corrosion rate in highly deaerated (nitrogen-sparged) acid environments is essentially nil. The conventional polarization curve of an active-passive alloy showing a current density maximum is not observed. Rather, the current density initially increases rapidly from near the half-cell potential for copper in contact with a solution very dilute in copper ions (160 mV (SHE) at $a_{Cu^{2+}} = 10^{-6}$). This is followed by a rapid transition to high current densities essentially independent of potential indicating a diffusion limiting mechanism. This limit is associated with the very rapid dissolution of the copper and probable precipitation of copper sulfate.

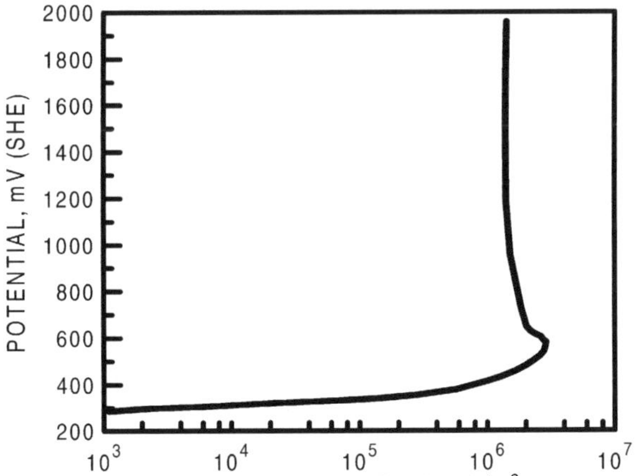


**Fig. 5.23** Potentiostatic anodic polarization curve for copper in deaerated 1 N $H_2SO_4$ at 25 °C. Redrawn from Ref 20

At potentials near 1800 mV (SHE), an increase in current density may be observed due to the oxidation of water to form oxygen gas.

## Anodic Polarization of Several Active-Passive Alloy Systems

The anodic polarization of a given alloy base metal such as iron or nickel is sensitive to alloying element additions and to heat treatments if the latter influences the homogeneity of solid solutions or the kinds and distribution of phases in the alloy. The effect of chromium in iron or nickel is to decrease both $E_{pp}$ and $i_{crit}$ and hence to enhance the ease of placing the alloy in the passive state. The addition of chromium to iron is the basis for a large number of alloys broadly called stainless steels, and chromium additions to nickel lead to a series of alloys with important corrosion-resistant properties.

### *Anodic Polarization Curves for Iron-Chromium Alloys*

Polarization curves for iron, chromium, and alloys with 1, 6, 10, and 14 weight percent (wt%) chromium in iron are shown in Fig. 5.24; the environment is 1 N $H_2SO_4$ at 25 °C (Ref 21). Iron and chromium are body-centered-cubic metals, and the alloys are solid solutions having this structure. The passivation potential ($E_{pp}$), the active peak current density ($i_{crit}$), and the passive state current density ($i_p$) are decreased

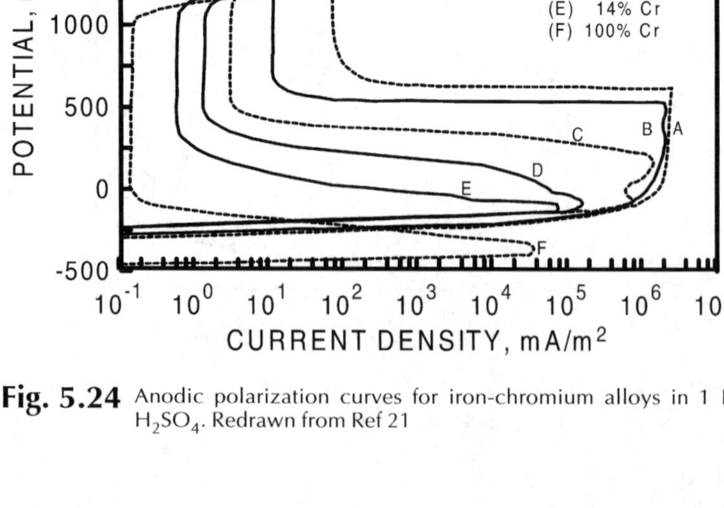


**Fig. 5.24** Anodic polarization curves for iron-chromium alloys in 1 N $H_2SO_4$. Redrawn from Ref 21

significantly as the chromium concentration is increased up to 10 to 14 wt% Cr. The rate of the effect of chromium on these parameters characterizing the polarization curve decreases at higher chromium concentrations (Ref 22). The passive films on these alloys are complicated in terms of the crystalline structure, chromium concentration, and thickness; these features also depend on the time that the alloys are held in the passive potential range (Ref 21). The passive oxide films are related to a spinel structure with the general formula, $FeFe_{(2-x)}Cr_xO_4$, in which the chromium concentration varies within the film. At low chromium concentrations, the crystalline structure is essentially that of $\gamma$-$Fe_2O_3$, about 3 nm thick, and with a crystal lattice orientation relationship to the metal substrate at the metal/oxide interface (Ref 23). With increasing chromium concentration, the films are less crystalline, becoming completely amorphous at 19 to 24 wt% Cr in the alloy and about 2 nm thick. At 18 wt% Cr, about 70 wt% of the metal ions in the oxide film are chromium (Ref 21).

## *Anodic Polarization of Iron-Chromium-Molybdenum Alloys*

Alloys containing 10 to 25 wt% chromium span the compositions of the commercial ferritic stainless steels. The effect of chromium in decreasing $E_{pp}$ and $i_{crit}$, and in changing the properties of the passive film, are important factors in relating alloy composition to corrosion resistance when maintenance of a passive state is critical to satisfactory performance in a particular environment. The corrosion resistance of these ferritic alloys is improved by additions of 0 to 6 wt% molybdenum. The major effect of the molybdenum on the polarization curve is to significantly decrease the active peak current density, $i_{crit}$. Polarization curves in the vicinity of the active peak of an Fe-18 wt% Cr alloy with additions of 0, 2, 4, and 6 wt% Mo in 1 N $H_2SO_4$ are shown in Fig. 5.25 (Ref 24). It is evident that the active peak is decreased progressively to about $10^3$ mA/m$^2$, which indicates that the molybdenum has enhanced the ability to establish the passivated state. Again, the corrosion resistance of the alloy, particularly with respect to localized corrosion, will depend on the stability of the passive film, once formed. The significance of increasing the chromium concentration above 12 wt% Cr and adding 0 to 6 wt% Mo on improved resistance to localized corrosion of stainless steels is discussed in greater detail in Chapter 7.

## *Anodic Polarization of Iron-Chromium-Nickel Alloys*

Nickel (face-centered cubic) is a major addition to iron-chromium alloys and with 8 to 22 wt% Ni forms the basis of the austenitic stainless steels. The major influence of nickel is to permit the formation of face-centered-cubic solid solution alloys, which generally have more favorable metallurgical properties than the body-centered-cubic, iron-chromium alloys. The corrosion resistance of these alloys, however, is still due to the presence of chromium in the passive film. This influence is shown by the polarization curves in Fig. 5.26 where addi-

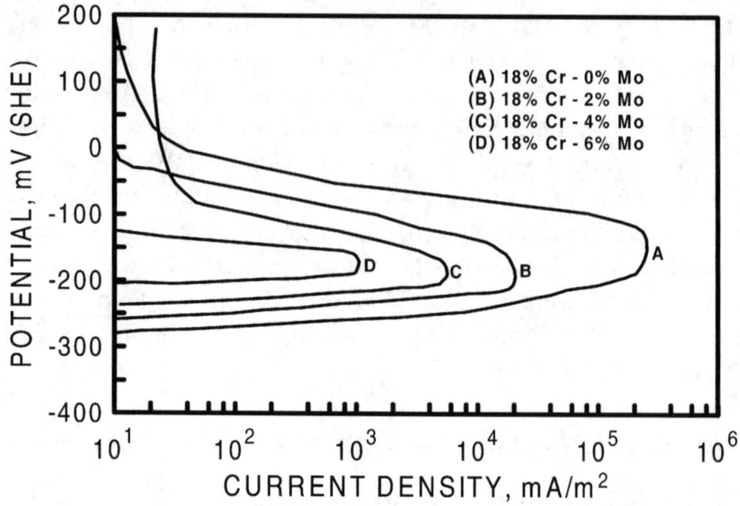


**Fig. 5.25** Polarization curves of Fe-18 wt% Cr alloys showing the shift in the active peak current density by 0–6 wt% Mo. Redrawn from Ref 24

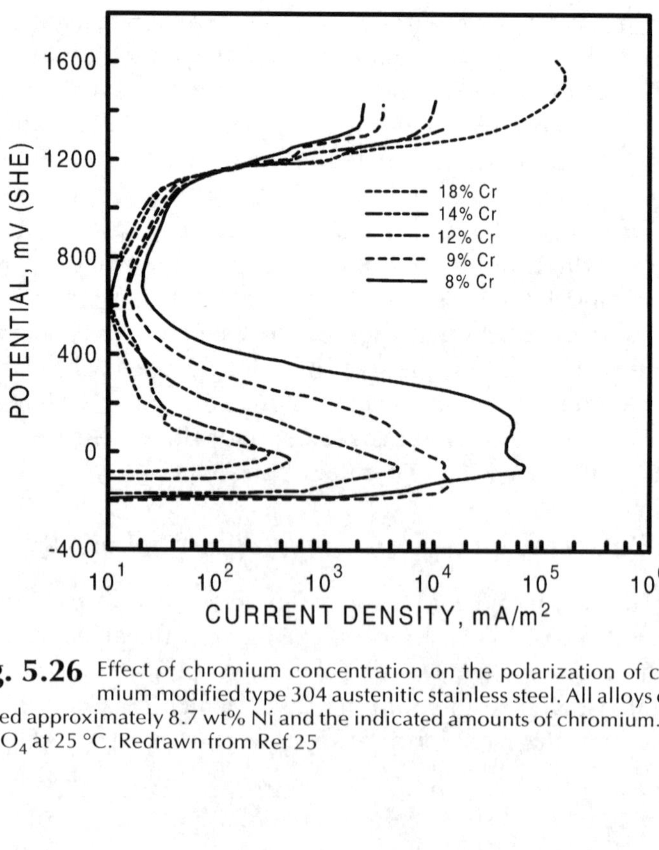


**Fig. 5.26** Effect of chromium concentration on the polarization of chromium modified type 304 austenitic stainless steel. All alloys contained approximately 8.7 wt% Ni and the indicated amounts of chromium. 1 N $H_2SO_4$ at 25 °C. Redrawn from Ref 25

tions of chromium to an Fe-8.7 wt% Ni-base alloy results in progressive decreases in $E_{pp}$, $i_{crit}$, and $i_p$ (Ref 25).

## *Anodic Polarization of Nickel-Chromium Alloys*

Polarization curves for nickel-rich nickel-chromium alloys in 1 N $H_2SO_4$ are shown in Fig. 5.27 and for chromium-rich alloys in Fig. 5.28

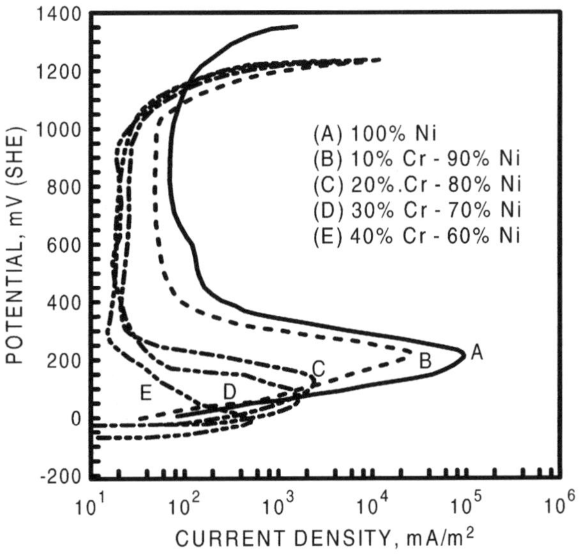


**Fig. 5.27** Anodic polarization curves for nickel-chromium alloys in 1 N $H_2SO_4$. Redrawn from Ref 13

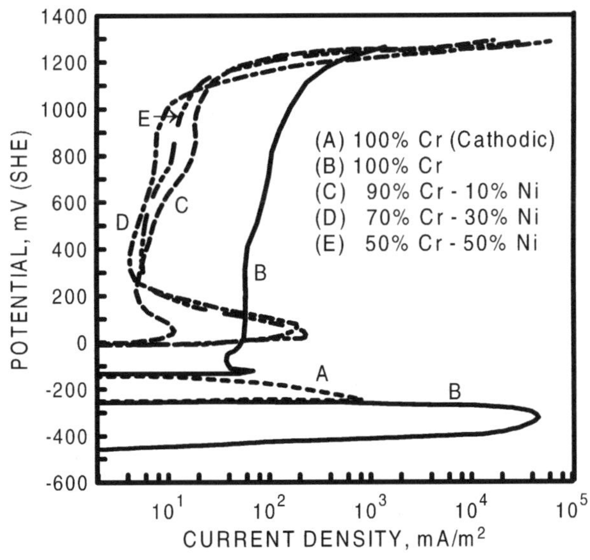


**Fig. 5.28** Anodic polarization curves for chromium-nickel alloys in 1 N $H_2SO_4$. Redrawn from Ref 13

(Ref 13). These alloys are face-centered-cubic solid solutions from 0 to approximately 40 wt% chromium and body-centered-cubic from approximately 90 to 100 wt% chromium. The intermediate alloys are two-phase structures. The progressive influence of chromium in nickel in decreasing $E_{pp}$, $i_{crit}$, and $i_p$ is evident, and, hence the higher chromium alloys are more easily passivated. An exception is that the polarization curve for pure chromium occurs at larger current densities than for the 90 wt% chromium alloy.

## *Anodic Polarization of Nickel-Molybdenum Alloys*

Nickel dissolves up to 35 wt% molybdenum forming a face-centered-cubic solid solution (rapid cooling is required for alloys with >20 wt% Mo). Polarization curves for a series of alloys of 0 to 22 wt% molybdenum are shown in Fig. 5.29 (Ref 26). These curves illustrate an alloying effect in which the passivating potential, $E_{pp}$, and the anodic-peak current density, $i_{crit}$, are relatively unchanged, and the passive current density, $i_p$, is significantly increased with increasing molybdenum content. The potentials in the active polarization potential range, however, are progressively raised as the molybdenum concentration is increased. As a consequence, environmental conditions (dissolved oxygen and ferric ions) that place the corrosion potential in the passive potential range will be associated with larger current densities and hence higher corrosion rates for the alloys than for the base metal, nickel. In contrast, it is shown later in this chapter that when these oxi-

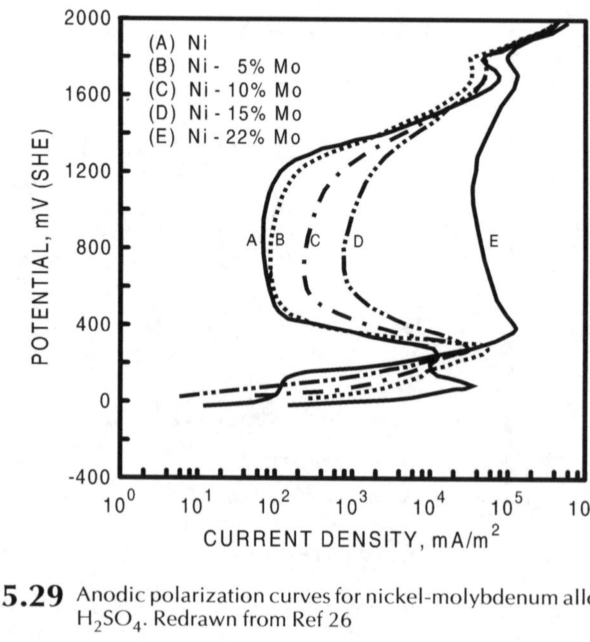


**Fig. 5.29** Anodic polarization curves for nickel-molybdenum alloys in 1 N $H_2SO_4$. Redrawn from Ref 26

dizing species are not present, the increased potential in the active potential range for the alloys is beneficial in decreasing the corrosion rate.

## Representative Polarization Behavior of Several Commercial Alloys

In the following section, polarization curves for several commercial alloys in different environments are presented along with discussions of the relationships between the curves and the corrosion behaviors of the alloys. As alloys are developed commercially and the range of their application expanded, publication of polarization curves in corrosion and manufacturer's literature becomes not only a basis for understanding the corrosion behavior of a specific environment/alloy combination but also a guide for understanding how reasonable changes in composition of both the alloy and its environment may change the corrosion response. Bases for making these approximations are illustrated in previous sections of this chapter and are used in the following with respect to commercial alloys.

Type 430 stainless steel (Fe, 16 to 18 wt% Cr, 0.12 wt% C maximum) is used as an ASTM standard material to certify the performance of potentiostats in accurately and reproducibly determining polarization curves (Ref 27). The environment is 1 N $H_2SO_4$ at 30 °C, and the scan rate is specified as 600 mV/h. To meet the standard, a measured polarization curve determined using the reference standard must fall within the band shown in Fig. 5.30. An advantage in using this alloy is the large

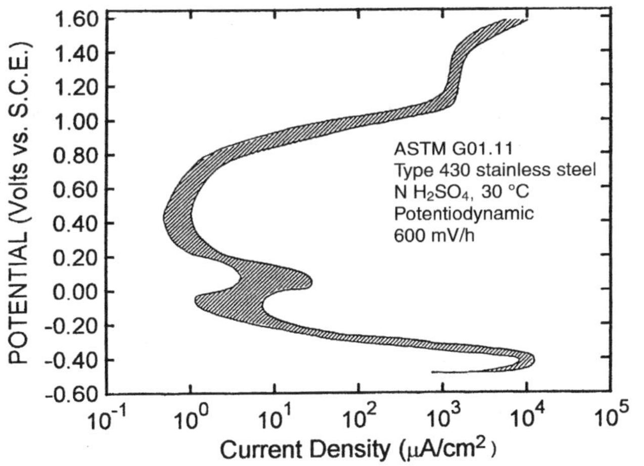


**Fig. 5.30** Standard ASTM potentiodynamic anodic polarization plot for certification of potentiostat performance. Type 430 stainless steel in 1 N $H_2SO_4$ at 30 °C. Scan rate of 600 mV/h. Test curve is to lie within the shaded region. Redrawn from Ref 27

difference in the current density at the end of the active dissolution range (approximately $10^4$ μA/cm$^2$ at –0.40 V (SCE)) and the low current density in the passive potential range (approximately 1 μA/cm$^2$ at 0.40 V (SCE)). The open circuit corrosion potential, $E_{corr}$, is approximately –0.50 V (SCE), and the increase in current density above 0.80 V (SCE) is associated with a change to the transpassive region of the polarization curve. It should be noted that environments changing the corrosion potential from near 0.40 V (SCE) (the passive range) to near –0.40 V (SCE) (the anodic peak current density) would correspond to an increase in corrosion rate by a factor of about $10^4$. Hence, the corrosion rate of this alloy can be very sensitive to environmental conditions.

The effect of pH on the polarization of iron is shown in Fig. 5.6. The effect of pH on the polarization of type 304 stainless steel (nominally 18 to 20 wt% Cr, 8 to 10.5 wt% Ni, 0.08 wt% C maximum) in environments based on 1 M $Na_2SO_4$ with additions of $H_2SO_4$ and NaOH to control the pH is shown in Fig. 5.31 (Ref 28). The influence of chromium and nickel in moving the anodic polarization curve of iron to lower current densities persists over the indicated pH range with the corrosion rates being very low for pH > 4.0.

The effects of acid concentration and temperature on the anodic polarization of a commercial nickel-base alloy (Hastelloy C, nominal composition: 54 wt% Ni, 2.5 wt% Co, 15.5 wt% Cr, 16 wt% Mo, 4 wt% W, 5.5 wt% Fe, 0.06 wt% C maximum) are shown in Fig. 5.32 (Ref 29). Qualitative conclusions from these curves indicate that the changes in corrosion rate on increasing the acid concentration from 1 to 10 N should be relatively small but that the effect of increasing the tempera-

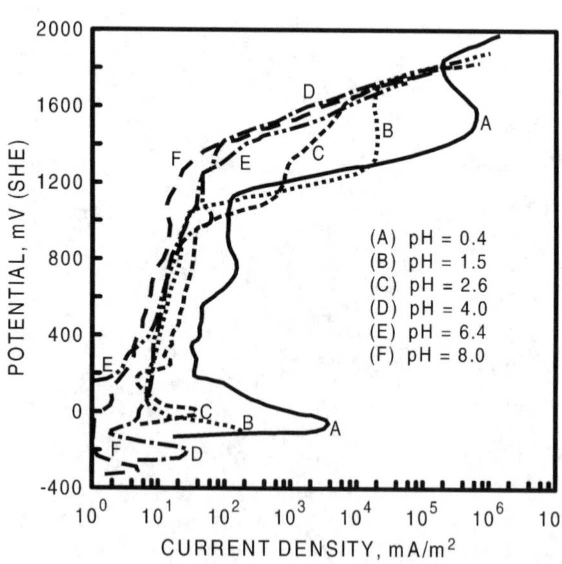


**Fig. 5.31** Changes in anodic polarization curves with pH for type 304 stainless steel in 1 M $Na_2SO_4$ solutions. Redrawn from Ref 28

ture from room temperature to 90 °C should be significant. These orders of changes are substantiated by weight-loss corrosion-rate measurements.

Figure 5.33 shows the very large difference in polarization behavior of three nickel-base alloys in 1 N HCl (Ref 29). Environments maintaining the potential near 600 mV (SHE) will clearly maintain passivity on the Hastelloy F alloy (22.34 wt% Cr, 7.07 wt% Mo, 0.07 wt% C, 1.35 wt% Nb) with a corrosion rate corresponding to the passive current density of 60 mA/m$^2$ (≈75 μm/year or 3 mpy), whereas the corrosion rate for Hastelloy C is somewhat higher, and the corrosion rate for Hastelloy B (0.75 wt% Cr, 26.5 wt% Mo, 0.02 wt% C, 5.2 wt% Fe) is prohibitively

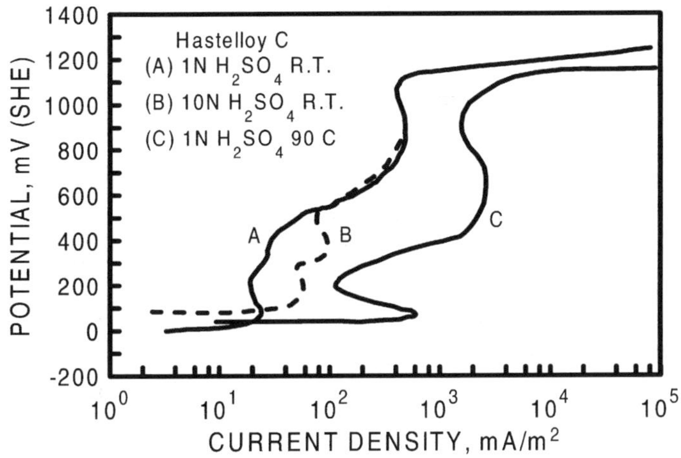


**Fig. 5.32** Anodic polarization curves for Hastelloy C (54 Ni, 2.5 Co, 15.5 Cr, 16 Mo, 5.5 Fe, and 0.06 C max, wt%), in the indicated environments. R.T., room temperature. Redrawn from Ref 29

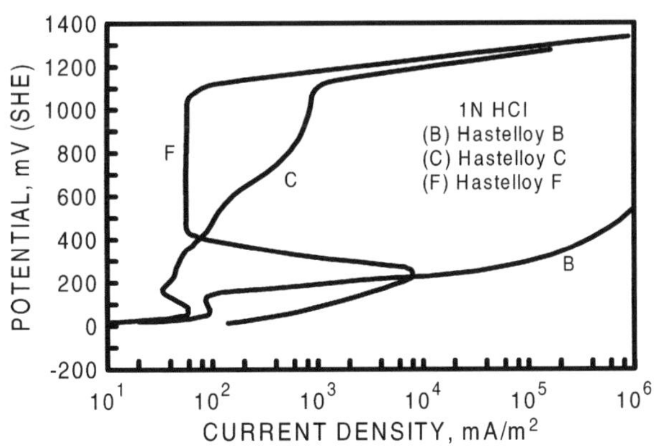


**Fig. 5.33** Anodic polarization curves for three nickel-base alloys in 1 N HCl at 25 °C. Alloy F, 22.34 Cr, 7.07 Mo, 0.07 C (wt%); alloy C, 54 Ni, 2.5 Co, 15.5 Cr, 16 Mo, 4 W, 5.4 Fe, 0.08 C (wt%); alloy B, 26.5 Mo, 0.75 Cr, 5.2 Fe, 0.02 C (wt%). Redrawn from Ref 29

high. In contrast, an environment maintaining a potential of 100 mV (SHE) would result in lower corrosion rates for Hastelloys C and B relative to Hastelloy F. Furthermore, Hastelloy F would have changed from corroding at a low rate in the passive state to corroding in the active potential range.

## Additional Examples of the Influence of Environmental Variables on Anodic Polarization Behavior

Reference has been made to the observation that both anionic and cationic species in the environment can influence the anodic polarization of active-passive types of metals and alloys. Specific examples have related to the effect of pH as it influences the stability and potential range of formation of oxide and related corrosion product films. The effect of pH, however, cannot be treated, even with single chemical species, independent of the accompanying anions. For example, chloride, sulfate, phosphate, and nitrate ions accompanying acids based on these ionic species will influence both the kinetics and thermodynamics of metal dissolution in addition to the effect of pH. Major effects may result if the anion either enhances or prevents formation of protective corrosion product films, or if an anion, both thermodynamically and kinetically, is an effective oxidizing species (easily reduced), then large changes in the measured anodic polarization curve will be observed.

### *Effects of Sulfide and Thiocyanate Ions on Polarization of Type 304 Stainless Steel*

The effects of sulfide, $S^{=}$, and thiocyanate, $SCN^-$, ions on the anodic polarization of type 304 stainless steel in 1 N $H_2SO_4$ are shown in Fig. 5.34 (Ref 30). It is evident that the major influence of these ions is to increase the active peak current density, $i_{crit}$, with relatively smaller effects in the passive potential range. Thus, the stainless steel is more difficult to passivate in the presence of these ions, or a pre-existing state of passivity established in the absence of the ions may be destroyed if they become present. A consequence of this influence of sulfide ions is initiation of localized corrosion in stainless steels at sites of pre-existing manganese sulfide inclusions (Ref 31–33). In acid environments, these inclusions are dissolved leading to local cavities high in sulfide-ion concentration. The formation of a protective passive film within the cavity is prevented, and the passive film in the vicinity of the initial inclusion may be destroyed. The local corrosion rate becomes much higher than exists over the otherwise passivated surface, and a pitting type corrosion results.

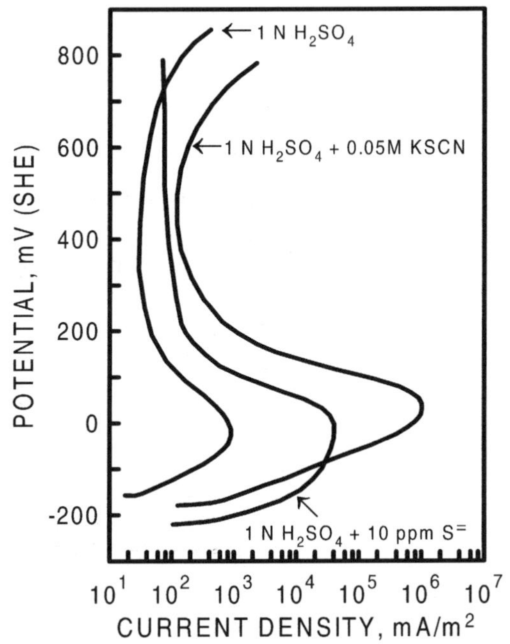


**Fig. 5.34** Effect of 0.05 M KSCN and 10 ppm $S^{=}$ in 1 N $H_2SO_4$ on the polarization of American Iron and Steel Institute (AISI) 304 stainless steel. Redrawn from Ref 30

## *Effects of Chloride Ions*

Chloride ions have a significant effect on the polarization and, hence, corrosion behavior of many metals and alloys over a wide range of pH and independent of other ionic species. Figure 5.35 is a schematic representation of the polarization curve of an active-passive alloy such as type 304 stainless steel in deaerated 1 N $H_2SO_4$ in the absence of chloride ions. An upscan potential traverse from the cathodic potential range passes through $E_{corr}$, then through the active peak into the passive region. Transition to the transpassive region occurs near 1200 mV (SHE). In the presence of chloride ions, the passive film breaks down at a specific potential identified as $E_{b,pit}$ (i.e., the breakdown potential for pitting corrosion) at which there is a rapid increase in current density (small-dashed curve). If the chloride concentration is sufficiently high to completely prevent passivation, the polarization curve follows the large-dashed curve, and very high current densities are observed with increasing potential. The interpretation of the increase in current density at $E_{b,pit}$ is that of a composite surface consisting of passive film with a low current density and pits, essentially free of protective film, corroding at the high current density given by the large-dashed curve at the pitting potential. With time, the current density increases as a larger fraction of the surface becomes pitted. For a given material, the potential at which pitting occurs is lower for a higher chloride ion concentration. Progressive local breakdown of the passive film will result in the entire surface approaching a condition of rapid active dissolution. An example of the effect of a range of chloride ion concentrations in near-neutral water on the polarization behavior of type 304 stainless

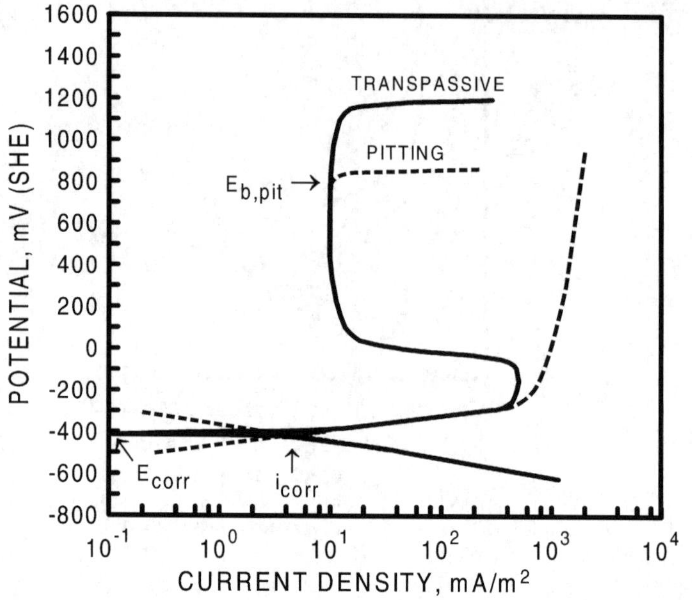


**Fig. 5.35** Schematic polarization curve for an active-passive alloy having susceptibility to localized corrosion (pitting) due to chloride ions. Pitting initiates at $E_{b,pit}$. Small-dashed section is observed when chloride ion concentration initiates penetration of the passive film.

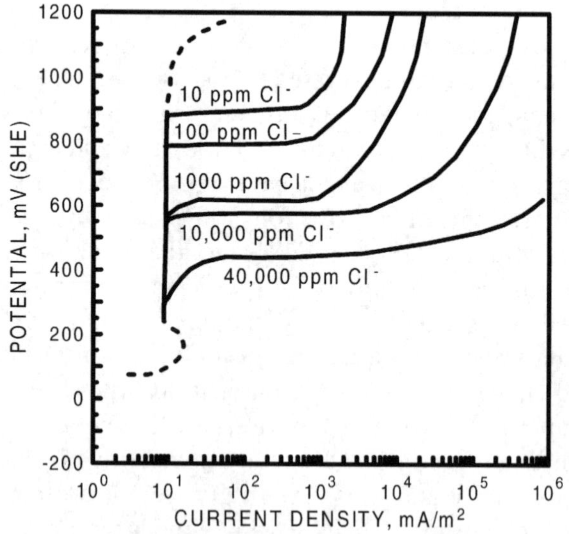


**Fig. 5.36** Effect of chloride-ion concentration in near-neutral water on anodic polarization of type 304 stainless steel. Dashed lines added to show approximate locations of transpassive and anodic-peak sections of the curve. Based on Ref 34

steel is shown in Fig. 5.36 (Ref 34). The uppermost dashed curve corresponds to the transition into the transpassive potential region where, as previously described, higher valent metal ions in solution are more stable than the passive film. Even 10 ppm chloride ion causes rupture of the passive film at 300 mV below the transpassive potential. Progressively increasing the chloride ion concentration has the effect shown.

Figures 5.37 and 5.38 show the effects of adding 1 N NaCl to the 1 N $H_2SO_4$ environment of the same set of alloys in Fig. 5.27 and 5.28 (Ref

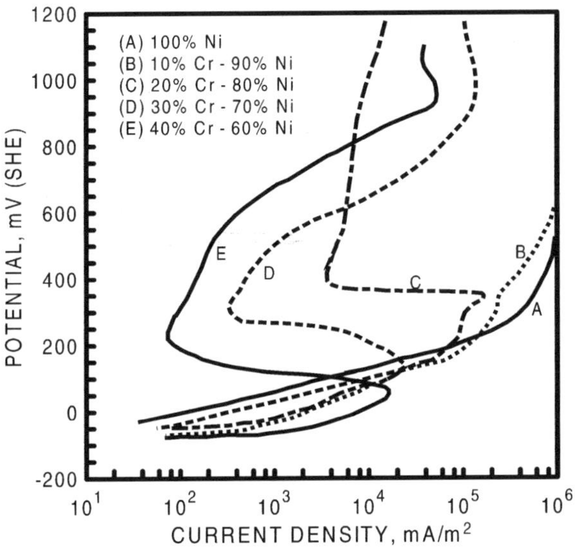


**Fig. 5.37** Anodic polarization of nickel-chromium binary alloys in 1 N $H_2SO_4$ + 1 N NaCl. Redrawn from Ref 13

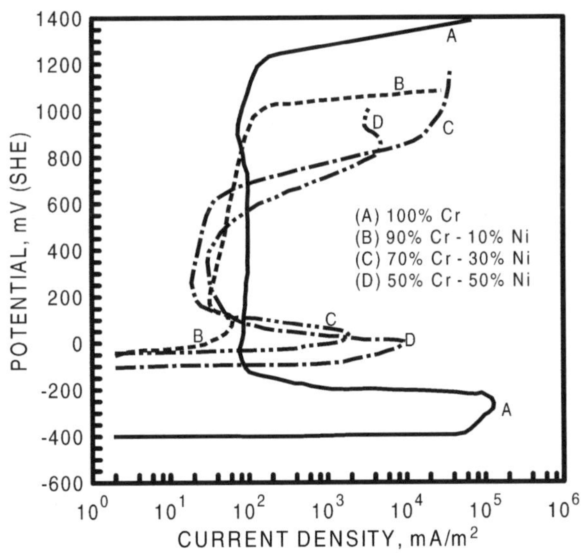


**Fig. 5.38** Anodic polarization of chromium-nickel binary alloys in 1 N $H_2SO_4$ + 1 N NaCl. Redrawn from Ref 13

13). The extents of the passive potential regions have been reduced for all materials except pure chromium, and the curves for 90 and 100 wt% nickel indicate that an active-to-passive state transition no longer occurs. The magnitude of the influence of the chloride ions is emphasized by comparing the current densities for each alloy at 200 mV (SHE) with and without chloride ions present.

## Polarization of Admiralty Brass

An example of the very large influence that different species in the environment can have on anodic polarization is shown in Fig. 5.39 for the copper-base alloy, admiralty brass (nominal composition: 71 wt% Cu, 28 wt% Zn, 1 wt% Sn, 0.06 wt% As) (Ref 35). The polarization curves show roughly two types of response depending on the species in solution. The polarization curves determined in the presence of $HPO_4^{=}$, $B_4O_7^{=}$, $MoO_4^{=}$, $CrO_4^{=}$, and $WO_4^{=}$ are characteristic of a passive film present on the metal surface at initiation of an increasing potential scan from the corrosion potential. The approximately linear initial portions of the polarization curves for the other environments is characteristic of Tafel behavior and implies active corrosion over this potential range. There is a tendency toward passivation in the $Cl^-$, $ClO_3^-$, and $NO_2^-$ environments immediately followed in the latter two cases by rapidly increasing cur-

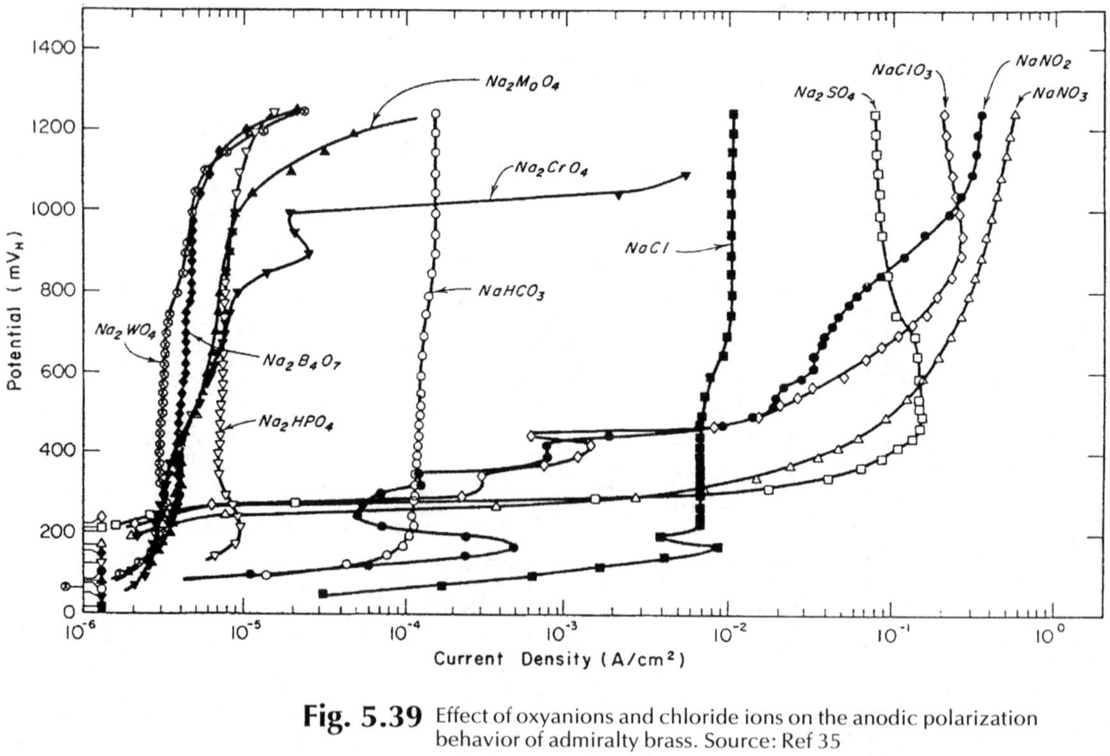


**Fig. 5.39** Effect of oxyanions and chloride ions on the anodic polarization behavior of admiralty brass. Source: Ref 35

rent density. There is a high limiting current density with no tendency for passivation in the $SO_4^=$ and $NO_3^-$ containing solutions. The wide range of positions of these polarization curves indicates potentially large differences in corrosion rates depending on the environment, not only depending on which of the species shown are present, but equally, if not more importantly, on the oxidizing species present (such as dissolved oxygen), which provide the cathodic reactant and contribute significantly in establishing the corrosion potential. It is evident from Fig. 5.39 that if an environment establishes a corrosion potential of 400 mV (SHE), corrosion rates extending from near $3 \times 10^{-6}$ A/cm$^2$ to $10^{-1}$ A/cm$^2$ could occur.

## *Effect of Temperature on the Polarization of Titanium*

Brief reference was made to the polarization curve for titanium in Fig. 5.20. The environment was 1 N $H_2SO_4$ at room temperature. Figure 5.40 shows the effect of temperature, 25 to 90 °C, on the polarization of titanium in 40% $H_2SO_4$ (Ref 36). The effect is to increase the active peak current density by a factor of about 100 with a much smaller effect in the passive potential range. These curves also emphasize that the passive potential range for titanium is very large starting near 0 mV (SHE). The passive film, $TiO_2$, is very protective, and because of its high ohmic resistivity, the passive range may extend to very high potentials. As is discussed in Chapter 7, this passive film can become unstable in the presence of chloride ions, and pitting can become a mode of corrosion failure.

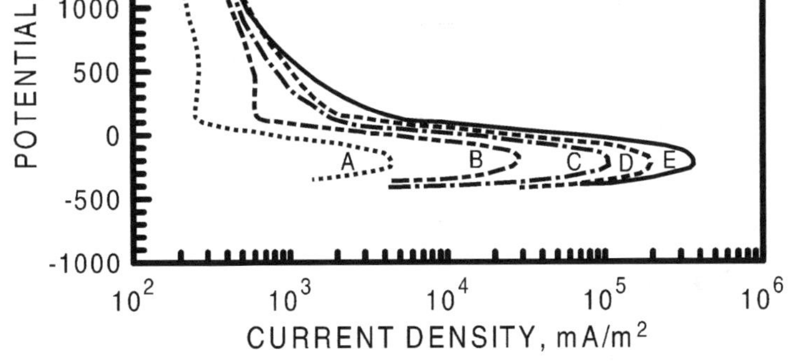


**Fig. 5.40** Anodic polarization curves for titanium in 40% $H_2SO_4$ solutions as a function of temperature. Redrawn from Ref 36

# Prediction of Corrosion Behavior of Active-Passive Type Metals and Alloys in Specific Environments

In principle, if the anodic polarization curve of a metal is known for a given environment and the cathodic reduction curves of reducible species in the environment are known, superposition of these curves should permit prediction of the corrosion behavior of the metal/environment system. This follows from the earlier discussion of the relationship of anodic and cathodic polarization curves to the net or experimentally determined curves. The obvious limitation of the procedure is the problem of establishing by experiment the individual anodic and cathodic polarization curves. The polarization curves for cathodic reactants such as dissolved oxygen, water, and hydrogen ions, which are inherent to all natural aqueous environments, as well as other species such as $Fe^{3+}$, $NO_2^-$, and $Cr_2O_7^=$, may be determined on inert surfaces such as platinum. The extent to which these curves are applicable when the reactions occur on active metal surfaces must be questioned. The exchange current densities will almost always be lower on the active metals and will differ depending on whether the surface of the metal contains a passive film. Theoretically, the limiting diffusion current density should be the same since the current limiting factor is the diffusion rate of cathodic reactant species through the boundary layer. This should be independent of the metal substrate. However, corrosion product films may limit the diffusion rate, of oxygen, for example, and establish a lower limiting current density. In these circumstances, it is necessary to use qualitative estimates of the positions of the polarization curves. It is more difficult to establish the individual anodic polarization curves since measurements cannot always be made independent of species that are inherent to the aqueous environment. Careful deaeration can obviously remove or, at least, significantly reduce the concentration of dissolved oxygen. However, the effect of pH on the anodic polarization curve cannot be determined independent of the coexistence of the hydrogen reduction reaction if the latter can occur in the potential range over which the measurements are being made.

Some examples of how these predictions are made and some of the limitations and precautions that must be recognized are presented in this section.

## *Corrosion of Iron at pH = 7 in Deaerated and Aerated Environments and with Nitrite Additions*

A representative anodic polarization curve for iron in a buffered solution of pH = 7 is shown in Fig. 5.41. Also shown are cathodic polarization curves for dissolved oxygen and nitrite ions on platinum under aerated conditions (dissolved oxygen = 8.5 ppm) and under deaerated conditions (dissolved oxygen ≈ 1 ppb). Approximations also have been made to illustrate the effect of the formation of a corrosion product layer (the $Fe_3O_4/Fe_2O_3$ rust layer on iron) in shifting the oxygen reduction curve to lower current densities. The polarization curve for nitrite ion reduction is related to the reaction:

$$NO_2^- + 8H^+ + 6e \rightarrow NH_4^+ + 2H_2O \qquad \text{(Eq 5.6)}$$

The exact reaction involving the nitrite ion is uncertain, but the curve is the experimental result of polarizing a platinum electrode in a deaerated solution containing 1000 ppm $NO_2^-$. The intersection between the appropriate cathodic curves and the anodic curve for iron is identified by pairs of values of $E_{corr}$ and $i_{corr}$. The corrosion rates in terms of $i_{corr}$ for the three environments are:

- *C1, aerated (clean surface):* 500 mA/m$^2$
- *C2, aerated (rust surface):* 60
- *C3, deaerated:* 5
- *C4, deaerated with nitrite ion:* 2
- *C5, aerated with nitrite ion:* 1.4

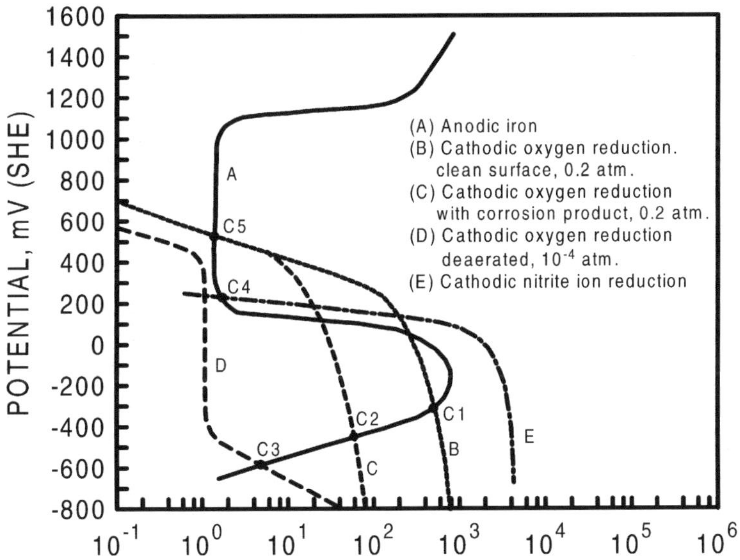


**Fig. 5.41** Approximate anodic polarization curve for iron and cathodic polarization curves for oxygen under several conditions and for nitrite ions. The polarization curves are used to estimate the effects of these environments on corrosion rate. Estimated $E_{corr}$ and $i_{corr}$ for the several environments are C1, clean surface, aerated; C2, surface with corrosion product, aerated; C3, clean surface, deaerated; C4, clean surface, deaerated plus nitrite ions, passivated; C5, clean surface, aerated plus nitrite ions, passivated

The corrosion occurs in the active potential range of the anodic curve for both the aerated and deaerated conditions without the nitrite ions. For the aerated environments, the major cathodic reaction is oxygen reduction with the rate much lower when the surface is covered by a corrosion product layer that reduces access of oxygen to the surface. In the deaerated environment, the major cathodic reaction is the direct reduction of water. The thermodynamics and kinetics of the nitrite ion reaction are such that the polarization curve for the reduction of this ion intersects the iron curve only in the passive region. The combined effect of the nitrite and oxygen is to move the corrosion potential into the passive range. The iron is, therefore, passivated by the nitrite ion, which is referred to as a passivating type inhibitor. It should be noted, however, that its use in inhibiting corrosion is significant only in the aerated environment where the rate is reduced by a factor of about 43 over the aerated environment with corrosion product layer. In the deaerated environment, the rate is already sufficiently small so as not to require the nitrite inhibitor to usefully decrease the rate. Additions of the order of 100 ppm chloride ion to the aerated nitrite environment will cause the corrosion potential to decrease into the active range and the corrosion rate to increase. In the presence of chloride ions, the anodic polarization curve in the vicinity of $i_{crit}$ is increased. The net cathodic curve now intersects the anodic curve in the active range and at a higher current density than in the absence of the nitrite ion, in which case addition of nitrite increases rather than inhibits corrosion.

## *Corrosion of Iron, Nickel, Chromium, and Titanium in Sulfuric and Nitric Acids*

The approximate anodic polarization curves for iron, nickel, chromium, and titanium in 1 N $H_2SO_4$ are shown in Fig. 5.42. The cathodic reactions are for the environments shown and are representative of curves obtained on platinum. Since they may be displaced significantly when the reactions occur on the other metal surfaces, particularly the shift of the oxygen curves to lower potentials and current densities, the following discussion is qualitative. The conclusions drawn, however, are consistent with observations on the actual metal/environment systems.

In deaerated 1 N $H_2SO_4$ (pH = 0.56), hydrogen-ion reduction is the cathodic reaction with the cathodic polarization curve intersecting the iron, nickel, and chromium curves in the active potential region. Hence, active corrosion occurs with hydrogen evolution, and the corrosion rates would be estimated by the intersections of the curves. The curves predict that the titanium will be passivated. However, the position of the cathodic hydrogen curve relative to the anodic curves for titanium and chromium indicates that if the exchange current density for the hydrogen reaction were higher (e.g., 10 mA/m$^2$), both titanium and chromium would exist in the passive state with low corrosion rates. Conversely, if the exchange current density were lower (e.g., 0.01 mA/m$^2$), both metals would corrode in the active state with the rate being much larger for chromium. As a consequence, the corrosion behavior of these metals can be very sensitive to small changes in the environment, metal composition, and surface condition, all of which may influence the exchange current density for the hydrogen reaction. This sensitivity has been demonstrated by showing that small additions of platinum to both titanium and chromium result in large decreases in corrosion rate in boiling sulfuric acid (Ref 37). The platinum increases the hydrogen exchange current density and brings about the decrease in corrosion rate as just described.

Limited information is available on the anodic polarization of the four metals in Fig. 5.42 in nitric acid. As an approximation, the behavior in sulfuric acid is assumed to apply in nitric acid. The overall reaction for the reduction of nitric acid is:

$$2H^+ + NO_3^- + 2e \rightarrow NO_2^- + H_2O \quad \text{(Eq 5.7)}$$

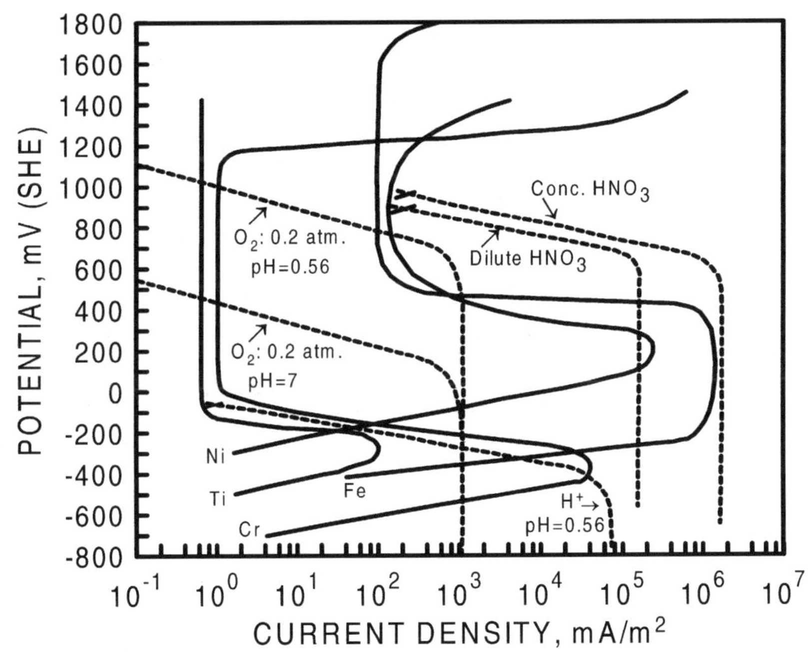


**Fig. 5.42** Approximate polarization curves for iron, nickel, chromium, and titanium in 1 N $H_2SO_4$. Approximate cathodic polarization curves for reduction of nitric acid, dissolved oxygen, and hydrogen ions. An explanation for predicting corrosion behavior based on intersection of anodic and cathodic curves can be found in the text.

The sequence of reactions involved in the overall reduction of nitric acid is complex, but direct measurements confirm that the acid has a high oxidation/reduction potential, ~940 mV (SHE), a high exchange current density, and a high limiting diffusion current density (Ref 38). The cathodic polarization curves for dilute and concentrated nitric acid in Fig. 5.42 show these thermodynamic and kinetic properties. Their position relative to the anodic curves indicate that all four metals should be passivated by concentrated nitric acid, and this is observed. In fact, iron appears almost inert in concentrated nitric acid with a corrosion rate of about 25 μm/year (1 mpy) (Ref 8). Slight dilution causes a violent iron reaction with corrosion rates >25 × $10^6$ μm/year ($10^6$ mpy). Nickel also corrodes rapidly in the dilute acid. In contrast, both chromium and titanium are easily passivated in dilute nitric acid and corrode with low corrosion rates.

## *Corrosion of Type 304 Stainless Steel in Sulfuric Acid*

Type 304 stainless steel is basically an alloy of 18 to 19 wt% Cr and 8 to 10 wt% Ni. Its corrosion behavior in sulfuric acid is sensitive to both alloy composition and the sulfuric acid environment. Variables with respect to alloy composition include whether the Cr and Ni concentrations are high or low within the allowed range and the concentrations of residual elements such as sulfur, phosphorus, copper, and molybdenum. Thermal and mechanical treatments are also variables but are not considered in the following. Important variables with respect to the sulfuric acid environment include degree of aeration and agitation (velocity effect) and small concentrations of species such as nitric acid, cupric ions, and ferric ions. The net influence of these variables is to find corrosion rates varying from <25 μm/year (1 mpy) to >2500 μm/year (100 mpy) (Ref 39).

This wide range of corrosion behavior can be understood by analyzing how the positions of the individual anodic and cathodic polarization curves lead to significant differences in $E_{corr}$ and $i_{corr}$. Figure 5.43 is an approximate representation of the individual polarization curves of reactions to be considered in an analysis of the corrosion behavior. The peaks of the anodic curves (L and H) are representative of the limits, $i_{crit}$, that have been observed for type 304 stainless steels in *deaerated* 1 N sulfuric acid. Values range from 100 to 30,000 mA/$m^2$ (Ref 40). High values have been associated with alloys having abnormally high sulfur concentrations with the sulfur concentrated in nonmetallic inclusions. These inclusions dissolve to give high local concentrations of sulfide ions that increase the active-peak current density as discussed in relation to Fig. 5.34. The hydrogen-ion reduction curve is representative of 1 N $H_2SO_4$ (pH ≈ 0.6), and the oxygen reduction curve is representative of this acid saturated with air. Under deaerated conditions, the only cathodic reaction is hydrogen-ion reduction. Under aerated conditions, the effective cathodic curve is the sum of the oxygen and hydrogen-ion reduction curves; this sum curve is shown by the crosses and is used in the analysis of corrosion under aerated conditions.

Intersections of anodic and cathodic polarization curves define the electrochemical parameters, $E_{corr}$ and $i_{corr}$, for corrosion. In Fig. 5.43, four intersections occur; two occur between the cathodic hydrogen reduction curve and the anodic curves, (L) and (H), and two between the cathodic sum curve and each of the two anodic curves. The former two intersections apply to deaerated conditions and the latter to aerated conditions. Figure 5.44 shows the two polarization curves predicted for the two alloys under deaerated conditions. The shift in the active-peak, current-density maximum results in a change in intersection of the anodic and cathodic curves such that alloys with the high $i_{crit}$ have a lower $E_{corr}$ and a higher $i_{corr}$. These differences correlate with direct measurements of corrosion potentials and corrosion rates of stainless steels. It is important to recognize that in the deaerated acid, corrosion occurs in the active range of the polarization curve for alloys of both low and high anodic-peak current density.

Figure 5.45 shows the two polarization curves predicted for the two alloys under aerated conditions. The solid curve is predicted for the alloy with the higher (H) anodic-peak current density, and the curve defined by the crosses is predicted for the alloy with the lower (L) anodic-

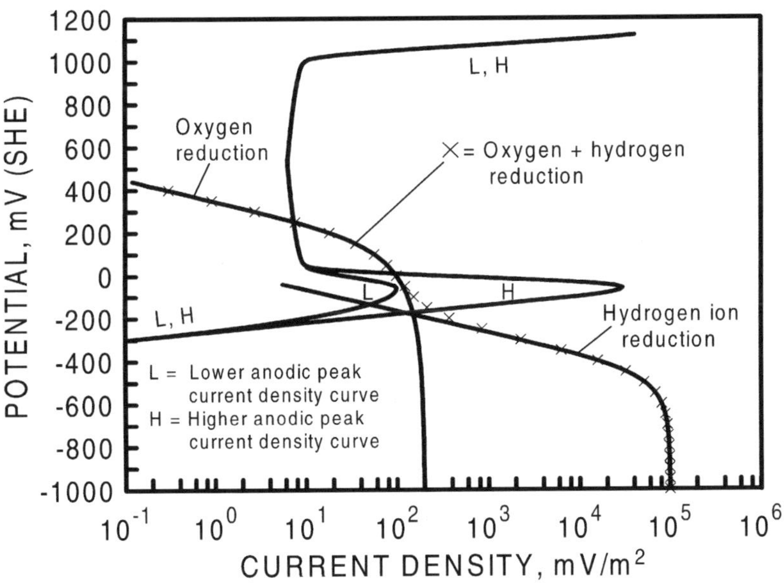


**Fig. 5.43** Schematic polarization curves for type 304 stainless steel in 1 N $H_2SO_4$. L (low) and H (high) distinguish the effects that minor composition variables can have on the position of the active peak current density ($i_{crit}$) in the stainless steel polarization curve.

-peak current density. The curves indicate that the alloy with the lower anodic peak would be passivated by the aeration; the anodic and cathodic polarization curves cross in the passive potential range of the alloy.

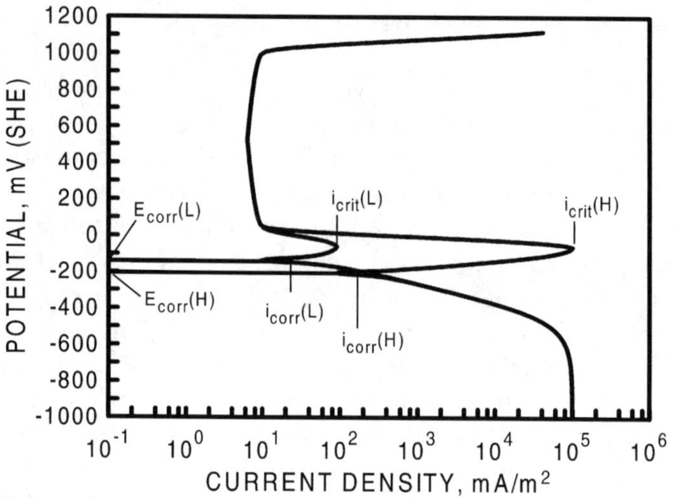


**Fig. 5.44** Schematic polarization curves for type 304 stainless steel in deaerated 1 N $H_2SO_4$. L (low) and H (high) distinguish the effects that minor composition variables can have on the position of the active peak current density ($i_{crit}$) in the stainless steel polarization curve. Estimated corrosion potentials and corrosion current densities are shown.

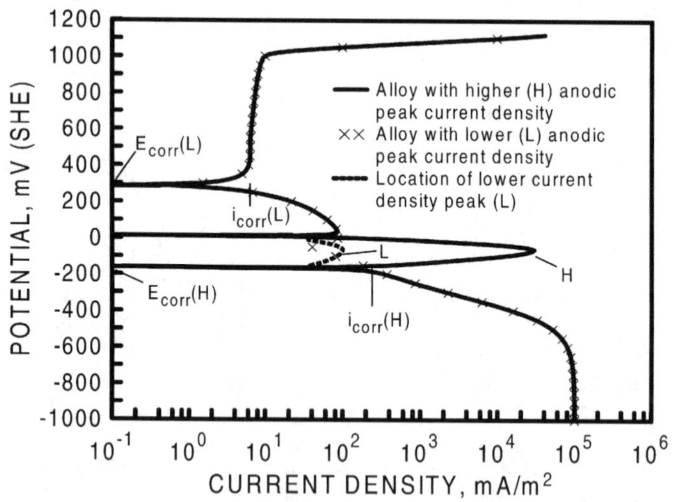


**Fig. 5.45** Schematic polarization curves for type 304 stainless steel in aerated 1 N $H_2SO_4$. L (low) and H (high) distinguish the effects that minor composition variables can have on the position of the active peak current density ($i_{crit}$) in the stainless steel polarization curve. Estimated corrosion potentials and corrosion current densities are shown. In particular, note that corrosion can occur in the active or passive potential range depending on the position of $i_{crit}$.

The result is a corrosion rate of about 10 mA/m$^2$, $i_{corr}$ (L). In contrast, the alloy with the higher anodic peak would not be passivated. The polarization curves cross in the active potential range of the alloy resulting in an active corrosion rate corresponding to about 250 mA/m$^2$.

This analysis provides explanations of observations that slight increases in oxidizing power of the environment can significantly decrease the corrosion rate by changing the corrosion mode from active to passive. For example, increasing the amount of dissolved oxygen in the environment or increasing fluid velocity to increase the limiting-diffusion current density can move the cathodic curve beyond the anodic-peak current density. Other examples are the decrease in corrosion rate with small additions of nitric acid, ferric ions, and cupric ions to the environment, all of which result in a net cathodic curve at higher current densities, thereby placing the alloy in the passive state.

## Chapter 5 Review Questions

1. a. Sketch an anodic polarization curve (E versus log i) for an active-passive metal, starting at E′, $i_o$. Identify $i_{crit}$, $E_{pp}$, and $i_p$.
   b. In developing a new corrosion-resistant, active-passive alloy, discuss why it is desirable to have values of $i_{crit}$, $E_{pp}$, and $i_p$ as low as possible. (Hint: Consider the intersections of anodic and cathodic polarization curves.)
2. Based on the data presented in Fig. 5.42, for each element/electrolyte listed below, state whether active or passive corrosion occurs and give the corrosion current density, $i_{corr}$. In each situation, assume the worst-case condition.

   a. Fe in concentrated nitric acid
   b. Fe in dilute nitric acid
   c. Ni in aerated neutral solution
   d. Cr in aerated neutral solution
   e. Cr in aerated acidic (pH = 0.56) solution
   f. Cr in deaerated acidic (pH = 0.56) solution
   g. Ti in aerated acidic (pH = 0.56) solution
   h. Ti in deaerated acidic (pH = 0.56) solution

3. With reference to the polarization curves in Fig. 5.42:

   a. Determine the values of $i_{crit}$, $E_{pp}$, and $i_p$ for iron.
   b. Give the approximate potential ranges for active, passive, and transpassive corrosion of chromium.
   c. Could an increase in fluid velocity for an aerated acid solution at pH = 0.56 result in the passivation of either iron or chromium? Explain.

d. Does the contribution of dissolved oxygen to the corrosion of iron change significantly when the acidity is decreased from a pH of 0.56 to 7.0? Explain.

4. When ferric chloride ($FeCl_3$) is progressively added to deaerated water in contact with stainless steel, the following observations are made: (a) for small additions, the corrosion rate increases; (b) for intermediate additions, the corrosion rate suddenly decreases; and (c) for larger additions, pitting corrosion occurs. Use appropriate polarization curves to explain these observations.
5. A stainless steel undergoes pitting corrosion in a chloride-ion-containing environment. If the oxidizing potential of the environment could be changed, should it be increased or decreased in order to minimize or eliminate the pitting corrosion? Explain.
6. When pitting corrosion is a problem with type 304 stainless steel, the problem can frequently be solved by changing to type 316 stainless steel, which contains molybdenum. To explain this effect, which of the following would represent the major influence of molybdenum: decreases $i_{crit}$, decreases $i_p$, or increases $E_{b,pit}$? Explain.

## Answers to Chapter 5 Review Questions

2. (a) Passive, 100 mA/m$^2$; (b) Active, 160,000 mA/m$^2$; (c) Active, 1000 mA/m$^2$; (d) Active, 1000 mA/m$^2$; (e) Active, 40,000 mA/m$^2$; (f) Active, 40,000 mA/m$^2$; (g) Passive, 0.6 mA/m$^2$; (h) Passive, 0.6 mA/m$^2$
3. (c) More likely for chromium; (d) Does not change at all ($O_2$ diffusion control)
5. Decreased to lower $E_{corr}$ relative to $E_{b,pit}$
6. Increases $E_{b,pit}$

## References


1. H.H. Uhlig, History of Passivity, Experiments and Theories, *Passivity of Metals,* R.P. Frankenthal and J. Kruger, Ed., The Electrochemical Society, 1978, p 1–28
2. M. Pourbaix, "Atlas of Electrochemical Equilibria in Aqueous Solutions," Pergamon Press, 1974, p 307–321
3. M. Pourbaix, Corrosion, *Atlas of Electrochemical Equilibria in Aqueous Solutions,* M. Pourbaix, Ed., National Association of Corrosion Engineers, 1974, p 70–83

4. M. Nagayama and M. Cohen, The Anodic Oxidation of Iron in a Neutral Solution I. The Nature and Composition of the Passive Film, *J. Electrochem. Soc.,* Vol 109, 1962, p 781–790
5. N. Sato, The Passivity of Metals and Passivating Films, *Passivity of Metals,* R.P. Frankenthal and J. Kruger, Ed., The Electrochemical Society, 1978, p 29–58
6. E.J. Kelly, The Active Iron Electrode I. Iron Dissolution and Hydrogen Evolution Reactions in Acidic Sulfate Solutions, *J. Electrochem. Soc.,* Vol 112, 1965, p 124–131
7. M.J. Humphries and R.N. Parkins, Stress-Corrosion Cracking of Mild Steels in Sodium Hydroxide Solutions Containing Various Additional Substances, *Corros. Sci.,* Vol 7, 1967, p 747–761
8. T.P. Sastry and V.V. Rao, Anodic Protection of Mild Steel in Nitric Acid, *Corrosion,* Vol 39, 1983, p 55
9. B.E. Wilde and F.G. Hodge, The Cathodic Discharge of Hydrogen on Active and Passive Chromium Surfaces in Dilute Sulphuric Acid Solutions, *Electrochim. Acta,* Vol 14, 1969, p 619–627
10. W.A. Mueller, Derivation of Anodic Dissolution Curve of Alloys from Those of Metallic Components, *Corrosion,* Vol 18, 1962, p 73t–79t
11. K. Sugimoto and Y. Sawada, The Role of Molybdenum Additions to Austenitic Stainless Steels in the Inhibition of Pitting in Acid Chloride Solutions, *Corros. Sci.,* Vol 17, 1977, p 425–445
12. T.M. Devine and B.J. Drummond, Use of Accelerated Intergranular Corrosion Tests and Pitting Corrosion Tests to Detect Sensitization and Susceptibility to Intergranular Stress Corrosion Cracking in High Temperature Water of Duplex 308 Stainless Steel, *Corrosion,* Vol 37, 1981, p 104–115
13. F.G. Hodge and B.E. Wilde, Effect of Chloride Ion on the Anodic Dissolution Kinetics of Chromium-Nickel Binary Alloys in Dilute Sulfuric Acid, *Corrosion,* Vol 26, 1970, p 146–150
14. E.J. Kelly, Anodic Dissolution and Passivation of Titanium in Acidic Media III. Chloride Solutions, *J. Electrochem. Soc.,* Vol 126, 1979, p 2064–2075
15. C.J. Mauvais, R.M. Latanision, and A.W. Ruff, Jr., On the Anisotropy Observed During the Passivation of Nickel Monocrystals, *J. Electrochem. Soc.,* Vol 117, 1970, p 902
16. R.T. Foley, Localized Corrosion of Aluminum Alloys—A Review, *Corrosion,* Vol 42, 1986, p 277–288
17. H. Kaesche, Pitting Corrosion of Aluminum and Intergranular Corrosion of Aluminum Alloys, *Localized Corrosion NACE 3,* R.W. Staehle, B.F. Brown, J. Kruger, and A. Agrawal, Ed., National Association Corrosion Engineers, 1974, p 516–525
18. F.H. Haynie and S.J. Ketcham, Electrochemical Behavior of Aluminum Alloys Susceptible to Intergranular Corrosion. II. Electrode

Kinetics of Oxide-Covered Aluminum, *Corrosion,* Vol 19, 1963, p 403t–407t

19. A.P. Bond, G.F. Bolling, H.A. Domian, and H. Bilon, Microsegregation and the Tendency for Pitting Corrosion in High-Purity Aluminum, *J. Electrochem. Soc.,* Vol 113, 1966, p 773–778
20. F. Mansfeld and H.H. Uhlig, Passivity in Copper-Nickel-Aluminum Alloys–A Confirmation of the Electron Configuration Theory, *J. Electrochem. Soc.,* Vol 115, 1968, p 900–904
21. R. Kirchheim, B. Heine, H. Fischmeister, S. Hofmann, H. Knote, and U. Stolz, The Passivity of Iron-Chromium Alloys, *Corros. Sci.,* Vol 29, 1989, p 899–917
22. P.F. King and H.H. Uhlig, Passivity in the Iron-Chromium Binary Alloys, *J. Electrochem. Soc.,* Vol 63, 1959, p 2026–2032
23. C.L. McBee and J. Kruger, Nature of Passive Films on Iron-Chromium Alloys, *Electrochim. Acta,* Vol 17, 1972, p 1337–1341
24. M.B. Rockel, The Effect of Molybdenum on the Corrosion Behavior of Iron-Chromium Alloys, *Corrosion,* Vol 29, 1973, p 393–395
25. W.Y.C. Chen and J.R. Stephens, Anodic Polarization Behavior of Austenitic Stainless Steel Alloys with Lower Chromium Content, *Corrosion,* Vol 35, 1979, p 443–450
26. K. Tachibana and M.B. Ives, Selective Dissolution Measurements to Determine the Nature of Films on Nickel-Molybdenum Alloys, *Passivity of Metals,* The Electrochemical Society, 1978, p 878–897
27. "Standard Reference Test Method for Making Potentiostatic and Potentiodynamic Anodic Polarization Measurements," G 5-94, *Annual Book of ASTM Standards,* Vol 03.02, ASTM, 1995, p 48–58
28. K. Sugimoto and Y. Sawada, Interfacial Impedance of Austenitic Steel under Anodic Polarization, *Proceedings of the Fifth International Congress on Metallic Corrosion,* National Association of Corrosion Engineers, 1974, p 290–297
29. N.D. Greene, The Passivity of Nickel and Nickel-base Alloys, *First International Congress on Metallic Corrosion,* National Association of Corrosion Engineers, 1961, p 113–117
30. E.E. Stansbury, Potentiostatic Etching, *Applied Metallography,* G.F. Vander Voort, Ed., Van Nostrand Reinhold, New York, 1988, p 21–39
31. Z. Smialowska, Influence of Sulfide Inclusions on the Pitting Corrosion of Steels, *Corrosion,* Vol 28, 1972, p 388–396
32. G. Wranglen, Pitting and Sulphide Inclusions in Steel, *Corros. Sci.,* Vol 14, 1974, p 331–349
33. Z. Smialowska, *Pitting Corrosion of Metals,* National Association of Corrosion Engineers, 1986, p 69–97
34. M.J. Johnson, Relative Critical Potentials for Pitting Corrosion of Some Stainless Steels, *Localized Corrosion,* STP 516, ASTM, 1972, p 262–272

35. A. Kawashima, A.K. Agrawal, and R.W. Staehle, Effect of Oxyanions and Chloride Ion on the Stress Corrosion Cracking Susceptibility of Admiralty Brass in Nonammonical Aqueous Solutions, *Stress Corrosion Cracking—The Slow Strain-Rate Technique,* STP 665, G.M. Uglansky and J.H. Payer, Ed., ASTM, 1979, p 266–278
36. N.D. Tomashov, G.P. Chernova, Y.S. Ruskol, and G.A. Ayuyan, Passivation of Alloys on Titanium Base, *Proceedings of the Fifth International Conference on Metallic Corrosion,* National Association of Corrosion Engineers, 1974, p 248–252
37. N.D. Tomashov, Methods of Increasing the Corrosion Resistance of Metal Alloys, *Corrosion,* Vol 14, 1958, p 229t–236t
38. K.J. Vetter, *Electrochemical Kinetics,* Academic Press, 1967, p 490–493
39. B.E. Wilde and N.D. Greene, Jr., The Variable Corrosion Resistance of 18Cr-8Ni Stainless Steels: Behavior of Commercial Alloys, *Corrosion,* Vol 25, 1969, p 300–304
40. N.D. Greene and B.E. Wilde, Variable Corrosion Resistance of 18 Chromium-8 Nickel Stainless Steels: Influence of Environmental and Metallurgical Factors, *Corrosion,* Vol 26, 1970, p 533–538

CHAPTER 6

# Electrochemical Corrosion-Rate Measurements

Electrochemical corrosion studies to determine both corrosion rates and behaviors frequently employ a potentiostatic circuit, which includes a polarization cell, as schematically shown in Fig. 6.1. The working electrode (WE) is the corrosion sample (i.e., the material under evaluation). The auxiliary electrode (AE), or counter electrode, is ideally made of a material that will support electrochemical oxidation or reduction reactions with reactants in the electrolyte but will not itself undergo corrosion and thereby contaminate the electrolyte. The AE is usually made of platinum or high-density graphite. The reference electrode (RE) maintains a constant potential relative to which the potential of the WE is measured with an electrometer, a high-impedance ($>10^{14}$ ohms) voltmeter that limits the current through the electrometer to extremely small values that negligibly influence either the RE or WE potential. The potentiostat is a rapid response direct-current (dc) power supply that will maintain the potential of the WE relative to the RE at a constant (preset or set point) value even though the external circuit current, $I_{ex}$, may change by several orders of magnitude. When the potentiostat is disconnected from the corrosion sample (WE), the open-circuit or open-cell condition exists, the WE is freely corroding, the potential measured is the open-circuit corrosion potential, $E_{corr}$, and, of course, $I_{ex} = 0$.

The potentiostat can be set to polarize the WE either anodically, in which case the net reaction at the WE surface is oxidation (electrons removed from the WE), or cathodically, in which case the net reaction at the WE surface is reduction (electrons consumed at the WE). With reference to the potentiostatic circuit in Fig. 6.1, determination of a polarization curve is usually initiated by measuring the open-circuit corrosion potential, $E_{corr}$, until a steady-state value is achieved (e.g., less than 1.0 mV change over a five-minute period). Next, the potentiostat is set to control at $E_{corr}$ and connected to the polarization cell. Then, the set-point potential is reset continuously or stepwise to control the potential-time history of the WE while $I_{ex}$ is measured. If the set-point potential is continuously increased (above $E_{corr}$), an anodic polarization curve is generated; conversely, if the potential is continuously decreased (below $E_{corr}$), a cathodic polarization curve is produced.

Interpretation of an experimentally determined polarization curve, including an understanding of the information derivable therefrom, is based on the form of the polarization curve, which results from the polarization curves for the individual anodic and cathodic half-cell reactions occurring on the metal surface. These individual polarization curves, assuming Tafel behavior in all cases, are shown in Fig. 6.2 (dashed curves) with $E_{corr}$ and the corrosion current, $I_{corr}$, identified. It is assumed that over the potential range of concern, the $I_{ox,X}$ and $I_{red,M}$ contributions to the sum-anodic and sum-cathodic curves are negligible; consequently, $\Sigma I_{ox} = I_{ox,M}$ and $\Sigma I_{red} = I_{red,X}$. At any potential of the

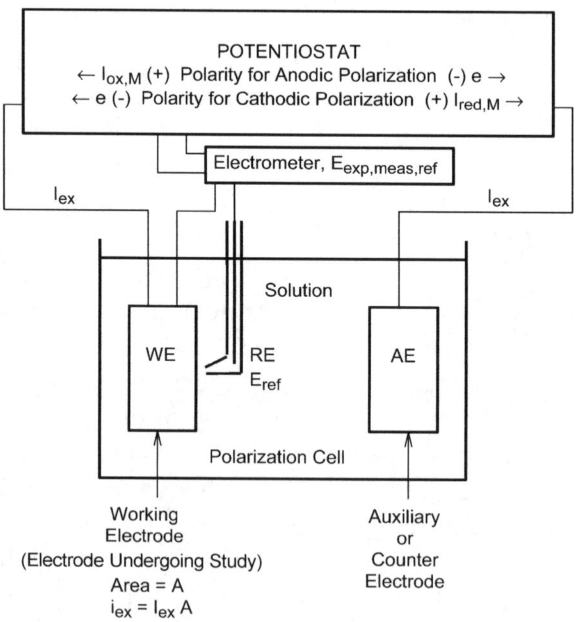


**Fig. 6.1** The potentiostatic circuit

WE established by the potentiostat, the external current, $I_{ex}$, is the difference between $I_{ox,M}$ and $I_{red,M}$. This difference, in terms of the Tafel expressions for the individual reactions (see Eq 4.69), is:

$$I_{ex} = I_{ox,M} - I_{red,X} = I_{o,M}\, e^{2.3(E-E'_M)/\beta_{ox,M}} - I_{o,X}\, e^{-2.3(E-E'_X)/\beta_{red,X}} \quad \text{(Eq 6.1)}$$

It is evident that $I_{ex}$ changes from positive to negative when $I_{red,X}$ becomes greater than $I_{ox,M}$. This change in sign occurs as $I_{ex}$ passes through $I_{ex} = 0$, at which point, $E = E_{corr}$ and $I_{ox,M} = I_{red,X} = I_{corr}$. Thus, two current ranges can be identified: $I_{ex} = I_{ex,a} > 0$, over which the anodic or oxidation reaction is dominant, and $I_{ex} = I_{ex,c} < 0$, over which the cathodic or reduction reaction is dominant. The properties of these two ranges are summarized below.

In the current range, $I_{ex} = I_{ex,a} > 0$, the WE potential set by the potentiostat is greater than $E_{corr}$. The electrons produced per unit time by the $M \rightarrow M^{m+} + me$ reaction exceed those consumed per unit time by the $X^{x+} + xe \rightarrow X$ reaction, and net oxidation occurs at the WE. A positive current is consistent with the sign convention that assigns a positive value to the external circuit current when net oxidation occurs at the WE. A plot of E versus log $I_{ex,a}$ takes the form of the upper solid curve in Fig. 6.2, the anodic branch of the experimental polarization

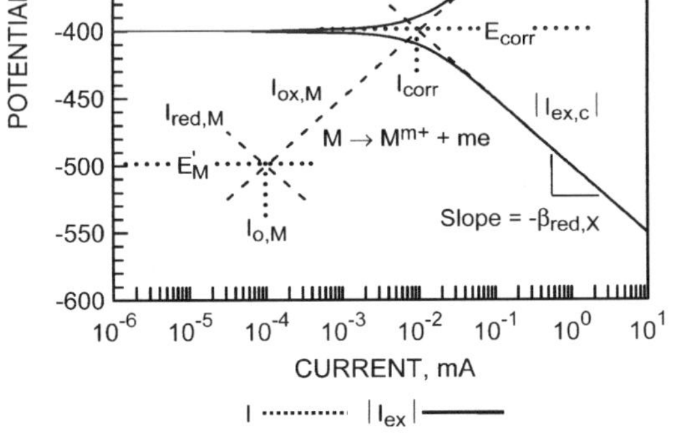


**Fig. 6.2** Schematic experimental polarization curves (solid curves) assuming Tafel behavior for the individual oxidation and cathodic-reactant reduction polarization curves (dashed curves)

curve. When E is increased sufficiently above $E_{corr}$ to cause $I_{red,X}$ to become negligible with respect to $I_{ox,M}$ (normally 50 to 100 mV):

$$I_{ex,a} = I_{ox,M} \quad \text{(Eq 6.2)}$$

and $I_{ex,a}$ becomes a direct measure of the oxidation rate, $I_{ox,M}$, of the metal in this potential range. This linear portion of an experimental curve reveals the Tafel curve of the anodic metal reaction, and extrapolation of the Tafel curve to $E'_M$ provides an estimate from experiment of the metal exchange current density, $I_{o,M}/A_a$, where $A_a$ is the area of the WE.

In the current range, $I_{ex} = I_{ex,c} < 0$, the WE potential set by the potentiostat is less than $E_{corr}$. At the metal surface, electrons consumed per unit time by the $X^{x+} + xe \rightarrow X$ reaction exceed those produced per unit time by the $M \rightarrow M^{m+} + me$ reaction. Net reduction is occurring, and electrons must be supplied to the WE by the external circuit; the external circuit current ($I_{ex,c}$) will be negative. A plot of E versus log $|I_{ex,c}|$ takes the form of the lower solid curve in Fig. 6.2. When E is decreased sufficiently below $E_{corr}$ to cause $I_{ox,M}$ to become negligible (normally 50 to 100 mV):

$$I_{ex,c} = -I_{red,X}$$

or

$$|I_{ex,c}| = I_{red,X} \quad \text{(Eq 6.3)}$$

and $I_{ex,c}$ becomes a direct measure of the rate of the cathodic reaction, $I_{red,X}$, on the metal. This linear portion of an experimental curve reveals the Tafel curve of the cathodic reaction, and extrapolation of the Tafel curve to $E'_X$ provides an estimate from experiment of the cathodic reaction exchange current density, $I_{o,X}/A_c$, where $A_c$ is the area of the WE.

The net (or experimental) anodic and cathodic polarization curves in Fig. 6.2 also can be expressed with $E_{corr}$ and $I_{corr}$ as parameters. This form is used in establishing expressions that provide the basis of one of the experimental techniques for determination of $I_{corr}$. At the specific condition that $E = E_{corr}$ and $I_{ex} = 0$, $I_{ox,M} = I_{red,X} = I_{corr}$; therefore, the Tafel expressions for the currents of the individual anodic and cathodic reactions can be equated, or

$$I_{corr} = I_{o,M}\, e^{2.3(E_{corr} - E'_M)/\beta_{ox,M}} = I_{o,X}\, e^{-2.3(E_{corr} - E'_X)/\beta_{red,X}} \quad \text{(Eq 6.4)}$$

Division of Eq 6.4 into Eq 6.1 results in the desired expression with $E_{corr}$ and $I_{corr}$ as parameters:

$$I_{ex} = I_{corr}\left[e^{2.3(E - E_{corr})/\beta_{ox,M}} - e^{-2.3(E - E_{corr})/\beta_{red,X}}\right] \quad \text{(Eq 6.5)}$$

When $E > E_{corr}$, the first exponential term is greater than the second exponential term and $I_{ex}$ is positive. Plotted as E versus log $I_{ex}$, Eq 6.5 plots as the upper solid curve in Fig. 6.2. For $E < E_{corr}$, $I_{ex}$ is negative, and a plot of E versus log $|I_{ex}|$ plots as the lower solid curve in Fig. 6.2. These equations will be used in establishing relationships for the analysis of corrosion rates by the experimental techniques of Tafel-curve modeling and polarization resistance.

It is emphasized that more generally, $I_{ex}$ is the experimentally measured current representing the net difference between the sum of all oxidation-reaction currents and the sum of all reduction-reaction currents at the interface:

$$I_{ex} = \Sigma I_{ox} - \Sigma I_{red} \qquad \text{(Eq 6.6)}$$

For the two half-cell reactions under consideration:

$$I_{ex} = (I_{ox,M} + I_{ox,X}) - (I_{red,X} + I_{red,M}) \qquad \text{(Eq 6.7)}$$

Under the condition that $I_{ox,X}$ and $I_{red,M}$ are negligible:

$$I_{ex} = I_{ox,M} - I_{red,X} \qquad \text{(Eq 6.8)}$$

The above relationship is equally applicable if either the metal oxidation-rate curve or the reduction-rate curve for the cathodic reactant does not obey Tafel behavior. To illustrate this point, three additional schematic pairs of individual anodic and cathodic polarization curves are examined. In Fig. 6.3, the metal undergoes active-passive oxidation behavior and $E_{corr}$ is in the passive region. In Fig. 6.4, where the total re-

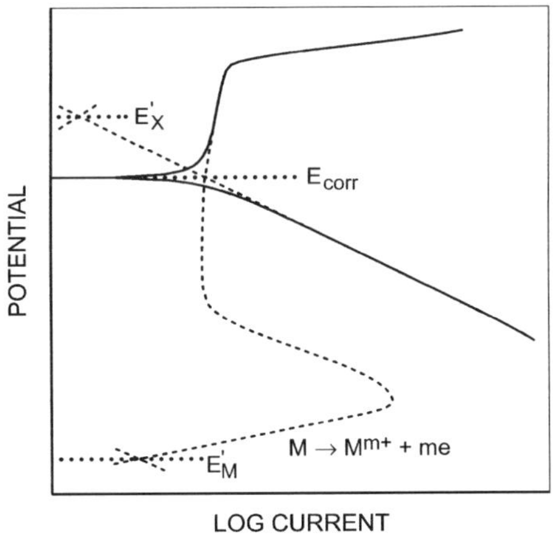


**Fig. 6.3** Schematic experimental polarization curves (solid curves) assuming active-passive behavior for the individual metal-oxidation curve and Tafel behavior for the individual cathodic-reactant reduction curve (dashed curves)

duction-rate curve involves reduction of both dissolved oxygen and hydrogen ions, and their respective limiting diffusion currents, the metal shown undergoes active-passive oxidation behavior, and $E_{corr}$ is in the passive region. It is to be noted for the example in Fig. 6.4 that if the dissolved oxygen were removed from the electrolyte, $E_{corr}$ would be in the active region, $I_{corr}$ would be considerably larger, and the experimental polarization curves would appear as in Fig. 6.5.

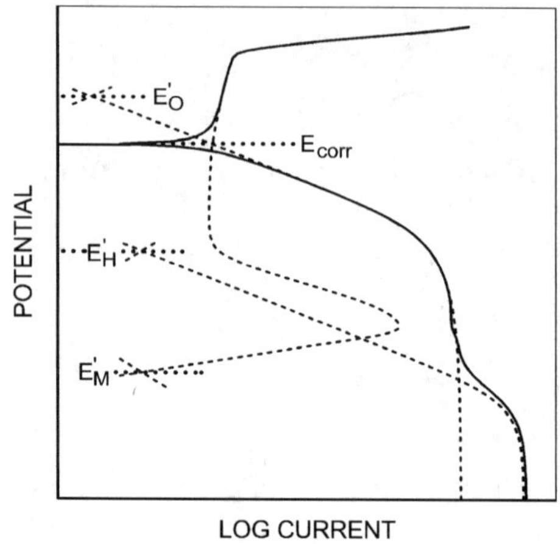


**Fig. 6.4** Schematic experimental polarization curves (solid curves) assuming active-passive behavior for the individual metal-oxidation curve and Tafel behavior plus limiting diffusion for the individual dissolved-oxygen and hydrogen-ion reduction curves (dashed curves)

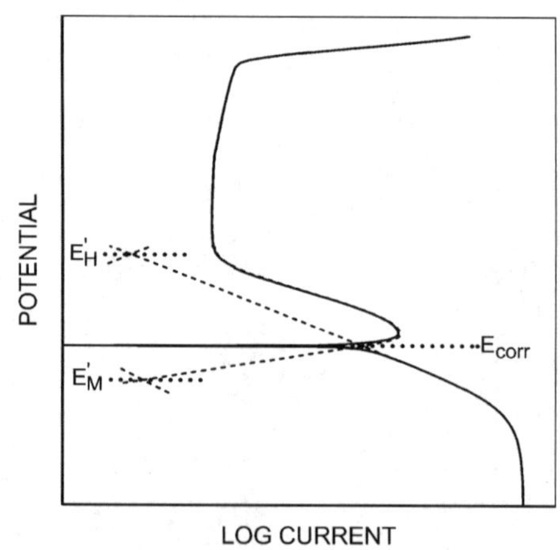


**Fig. 6.5** Schematic experimental polarization curves (solid curves) assuming active-passive behavior for the individual metal-oxidation curve and Tafel behavior plus limiting diffusion for the individual hydrogen-ion reduction curve in deaerated aqueous solution (dashed curves)

## Potential Measurement: Reference Electrodes and Electrometers (Ref 1)

Reference half cells, or reference electrodes, are used to establish the relative potentials of metals in contact with aqueous environments. The metal/aqueous-environment systems of concern may extend from pure metals in contact with electrolytes containing only the ions of that metal, to complex alloys in contact with complex electrolytes. In the latter case, the reference half cell measures the corrosion potential. The source of the potential to be measured is discussed in Chapter 2 as the difference of electrical potential between the metal and its aqueous environment. It is emphasized that this difference cannot be measured directly because the introduction of a measuring probe into the aqueous medium introduces another metal/liquid interface, across which an additional potential difference exists. Thus, any potential measuring instrument connected between the metal and the probe will indicate only a difference in potential, and absolute values of the individual half-cell potentials cannot be determined. The discussion shows that relative values of half-cell potentials are established if the measuring probe is a highly reproducible second half cell. These half cells are referred to as reference half cells or, frequently, as reference electrodes. The accepted primary reference half cell is the standard hydrogen electrode (SHE) consisting of platinum simultaneously in contact at 25 °C with a solution of hydrogen ions at unit activity and hydrogen gas at one atmosphere pressure. This half-cell potential is assigned the value, E(SHE) = 0.

Arrangements for making potential measurements are illustrated in Fig. 6.6. The working-electrode potential is measured relative to the reference electrode. The working electrode may be a pure metal, and it may be immersed in a solution containing its own ions, in which case the half-cell potential, $E'_{M,M^{m+}}$, is measured, assuming that another possible half-cell reaction (e.g., $O_2 + 2H_2O + 4e = 4OH^-$) does not significantly polarize the potential away from $E'_{M,M^{m+}}$. In corrosion investigations, both the working electrode and the solution are typically complex in composition, and the corrosion potential, $E_{corr}$, established by simultaneous anodic and cathodic reactions at the metal surface, is measured. The reference electrode contacts the working solution through a small opening. In Fig. 6.6(a), contact is made through a salt bridge, a tube frequently containing KCl solution or the electrolyte of the electrochemical cell. This salt bridge minimizes cross contamination by species in the two solutions that could alter the potentials of the individual electrodes. If cross contamination is not a problem, the reference electrode is frequently placed directly into the working solution, as shown in Fig. 6.6(b).

The potential of the working electrode relative to the reference electrode is measured with an electrometer or high-impedance voltmeter. This instrument must have an internal impedance large enough to limit the measuring current to values less than currents that can significantly affect processes occurring at the electrodes. For example, if sufficient current passes through the reference electrode interface, the reference electrode can polarize and shift its potential from the reversible value. For corroding metals, if the measuring current is comparable to the corrosion current, anodic and cathodic reactions will be affected, and the measured potential will not be representative of the corrosion potential. Potential-measuring instruments with measuring currents sufficiently small so as not to influence the reference-electrode or working-electrode potentials are generally called electrometers. These instruments should have an internal impedance $>10^{10}$ ohms, and frequently they have values $>10^{14}$ ohms. The importance of having an internal impedance of this magnitude is illustrated by considering the measurement of the corrosion behavior of specimens having an area of 1 $cm^2$, a size frequently used in laboratory measurements. A reasonably low corrosion rate of 25 μm/year (1 mpy) corresponds to a corrosion current density of approximately $10^{-6}$ $A/cm^2$ for most metals. An externally imposed current should be $<10^{-8}$ A to be only 1% of the corrosion current. If the potential between the reference electrode and the corroding metal is 1 V, the resistance of an electrometer should be greater than $1/10^{-8} = 10^8$ ohms. This rough estimate, plus other factors, confirms that the internal

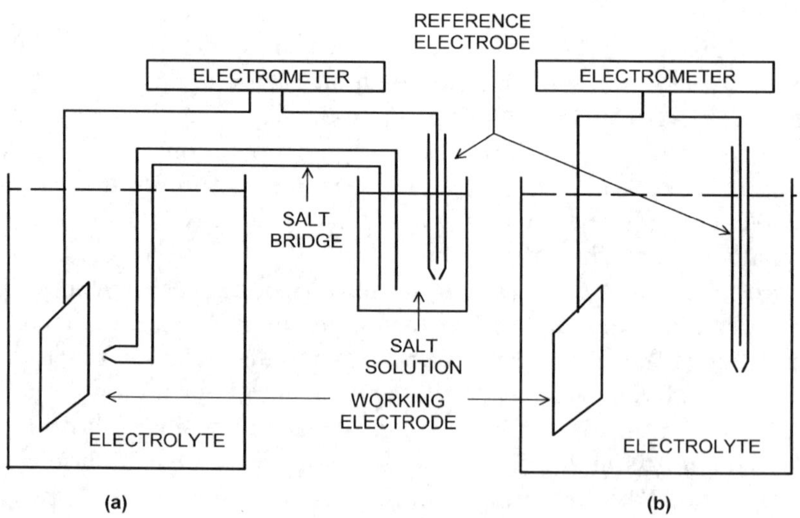


**Fig. 6.6** Schematic arrangements for measurements of electrode potentials using an electrometer. (a) Circuit completed through a salt bridge between the reference electrode and the specimen electrolyte. (b) Circuit completed through the specimen electrolyte

impedance of an electrometer needs to be of the magnitude indicated here.

The potentials, relative to the standard hydrogen electrode, of several half cells used as reference electrodes are given in Table 6.1 (the section "Examples of Half-Cell Reactions and Nernst-Equation Calculations" in Chapter 2 provides discussions of half-cell-potential calculations). There are a number of factors that contribute to the selection of components and to the design of a satisfactory reference half cell:

- The metals must have sufficiently positive half-cell potentials that corrosive reactions that would change the potential to a corrosion potential, $E_{corr}$, do not occur. This requirement, with few exceptions, restricts the metal component of the half cell to silver, mercury, and copper. For these metals, appearing in Table 6.1, corrosion due to hydrogen evolution will not occur since the metal half-cell potentials are above potentials for hydrogen evolution. Also, the kinetics of the reduction of any dissolved oxygen are sufficiently slow that the potential is shifted negligibly from that of the metal half cell.
- The half cells in Table 6.1 identified as saturated (e.g., "in saturated KCl" as compared to "in 1 N KCl") have the advantage of more easily maintaining a constant anion concentration and, hence, half-cell potential. The cells saturated in KCl maintain a constant $Cl^-$ ion concentration and, therefore, a constant $Ag^+$ ion concentration

**Table 6.1 Potentials of selected reference half cells**

| Type of half cell | mV (SHE) |
|---|---|
| **Silver/silver-chloride** | |
| Ag/AgCl (sat.)(a) in sat. KCl | +196 |
| Ag/AgCl (sat.) in 1 N KCl | +234 |
| Ag/AgCl (sat.) in 0.1 N KCl | +289 |
| **Mercury/mercurous-chloride (calomel)** | |
| $Hg/Hg_2Cl_2$ (sat.) in sat. KCl | +241 |
| $Hg/Hg_2Cl_2$ (sat.) in 1 N KCl | +288 |
| $Hg/Hg_2Cl_2$ (sat.) in 0.1 N KCl | +334 |
| **Copper/copper-sulfate** | |
| $Cu/CuSO_4$ (sat.) in sat. $CuSO_4$ | +316 |
| **Mercury/mercurous-sulfate** | |
| $Hg/HgSO_4$ (sat.) in sat. $HgSO_4$ | +621 |
| **Silver/silver-sulfate** | |
| $Ag/AgSO_4$ (sat.) in sat. $Ag_2SO_4$ | +654 |
| **Mercury/mercuric-oxide** | |
| Hg/HgO (sat.) in sat. HgO (1 N $OH^-$) | +98 |

(a) sat., electrolyte saturation with respect to the indicated salt through equilibrium with AgCl or a constant $Hg_2^{2+}$ ion concentration through equilibrium with $Hg_2Cl_2$. In saturated cells such as $Cu/CuSO_4$, a constant metal-ion concentration is maintained by an excess of solid salt in contact with the electrolyte. These cells have advantages in that: (a) evaporation of water from the cell is compensated by precipitation of additional solid salt; (b) although rarely a factor, addition or removal of metal ions by current flow to or from the metal electrode is compensated by precipitation or dissolution of solid salt; and (c) preparation of the electrolyte requires only that an excess of solid salt be present. In contrast, the potential of unsaturated half cells is affected by evaporation and by transfer of metal ions at the metal interface. Also, greater care is required in establishing the electrolyte concentration. The major disadvantage of the saturated-salt half cell is that the potential is more sensitive to changes in temperature because of the temperature dependence of the solubility of the salt and, hence, the temperature dependence of the concentration of metal ions in contact with the metal electrode. In this respect, the potentials of unsaturated half cells are less temperature sensitive.

- Reference electrodes are usually constructed as glass tubes containing the reference metal electrode and electrolyte. An opening must be provided in the end of the tube to allow contact between the reference-cell electrolyte and the aqueous environment of the system being measured. This opening must be large enough to allow the measuring current to flow, usually $<10^{-12}$ A, but as small as possible to minimize cross contamination of the electrolytes by diffusion. Openings to the reference electrode have been made by fusing the glass around asbestos fiber, by producing controlled cracks in the end of a glass tube, and by very-fine-pore fritted glass plugs. The salt bridge shown in Fig. 6.6(a) provides a long diffusion path and also can be terminated as just described. Such a tube also provides a method for separating a test electrode at an elevated temperature from a standard reference electrode at ambient temperature. The importance of minimizing leakage from the reference electrode can be illustrated by problems encountered in determining the corrosion behavior of stainless steels in small test containers. Contamination of the test environment by chloride ions from the reference cell, even by 10 ppm $Cl^-$ ion, may cause serious error in interpretation of potential measurements of the stainless steel.
- A potential difference develops at the junction between the electrolytes of the reference half cell and the WE being measured. This potential difference contributes to the potential difference between the two electrodes and should be considered when precise measurements are required. The junction potential can frequently be de-

creased by using a salt bridge containing KCl between the two electrolytes.

- The chloride ion contamination problem can be avoided by using reference electrodes that do not contain these ions. Examples are the sulfate and oxide types of reference electrodes in Table 6.1.

## The IR Correction to Experimentally Measured Potentials (Ref 2, 3)

It is noted in Fig. 6.1 that the experimentally measured potential, as measured against any given reference electrode (e.g., the saturated calomel electrode, SCE), is denoted as $E_{exp,meas,ref}$. When converting this potential to the standard hydrogen electrode scale (SHE), the following relationship applies:

$$E_{exp} = E_{exp,meas,ref} + E_{ref} \qquad \text{(Eq 6.9)}$$

where $E_{ref}$ is the potential of the reference electrode on the SHE scale, for example, +241 mV (SHE) for the SCE as shown in Table 6.1. It is further noted that the interface potential, E, as used throughout this text, is on the SHE scale. Under consideration at this point is how the experimentally measured interface potential, $E_{exp}$, is related to the actual interface potential, E, which is the desired quantity.

In Chapter 4, in the section "Relationship between Interface Potentials and Solution Potentials," E (or $E_M$) is defined relative to the potentials, $\phi$, as follows:

$$E = (\phi_{M^o} - \phi_{M^{m+}}) - (\phi'_{H_2} - \phi'_{H^+}) \qquad \text{(Eq 6.10)}$$

$$E = (\phi_{M^o} - \phi_{M^{m+}}) - \Delta\phi_{SHE} \qquad \text{(Eq 6.11)}$$

where $\phi_{M^o}$ is the potential in the metal, $\phi_{M^{m+}}$ is the potential of the metal ion in the solution adjacent to the metal surface, and $\phi'_{H_2}$ and $\phi'_{H^+}$ have similar meanings relative to the SHE. In terms of the potentiostatic circuit of Fig. 6.1, Eq 6.11 may be rewritten as:

$$E = (\phi_{M,WE} - \phi_{S,WE}) - \Delta\phi_{SHE} \qquad \text{(Eq 6.12)}$$

where $\phi_{M,WE}$ is the potential in the metal at the WE, and $\phi_{S,WE}$ is the potential in the solution adjacent to the WE. On the other hand, the experimentally measured potential, $E_{exp}$, is given by:

$$E_{exp} = (\phi_{M,WE} - \phi_{S,RE}) - \Delta\phi_{SHE} \qquad \text{(Eq 6.13)}$$

where $\phi_{S,RE}$ is the potential in the solution at the RE position. Subtracting Eq 6.13 from 6.12 yields:

$$E - E_{exp} = \phi_{S,RE} - \phi_{S,WE} \quad \text{(Eq 6.14)}$$

Obviously, if the RE could be placed at the WE surface, then $\phi_{S,RE} = \phi_{S,WE}$ and $E = E_{exp}$. Otherwise, $E \neq E_{exp}$.

Now consider the relationship between $\phi_{S,RE}$ and $\phi_{S,WE}$. With reference to Fig. 6.7(a), consider an anodic external current, $I_{ex,a}$. In the solution, this current flows from the higher solution potential at the WE surface, $\phi_{S,WE}$, past the RE, to the lower solution potential at the AE surface. The solution potential at the RE location is $\phi_{S,RE}$. A simple case is assumed in which the current distribution in the solution is uniform, leading to a linear solution-potential gradient. The potential difference in the solution between the WE surface and the RE position is $I_{ex,a}R'_S$ where $R'_S$ is the solution resistance (ohms) between the WE and RE. From the geometry in Fig. 6.7(a):

$$\phi_{S,WE} = \phi_{S,RE} + I_{ex,a}R'_S \quad \text{(Eq 6.15)}$$

where it is noted that, according to sign convention, $I_{ex,a}$ is a positive quantity. On substituting into Eq 6.14, the desired relationship between E and $E_{exp}$ is produced:

$$E = E_{exp} - I_{ex,a}R'_S \quad \text{(Eq 6.16)}$$

Therefore, for an anodic polarization curve, the true potential, E, is less than the experimentally measured potential, $E_{exp}$, by an amount equal to the "IR correction," $I_{ex,a}R'_S$, as indicated in Fig. 6.8. This correction becomes smaller as $I_{ex,a}$ becomes smaller and as $R'_S$ becomes smaller (i.e., lower solution resistivity and/or shorter distance between the WE and RE).

Now, with reference to Fig. 6.7(b), consider a cathodic external current, $I_{ex,c}$. In the solution, this current flows from the higher solution po-

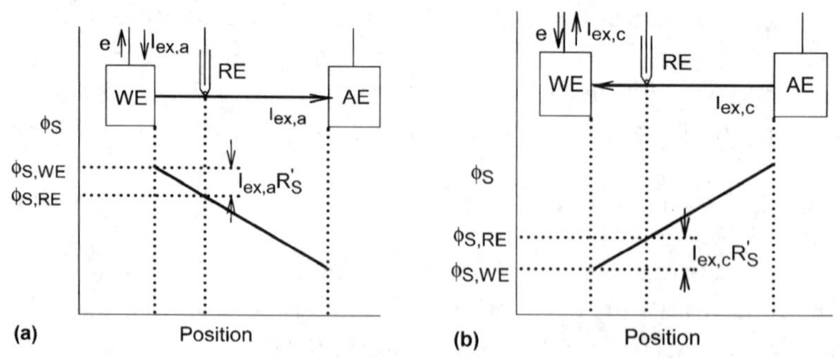


**Fig. 6.7** The IR connection. (a) Anodic external current. (b) Cathodic external current

tential at the AE, past the RE, to the lower solution potential at the WE. The potential difference in the solution between the WE surface and the RE position is $I_{ex,c}R'_S$, where it is noted that, according to sign convention, $I_{ex,c}$ is a negative quantity. From the geometry in Fig. 6.7(b):

$$\phi_{S,WE} = \phi_{S,RE} + I_{ex,c}R'_S \quad \text{(Eq 6.17)}$$

and on substituting into Eq 6.14:

$$E = E_{exp} - I_{ex,c}R'_S \quad \text{(Eq 6.18)}$$

Thus, for a cathodic polarization curve, the true potential, E, is greater than the experimentally measured potential, $E_{exp}$, by an amount equal to the IR correction, $|I_{ex,c}R'_S|$, as indicated in Fig. 6.8.

On comparison of Eq 6.16 and 6.18, it is seen that a single IR correction equation may be written for both anodic and cathodic polarization:

$$E = E_{exp} - I_{ex}R'_S \quad \text{(Eq 6.19)}$$

where $I_{ex}$ is positive for anodic external currents and negative for cathodic external currents. Division of $I_{ex}$ by specimen (WE) area (A) to produce external current density ($i_{ex}$), and simultaneous multiplication of $R'_S$ by specimen area to produce a normalized solution resistance, yields the equivalent relationship:

$$E = E_{exp} - i_{ex}R_S \quad \text{(Eq 6.20)}$$

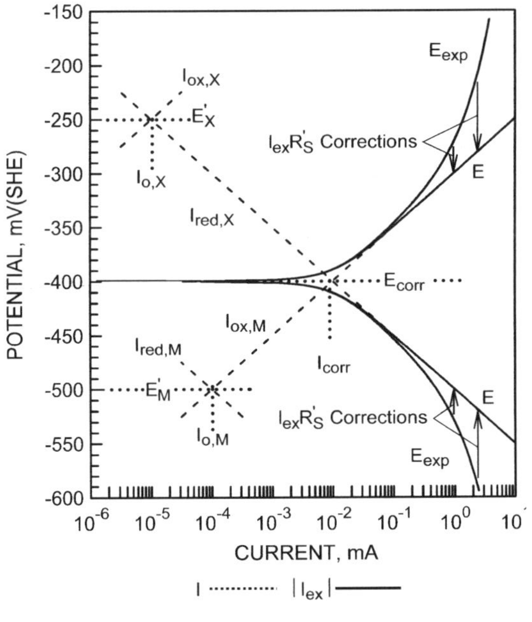


**Fig. 6.8** Polarization curves with IR corrections

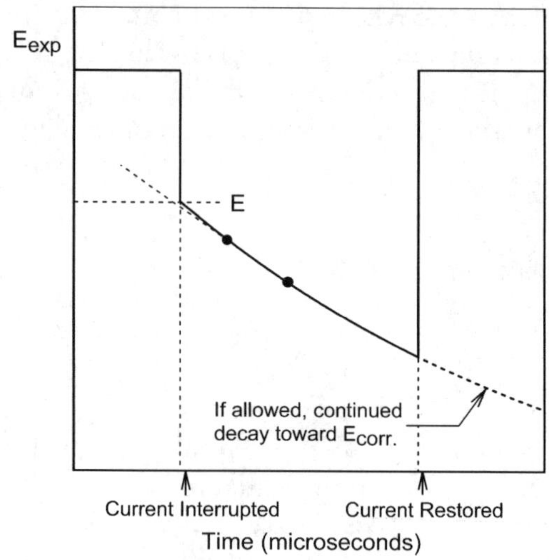


**Fig. 6.9** The current-interrupt IR-correction method. Based on Ref 2

where the dimensions of the solution resistance between WE and RE, $R_S$, are electrical resistance times area (e.g., ohms-$m^2$).

One technique for continuously making IR corrections while generating experimental polarization curves is the "current-interrupt" method (Ref 2). The circuitry necessary to perform this method is currently incorporated in some commercial potentiostats. The operational principles are illustrated in Fig. 6.9. At a given control potential, $E_{exp}$, the external current, $I_{ex}$, is interrupted by electronically opening an appropriate switch. $I_{ex}$ instantly goes to zero, and, therefore, $E_{exp}$ instantly goes to E (see Eq 6.19). During the very short current-interrupt time period (on the order of microseconds), the potential is recorded as a function of time. Data points are selected in this time period (as illustrated in Fig. 6.9) to allow a linear extrapolation to the potential corresponding to the time of current interruption. This potential is an excellent approximation to the true potential, E, and, therefore, is used in the polarization-curve data output from the potentiostat. It is noted in Fig. 6.9 that after current interruption, if the current were not restored, the potential would continue to exponentially decay toward the open-circuit potential, $E_{corr}$.

## Electrochemical Corrosion-Rate Measurement Methods and the Uniform-Corrosion Consideration

The thermodynamic and kinetic principles along with measurement techniques described in previous sections provide the basis for both predicting and measuring rates of corrosion. All electrochemical techniques for corrosion-rate determination are directed to measurement of the corrosion current, $I_{corr}$, from which the corrosion current density ($i_{corr} = I_{corr}/A_a$), the corrosion intensity, and the corrosion penetration rate are calculated, providing the area of the anodic sites ($A_a$) also can be determined. In the limit, these sites are assumed to be uniformly distributed on a scale approaching atom dimensions and indistinguishable from sites of the cathodic reaction supporting the corrosion. In this limit, the corrosion is uniform, and the area of the anodic sites ($A_a$) is taken to be the total specimen area (A). From this limit, anodic sites can vary from microscopic to macroscopic dimensions, thus leading to localized corrosion. Hence, polarization measurements leading to a value for the corrosion current density by dividing the corrosion current by the total specimen area ($i_{corr} = I_{corr}/A$) must be accompanied by a surface examination to determine the actual anodic areas. Further, if there is a distribution over both anodic and cathodic sites with respect to the current density of these respective reactions, the calculations are obviously more difficult. Frequently, the heterogeneity of these reactions over the surface must be evaluated qualitatively, recognizing that the calculated corrosion current density, $i_{corr} = I_{corr}/A$, gives only a lower limit to the actual current density and hence that local corrosion intensities and penetration rates can be much higher. Assuming that a specimen surface undergoing measurement contains at least a statistical distribution of anodic and cathodic sites and that the intersite electrical resistance is small, previous discussions (Chapter 4) have shown that the intersection of the extrapolated Tafel regions of the anodic and cathodic polarization curves gives $I_{corr}$. To establish this intersection experimentally requires determination of the anodic and cathodic polarization curves in the vicinity of the intersection. Since the data analysis techniques involve extrapolations and measurements of slopes of these curves, the accuracy of their experimental determination is important. Thus, the experimental methods must be critically evaluated with respect to their sensitivity to the polarization variables and how various conditions established at the interface by the variables contribute to an electrochemical measurement. These variables include exchange current densities, Tafel slopes, diffusion of species to and from the interface, corrosion-product formation, and the potential scan rate.

Some of the experimental methods for determining $I_{corr}$, and in some cases, additional information, are discussed in the following sections. The methods are used to not only establish the corrosion characteristics of a metal/environment system of immediate interest but also as tools to investigate the wide range of variables that can be imposed on a system and the intercomparison of different metal/environment systems. Some obvious variables include pH; temperature; types and concentrations of oxidizing agents supporting the cathodic reaction, including the frequently important variable of dissolved oxygen; corrosion inhibitors that may selectively affect either the anodic or cathodic reaction; mechanical and thermal treatments applied to the metal or alloy, including comparisons of cast, wrought, and welded structures; surface finishes, including chemical modifications and coatings; and galvanic coupling between metals or between metals and conducting surface layers such as scales.

## *Tafel Analysis*

It is shown in Chapter 3 that a simple kinetic model of half-cell reactions leads to Tafel equations in which the overpotentials ($\eta$) or polarizations of the oxidation and reduction components of a half-cell reaction are linearly dependent on the logarithm of the oxidation and reduction currents ($I_{ox}$ and $I_{red}$), respectively, or

$$\eta = E - E' = +\beta_{ox} \log \frac{I_{ox}}{I_o} \quad \text{(oxidation)} \qquad \text{(Eq 6.21)}$$

$$\eta = E - E' = -\beta_{red} \log \frac{I_{red}}{I_o} \quad \text{(reduction)} \qquad \text{(Eq 6.22)}$$

where $E'$ is the equilibrium potential of the half-cell reaction, $\beta_{ox}$ and $\beta_{red}$ are the Tafel constants for the oxidation and reduction components, and $I_o$ is the exchange current for the half-cell reaction. In Chapter 4, simple corrosion processes (controlled only by charge transfer at the metal interface) are analyzed as coupled half-cell reactions between the oxidation reaction of a metal and the reduction reaction of the species causing corrosion (e.g., $H^+$ ions or dissolved $O_2$). The result is illustrated graphically in Fig. 6.2 in which the corrosion potential, $E_{corr}$, and the corrosion current, $I_{corr}$, are given by the intersection of the Tafel curves for the metal oxidation and cathodic-reactant reduction reactions. Therefore, if the parameters governing the anodic and cathodic reactions ($E'$, $I_o$, and $\beta$ values) of a corroding system are available, the corrosion current density ($I_{corr}$/area) can be calculated, and hence the corrosion intensity or penetration rate can be determined by application of Faraday's law (see Chapter 4 and Table 6.2). Since even approximate values of the parameters usually are not known, reliable calculations cannot be made. However, the forms of the curves in Fig. 6.2 become the basis of several experimental procedures referred to as Tafel analysis.

Although the primary objective of Tafel analysis based on experimental measurements is the determination of the corrosion current density, $i_{corr}$, the measurements also can give values for the cathodic and anodic Tafel constants, $\beta_{red,X}$ and $\beta_{ox,M}$, and estimates of the exchange current densities, $i_{o,X}$ and $i_{o,M}$. The values of these parameters can provide information on the kinetic mechanisms of the electrochemical reactions, particularly by observing changes in the parameters with changes in the electrolyte composition observed to influence the corrosion rate.

If the experimentally determined anodic and cathodic polarization curves have the forms shown in Fig. 6.2 (solid curves), the linear sections can be extrapolated to $E_{corr}$ to give values of $I_{corr}$, from which the corrosion rate (corrosion intensity or corrosion penetration rate) can be calculated. Unfortunately, the curves may not show sufficiently distinct linear sections to allow acceptable extrapolation. A potential scan greater than ±(50 to 100) mV about $E_{corr}$ is generally required to reach potentials at which the anodic-Tafel or cathodic-Tafel behavior dominates and linear polarization is expected. As this deviation from $E_{corr}$ occurs, conditions at the metal/solution interface may change progressively and prevent linear behavior. Changing interface conditions may include corrosion-product buildup along the anodic branch and corrosion-product reduction along the cathodic branch, diffusion of species to and from the interface, and IR potential drops between the working and reference electrodes in the potentiostat circuit (Fig. 6.8).

**Tafel Extrapolation.** The most fundamental procedure for experimentally evaluating $I_{corr}$ is by Tafel extrapolation. This method requires the presence of a linear or Tafel section in the E versus log $I_{ex}$ curve. A potential scan of ±300 mV about $E_{corr}$ is generally required to determine whether a linear section of at least one decade of current is present such that a reasonably accurate extrapolation can be made to the $E_{corr}$ potential. Such linear sections are illustrated for the cathodic polarization curves in Fig. 6.2 to 6.5. The current value at the $E_{corr}$ intersection is the corrosion current, $I_{corr}$, as shown in Fig. 6.10. Assuming uniform corrosion, the corrosion current density is obtained by dividing $I_{corr}$ by the specimen area (i.e., $i_{corr} = I_{corr} / A$). Anodic polarization curves are not often used in this method because of the absence of linear regions over

**Table 6.2 Faraday's law expressions**

**Corrosion intensity (CI)**

$$CI\ (g/m^2 \cdot y) = 0.327\ \frac{M\ i_{corr}}{m}$$

$$CI\ (m\ g/cm^2 \cdot y) = 0.0327\ \frac{M\ i_{corr}}{m}$$

**Corrosion penetration rate (CPR)**

$$CPR\ (\mu m/y) = 0.327\ \frac{M\ i_{corr}}{m\rho}$$

$$CPR\ (mm/y) = 0.327 \times 10^{-3}\ \frac{M\ i_{corr}}{m\rho}$$

$$CPR\ (mpy) = 0.0129\ \frac{M\ i_{corr}}{m\rho}$$

Note: M, g/mol; m, oxidation state or valence; ρ, g/cm$^3$; $i_{corr}$, mA/m$^2$; and mpy, mils (0.001 in.) per year. For alloys, use atomic-fraction-weighted values for M, m, and ρ. This procedure assumes nonselective corrosion of the elemental constituents of the alloy.

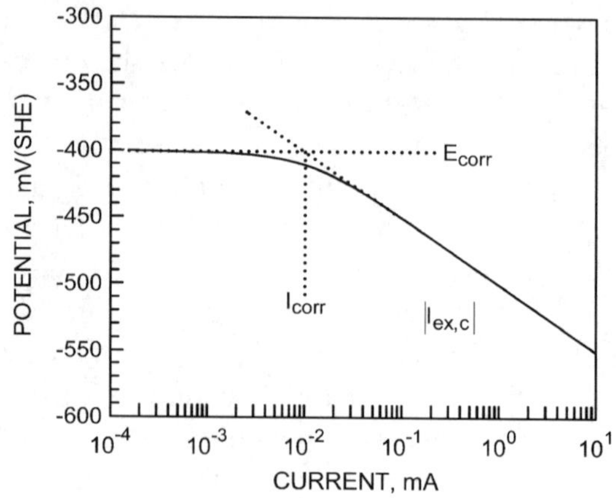


**Fig. 6.10** The Tafel extrapolation method

at least one decade of current for many metals and alloys exhibiting active-passive behavior. For example, inspection of Fig. 6.5 shows that extrapolation of the linear portion of the cathodic curve would yield more accurate results than attempted extrapolation of the anodic curve. In many cases, a linear region may not be observed even in the cathodic curve. This can be a result of the corrosion being under diffusion control or, on decreasing the potential, entering into the diffusion-control region, or even that the nature of the interface changes with changing potential.

The time required to determine $I_{corr}$ by Tafel extrapolation is approximately 3 h, which corresponds to the approximate time required for experimental setup and generation of a cathodic polarization curve at a commonly employed, slow scan rate of 600 mV/h. In comparison, a comparable gravimetric evaluation (mass-loss measurement) on a corrosion-resistant metal or alloy could take months, or longer. A limitation of the Tafel extrapolation method is the rather large potential excursion away from $E_{corr}$, which tends to modify the WE surface, such that if the measurement is to be repeated, the sample should be re-prepared following initial procedures and again allowed to stabilize in the electrolyte until a steady-state $E_{corr}$ is reached. Consequently, the Tafel extrapolation method is not amenable to studies requiring faster, or even continuous, measurements of $I_{corr}$.

**Tafel Curve Modeling (Ref 4, 5).** Equation 6.5 provides the form of the experimental polarization curve when the anodic and cathodic reactions follow Tafel behavior. The equation accounts for the curvature near $E_{corr}$ and $I_{corr}$, which is observed experimentally. Physically, the curvature is a consequence of both the anodic and cathodic reactions having measurable effects on $I_{ex}$ at potentials near $E_{corr}$. Tafel-curve modeling uses experimental data taken within approximately ±25 mV of $E_{corr}$ where the corrosion process is less disturbed by induced corro-

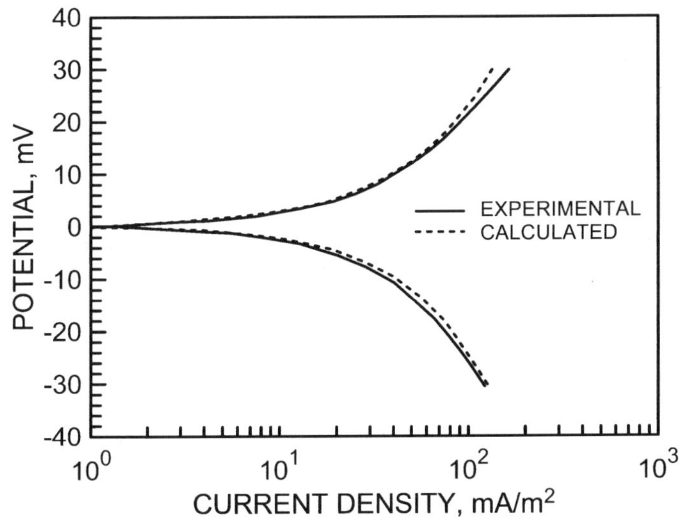


**Fig. 6.11** Comparison of experimental and calculated polarization curves for low-carbon steel in boric acid at 10% saturation, where the potential is $E - E_{corr}$

sion-product formation or diffusion-limiting processes. The procedure is to take the experimentally measured $E_{corr}$ value, and the pairwise values of E and $I_{ex}$ along the experimental polarization curve, and use mathematical techniques to determine the values of $I_{corr}$, $\beta_{ox,M}$, and $\beta_{red,X}$, which provide the best fit of Eq 6.5 to the experimental data. Statistical methods are used to determine the goodness of fit. The corrosion current density can then be calculated knowing the specimen area, and hence any measure of corrosion rate, such as the corrosion penetration rate, can be determined through Faraday's law (see Chapter 4 and Table 6.2). In addition, Tafel-curve modeling gives experimentally determined values of the slopes of the Tafel curves ($\beta_{ox}$ and $-\beta_{red}$). These can then be examined for information on the nature of the interface reactions and can be used in the polarization-resistance technique for determining corrosion rates. The agreement that can be obtained between experimental and calculated polarization curves by the method just described is illustrated in Fig. 6.11 for the polarization of low-carbon steel over ±25 mV about $E_{corr}$ in saturated boric-acid solution at 49 °C.

## *Polarization Resistance (Ref 6–11)*

The polarization-resistance, or Stern-Geary (Ref 12), method allows faster corrosion-rate measurements. The theoretical justification for this method is based on the expression for the external current given by Eq 6.5, which, on division by the specimen area to convert to current density, has the form:

$$i_{ex} = i_{corr}\left[e^{+2.3(E-E_{corr})/\beta_{ox,M}} - e^{-2.3(E-E_{corr})/\beta_{red,X}}\right] \quad \text{(Eq 6.23)}$$

This equation has the form of the solid curve in Fig. 6.12 when plotted near $E_{corr}$ (usually within ±25 mV of $E_{corr}$).

Differentiation of Eq 6.23 with respect to E yields:

$$\frac{di_{ex}}{dE} = i_{corr}\left[\frac{2.3}{\beta_{ox,M}} e^{2.3(E-E_{corr})/\beta_{ox,M}} + \frac{2.3}{\beta_{red,X}} e^{-2.3(E-E_{corr})/\beta_{red,X}}\right] \quad \text{(Eq 6.24)}$$

At $E = E_{corr}$, the exponential terms are unity, and upon rearrangement, Eq 6.24 reduces to:

$$\left(\frac{dE}{di_{ex}}\right)_{E_{corr}} = R_p = \frac{\beta_{ox,M}\,\beta_{red,X}}{2.3\, i_{corr}\,(\beta_{ox,M} + \beta_{red,X})} \quad \text{(Eq 6.25)}$$

where $(dE/di_{ex})_{E_{corr}}$ is known as the polarization resistance, $R_p$. It has dimensions of resistance times area (i.e., total specimen area) (e.g., ohms-m$^2$). As seen by Eq 6.25 and indicated in Fig. 6.12, $R_p$ is the slope of the experimental E versus $i_{ex}$ curve at $E_{corr}$. The curve tends to be linear near $E_{corr}$, which facilitates determination of the slope. A linear relationship is shown by applying the following series expansion to Eq 6.23:

$$e^x = 1 + x + \frac{x^2}{2!} + \ldots + \frac{x^n}{n!} \quad \text{(Eq 6.26)}$$

Assuming that the third and higher terms in the series are negligible (i.e., x is small), Eq 6.23 takes the form:

$$\frac{E - E_{corr}}{i_{ex}} = \frac{\beta_{ox,M}\,\beta_{red,X}}{2.3\, i_{corr}\,(\beta_{ox,M} + \beta_{red,X})} \quad \text{(Eq 6.27)}$$

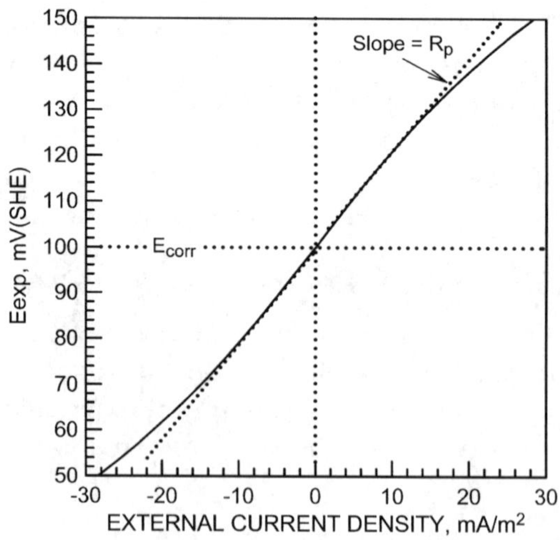


**Fig. 6.12** The polarization resistance method

This equation shows that E versus $i_{ex}$ is linear (i.e., the term on the right side of the equation is a constant) provided the quantities $(E - E_{corr})/\beta_{ox,M}$ and $(E - E_{corr})/\beta_{red,X}$ are small (i.e., provided the x values in the series expansion are small). Since a typical value for the Tafel constants is 100 mV, the condition is generally considered to be met when $(E - E_{corr})$ is less than about 10 mV.

Equation 6.25 may be rewritten in the following form, since the desired quantity in the polarization-resistance analysis is the corrosion current density:

$$i_{corr} = \frac{\beta_{ox,M}\,\beta_{red,X}}{2.3\,R_p\,(\beta_{ox,M} + \beta_{red,X})} = \frac{B}{R_p} \qquad \text{(Eq 6.28)}$$

This equation is used directly to determine $i_{corr}$, providing that the experimentally measured potential, $E_{exp}$, is the actual potential at the WE/electrolyte interface, E (i.e., no IR correction is needed). Under these conditions, the analysis procedure involves evaluating the slope of the E versus $i_{ex}$ curve at $E_{corr}$, as shown in Fig. 6.12, to determine $R_p$. From $R_p$, and known or experimentally determined Tafel constants (β values), $i_{corr}$ is calculated. If an IR correction is necessary, then, because $E = E_{exp} - i_{ex}R_S$ (Eq 6.20):

$$R_p = \left(\frac{dE}{di_{ex}}\right)_{E_{corr}} = \left(\frac{dE_{exp}}{di_{ex}}\right)_{E_{corr}} - R_S \qquad \text{(Eq 6.29)}$$

Therefore, in this case, the experimental slope must be corrected by the value of $R_S$ to obtain $R_p$.

As previously stated, once $R_p$ is determined, calculation of $i_{corr}$ requires knowledge of the Tafel constants. These constants can be determined from experimental anodic and cathodic polarization curves, or by Tafel-curve modeling, for the material and solution of interest as discussed earlier. In the absence of these values, an approximation is sometimes used.

In terms of rationalizing an approximation for B in Eq 6.28, it is convenient to express B as:

$$B = \frac{1}{2.3(1/\beta_{ox,M} + 1/\beta_{red,X})} \qquad \text{(Eq 6.30)}$$

It has been observed that experimental values of $\beta_{ox,M}$ normally range between 60 and 120 mV, whereas values of $\beta_{red,X}$ normally range between 60 mV and infinity (the latter corresponding to diffusion control for the cathodic reaction) (Ref 11, 13). Given the ranges in β values, the extreme values of B are 13 and 52 mV, corresponding to $\beta_{ox,M} = \beta_{red,X} = 60$ mV and $\beta_{ox,M} = 120$ mV, $\beta_{red,X}$ = infinity, respectively. If $\beta_{ox,M} = \beta_{red,X} = 120$ mV is used as an approximation, then

$B = 26$ mV. The expected error in the calculated value of $i_{corr}$ (Eq 6.28) when using $B = 26$ mV as an approximation (as compared with extreme values of 13 and 52 mV) should be less than a factor of two. Therefore, the following approximation provides a reasonably good estimate of $i_{corr}$ from polarization-resistance measurements:

$$i_{corr} \cong \frac{26 \text{ mV}}{R_p} \tag{Eq 6.31}$$

In generating an $E_{exp}$ versus $i_{ex}$ curve for polarization-resistance analysis, only very small potential excursions about $E_{corr}$ are employed, normally ±10 mV or less. The general assumption is that on scanning through this small potential range, the material surface remains unchanged. Consequently, repeat measurements may be made as a function of time without removing the sample and re-preparing the surface.

## *Electrochemical Impedance Spectroscopy (EIS) (Ref 14–18)*

This method for evaluating the corrosion rate is based on measurement of alternating current (ac) impedance over a range of applied frequencies. The method is rapidly expanding due to the development of applicable instruments and the capability of the method in providing additional information on the corrosion process and electrochemical-cell performance. The method normally involves application of time-varying, small, potential excursions about $E_{corr}$, measurement of $I_{ex}$, and determination of the system impedance, Z, and the impedance phase angle, δ. The applied ac potential, $(E_{exp} - E_{corr})$, is normally sinusoidal, as indicated in Fig. 6.13. At each frequency, the measured external current, $I_{ex}$, may be out of phase with the applied potential, as indicated in Fig. 6.13, due to physical processes that behave as capacitive or inductive elements (rather than resistive elements) in an electrical circuit considered to be "equivalent" to the electrochemical cell. Consequently, the system impedance also may be out of phase with the applied potential. In the EIS analysis, the impedance of the system, Z, and the phase angle between the impedance and the applied potential, δ, are

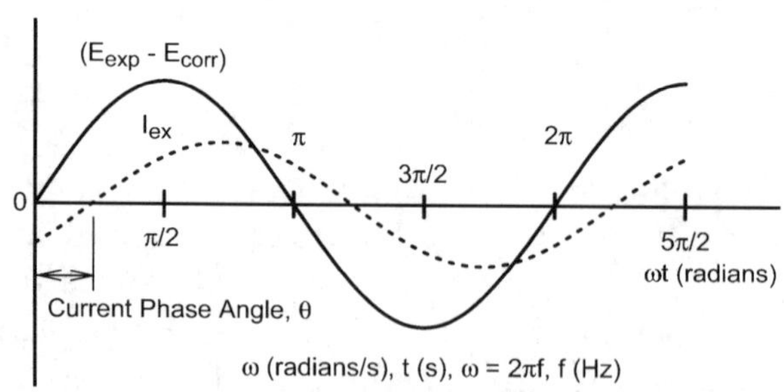


**Fig. 6.13** Applied potential and resultant external current relative to the electrochemical impedance spectroscopy method

determined as a function of applied frequency. These quantities are then interpreted in relationship to the electrochemical, chemical, and physical processes associated with the cell. To obtain maximum information, the impedance and phase angle must be determined over a wide range of frequencies.

In contrast to the EIS method, the Tafel-extrapolation, Tafel-curve-modeling and polarization-resistance methods are conducted under essentially dc conditions. In these cases, in generating the appropriate $E_{exp}$ versus log $i_{ex}$ or $i_{ex}$ curve, the potentiodynamic potential scan rate is very slow, or the time between potentiostatic potential steps is very long. The common practice is a potential scan rate of 600 mV/h or an equivalent step rate of 50 mV every 5 min. Under these conditions, it is assumed that a steady-state, external-current-density results at every discrete potential. Consequently, every element in the electrical circuit is purely resistive in nature, and therefore, the applied potential and resultant external-current-density are exactly in phase. Since the impedance (normalized with respect to specimen area) is $dE_{exp}/di_{ex}$, under these conditions, the impedance, Z, at $E_{corr}$ is given by (see Eq 6.29):

$$Z = \left(\frac{dE_{exp}}{di_{ex}}\right)_{E_{corr}} = R_p + R_S \qquad \text{(Eq 6.32)}$$

with a phase angle equal to zero. However, even under these conditions, since the potential has to be scanned or stepped at some rate, the effective frequency is not zero. In the polarization-resistance method, if a 600 mV/h rate is used to go from $E_{corr}$ to $(E_{exp} - E_{corr}) = 10$ mV, and this $E_{exp}$ versus time variation is assumed to be $\frac{1}{4}$ of a triangular wave, the effective frequency is $4.17 \times 10^{-3}$ Hz.

**Some Basic Relationships in ac Circuit Analysis.** Consider an ac voltage, V, applied to the circuit shown in Fig. 6.14(a), which consists of a resistor and a capacitor in parallel. For the resistor, the voltage, V, and current, $I_1$, as a function of time, t(s), at a given frequency, $\omega_1$(radians/s), are illustrated in Fig. 6.14(b) and are given by the equations:

$$V = V_{max} \sin \omega_1 t \qquad \text{(Eq 6.33)}$$

$$I_1 = I_{1,max} \sin \omega_1 t \qquad \text{(Eq 6.34)}$$

The current and voltage are exactly in phase, and therefore, the phase angle of the current relative to the voltage, $\theta$, is zero. Next, consider the capacitor, where the same ac voltage, V, is applied at the same frequency, $\omega_1$. The capacitance, C, is given by:

$$C = q/V \qquad \text{(Eq 6.35)}$$

where C is in farads, the charge, q, in coulombs, and V in volts. Substitution for V and rearrangement yields:

$$q = C\,V_{max} \sin \omega_1 t \qquad \text{(Eq 6.36)}$$

The capacitor current, $I_2$, is equal to dq/dt; thus:

$$I_2 = \omega_1 C V_{max} \cos \omega_1 t \qquad \text{(Eq 6.37)}$$

$$I_2 = \frac{V_{max}}{(1/\omega_1 C)} \sin(\omega_1 t + \pi/2) \qquad \text{(Eq 6.38)}$$

$$I_2 = \frac{V_{max}}{X_c} \sin(\omega_1 t + \pi/2) = I_{2,max} \sin(\omega_1 t + \pi/2) \qquad \text{(Eq 6.39)}$$

where $I_{2,max} = V_{max}/X_c$ and $X_c = 1/\omega_1 C$ is the capacitive reactance. Therefore, the capacitive current is out of phase relative to the voltage by the phase angle, $\theta = \pi/2$. The voltage and current for the capacitor as a function of time at the frequency, $\omega_1$, are shown in Fig. 6.14(c).

The resultant current, $I_r$ (Fig. 6.14a), and the phase angle, $\theta_r$, between the resultant current and the voltage are obtained by adding the resistive current, $I_1$, and the capacitive current, $I_2$. This operation can be accomplished by treating the currents and voltage as rotating vectors, $\mathbf{I_1}$, $\mathbf{I_2}$, and **V** (indicated by bold type), as shown in Fig. 6.15(a). The vectors have magnitudes of $I_{1,max}$, $I_{2,max}$, and $V_{max}$, and all rotate at the same angular frequency, $\omega_1$. At any time (e.g., $t_1$) the V, $I_1$, and $I_2$ values are the vertical components of the corresponding vectors; that is, $V = V_{max} \sin \omega_1 t_1$, $I_1 = I_{1,max} \sin \omega_1 t_1$ and $I_2 = I_{2,max} \sin(\omega_1 t_1 + \pi/2)$. In Fig. 6.15(a), at $t_1$, the vertical components of the vectors are projected across to the corresponding points on the sine waves in Fig. 6.15(b). The resultant current at $t_1$, $I_r$, may be determined by adding the $I_1$ and $I_2$ sine-wave values in Fig. 6.15(b) or by using the parallelogram law to add the $\mathbf{I_1}$ and

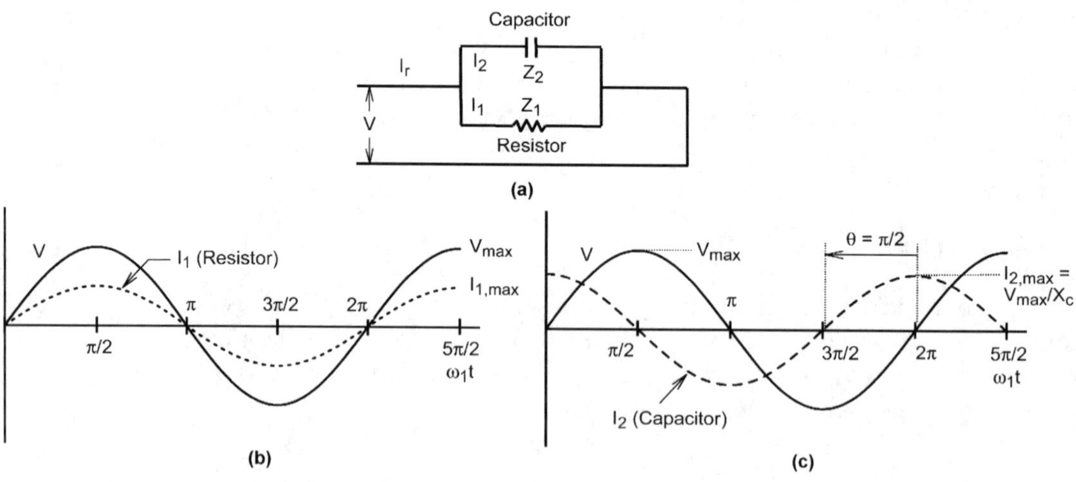


**Fig. 6.14** (a) A simple resistor/capacitor parallel circuit and the corresponding voltage and current variations for the (b) resistor and (c) capacitor

$\mathbf{I_2}$ vectors in Fig. 6.15(a) to obtain $\mathbf{I_r}$, then recognizing that the vertical component of $\mathbf{I_r}$ is the value of $I_r$ at $t_1$; that is, $I_r = I_{r,max} \sin(\omega_1 t_1 + \theta_r)$, where $\theta_r$ is the phase angle between the resultant current and the voltage. It is noted that the vector addition easily provides both the maximum value and the phase angle for the resultant current; these quantities are all that are needed to fully describe the resultant current.

Since all of the vectors in Fig. 6.15(a) are rotating at the same angular frequency ($\omega_1$) and thus always are maintaining the same relative positions, it is convenient to treat them as stationary vectors, as indicated in Fig. 6.16. In this figure, the vector magnitudes are the same as in Fig. 6.15(a), but the angles are the phase angles, θ, between the currents and the voltage. Furthermore, for ease of mathematical analysis of the vectors, it is convenient to employ complex notation with "real" and "imaginary" components. These components are defined relative to the reference waveform (i.e., the applied voltage, **V**). The "real" component is exactly in phase with **V**, whereas the "imaginary" component is exactly

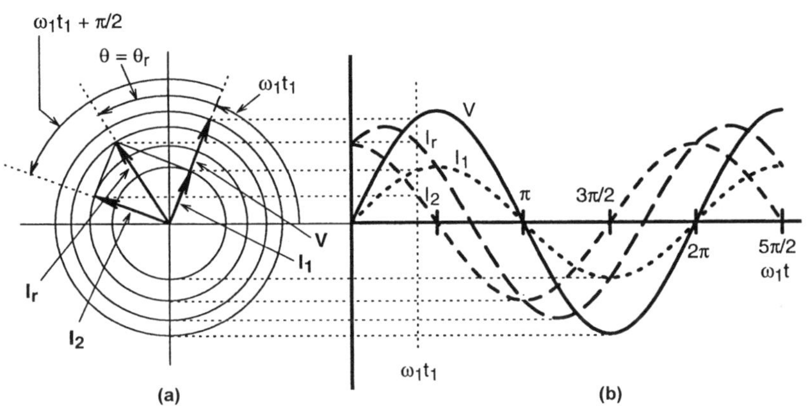


**Fig. 6.15** (a) Rotating-vector and (b) sine-wave descriptions of the voltage and current variations in the ac circuit of Fig. 14(a)

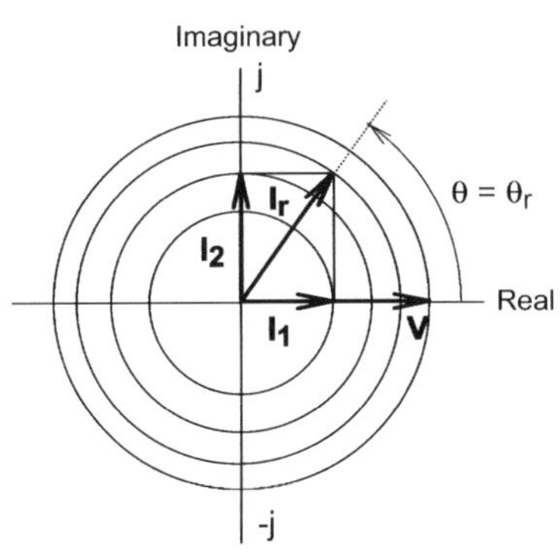


**Fig. 6.16** Stationary-vector descriptions of the voltage and currents in the ac circuit of Fig. 14(a)

90° out of phase with **V**. For example, consider an arbitrary current vector, **I**, with a phase angle, θ, relative to **V**. The "real" component is $I_{max}$ cos θ, the "imaginary" component is $I_{max}$ sin θ, and the complex-notation description is $\mathbf{I} = I_{max}(\cos\theta + j\sin\theta)$, where $j = \sqrt{-1}$. Therefore, with reference to Fig. 6.16, the vectors **V**, $\mathbf{I_1}$, and $\mathbf{I_2}$ may be expressed as:

$$\mathbf{V} = V_{max}(\cos\theta + j\sin\theta) = V_{max}(\cos 0 + j\sin 0) = V_{max}(1 + j0) \quad (6.40)$$

$$\mathbf{I_1} = I_{1,max}(\cos\theta + j\sin\theta) = I_{1,max}(\cos 0 + j\sin 0) = I_{1,max}(1 + j0) \quad (6.41)$$

$$\mathbf{I_2} = I_{2,max}(\cos\theta + j\sin\theta) = I_{2,max}(\cos\pi/2 + j\sin\pi/2) = I_{2,max}(0 + j1) \quad (6.42)$$

The resultant-current vector, $\mathbf{I_r}$, is then determined as the sum of $\mathbf{I_1}$ and $\mathbf{I_2}$:

$$\mathbf{I_r} = \mathbf{I_1} + \mathbf{I_2} = I_{1,max} + jI_{2,max} \quad \text{(Eq 6.43)}$$

The magnitude of the $\mathbf{I_r}$ vector (i.e., $I_{r,max}$) is given by the square root of the sum of the squares of its "real" component, $I_{1,max}$, and its "imaginary" component, $I_{2,max}$:

$$|\mathbf{I_r}| = I_{r,max} = \left[(I_{1,max})^2 + (I_{2,max})^2\right]^{1/2} \quad \text{(Eq 6.44)}$$

Also, as seen in Fig. 6.16, the tangent of the phase angle of $\mathbf{I_r}$ relative to the voltage is given by the "imaginary" component divided by the "real" component:

$$\tan\theta_r = I_{2,max}/I_{1,max} \quad \text{(Eq 6.45)}$$

And finally, the equation for the resultant-current sine wave is:

$$I_r = I_{r,max}\sin(\omega_1 t + \theta_r) \quad \text{(Eq 6.46)}$$

Now, with reference to Fig. 6.17, consider the impedances relative to the above situation, while continuing to employ vectors and complex notation. For the resistor, the impedance, $\mathbf{Z_1}$, is given by:

$$\mathbf{Z_1} = \mathbf{V}/\mathbf{I_1} = \frac{V_{max}(1 + j0)}{I_{1,max}(1 + j0)} = \frac{V_{max}}{I_{1,max}} = R \quad \text{(Eq 6.47)}$$

where R is the resistance. Since the current and voltage are exactly in phase, the impedance is correspondingly in phase. Therefore, the impedance phase angle, δ, for the resistor is equal to zero (i.e., the phase angle between the impedance, $\mathbf{Z_1}$, and the voltage, **V**, which serves as the reference, is zero). For the capacitor, the impedance, $\mathbf{Z_2}$, is given by:

$$\mathbf{Z_2} = \mathbf{V}/\mathbf{I_2} = \frac{V_{max}(1 + j0)}{I_{2,max}(0 + j1)} = \frac{-jV_{max}}{I_{2,max}} = \frac{-jV_{max}}{(V_{max}/X_c)} = -jX_c \quad \text{(Eq 6.48)}$$

The tangent of the impedance phase angle for the capacitor is given by the "imaginary" component divided by the "real" component, tan $\delta = -X_c/0 = -\infty$. Therefore, the impedance phase angle for the capacitor is $-\pi/2$.

The equivalent impedance of the resistor/capacitor parallel circuit in Fig. 6.14(a), **Z**, must be determined by application of Kirchhoff's rule (i.e., the algebraic sum of the voltages of the voltage sources in any circuit loop must equal the algebraic sum of the voltage drops in the same loop). Thus:

$$\mathbf{V} = \mathbf{I_2 Z_2} \quad \text{(Eq 6.49)}$$

$$\mathbf{V} = \mathbf{I_1 Z_1} \quad \text{(Eq 6.50)}$$

Also:

$$\mathbf{I_r} = \mathbf{I_1} + \mathbf{I_2} \quad \text{(Eq 6.51)}$$

Elimination of $\mathbf{I_1}$ and $\mathbf{I_2}$ yields:

$$\mathbf{I_r} = \mathbf{V/Z_1} + \mathbf{V/Z_2} = \mathbf{V/[Z_1Z_2/(Z_1 + Z_2)]} = \mathbf{V/Z} \quad \text{(Eq 6.52)}$$

where **Z** is the equivalent circuit impedance, that is:

$$\mathbf{Z} = (\mathbf{Z_1Z_2})/(\mathbf{Z_1} + \mathbf{Z_2}) = \frac{R(-jX_c)}{R - jX_c} = \frac{RX_c^2}{(R^2 + X_c^2)} - j\frac{R^2X_c}{(R^2 + X_c^2)} \quad \text{(Eq 6.53)}$$

The magnitude of **Z** is:

$$|\mathbf{Z}| = \left[\frac{R^2X_c^4}{(R^2 + X_c^2)^2} + \frac{R^4X_c^2}{(R^2 + X_c^2)^2}\right]^{1/2} = \frac{RX_c}{(R^2 + X_c^2)^{1/2}} \quad \text{(Eq 6.54)}$$

The phase angle, δ, between **Z** and **V** is given by:

$$\tan\delta = \frac{-R^2X_c/(R^2 + X_c^2)}{RX_c^2/(R^2 + X_c^2)} = \frac{-R}{X_c} \quad \text{(Eq 6.55)}$$

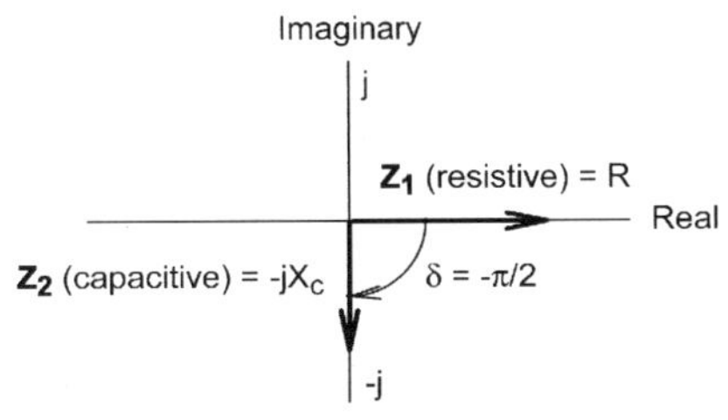


**Fig. 6.17** Stationary-vector descriptions of the impedances in the ac circuit of Fig. 14(a)

**Details of the EIS Method (Ref 14–18).** In the EIS method, potentials are applied over the frequency range of approximately $10^{-3}$ to $10^4$ Hz in order to provide full information on the corrosion process and cell performance. Since the amplitude of the potential wave is small, on the order of 10 mV, the assumption is made (as with the polarization-resistance method) that the surface of the material is not disturbed.

In the EIS method, the electrochemical cell is modeled by an equivalent electrical circuit with each element in the circuit corresponding to an electrochemical, chemical, or physical process taking place in the cell. If the impedance of the assumed model differs from the observed impedance, the model is changed until reasonable agreement results. The simplest equivalent circuit for EIS analysis is shown in Fig. 6.18, where $R_p$ is the polarization resistance, and $R_S$ is the solution resistance (normalized with respect to specimen area) between the working electrode (corrosion specimen) and reference electrode. C is the capacitance (normalized with respect to specimen area) associated with the specimen/electrolyte interface, in simple cases the double-layer capacitance, and must be considered when alternating, higher frequency potentials are employed. Physically, C relates to ions and polar molecules in the electrolyte that undergo charge redistribution and hence produce a current under time-varying potentials, but not after a decay time at constant potentials (steady-state condition).

It is noted that the equivalent circuit in Fig. 6.18 has been normalized with respect to specimen (WE) area. Therefore, external current density is shown, $i_{ex} = I_{ex}/A$, and the dimensions of the impedances are electrical resistance times specimen area (e.g., ohm-m$^2$). To analyze this ac circuit, the potential, current density, and impedances will be treated as vectors (again indicated by bold type), which will be represented by complex numbers. The analysis first will provide the equivalent impedance of the circuit, **Z**, and then the phase angle, δ, of the impedance with respect to the applied potential, $(\mathbf{E_{exp}} - E_{corr})$.

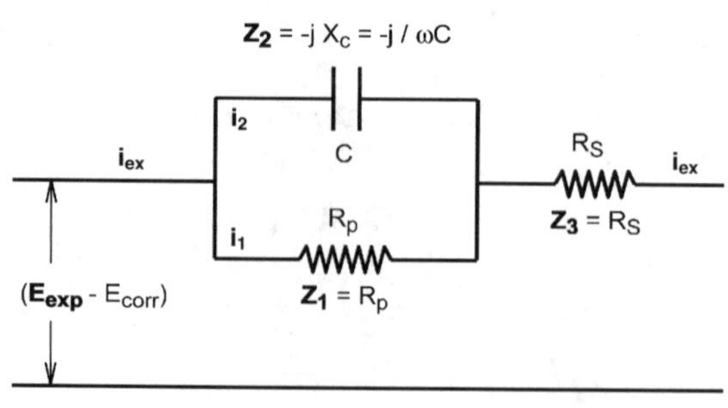


**Fig. 6.18** Electrochemical impedance spectroscopy (EIS): simplest electrical-circuit model

Application of Kirchhoff's rule to the equivalent circuit yields:

$$(\mathbf{E_{exp}} - E_{corr}) = \mathbf{i_1}\,\mathbf{Z_1} + \mathbf{i_{ex}}\,\mathbf{Z_3} \qquad \text{(Eq 6.56)}$$

$$(\mathbf{E_{exp}} - E_{corr}) = \mathbf{i_2}\,\mathbf{Z_2} + \mathbf{i_{ex}}\,\mathbf{Z_3} \qquad \text{(Eq 6.57)}$$

Also:

$$\mathbf{i_{ex}} = \mathbf{i_1} + \mathbf{i_2} \qquad \text{(Eq 6.58)}$$

Elimination of $\mathbf{i_1}$ and $\mathbf{i_2}$ from the above equations yields:

$$\mathbf{i_{ex}} = (\mathbf{E_{exp}} - E_{corr})/\mathbf{Z} \qquad \text{(Eq 6.59)}$$

where

$$\mathbf{Z} = (\mathbf{Z_1 Z_2} + \mathbf{Z_2 Z_3} + \mathbf{Z_1 Z_3}) \,/\, (\mathbf{Z_1} + \mathbf{Z_2}) \qquad \text{(Eq 6.60)}$$

In complex notation, the impedances are given as:

$$\mathbf{Z_1} = R_p \qquad \text{(Eq 6.61)}$$

$$\mathbf{Z_2} = -jX_c = -j/\omega C \qquad \text{(Eq 6.62)}$$

$$\mathbf{Z_3} = R_S \qquad \text{(Eq 6.63)}$$

Recall that in a purely capacitive circuit element, the phase angle between the current and applied potential is $\theta = \pi/2$ and $\mathbf{Z_2} = -jX_c$, where $X_c$ is the capacitive reactance, in this case normalized with respect to the specimen area (ohm-m$^2$). $X_c$ is equal to $1/\omega C$, where $\omega$ is the angular frequency (radians/s, that is, $\omega = 2\pi f$, where f is frequency in cycles/s or Hertz) and C is the normalized capacitance (farad/m$^2$). In a purely resistive circuit element (e.g., $R_p$ and $R_S$), the current is exactly in phase with the applied potential ($\theta = 0$); thus, $\mathbf{Z_1} = R_p$ and $\mathbf{Z_3} = R_S$, where, again, $R_p$ and $R_S$ are normalized with respect to specimen area (ohm-m$^2$). Upon substitution of Eq 6.61 to 6.63 into Eq 6.60, the equivalent circuit impedance is determined to be:

$$\mathbf{Z} = R_S + \frac{R_p}{(\omega^2 C^2 R_p^2 + 1)} - \frac{j\omega C R_p^2}{(\omega^2 C^2 R_p^2 + 1)} \qquad \text{(Eq 6.64)}$$

or

$$\mathbf{Z} = Z' + jZ'' \qquad \text{(Eq 6.65)}$$

where

$$Z' = R_S + \frac{R_p}{(\omega^2 C^2 R_p^2 + 1)} \qquad \text{(Eq 6.66)}$$

$$Z'' = \frac{-\omega C R_p^2}{(\omega^2 C^2 R_p^2 + 1)} \qquad \text{(Eq 6.67)}$$

Z′ and Z″ are the so-called real and imaginary components of the equivalent impedance. The absolute magnitude of the impedance, obtained as $|\mathbf{Z}| = ((Z')^2 + (Z'')^2)^{1/2}$, is:

$$|\mathbf{Z}| = \left[ R_S^2 + \frac{2R_S R_p}{(\omega^2 C^2 R_p^2 + 1)} + \frac{R_p^2}{(\omega^2 C^2 R_p^2 + 1)} \right]^{1/2} \qquad \text{(Eq 6.68)}$$

The impedance phase angle, δ, defined by the relationship $\tan \delta = Z''/Z'$, is given by:

$$\tan \delta = \frac{-\omega C R_p^2}{R_S + R_p + R_S(\omega C R_p)^2} \qquad \text{(Eq 6.69)}$$

EIS data often are plotted in the complex plane as –Z″ (j axis) versus Z′, the so-called Nyquist plot. With this data-presentation format, it is instructive to obtain a relationship between Z″ and Z′ by use of Eq 6.66 and 6.67 and elimination of ω. The result is:

$$(Z'')^2 = 2R_S Z' + R_p Z' - (Z')^2 - R_S^2 - R_S R_p \qquad \text{(Eq 6.70)}$$

Upon rearrangement, Eq 6.70 has the form:

$$\left[ Z' - \left( R_S + \frac{R_p}{2} \right) \right]^2 + (-Z'')^2 = \left( \frac{R_p}{2} \right)^2 \qquad \text{(Eq 6.71)}$$

which is the equation of a circle with the center on the Z′ axis at $Z' = R_S + R_p/2$, and a radius equal to $R_p/2$, as shown in Fig. 6.19. At the apex of the semicircle (i.e., at the maximum |–Z″| value), it can be

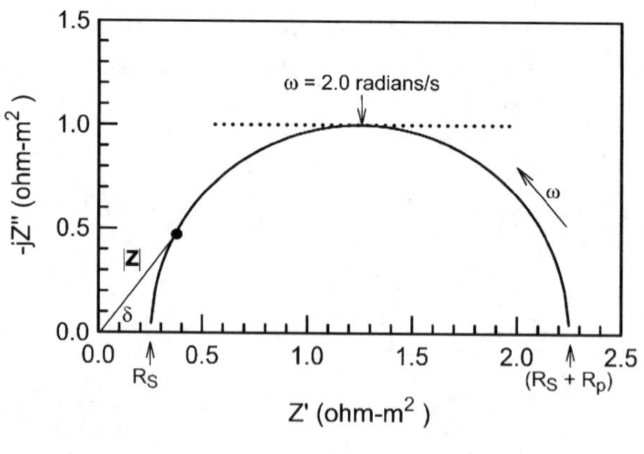


**Fig. 6.19** Electrochemical impedance spectroscopy method, Nyquist data-presentation format

shown by differentiating $-Z''$ with respect to $Z'$ and setting it equal to zero that:

$$C = \frac{1}{\omega R_p} \quad \text{at } |-Z''|_{max} \qquad \text{(Eq 6.72)}$$

Therefore, if the ac impedance data collected fit the particular model described, the data points will fit a semicircle on the Nyquist plot. From the semicircle, all three parameters can be determined directly: $R_S$ corresponds to the $Z'$ value at $-Z'' = 0$ at the highest frequency, $R_p$ corresponds to the diameter of the semicircle, and C may be calculated from Eq 6.72 using the frequency at the apex of the semicircle.

Another method often used for plotting and evaluating EIS data involves plots of log $|\mathbf{Z}|$ and $\delta$ versus log $\omega$. These data presentations are known as Bode plots and are illustrated by the example in Fig. 6.20, again for the simplest equivalent circuit of Fig. 6.18. Bode plots have advantages in that the impedance and impedance phase angle are shown as explicit functions of the frequency, which is the independent experimental variable. Reference to Eq 6.68 shows that at very high $\omega$ values, $|\mathbf{Z}|$ approaches $R_S$, and at very low frequencies, $|\mathbf{Z}|$ approaches $(R_S + R_p)$. These limits are indicated in Fig. 6.20. In analyzing intermediate frequencies, that is, when $(R_S + R_p) > |\mathbf{Z}| > R_S$, it first is convenient to rewrite Eq 6.68 in the form:

$$|\mathbf{Z}|^2 = R_S^2 + \frac{2R_S R_p}{(\omega^2 R_p^2 C^2 + 1)} + \frac{R_p^2}{(\omega^2 R_p^2 C^2 + 1)} \qquad \text{(Eq 6.73)}$$

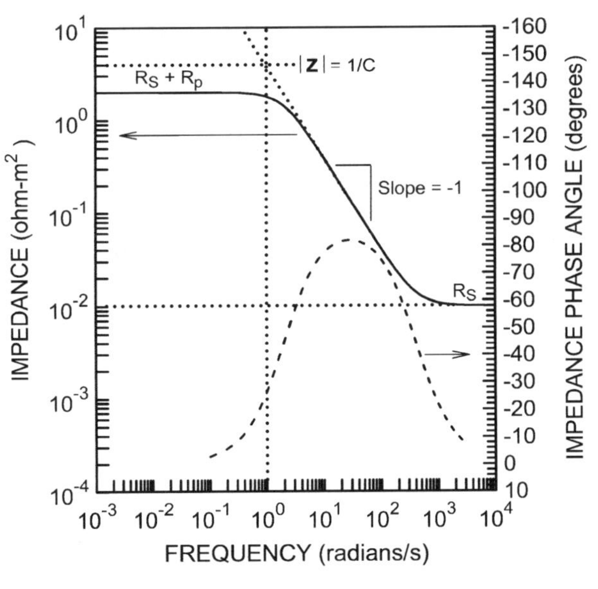


**Fig. 6.20** Electrochemical impedance spectroscopy, Bode data-presentation format

When $(R_S + R_p) > |\mathbf{Z}| > R_S$, $(R_S + R_p)^2 >> |\mathbf{Z}|^2 >> R_S^2$. Furthermore, assuming that $R_S << R_p$, then $R_p^2 >> |\mathbf{Z}|^2 >> R_S^2$. From the conditions that $|\mathbf{Z}|^2 >> R_S^2$ and $R_S << R_p$, Eq 6.73 becomes:

$$|\mathbf{Z}|^2 = \frac{R_p^2}{(\omega^2 R_p^2 C^2 + 1)} \qquad \text{(Eq 6.74)}$$

From the condition, $|\mathbf{Z}|^2 << R_p^2$, the denominator in Eq 6.74 is much greater than one, and therefore, $(\omega^2 R_p^2 C^2 + 1) \approx \omega^2 R_p^2 C^2$. Consequently, corresponding to the condition of intermediate frequencies, Eq 6.74 becomes:

$$|\mathbf{Z}| = \frac{1}{\omega C} \qquad \text{(Eq 6.75)}$$

or

$$\log |\mathbf{Z}| = -\log \omega - \log C \qquad \text{(Eq 6.76)}$$

Therefore, as shown in Fig. 6.20, extrapolation of the intermediate-frequency portion of the log $|\mathbf{Z}|$ versus log $\omega$ curve to log $\omega = 0$ (i.e., $\omega = 1$) yields:

$$\log |\mathbf{Z}| = -\log C \quad \text{at } \log \omega = 0 \qquad \text{(Eq 6.77)}$$

or

$$|\mathbf{Z}| = 1/C \quad \text{at } \omega = 1 \qquad \text{(Eq 6.78)}$$

Furthermore, by inspection of Eq 6.76, it is seen that the slope of the intermediate-frequency portion of the curve, $d(\log |\mathbf{Z}|)/d(\log \omega)$, is equal to $-1$.

If the data collected do not fit the simplest equivalent-circuit model (Fig. 6.18), more complex models are analyzed. A number of equivalent circuits have been developed to model corrosion processes involving diffusion control, porous films or coatings, pseudoinductive mechanisms, simultaneous electrochemical and chemical reactions, and pitting corrosion (Ref 14–18).

Once $R_p$ is determined by the EIS method, $i_{corr}$ is evaluated in the same way as with the polarization-resistance method (i.e., with Eq 6.28). Therefore, the Tafel constants still must be experimentally determined. The intrinsic value of the EIS method lies in the fact that extensive information is extracted (i.e., $R_p$, $R_S$, and C are all determined) and, ideally, interpreted to not only determine the corrosion rate but also the rate-controlling mechanisms at the material surface and within the electrolyte.

## *Two-Electrode Method (Ref 19–20)*

This method employs the basic principles previously described for the EIS method but with the use of two identical working electrodes. The method does not use an auxiliary electrode nor a reference electrode. With reference to Fig. 6.21, the two working electrodes (A and B), ideally, are identical in all aspects—geometry, chemical composition, microstructure, surface condition, etc. The method involves application of a low-amplitude (e.g., 20 mV) AC potential across the two electrodes, at a very low frequency (lf) and at a very high frequency (hf), and measurement of the impedance of the system at each frequency, $|\mathbf{Z}|_{lf}$ and $|\mathbf{Z}|_{hf}$. The assumed equivalent electrical circuit for the system also is indicated in Fig. 6.21. This circuit assumes that the simplest equivalent electrical circuit, as shown in Fig. 6.18, is applicable to each of the electrodes in the two-electrode method. In this case, $R_S$ is the solution resistance (normalized with respect to specimen area, for example, ohms-m$^2$) between the two electrodes. With reference to Fig. 6.21 (and also with reference to the previous discussion of the EIS method), it is seen that:

$$|\mathbf{Z}|_{lf} = 2R_p + R_S \qquad \text{(Eq 6.79)}$$

and

$$|\mathbf{Z}|_{hf} = R_S \qquad \text{(Eq 6.80)}$$

Substitution of Eq 6.80 into Eq 6.79 yields:

$$R_p = (|\mathbf{Z}|_{lf} - |\mathbf{Z}|_{hf})/2 \qquad \text{(Eq 6.81)}$$

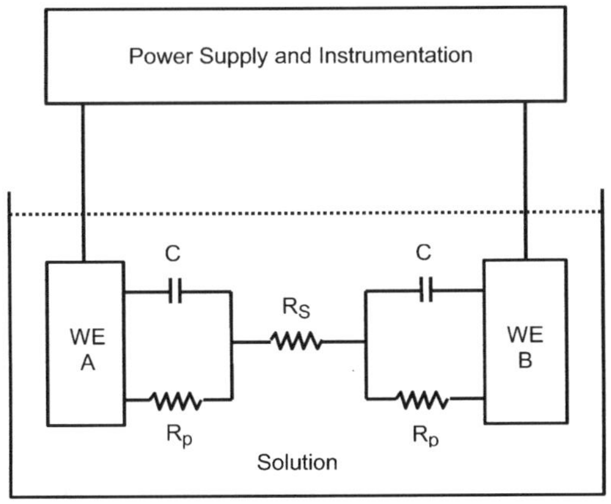


**Fig. 6.21** Two-electrode method

Once $R_p$ is determined, $i_{corr}$ is evaluated with the polarization-resistance (or Stern-Geary) equation, Eq 6.28. The two-electrode method is a relatively simple and fast method for evaluating $R_p$ when compared with the standard polarization-resistance and electrochemical-impedance methods.

### *Reminder of the Uniform-Corrosion Consideration*

After $i_{corr}$ is evaluated by any one of the foregoing methods, use of one of the Faraday-law expressions (Table 6.2 and Chapter 4) leads to either the average corrosion intensity (CI) or average corrosion penetration rate (CPR). If the corrosion process is uniform, these average values relate directly to the uniform surface dissolution rate. If, on the other hand, the corrosion process is localized, the actual corrosion intensity and corrosion penetration rate at local areas can be orders of magnitude greater than the average values.

## Chapter 6 Review Questions

1. Describe the function of each component in the potentiostatic circuit.
2. Derive Eq 6.5, which expresses the external current as a function of the potential, corrosion current, corrosion potential, and Tafel constants.
3. In measuring the potential of a metal/electrolyte system, why should the potential-measuring instrument have a high impedance, on the order of $10^{10}$ ohms or greater?
4. Why is the metal component of a reference electrode generally restricted to either silver, mercury, or copper?
5. Discuss the advantages and disadvantages of "saturated" reference half cells.
6. Give some ways in which reference-electrode openings are constructed in order to minimize cross contamination between the reference-electrode electrolyte and the electrochemical-cell electrolyte. Can the openings be made too small? Explain.
7. For a given electrolyte resistivity, and relative to positioning the reference-electrode or salt-bridge tip, how can the magnitude of the IR correction be reduced?
8. For the data presented in Fig. 6.8, evaluate the solution resistance, $R'_S$, between the working electrode and the reference electrode.
9. Briefly describe how the current-interrupt IR-correction is performed.

10. If localized corrosion is occurring (e.g., pitting corrosion), and the experimentally determined value of $I_{corr}$ is divided by the total specimen area, A, to determine $i_{corr}$, then Faraday's law is used to calculate the corrosion intensity, CI, or corrosion penetration rate, CPR, why are the resultant values to be regarded as *minimum* values (i.e., the actual local values will be considerably higher)?
11. Why should the specimen surface be carefully examined for localized corrosion after an electrochemical corrosion-rate test before calculating the corrosion intensity or corrosion penetration rate based on the total exposed area of the specimen, the experimentally determined corrosion current, and Faraday's law?
12. In the Tafel-extrapolation method for evaluation of $I_{corr}$, why is the cathodic polarization curve generally analyzed rather than the anodic polarization curve?
13. Why should the corrosion-specimen surface be re-prepared after a Tafel-extrapolation corrosion-rate measurement before conducting a subsequent measurement?
14. Starting with Eq 6.1, derive the polarization-resistance equation, Eq 6.25.
15. From the polarization-resistance data in Fig. 6.12, evaluate $i_{corr}$. Assume $\beta_{ox} = \beta_{red} = 100$ mV.
16. Use the approximate equation, Eq 6.31, to determine $i_{corr}$ from the polarization-resistance data in Fig. 6.12.
17. In the polarization-resistance method, why is it generally assumed that repeat measurements may be made without removing the sample from the electrolyte and re-preparing the surface?
18. Describe the experimental procedures employed in collecting electrochemical-impedance-spectroscopy (EIS) data.
19. With reference to the EIS method, prove that for the electrical-circuit model of Fig. 6.18, the equivalent circuit impedance is given by Eq 6.64.
20. When using the Nyquist data-presentation format in the EIS method (Fig. 6.19) and assuming the simplest equivalent electrical-circuit model of Fig. 6.18, prove that the data points will fit on a semicircle, that the $Z'$ value at the $-Z'' = 0$ high-frequency intersection corresponds to $R_S$, that the $Z'$ value at the $-Z'' = 0$ low-frequency intersection corresponds to $R_S + R_p$, and that C is calculated from $C = 1/\omega R_p$, where $\omega$ is the angular frequency at the apex of the semicircle.
21. From the EIS Nyquist-format data in Fig. 6.19, determine $R_S$, $R_p$, C, and $i_{corr}$. Assume $\beta_{ox} = \beta_{red} = 100$ mV.
22. From the EIS data in Fig. 6.20, plotted using the Bode data-presentation format, evaluate $R_S$, $R_p$, C, and $i_{corr}$. Assume $\beta_{ox} = \beta_{red} = 100$ mV.

23. Give examples of corrosion processes that are not adequately modeled by the simplest equivalent electrical circuit of Fig. 6.18.
24. With reference to the two-electrode method (Fig. 6.21), why is the low-frequency impedance equal to $2R_p + R_S$ and the high-frequency impedance equal to $R_S$?
25. In the two-electrode method, why should the two working electrodes be identical in every way?

## Answers to Chapter 6 Review Questions

8. 30 ohms
15. 10 $mA/m^2$
16. 12 $mA/m^2$
21. $R_S$ = 0.25 ohm-$m^2$; $R_p$ = 2.00 ohm-$m^2$; C = 0.25 farad/$m^2$; $i_{corr}$ = 10.9 $mA/m^2$
22. $R_S$ = 0.01 ohm-$m^2$; $R_p$ = 2.00 ohm-$m^2$; C = 0.25 farad/$m^2$; $i_{corr}$ = 10.9 $mA/m^2$

## References


1. D.J.G. Ives and G.J. Janz, *Reference Electrodes,* Academic Press, 1961; NACE International, 1996
2. "Potential Error Correction (iR Compensation)," Technical Note 101, EG&G Princeton Applied Research, Princeton, NJ, 1986
3. D.K. Roe, Overcoming Solution Resistance with Stability and Grace in Potentiostatic Circuits, *Laboratory Techniques in Electroanalytical Chemistry,* P.T. Kissinger and W.R. Heineman, Ed., Marcel Dekker, Inc., 1984, p 193–234
4. N.D. Greene and R.H. Gandhi, Calculation of Corrosion Rates from Polarization Data with a Microcomputer, *Mater. Perform.,* Vol 21 (No. 7), 1982, p 34–39
5. N.D. Greene and R.H. Gandhi, Betacrunch Version 2.0, *Mater. Perform.,* Vol 26 (No. 7), 1987, p 52–53
6. "Basics of Corrosion Measurements," Application Note Corr 1, EG&G Princeton Applied Research, Princeton, NJ, 1980
7. "Standard Practice for Conducting Potentiodynamic Polarization Resistance Measurements," G 59-91, *Annual Book of ASTM Standards*, Vol 03.02, ASTM, 1995
8. M. Stern and R.M. Roth, *J. Electrochem. Soc.*, Vol 104, 1957, p 390
9. M. Stern, A Method for Determining Corrosion Rates from Linear Polarization Data, *Corrosion,* Vol 14, 1958, p 440t

10. F. Mansfeld, The Polarization Resistance Technique for Measuring Corrosion Currents, *Corros. Sci. Technol.,* Vol 6, Plenum Press, 1976, p 163
11. D.A. Jones, Polarization Methods to Measure Corrosion Rate, *Principles and Prevention of Corrosion,* Macmillan Publishing Co., 1992, p 142–166
12. M. Stern and A.L. Geary, Electrochemical Polarization I. A Theoretical Analysis of the Shape of Polarization Curves, *J. Electrochem. Soc.,* Vol 104, 1957, p 56–63
13. M. Stern and E.D. Weisert, Experimental Observations on the Relation between Polarization Resistance and Corrosion Rate, *Proc. ASTM,* Vol 59, 1959, p 1280
14. "Basics of Electrochemical Impedance Spectroscopy (EIS)," Application Note AC-1, EG&G Princeton Applied Research, Princeton, NJ, 1989
15. "Standard Practice for Verification of Algorithm and Equipment for Electrochemical Impedance Measurements," G 106-89, *Annual Book of ASTM Standards,* Vol 03.02, ASTM, 1995
16. D.C. Silverman, Primer on AC Impedance Technique, *Electrochemical Techniques for Corrosion Engineering,* R. Baboian, Ed., NACE International, 1986, p 73–79
17. F. Mansfeld, Recording and Analysis of AC Impedance Data for Corrosion Studies, I. Background and Methods of Analysis, *Corrosion,* Vol 36 (No. 5), 1981, p 301
18. F. Mansfeld, M.W. Kendig, and S. Tsai, Recording and Analysis of AC Impedance Data for Corrosion Studies, II. Experimental Approach and Results, *Corrosion,* Vol 38 (No. 1), 1982, p 570
19. *Model 9030 Corrater Corrosion Rate Monitor, User Manual,* Rohrback Cosasco Systems, Santa Fe Springs, CA, 1987
20. S. Haruyama and T. Tsuru, A Corrosion Monitor Based on Impedance Method, *Electrochemical Corrosion Testing,* STP 727, F. Mansfeld and U. Bertocci, Ed., ASTM, 1981, p 167–186

CHAPTER 7

# Localized Corrosion

## The Concept of Localized Corrosion

A concept of uniform corrosion should be defined as a basis to which localized corrosion can be compared. Idealized uniform corrosion occurs when the flux of metal ions from the surface and the flux of cathodic reactants to the surface are uniform to atomic dimensions. From a practical standpoint, uniform corrosion occurs when localized anodic and cathodic sites are sufficiently small and uniformly distributed so as not to lead to failure due to localization of the anodic reaction. Actually, any physical irregularity in the metal surface tends to form a local anode. This includes grain boundaries; crystal imperfections such as dislocations and surface steps; different phases; and rough surfaces from machining, grinding, scratches, etc. Also, different crystallographic planes of the crystal lattice of a metal have different atom arrangements and behave differently electrochemically, some becoming more anodic than others in aqueous environments. As a consequence, the grains of the exposed surface of a polycrystalline metal may exhibit different corrosion rates. Frequently, these differences in localized behavior are small, and on a practical macroscopic scale, the corrosion appears to be uniform, and effectively is uniform. In other cases, the attack may be very localized and lead to localized failure. Effective uniform corrosion also occurs when diffusion through a corrosion product layer is the controlling factor in the corrosion rate.

# Deviations from Strictly Uniform Corrosion

## *Surface Conditions Leading to Localized Corrosion*

It is evident from the preceding discussion that strictly uniform corrosion is a limiting concept. The critical factor is whether local regions of corrosion are few, or even singular, and lead to failure, or whether local attack is relatively widespread with each localized region corroding at about the same rate. As an example, which is discussed in greater detail subsequently, conditions can exist causing preferential corrosion at grain boundaries. If all grain boundaries are penetrated at about the same, but slow, rate, the corrosion is local, but may still appear to be uniform visually. In contrast, if conditions operate to cause rapid penetration, this corrosion frequently leads to failure.

Localized corrosion can be related to the microstructure of a metal or alloy, the physical condition of the surface, or coupling of the metal to a dissimilar metal or to conducting surface films, usually oxides. These conditions are listed in an order generally observed to be increasingly conducive to localized attack leading to failure.

- Preferred dissolution sites such as dislocations, grain boundaries, and localized cold working such as at scratches
- Dispersed phases such as carbides, sulfides, oxides, and intermetallic compounds
- Irregular surface coatings such as discontinuous oxide coatings, deposits of more noble metals (e.g., copper on iron), and deposits of conducting materials, such as graphite, resulting from fabrication processes
- Irregular deposits such as dirt, scale, and biological growths
- Areas from which protective coatings have been removed physically or by corrosion (e.g., chromium, copper, and nickel plates from steel)
- Junction of dissimilar metals

In all of these conditions, important variables are the cathode-to-anode area ratio, $A_c/A_a$, and the ability of the cathodic surfaces to support cathodic reactions and thereby enhance the corrosion of anodic sites.

## *Environmental Conditions Leading To Localized Corrosion*

Environmental conditions leading to localized corrosion are usually associated with nonuniform concentrations of cathodic reactant species or corrosion product ions. These conditions can be brought about by the following:

- *Nonuniform access of the cathodic species due to location of the source of the species:* For example, a gradient in the dissolved-oxygen concentration exists along the wall of an open tank, providing more oxygen near the top and inducing anodic regions just below the water line. Similar conditions may exist near entrance piping into tanks and heat exchangers where the concentration of cathodic reactants will be higher.
- *Nonuniform fluid velocity:* This condition can induce local anodic and cathodic regions, causing variations at the surface in the concentration of cathodic species supporting corrosion and by removing corrosion products. This condition is frequently found in piping systems and pumps.
- *Localized restriction of the cathodic reactant:* This condition is observed in crevices such as overlapping surfaces, incompletely sealed gaskets, and incomplete press fits, particularly where heat exchanger tubes are rolled into a tube sheet. It also is observed under accumulated porous scale, dirt, and corrosion product deposits. The attack is due to localized acidification from the hydrolysis of metal ions and to the products of the metabolism of microbiological organisms.

## Localized Corrosion Induced by Rupture of Otherwise Protective Coatings

Following are representative examples:

- *Rupture of Organic Protective Films:* This condition differs from other causes of localized corrosion since these protective films are nonconductors and, as such, do not support the cathodic reaction. After the rupture in the coating, corrosion may progress under the coating by crevice corrosion mechanisms, resulting in further damage.
- *Rupture of Passive Films on Active-Passive Type Alloys such as Stainless Steels:* Several conditions may cause rupture. Chemical species in solution can cause local breakdown of the passive film, particularly the presence of chloride ions in contact with stainless steels and other alloys. The result is usually pitting corrosion. The rupture of passive films may be due to the loss of oxidizing species in solution (e.g., dissolved oxygen, $Fe^{3+}$ ions, $NO_2^-$ ions, etc). Rupture also may be due to stresses and the presence of environmental conditions incapable of immediate film repair. The rupture propagates into the underlying metal and is sustained by the concentration of stress at the leading edge of the crack and by corrosion mechanisms associated with a crevice. The result is stress-corrosion cracking (SCC).

## Localized Corrosion due to Variations in Chemical Composition in Alloys

Following are representative examples:

- *Chemically homogeneous alloys:* Completely annealed solid solution alloys are single phase and show uniform chemical composition throughout. These alloys can corrode uniformly as with pure metals, although preferential corrosion of one or more of the alloying elements may occur, creating nonuniform compositions along the surface. For example, a type of corrosion called dezincification removes zinc from copper-zinc solid solution alloys (brasses), resulting in localized regions of copper on the surface.
- *Multiphase alloys:* Multiphase alloys inherently contain dispersions of phases of differing composition, and these dispersions may be uniform or nonuniform depending on the processing of the alloy. Some phases tend to preferentially support cathodic reactions inducing other phases to be anodic and corrode. The extent to which damaging localized corrosion occurs depends on the environment and the size, shape, and uniformity of the dispersed phases in the microstructure. For example, iron carbide in steels tends to act as a cathodic surface supporting cathodic hydrogen reduction. The iron matrix in which the carbide is dispersed becomes anodic and corrodes. In other cases, the dispersed phase is anodic, and the continuous phase supports the cathodic reaction.
- *Chemical segregation in castings:* The mechanism of solidification of alloys almost always leads to segregation on either a microscale or macroscale. This is particularly true of solid solution alloys that solidify with dendritic segregation (or coring) or with differences in composition between the center and surfaces of castings. These differences may be large and cause corrosion problems. The long-time high-temperature heat treatments required to make castings uniform in composition are usually not feasible in industrial practice. Therefore, if a corrosion problem exists for a given alloy, the solution to the corrosion problem may require a change in alloy composition or to a different alloy. Cast eutectic-type alloys usually do not show segregation. However, alloys having a primary dendritic phase mixed with a eutectic microconstituent may show local dendritic segregation and segregation between center and surface sections of castings. As with solid-solution alloys, this segregation may cause corrosion problems.
- *Chemical segregation in ingots and retention after processing:* Wrought products such as pipes, tubes, and plates are produced by the mechanical processing of ingots. The solidification of the ingots may result in dendritic segregation or center-to-surface segregation

of alloying elements as discussed for castings. In some instances, this segregation persists to the final fabricated product and appears as stringers of variations in composition. Depending on the degree of segregation and the corrosive environment, this condition may cause corrosion problems. If such is observed, the segregation can usually be reduced by careful attention to fabrication sequences of hot and cold working followed by annealing.

- *Chemical segregation resulting from precipitation of phases from thermodynamically unstable solid solutions:* A large number of alloys exhibit corrosion-resistant properties when placed in service in the rapidly cooled condition. The corrosion resistance is obtained only for alloy compositions existing as complete solid solutions at elevated temperatures and for which the solid solution is retained for practical rates of cooling. If cooled too slowly or reheated to lower temperatures following quenching, one or more phases precipitate from the solid solution, and local changes in composition associated with this precipitation may make the alloy susceptible to localized corrosion. Depending on the alloy, the time required for precipitation may extend from seconds to hours. The former is important in welding and the latter in stress-relief annealing.

## General Characterization of Pitting and Crevice Corrosion (Ref 1)

Pitting and crevice corrosion are two modes of localized corrosion generally associated with a type of occluded cell in which small metal/solution interface areas are restricted from the bulk environment. A major characteristic of these modes of corrosion is a large ratio of cathode area (areas in full contact with the bulk environment) to anode area (occluded region). As a consequence, the current density and, hence, the corrosion rate over the occluded area is very large. Exceptions to this generalization are those metals and alloys that form oxides that are poor electron conductors and, therefore, provide poor support for cathodic (reduction) reactions (e.g., oxides of aluminum, titanium, and tantalum). These differences and the observation that the occurrence of both pitting and crevice corrosion are frequently specific to each alloy and environment have limited development of a generally applicable theory for these corrosion modes.

Although the types of occluded regions vary over a wide range, three types are distinguished in the following discussion. The first type, generally associated with alloys forming very protective passive films, is restricted to localized penetration of a passive film, resulting in a sharply defined discontinuity in the surface with penetration into the metal, which may enlarge with depth. Pits of this type may be initiated

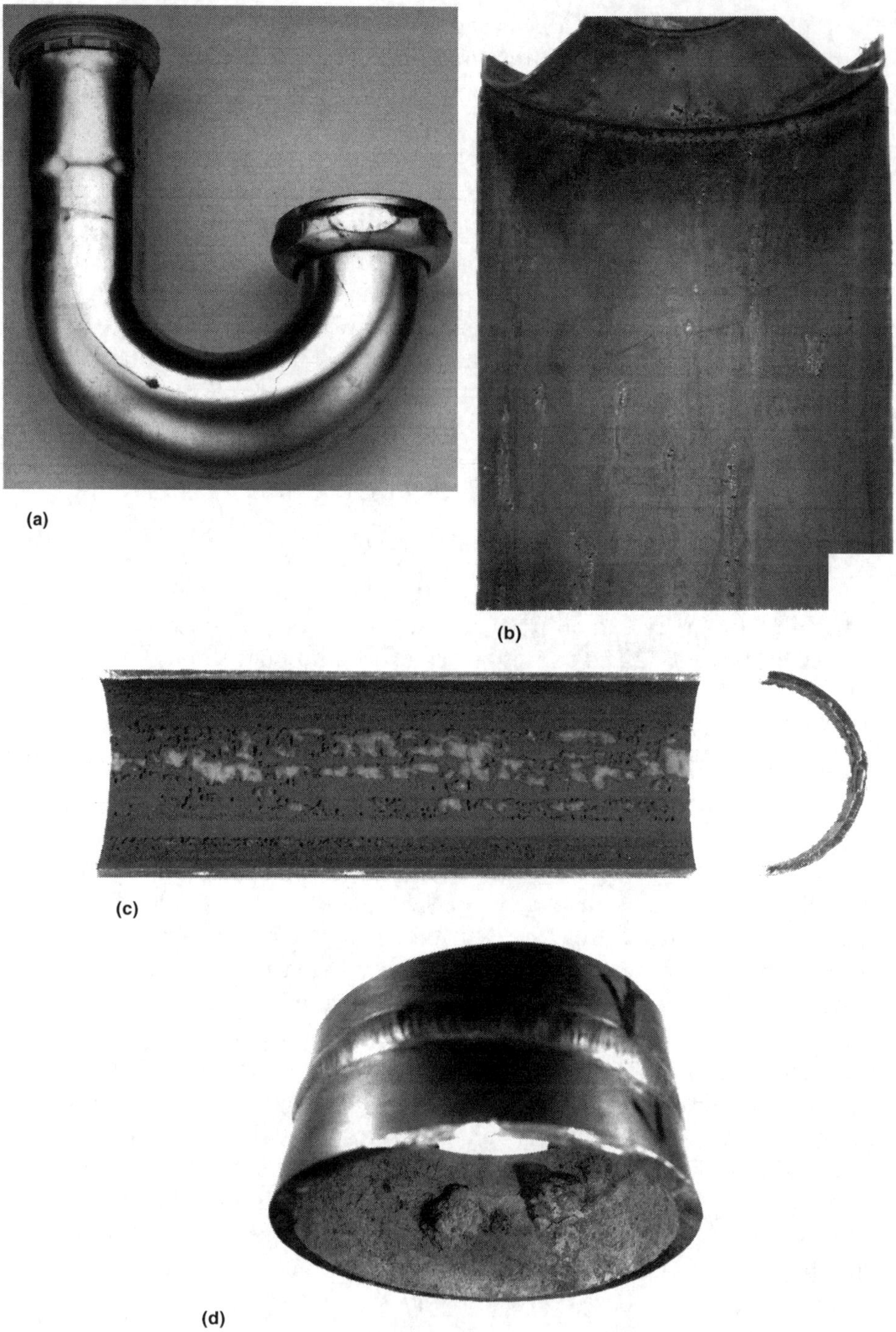


**Fig. 7.1** Examples of pitting corrosion. (a) Pitting and subsequent cracking in a chromium-plated copper sink-drain trap. (b) Pitting in a stainless steel thermos-bottle liner. (c) Pitting in a brass condensate line. (d) Mounds (or tubercles) associated with microbiologically influenced corrosion of a type 304 stainless steel pipe used for untreated fresh water. Underlying pits completely penetrate the wall thickness.

on a surface that is macroscopically free of deposits. Pits may also form under deposits of inert material that restrict access of a cathodic reactant or under deposits containing microbial species generating local acidic environments (Ref 2). A second type of occluded region, generally associated with alloys forming less protective passive films, is characterized by poorly defined, shallow, and rough penetrations of the surface. These regions also develop as a result of deposits of inert and biological materials, but the attack is more general under the deposit rather than having initiated at a very local region. Nonuniform corrosion product formation frequently leads to a rough underlying surface that may be referred to as pitted but is distinctly different from the morphology of sharply defined pits on highly passive alloys. These various types of nonuniform attack are illustrated in Fig. 7.1. A third type of occluded cell is associated with crevice corrosion. Types of crevices include overlapping surfaces, incompletely sealed gasket/metal interfaces, threaded joints, deep grooves and scratches, and irregular or incompletely penetrating welds, all of which preexist to contact with an aqueous environment. These crevices are preexisting occluded regions and are associated with the same mechanisms of propagation as in pitting corrosion. The modes differ, however, in that pitting is generally preceded by an initiation time followed by propagation. With crevice corrosion, the initiation times are either nonexistent or much shorter. As with pitting, crevice corrosion is generally encountered as a problem with active-passive-type alloys.

## Pitting of Typical Active-Passive Alloys

Pitting corrosion is usually associated with active-passive-type alloys and occurs under conditions specific to each alloy and environment. This mode of localized attack is of major commercial significance since it can severely limit performance in circumstances where, otherwise, the corrosion rates are extremely low. Susceptible alloys include the stainless steels and related alloys, a wide series of alloys extending from iron-base to nickel-base, aluminum, and aluminum-base alloys, titanium alloys, and others of commercial importance but more limited in use. In all of these alloys, the polarization curves in most media show a rather sharp transition from active dissolution to a state of passivity characterized by low current density and, hence, low corrosion rate.* As emphasized in Chapter 5, environments that maintain the corrosion potential in the passive potential range generally exhibit extremely low

*Aluminum alloys are an exception. The oxide film formed in air or on immediate contact with an aqueous environment places aluminum in a passive state and an active-to-passive transition is not observed experimentally in the polarization curve.

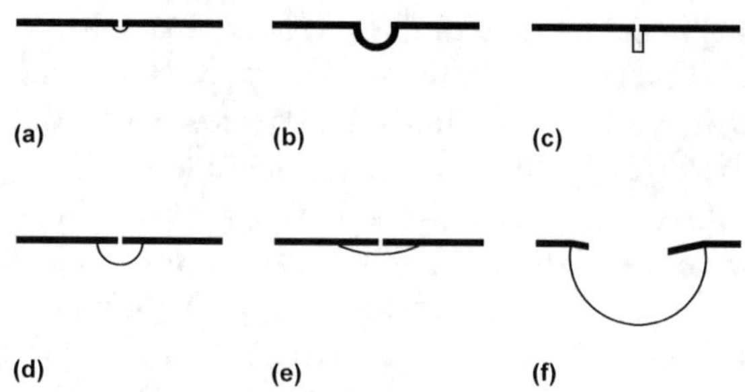


**Fig. 7.2** Schematic representation of shapes of pit initiation and propagation

corrosive attack rates. However, it is inherent with these materials that any sustained local loss of the passive film can lead to rapid local attack and possible failure.

Pits are initiated at preexisting conditions on a passive surface or as a consequence of local events such as physical or chemical damage to the passive surface. Pit propagation will not occur if conditions lead to immediate repassivation of the local region. Pitting is usually preceded by an induction time to activate the local region following which the pit propagates as an occluded cell.

The form of corrosion pits varies widely, reflecting a wide range of mechanisms of initiation and propagation that depend on the specific alloy and environment. Whether pits are initiated on an apparently uniformly passivated surface due to an aggressive bulk environment or under inert or active (microbial) foreign deposits that cause an aggressive environment to form, pit propagation results in different pit geometries. Penetration of the passive film immediately forms an occluded region, highly concentrated in corrosion product cations that hydrolyze to create a locally aggressive acidic environment. This initial stage is represented by Fig. 7.2(a). If the covering passive film breaks and the occluded region is cleared of corrosion products, the pit surface may repassivate, and propagation does not occur as illustrated in Fig. 7.2(b). Otherwise, the geometry of the enlarging cavity depends on the mechanical behavior of the covering passive film and the response of the specific alloy to the corrosive action of the occluded solution. Some observed geometries are represented schematically in Fig. 7.2(c) to 7.2(e) in which the covering passive film has partially remained in place, and cylindrical, spherical, and oblate cavities have formed. The faces of these cavities have been observed to be highly polished, faceted as a result of preferential attack associated with the crystal structure of the alloy, or very rough. In the case of aluminum, a complex network of corrosion tunnels may progress into the metal, leaving the pit surface very rough where these tunnels initiate. The stage at which the covering passive film breaks influences the subsequent propagation through compositional changes in the pit environment as it has access to the ex-

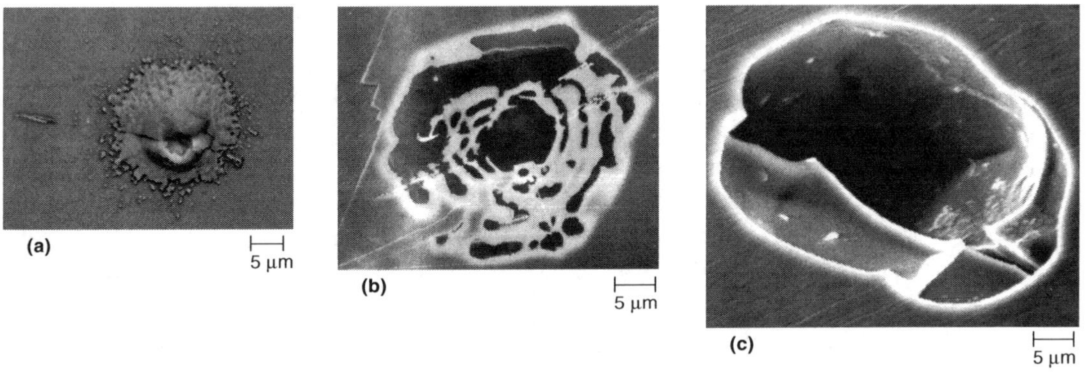


**Fig. 7.3** Stages of penetration of passive film leading to corrosion pit formation. (a) Initial stage of pit formation. (b) Partially perforated passive film on pit. (c) Fragment of passive film on edge of pit. Source: Ref 3

ternal environment. Since pits seldom remain covered beyond about 20 μm in diameter, pits visible even at low magnifications are open to the environment but may continue to propagate and form large cavities as illustrated in Fig. 7.2(f). Specific examples of pit morphologies representative of the schematic form shown in Fig. 7.2 are shown in Fig. 7.3 (Ref 3).

## *Pit Initiation*

Since the initiation of pitting is the localized penetration of the passive film, understanding of this step requires information on the structure of passive films and the mechanisms whereby they can be destroyed locally. Understanding of either of these is complicated by the thinness of the films and the question of the passive film structure when formed by and existing in the aqueous environment as compared with its structure when removed from this environment. The latter is necessary for the use of most of the surface analysis techniques applicable to structure evaluation. As a consequence, specific conclusions as to the structure are frequently inferred rather than more directly established.

**Chemical Structure of the Passive Film.** A metal surface on contact with an aqueous environment quickly develops a layer of adsorbed water molecules due to their dipole structure with the oxygen atom in the molecule tending to attach to the metal surface. One theory of passivity proposes that this layer is replaced by a film of adsorbed oxygen and that this film is sufficient to account for the passivity. Whether this film alone is responsible, in general, films thicken with increase in time usually to a steady value that is greater the higher the anodic potential. The steady-state thickness is observed to increase linearly with increase in potential, and for most active-passive metals, the maximum thickness is <10 nm (Ref 4). The film structures may be essentially those of the bulk oxides, although differences in interatomic distances may exist as a consequence of the conditions under which the oxide has formed. These conditions have also led to formation of oxides exclusively characteristic of passive films and not observed in bulk form. Some films exhibit semiconducting properties and as such contain metal-ion and oxygen-ion vacancies in the crystal structure (Ref 5). In some cases, for example, stainless steels of higher chromium concentration (>20%), the passive film tends to become amorphous; the film may also become effectively amorphous if the lattice defect concentration of ion vacancies becomes sufficiently large (Ref 6).

A model for the formation of a passive film on iron-base (stainless steels) and nickel-base alloys is shown schematically in Fig. 7.4 (Ref 6). In this generalized treatment, the anodic reaction is assumed to be $M \rightarrow M^{2+} + 2e$, and the cathodic reaction is $1/2O_2 + H_2O + 2e \rightarrow 2OH^-$. However, rather than the metal ions and hydroxyl ions immediately combining to form a solid product (i.e., $M^{2+} + 2OH^- \rightarrow M(OH)_2$), the following sequence of reactions is proposed for passive film development. First, the metal ion combines with a hydroxyl ion to form an intermediate complex ion, $M(OH)^+$ ( i.e., $M^{2+} + OH^- \rightarrow M(OH)^+$). The intermediate ion is then surrounded by water molecules and reacts to precipitate a solid film ($M(OH)^+ + H_2O \rightarrow M(OH)_2 + H^+$). The hydrogen ion produced by this last reaction combines with a hydroxyl ion left over from the cathodic reaction to form a water molecule ($H^+ + OH^- \rightarrow H_2O$). With time, and depending on the corrosion potential, an aging process occurs whereby the solid metal hydroxide is converted to the metal oxide ($M(OH)_2 \rightarrow MO + H_2O$)). Thus, the overall mechanism proposes that freshly formed films contain a large amount of bound water, and with time the film changes to a less hydrated structure. At any stage of aging, the film might contain the following types of bridges between metal ions: $H_2O$-M-$H_2O$, -HO-M-HO, and -O-M-O-. With loss of hydrogen ions, the structure progressively changes toward that of the metal oxide. The initial stage of film formation is represented in Fig. 7.4(a) (Ref 6) with an undeveloped region depicted near the center; the protective stage is shown in Fig. 7.4(b). Once the passive film has formed, metal ions slowly pass into the environment at a rate corresponding to the passive current density. The mechanism of film growth and maintenance of a steady-state thickness is transport of metal and/or oxygen ions by cation and anion vacancies. Which ion migrates fastest determines whether growth is predominantly at the metal/film or film/solution interface. At this steady-state condition, a balance between passive film formation and dissolution results in films that usually are <10 nm thick.

With the more active metals, such as aluminum, titanium, and tantalum, oxide films form immediately on contact with air and behave as passive films in aqueous solutions. In the case of tantalum, the passive film is protective over the entire pH range; in other cases, the films may become thermodynamically unstable at very high and/or low pH values, leading to high uniform corrosion rates (Ref 7). These films relate to the bulk oxide (e.g., $Al_2O_3$) but tend to be amorphous; they also exhibit very high resistance to electron conduction. This behavior is in contrast to that of the passive films formed on iron- and nickel-base alloys as described previously (i.e., for these alloys, passive films can be formed on the bare alloy surfaces in aqueous solutions and the films are electron conductors).

**Imperfections in Passive Films.** Physical flaws in the passive film, important to theories of pit initiation, are attributed to several factors. If the three-dimensional passive film develops by nucleation and lateral growth followed by thickening, impingement of growing regions may result in defects due to mismatch of crystal orientation; also, dimensional changes may lead to flaws on impingement as well as problems of epitaxial misfit with the metal substrate during growth. As proposed in the model described previously, the growing film may incorporate variable amounts of absorbed water and have a gradient of water concentration between the metal/film and film/solution interfaces. A number of surface and structural defects in the metal substrate have been suggested, and some substantiated, as causes of flaws in the passive film. These defects include grain boundaries, dislocations, surface

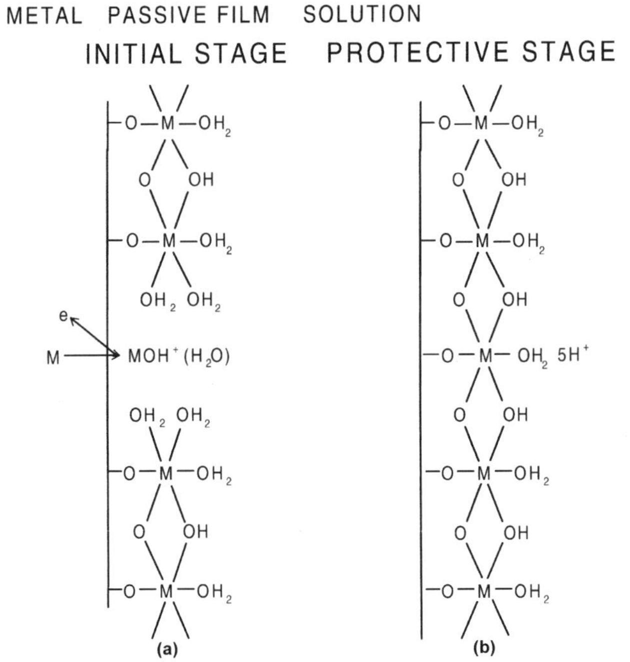


**Fig. 7.4** A mechanism for (a) initiation and (b) development of a passive film. Source: Ref 6

scratches, and, in particular, inclusions. As emphasized subsequently, inclusions can interfere or prevent local passive film formation, and if the inclusion is chemically attacked by the environment, the local chemical environment created by the dissolution, as well as the physical defect produced in the metal surface, may seriously affect the ability to form the passive film (Ref 8–12). If these are serious enough to prevent passivation at the site, active corrosion progresses locally and serious pitting occurs. In contrast, flaws in the preexisting, air-formed passive film of the active metals (aluminum, for example) have been associated with intermetallic compound particles in the substrate over which the passive film is less protective (Ref 13, 14).

**Interface Potential and Pit Initiation.** It is generally accepted that pit initiation occurs when the corrosion potential or potentiostatically imposed potential is above a critical value that depends on the alloy and environment. However, there is incomplete understanding as to how these factors (potential, material, and environment) relate to a mechanism, or more probably, several mechanisms, of pit initiation and, in particular, how preexisting flaws of the type previously described in the passive film on aluminum may become activated and/or when potential-driven transport processes may bring aggressive species in the environment to the flaw where they initiate local penetration. In the former case, the time for pit initiation tends to be very short compared with the initiation time on alloys such as stainless steels. Pit initiation is immediately associated with a localized anodic current passing from the metal to the environment driven by a potential difference between the metal/pit environment interface and sites supporting cathodic reactions. The latter may be either the external passive surface if it is a reasonable electron conductor or cathodic sites within the pit.

Several pit-initiation mechanisms, related to the potential of the passive film and to the potential gradient in the film that are more statistical in nature and compatible with pitting of otherwise flaw-free passive films, have also been proposed. It has been demonstrated, at least for some passive surfaces, that chloride ions are increasingly adsorbed to the surface (Ref 13, 15, 16). Exact mechanisms whereby the chloride ions penetrate the passive film and initiate pitting are uncertain. Suggested factors include the reaction of the chloride ion with metal cations in the passive film to form soluble metal-chloride complexes and substitution of chloride ions for water and/or $O^{2-}$ ions in the film. A proposed mechanism for the latter is represented in Fig. 7.5 (Ref 6). Chloride ions are absorbed into the passive film at local sites where soluble complex metal-chloride ions form and pass into solution. At the site of penetration, acidification due to hydrolysis of the metal ions reduces the stability of the oxide, and buildup of positive charge due to increased cation concentration attracts negative chloride ions. Local conditions are thereby enhanced for both the initiation and propagation of pitting.

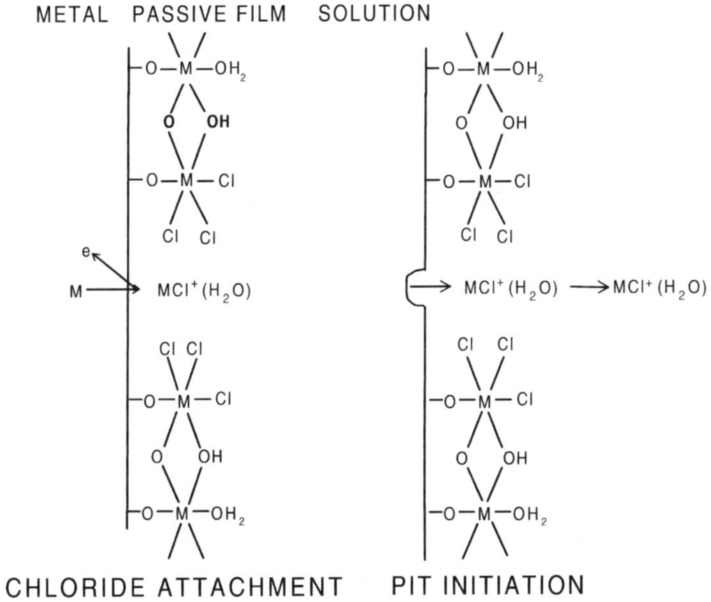


**Fig. 7.5** Schematic representation of pit initiation by chloride ion penetration into passive film. Source: Ref 6

Mechanisms of pit initiation that are associated with ion transport by cation and anion vacancies in the passive film have been proposed. If cation vacancies at the film/solution interface, formed by metal ions passing into the solution, migrate under the electrical field to the metal/film interface faster than metal ions pass from the metal to the film, then a supersaturation of cation vacancies could precipitate as a void at the interface (Ref 17). The resulting void becomes a flaw in the film and a site for pit initiation. If hydrogen ions or water are also diffusing to the metal/film interface, and the potential at the interface is sufficiently low, these can react to form hydrogen gas and further contribute to pit initiation. It has also been proposed that at the film/solution interface, fluctuations occur in the potential in the interface topology and in the electrolyte such that, in the vicinity of a critical potential, these fluctuations lead to pit initiation. This provides a statistical character to pitting in terms of pit initiation times and distribution that is observed experimentally and in service.

## *Pit Propagation*

Electrochemical, chemical, and physical processes associated with the anodic current determine the conditions leading either to local repassivation or to pit propagation. Since the particular set of processes determining repassivation or propagation is specific to each metal/environment combination, a generally applicable mechanism of propagation is probably not attainable. However, an overview of these mechanisms, with examples as to metal/environment combinations for which a process is or is not relevant, is useful. Detailed discussions of the pitting and crevice corrosion of representative classes of materials such as stainless steels, nickel-base alloys, and aluminum-base alloys are presented later in this chapter.

**Anodic Current and Cation Concentration in Occluded Regions.** The anodic current increases the local concentration of corrosion product cations ($M^{m+}$), which tend to hydrolyze according to reactions of the form $M^{m+} + xH_2O \rightarrow [M(OH)_x]^{(m-x)+} + xH^+$. Depending on the particular $M^{m+}$ concentration, the resulting pH is observed to range from <1 to ~5 (Ref 18, 19). Examples of the pH of saturated metal chloride solutions are given in Table 7.1 (Ref 18). If, within the pit, hydrolysis results in pH values that are less than the bulk environment pH, acidification within the pit occurs. Otherwise, the pit pH will increase. Since cations resulting from the dissolution of iron-, nickel- and aluminum-base alloys hydrolyze to pH < 3, occluded regions (pits and crevices) will become acidic when these alloys are in contact with bulk near-neutral environments. Two consequences of the lower pH are: (a) depending on the metal, the oxide may become soluble, and if so, repassivation is impossible; and (b) depending on the potential in the pit or crevice, hydrogen-ion reduction may become thermodynamically possible, resulting in local hydrogen-gas bubble formation. The first of these consequences is the equivalent of recognizing that decreased pH increases $i_{crit}$ and raises $E_{pp}$ of the anodic polarization curve (the section "Experimental Observations on the Anodic Polarization of Iron" in Chapter 5 provides more information), which makes it more difficult to form and maintain the passive state in the occluded region.

**Anion Migration into Occluded Regions.** Anions in the external environment, particularly chloride ions, will migrate into the occluded region as a consequence of the potential difference between the solution at the metal/environment interface in the pit and the solution at the external surface or, equivalently, in response to the increase in positive charge resulting from the increased cation concentration in the bottom of the pit. Chloride ions are known to stabilize the hydrolysis reactions and actually further lower the pH (Ref 19). If the increase in metal-ion concentration associated with the anodic current density at the pit inter-

**Table 7.1 Values of pH for concentrated chloride salt solutions at room temperature**

| Salt | pH at indicated salt concentrations | | |
|---|---|---|---|
| | 1 N | 3 N | Saturated |
| $FeCl_2$ | 2.1 | 0.8 | 0.2 |
| $NiCl_2$ | 3.0 | 2.7 | 2.7 |
| $CrCl_3$ | 1.1 | −0.3 | −1.4 |

Source: Ref 18

face and the migration of anions, such as $Cl^-$, from the external environment is such that the solubility of a salt is exceeded, then the salt deposits in the pit. This salt film provides an additional diffusion barrier for metal-ion migration and an associated increase in resistance to current flow. Thus, there is a competition within a pit for conditions allowing oxide-film formation (repassivation), salt-film formation, and maintenance of a bare-metal interface.

If the passive film cannot be reestablished and active corrosion occurs, a potential drop is established in the occluded region equal to IR where R is the electrical resistance of the electrolyte and any salt film in the restricted region. The IR drop lowers the electrochemical potential at the metal interface in the pit relative to that of the passivated surface. Fluctuations in corrosion current and corrosion potential (electrochemical noise) prior to stable pit initiation indicates that critical local conditions determine whether a flaw in the film will propagate as a pit or repassivate. For stable pit propagation, conditions must be established at the local environment/metal interface that prevents passive film formation. That is, the potential at the metal interface must be forced lower than the passivating potential for the metal in the environment within the pit. Mechanisms of pit initiation and propagation based on these concepts are developed in more detail in the following section.

## *An Analysis of Pitting Corrosion in Terms of IR Potential Changes in Occluded Regions and Relationship to Polarization Curves (Ref 20)*

For the active-passive-type metals, the current density and, hence, corrosion rate, in the active state may be $10^2$ to $10^5$ greater than in the passive state. As a consequence, the current density at any flaw exposing the substrate metal may be very large, leading to large localized penetration rates. With reference to Fig. 7.6 (Ref 20), the solid anodic curve is representative of the anodic polarization behavior of a stainless steel with passive film formation in an environment of pH = 1. The dashed extension of the active region represents the anodic polarization behavior in the absence of passive-film formation. A cathodic curve is shown resulting in a corrosion potential, $E_{corr}$, in the passive potential range. If a small flaw exists in the passive film, the very large passive/active area ratio tends to maintain the entire surface, including the small active region, at $E_{corr}$, and at this potential, the corrosion rate at the exposed active region is very high, $i_{corr,act} = i_{corr,pit}$. However, if the potential at the pit is still near $E_{corr}$, the flawed region should repassivate. Two factors operate to restrict this from occurring. If the flaw in the passive film is very small in cross section and depth, as assumed, then the resistance of the fluid path in the flaw is high, which, with the high current maintained by the cathodic reaction over the very large passivated surface, leads to an IR potential drop that decreases the potential at the bottom of the flaw. This potential changes progressively to more negative potentials as the flaw and, subsequently, pit depth increases. In addition, the corrosion reactions cause the concentration of metal ions in the pit environment to be high, and these will continue to undergo hydrolysis reactions, which lowers the pH.

These factors can be discussed with reference to the polarization curves for the initial and changing conditions within the occluded region. The combined effects of a potential drop into the pit and the effect of the lowered pH, which raises $E_{pp}$ and increases $i_{crit}$, are also analyzed by reference to Fig. 7.6 (Ref 20). As previously assumed, the solid anodic curve is taken as representative of a stainless steel in an environment of pH = 1. The dashed extension again represents the anodic polarization behavior in the absence of a passive film. At a potential, $E_{corr}$ (or $E_{pot}$ if the potential is maintained potentiostatically), the passive current density would be $i_{corr,pass}$ and the active corrosion current density would be $i_{corr,act}$. Assume that a small flaw through the passive film is associated with an $(IR)_1$ drop that lowers the potential in the bottom of the flaw to $E_1$. Since this potential is higher than the passivating potential, $E_{pp}$, this flaw should immediately repassivate and not propagate.

If the flaw in the passive film is smaller in cross section and greater in depth, then with reference to Fig. 7.6, the resulting increase in resistance can lead to an $(IR)_2$ potential drop that decreases the potential in the bottom of the flaw and/or pit to $E_2$. Then passivity cannot be maintained, and the corrosion current density increases to $i_2$ in the active range. The local corrosion rate is much higher, and a stable pit is initiated at the much higher current density. When the pH of the bulk envi-

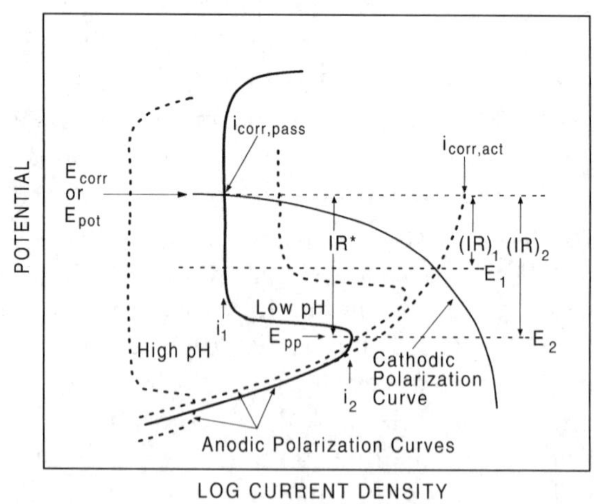


**Fig. 7.6** Schematic representation of polarization curves and variables relating to pit initiation and propagation. Based on Ref 20

ronment is higher (>9), the anodic polarization curve has the position at the extreme left in Fig. 7.6; the current density in the passive region is much smaller, $E_{pp}$ is lower, and the critical current density for passivation is less. All of these factors lead to less favorable conditions for pit initiation. In limiting cases, the passive state tends to be present over the entire potential range such that a lower potential associated with an IR drop in a defect may be of no consequence. Pitting may still occur, as appears to be the case with aluminum, in which case, at sufficiently high corrosion potentials, flaws at substrate intermetallic-compound particles allow an influx of chloride ions, which, when combined with hydrolysis of aluminum ions, provides an environment sustaining pit propagation (Ref 14).

This analysis leads to pit initiation as a consequence of flaws having an IR potential drop placing the bottom of the pit below $E_{pp}$. More generally, pits are initiated as a consequence of aggressive anions (e.g., chloride ion) concentrated at flaws or randomly on the passive film surface. Increasing potential increases this concentration, and at a critical potential, depending on material and environment, local dissolution of the passive film is initiated. Metal ions enter the local region where they undergo hydrolysis (e.g., $M^{2+} + H_2O \rightarrow M(OH)^+ + H^+$) resulting in a lower pH. This pH change moves the local polarization curve to higher $E_{pp}$ and greater $i_{crit}$, represented by the dashed anodic curve in Fig. 7.6, both of which contribute to sustaining the active corrosion at the base of the pit. The local environment created by the hydrolysis of the metal-ion corrosion products will vary with the specific ions as shown in Table 7.1.

This mechanism leads to the generalization that if the IR potential drop in the flaw/pit is greater than $[(E_{corr} \text{ or } E_{pot}) - E_{pp}]$, active pit propagation should occur. This critical condition is represented by the potential drop IR* in Fig. 7.6. Therefore, if local changes in the environment, such as a decrease in pH, results in the dashed polarization curve, IR* will be smaller and the probability of pitting greater. The conditions leading to an IR > IR* relate to a specific pit geometry (depth and cross section) and the environment within the pit, which determines the specific resistivity of the electrolyte, and to any solid corrosion products, such as precipitated metal salts, which contribute additional resistance.

Figure 7.6 also provides an understanding of a decrease in pit propagation rate as the pit depth increases and/or corrosion products accumulate. Both of these factors increase the IR potential drop and thereby decrease the electrochemical potential at the bottom of the pit. Since the potential and accompanying current density follow the polarization curve in the active potential range, a decreasing potential relates to a decreasing current density and, therefore, a decreasing corrosion rate in the pit.

A somewhat alternative analysis of pitting attributes pit initiation to the activation of defects in the passive film, defects such as those induced during film growth or those induced mechanically due to scratching or stress. The pit behavior is analyzed in terms of the product, xi, a parameter in which x is the pit or crevice depth (cm), and i is the corrosion current density ($A/cm^2$) at the bottom of the pit (Ref 21). Experimental measurements confirm that, for many metal/environment systems, the active corrosion current density in a pit is of the order of 1 $A/cm^2$. Therefore, numerical values for xi may be visualized as a pit depth in centimeters. A defect becomes a pit if the pH in the pit becomes sufficiently low to prevent maintaining the protective oxide film. Establishing the critical pH, for a specific oxide, will depend on the depth (metal ions trapped by diffusional constraints), the current density (rate of generation of metal ions) and the external pH. In turn, the current density will be determined by the local electrochemical potential established by corrosion currents to the passive external cathodic surface or by a potentiostat. Once the critical condition for dissolution of the oxide has been reached, the pit becomes deeper and develops a still lower pH by further hydrolysis.

During pit growth, current flows from the bottom of the pit and is distributed over the external passive film supporting the cathodic reaction, the current distribution depending on the specific conductivity of the environment. This current is carried by positive corrosion-product ions migrating from the pit and negative ions from the environment migrating into the pit. Chloride ions tend to dominate the negative ion contribution because of their high mobility. As a consequence, chloride ions build up in the pit until their back transfer by diffusion just balances the inward transfer. Thus, the concentration of chloride ions in the pit will depend on their concentration in the bulk environment, the concentration of corrosion product cations in the pit, and the pit or crevice geometry (area and depth). Deep narrow geometries favor high buildup of metal ions and, hence, low pH by hydrolysis, and high chloride ion concentration due to high rates of inward migration. This mechanism is supported by measurements in pits of pH $< 1$ and chloride concentrations on the order of 5 N when, in the bulk environment, the pH is near neutral and the chloride-ion concentrations are no greater than $10^{-3}$ N (100 ppm). It should be noted that this mechanism leads to local acidification by hydrolysis of metal ions as the critical factor in pit initiation and propagation. Simultaneously, the chloride ion concentration increases and thereby enhances the local dissolution rate. Pit initiation is not attributed to this increase in chloride concentration, a conclusion significantly different from proposals that pits are initiated by the incorporation of chloride ions into the passive film.

Examples of pit initiation and propagation at inclusions in a stainless steel are shown in Fig. 7.7 (Ref 3). The more acid-soluble inclusions

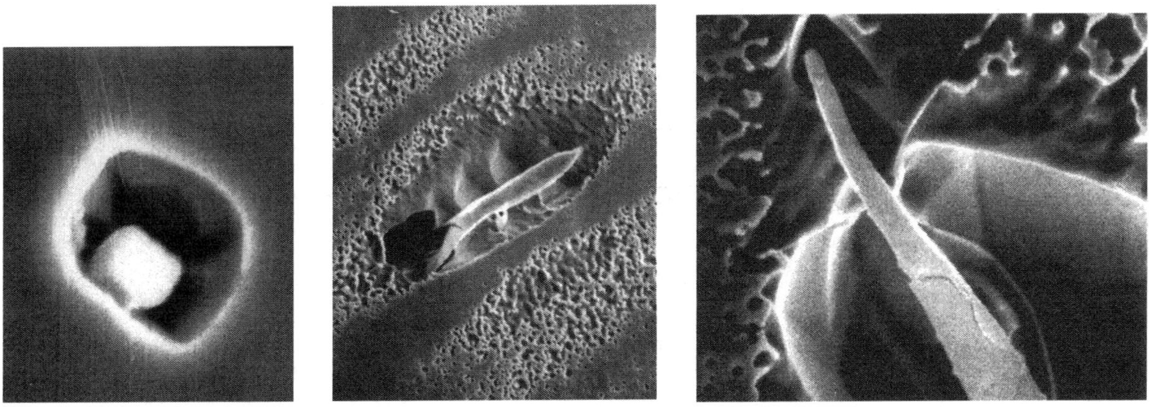

**Fig. 7.7** Pit formation at inclusions in type 304 stainless steel. Source: Ref 3

such as MnS or two-phase inclusions, one of which is soluble, are preferred sites for initiation (Ref 1, 8, 22, 23). The low pH and high sulfide-ion concentration resulting from dissolution of sulfide inclusions produces a local environment that prevents establishing passivity in the pit (the section "Effects of Sulfide and Thiocyanate Ions on Polarization of Type 304 Stainless Steel" in Chapter 5 describes the effect on anodic polarization behavior). If the inclusion and corrosion products are washed out, repassivation may occur; more generally, the environment in the pit remains aggressive and the pit continues to propagate. The condition is more severe with two-phase inclusions in which the MnS surrounds a nonreactive phase such as an insoluble oxide. The oxide particle effectively deepens the pit and allows retention of the aggressive surrounding solution, thus maintaining conditions for continued pit propagation. Pits formed on vertical surfaces may be accompanied by elongated downward attack on the passive film due to the gravitational flow of the more dense liquid from the pit that contains the acidic corrosion products. Since inclusions are elongated in the fabrication direction in products such as sheet and tubes, their geometries at exposed normal surfaces and at cut transverse and longitudinal sections frequently result in different susceptibilities to pitting (Ref 23).

## *Surface Instabilities during Pit Initiation*

The passive film contains or is susceptible to formation of a distribution of flaws that are potential sites for pitting, depending on the environment and the potential. The dynamic character of this surface under conditions conducive to pitting is illustrated in Fig. 7.8 (Ref 24) for a stainless steel in 0.4 M $FeCl_3$. In Chapter 4, it is shown that by scanning near the surface of a corroding metal with a reference electrode, the positions of local anodic and cathodic sites can be determined. In Fig. 7.8, the lines represent potential scans across the surface with the height proportional to the change in potential. The parallel displacement of the lines corresponds to a shift in the probe for successive scans. Peaks in the curves indicate local anodic sites. After 5 min of immersion, a large number of anodic sites are revealed. At 140 min, most of these have repassivated and a single site is observed, which has then repassivated at 160 min, and five new anodic sites have formed. At 380 min, a single stable pit has formed and is propagating. The single corrosion potentials at a distance from the interface are also listed for each time interval. They fluctuate by more than 150 mV and generally are more negative when the potential scans show greater anodic activity.

Distributions of current density over an iron surface exposed to an environment of 1 mM NaCl + 1 mM $Na_2SO_4$ (the cathodic reactant is dissolved oxygen, $O_2 + 2H_2O + 4e = 4OH^-$) are shown in Fig. 7.9 (Ref 25). In this case, the distribution of current density rather than the distribution of potential (Fig. 7.8) has been used to map the distribution of localized corrosion. Higher values of current density identify anodic areas and, therefore, areas of localized corrosion. In contrast to the pitting of stainless steel shown by Fig. 7.8, where pits formed and repassivated leading to a few regions of intense localized corrosion, localized corrosion on the iron surface remains active and spreads. In particular, the distribution of local corrosion on the iron after 20.2 h is distinctly different from that on the stainless steel in 380 min. A scan of variations of pH over the surface was consistent with the corrosion reactions. Anodic

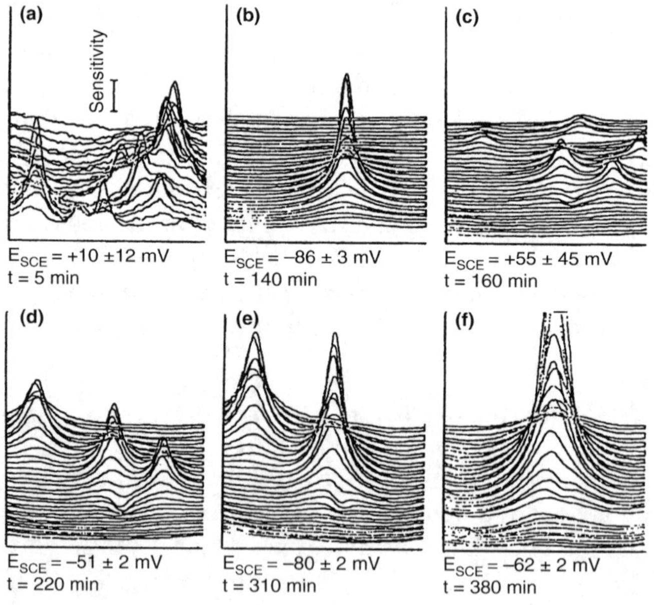


**Fig. 7.8** Potential distribution on the surface of a type 304 stainless steel in 0.4 M $FeCl_3$. Corrosion potentials (SCE) are indicated at each time period. Source: Ref 24

regions show a decrease in pH due to hydrolysis of the metal ions and the cathodic regions an increase in pH due to the formation of $OH^-$ ions. The cathodic surfaces are covered with a black oxide and the anodic areas by ferric hydroxides. Although the environments leading to the pitting behavior shown in Fig. 7.8 and 7.9 differ, the ferric chloride being more aggressive, the marked difference in behavior can be attributed to the greater stability of the passive film on the stainless steel.

The extremely local nature of pit initiation has been confirmed by observing the surface following initial contact with the aggressive environment. Slight local changes in surface morphology in the form of blisters are sometimes observed, on which an initial point of penetration may appear, or the entire blister surface may develop a lacelike appearance as a consequence of the uneven thinning and eventual penetration of the covering passive film. This latter stage is shown in Fig. 7.10(a) (Ref 3). Final rupture of the film may result in fragments of the passive film overlapping the edge of the pit as shown in Fig. 7.10(b). The geometry of the pit and the compositions of the occluded and external environments determine whether, on rupture of the film, pit propagation will occur or the pit surface will repassivate. Obviously, if deep pits remain under conditions preventing the loss of the pit environment, with

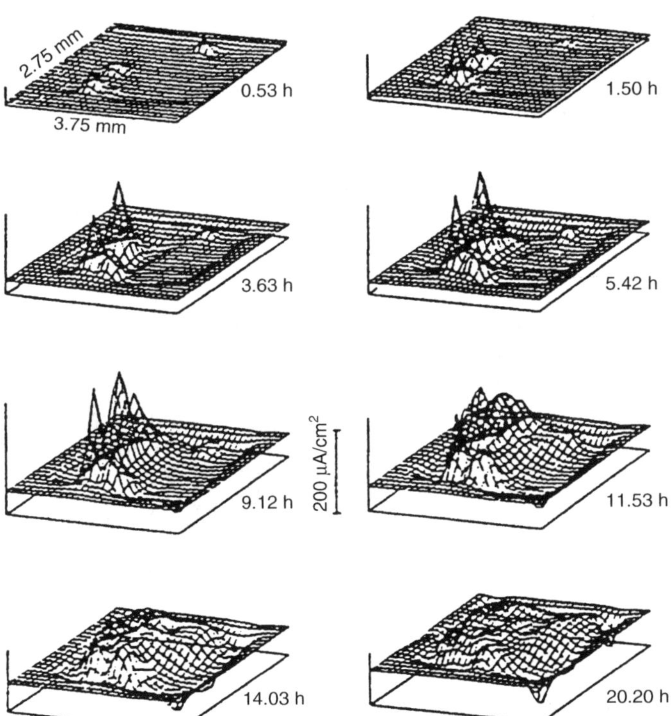


**Fig. 7.9** Distribution of current density over an iron surface exposed to 1 mM NaCl + 1 mM $Na_2SO_4$ at the times shown. Surface area 0.07 $cm^2$. Source: Ref 25

potentials in the pit in the active potential range of the polarization curve, pit propagation will occur. Thus, the early stages of pitting may be very probabilistic as to transition to permanent pits, which is consistent with the observations of the potential scans in Fig. 7.8.

Instability also is observed in the measurement of $E_{corr}$ as a function of time in pitting environments, as shown in Fig. 7.11 for type 304 stainless steel in 0.4 M $FeCl_3$ (Ref 26). The surface is initially passivated, and $E_{corr}$ remains essentially constant until rapid oscillations in poten-

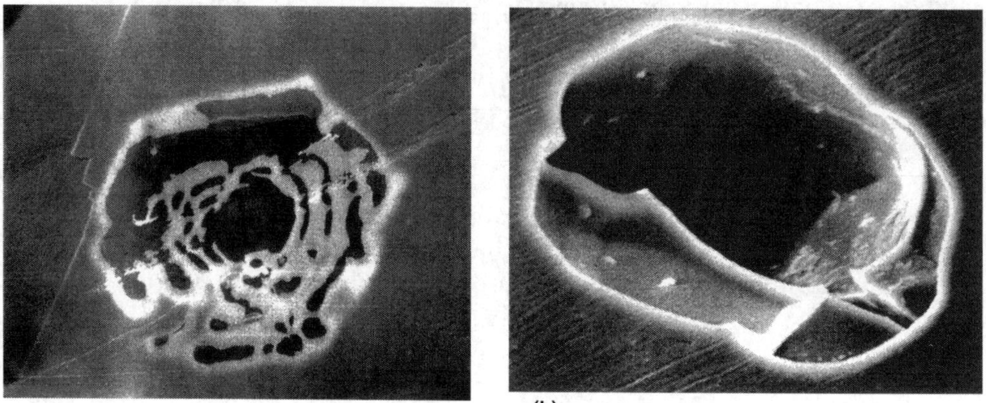

(a) (b)

**Fig. 7.10** (a) Partially perforated passive film on pit in type 304 stainless steel. (b) Fragment of passive film over edge of pit. 0.4 M $FeCl_3$. Source: Ref 3

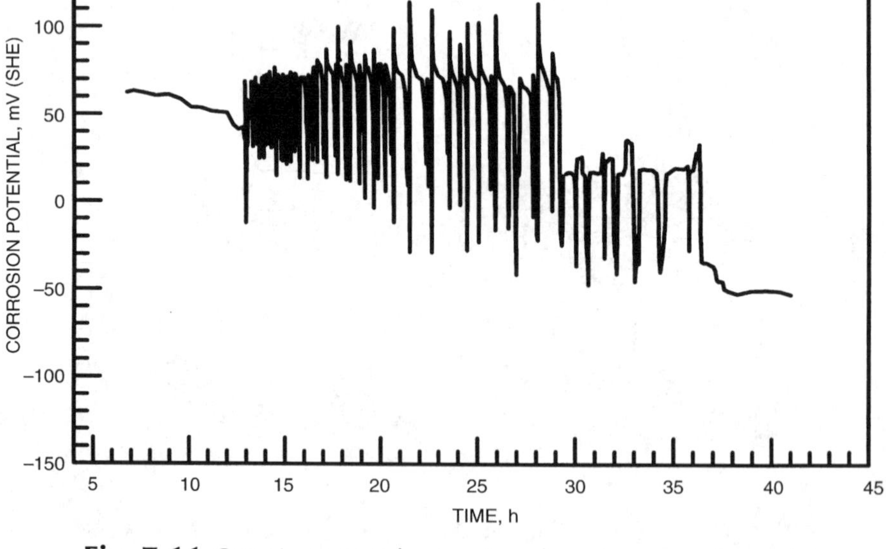


**Fig. 7.11** Corrosion potential versus time during exposure of type 304 stainless steel at 25 °C to 0.4 M $FeCl_3$. Source: Ref 26

tial (at 13 h) indicates initiation of pitting. The oscillations change in frequency and shape until, at 38 h, oscillations no longer occur, and the potential has decreased to values indicating active corrosion. Each time a pit opens, the surface becomes a galvanic couple between passivated and unpassivated regions with an IR potential drop between them; the measured $E_{corr}$ is that of the galvanic couple. Therefore, as the pit forms, $E_{corr}$ decreases and then increases if the pit repassivates. Otherwise, $E_{corr}$ progressively decreases as the size and number of pits increase. Thus, measuring the corrosion potential as a function of time can be an indicator of the initiation of pitting, providing that other factors changing $E_{corr}$, such as those affecting the cathodic reaction, are not responsible.

## *Pit Initiation and the Critical Pitting Potential*

A general discussion of passive film formation and structure has been given previously and then a mechanism described whereby defects in this film, if associated with sufficient IR potential drops, can lead to pit propagation. The tendency toward pitting of a material in a given environment can be investigated experimentally by increasing the potential using a potentiostat and observing the potential at which there is a significant increase in anodic current density. A schematic representation of this behavior is shown in Fig. 7.12. The solid curve is representative of the normal polarization curve, which at the higher potentials enters the transpassive region with increasing current density. Loss of the passive film at these potentials is an inherent characteristic of the alloy in that the passive film is no longer thermodynamically stable. If pitting is

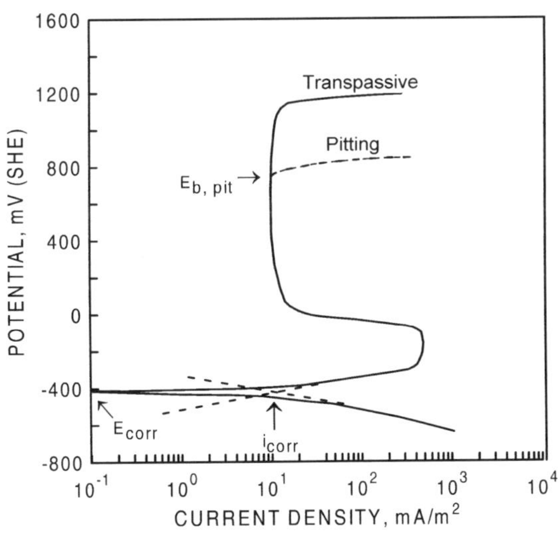


**Fig. 7.12** Schematic anodic polarization curve for a metal having susceptibility to pitting. Pitting is initiated at the breakdown potential $E_{b,pit}$.

initiated, an abrupt increase in current density (the dashed curve) occurs at a potential in the normally passive range, and pits are observed to form on the surface within which active corrosion is occurring. The potential at which pitting is initiated is referred to as the breakdown potential for pitting corrosion, $E_{b,pit}$ (or simply the pitting potential). Frequently, instabilities are observed as current fluctuations in the polarization curve as the breakdown potential is approached, as shown in Fig. 7.13 (see also Ref 11, 23). These instabilities are associated with initiation and repassivation of sites for pits prior to the propagation of stable pits as discussed previously. The instabilities in the potential with time, represented by Fig. 7.8 and 7.11, and the current fluctuations on increasing potential shown in Fig. 7.13, indicate metastable pitting (pit initiation followed by repassivation) terminating in statistically specific conditions required for the transition to stable pit propagation. For a given material and environment, the frequency of metastable pitting depends on surface conditions. Smooth surfaces exhibit less metastable pitting and have a slightly higher pitting potential (Ref 11, 23). The larger metastable pit activity associated with rougher surfaces has been attributed to gouging out and smearing of inclusions that are pit-initiation sites. In fact, this local mechanical effect has been considered a more important factor in pit initiation than just the presence of a rough surface (Ref 11, 23).

There are two major influences of the increasing potential that lead to pit initiation. First, as the potential is increased, the current density in preexisting flaws in the film increases such that IR* is exceeded, and active corrosion is initiated at the bottom of the flaw. Since the presence of certain anions in the environment are observed to lower $E_{b,pit}$, a second influence of increasing potential is to progressively attract these negative ions to the surface (which is becoming more positive) until local dissolution or penetration occurs. The potential at which this occurs

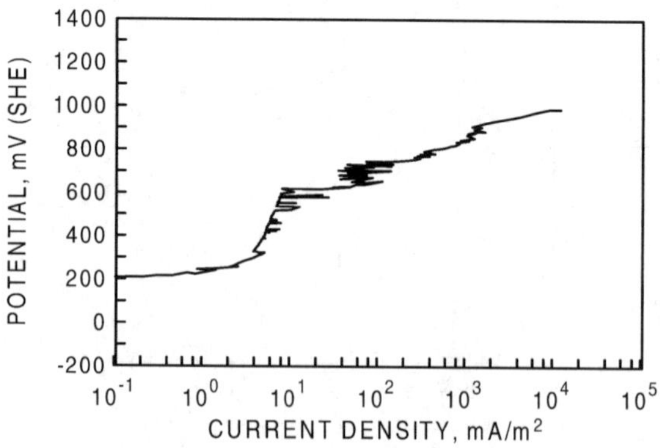


**Fig. 7.13** Anodic polarization curve showing current bursts at potentials below the breakdown potential. Type 304 stainless steel in 200 ppm chloride ion solution at room temperature, pH = 4

is lower than in the absence of the aggressive anions, and hence, the potential drop in a developing pit required to decrease the potential at the bottom of the pit to below $E_{pp}$ is less. The result is rapid dissolution at the bottom of the pit, and this, along with initiation of additional pits, results in an increase in the measured current density. The measured current may be expressed as $I = A_{pit}i_{pit} + A_{pass}i_{pass}$ where $A_{pit}$ and $A_{pass}$ represent the areas of the active (pit) and passive surfaces, respectively, at any time. The current density over the passive surface is $i_{pass}$ and in the pit is $i_{pit}$, with $i_{pit} >> i_{pass}$. As the total pitted area increases, the total current increases. If the potential is maintained by a cathodic reactant such as $O_2$ or $Fe^{3+}$, rather than potentiostatically, the corrosion potential decreases as pitting progresses as a consequence of the galvanic interaction between the passive surface and the active surface within the pit.

**Effect of Chloride Ions on Pit Initiation.** It is pointed out in the section "Interface Potential and Pit Initiation" that chloride ions are increasingly adsorbed and/or absorbed at the surface. Mechanisms whereby the chloride ions penetrate the passive film and initiate pitting are discussed. A representative example of the influence of progressive changes in chloride-ion concentration on polarization scans of type 304 stainless steel to reveal susceptibility to pitting is shown in Fig. 7.14 (Ref 27). It is evident that as the chloride-ion concentration increases, $E_{b,pit}$ decreases. It follows that an environment represented by cathodic curve A is predicted to induce pitting if the chloride concentration is greater than 200 ppm; whereas, an environment represented by cathodic curve B will not induce pitting even at a chloride-ion concentration of

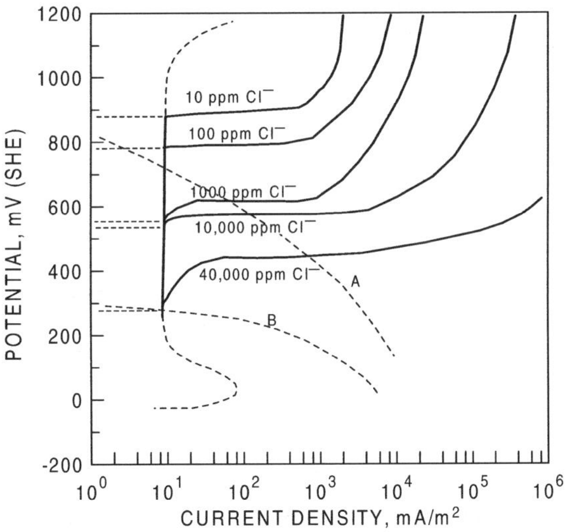


**Fig. 7.14** Effect of chloride-ion concentration on the anodic polarization of type 304 stainless steel. Dashed lines indicate breakdown potentials, $E_{b,\,pit}$. Curves A and B are schematic representations of polarization of cathodic reactions of relatively (A) high and (B) lower oxidizing strength. Based on Ref 27

40,000 ppm. For most alloys and environments, the chloride ion is most effective in initiating pitting (decreasing $E_{b,pit}$). The halide ions, $Br^-$ and $I^-$, are less aggressive, and $SO_4^=$ ions may have an inhibiting effect in chloride-containing environments (Ref 28). With respect to the chloride ion, three contributing factors are its high mobility; its small size, permitting incorporation into the passive film; and the predominant formation of soluble metal-chloride complexes. Two factors appear to contribute to the observation that the pitting potential is higher in more dilute chloride concentrations. First, lower chloride concentrations will contribute less to the conductivity of the pit environment, thus requiring higher external potentials to bring the potential in the pit to the critical value for pit propagation. The magnitude of this effect is uncertain since the concentration, and hence conductivity, of corrosion-product cations in the occluded region is already high. The more important factor may be that the lower bulk chloride concentration in the environment lowers the chloride-ion concentration in the occluded volume. The consequence is that, at the balance between migration into and diffusion from the occluded region, the hydrolysis reactions do not lower the pH sufficiently to initiate and/or maintain active corrosion. Thus, a higher potential is required to increase the chloride (and metal) ion concentrations, increasing the hydrolysis, and thereby lowering the pH to the critical value for active corrosion.

Extensive investigations have been reported covering the effects of single and mixed environments of anions on pitting behavior. A representative compilation of aggressive anions producing passivity breakdown on the listed metals is given in Table 7.2 (Ref 29). It should be noted that for those metals forming the more stable oxide films, such as Fe, Ni, Ti, and stainless steels, breakdown occurs for anions of strong acids. For the less-stable oxides, such as form on Zn and Mn, anions of weaker acids also cause breakdown.

**Table 7.2 Anions producing passivity breakdown**

| Metal | Aggressive anion |
|---|---|
| Iron | $Cl^-$, $Br^-$, $I^-$, $ClO_4^-$, $SO_4^=$ |
| Nickel | $Cl^-$, $Br^-$, $I^-$ |
| Stainless steel | $Cl^-$, $Br^-$, $SCN^-$ |
| Aluminum | $Cl^-$, $Br^-$, $I^-$, $ClO_4^-$, $NO_3^-$, $SCN^-$ |
| Titanium | $Cl^-$, $Br^-$, $I^-$ |
| Zirconium | $Cl^-$, $Br^-$, $I^-$, $ClO_4^-$ |
| Tantalum | $Br^-$, $I^-$ |
| Zinc | $Cl^-$, $Br^-$, $I^-$, $NO_3^-$, $SO_4^=$, $ClO_4^-$, $ClO_3^-$, $BrO_3^-$, $HCO_2^-$, $CH_3CO_2^-$ |
| Cadmium | $Cl^-$, $Br^-$, $ClO_4^-$, $SO_4^=$ |
| Manganese | $Cl^-$, $Br^-$, $ClO_4^-$, $SO_4^=$, $NO_3^-$, $CH_3CO_2^-$ |

Source: Ref 29

## *Cyclic Anodic Polarization Scans: the Protection Potential*

For most alloys, reversal of the anodic polarization scan, following the initiation and propagation of pitting, results in a "polarization loop" of the form shown in Fig. 7.15. When the potential scan is reversed at some potential above $E_{b,pit}$ where the current density has increased due to pitting, the downscan curve results in a loop of the form shown. The current density remains abnormally high and returns (if at all) to the passive current density at a lower potential. This lower-potential intersection is frequently referred to as the protection potential, $E_{prot}$, with the implication that if the potential is never raised above this value, pitting will not occur. This behavior is a direct consequence of the more aggressive environment generated in the pit during its propagation. On reversing the potential, the IR potential drop in the pit is decreased, allowing the potential in the pit to increase. $E_{prot}$ corresponds to the potential at which a stable passive film forms on the metal in the local pit environment; therefore, if the potential in the pit becomes or decreases below this value, repassivation should occur. This explanation is consistent with the observation that $E_{prot}$ is generally difficult to establish as a parameter characterizing the metal/environment. It also depends on such variables as the potential scan rate and the current density at which the scan is reversed. These variables influence the initial pit geometry, pit environment, and, hence, potential change within the pit. In some cases, the loop does not return within the passive potential range, suggesting that if the metal is held (with a potentiostat or by the environment) at any potential in the passive range, which is below $E_{b,pit}$, pitting will occur. Cyclic polarization scans, however, have been useful in the

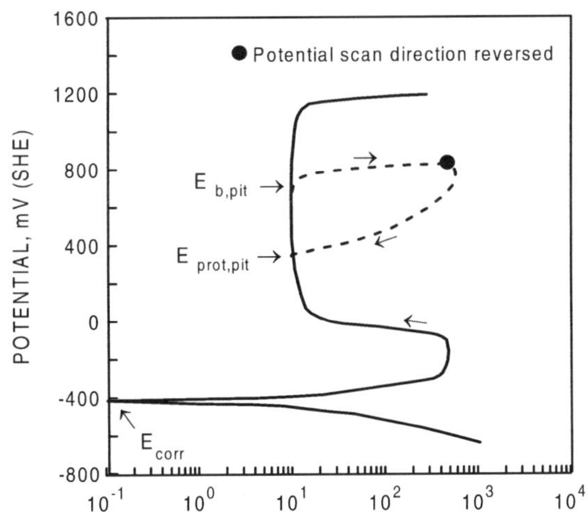


**Fig. 7.15** Schematic cyclic polarization curve for a metal showing susceptibility to pitting. Pitting is initiated at $E_{b,pit}$ and propagation stops at $E_{prot,pit}$.

study of pitting, allowing the several variables to be investigated and allowing classification of the relative resistance to pitting of alloys in terms of pitting potential, size of the anodic loop, and the corrosion potential.

## *Investigations of Pitting Corrosion Using Chemical Environments*

In the previous section, pitting of active-passive alloys is introduced in relationship to observations of potentiodynamic polarization scans. This leads to the concept of a breakdown potential for pit initiation and to a protection potential. In service, the environment induces a corrosion potential, and if this potential is above the protection potential, pitting is predicted to occur at some time (generally difficult to estimate). Ferric chloride solutions are frequently used as test environments for determining susceptibility of alloys to pitting corrosion. Four factors support the use of these solutions: (a) since the standard equilibrium potential for the $Fe^{3+}/Fe^{2+}$ half-cell reaction is 770 mV (SHE), the ferric-ion reduction reaction is highly oxidizing and is conducive to a high $E_{corr}$; (b) the exchange current density for the reduction of ferric ions is large, as also is the limiting diffusion current density, both of which make the reaction strong kinetically; (c) the ferric ions hydrolyze to lower the pH; and (d) $FeCl_3$ provides three chloride ions for every ferric ion. Factor (d) provides high chloride ion concentrations that are conducive to pitting. Factors (a), (b), and (c) are illustrated by the cathodic polarization curves for $FeCl_3$ previously shown in Fig. 3.19 and 3.20.

The effects of ferric chloride concentration on the pitting of type 304 stainless steel are shown in Fig. 7.16. Specimens were exposed for two weeks at room temperature to concentrations from 0.001 to 10 wt% ferric chloride. In this period of time, pitting was not observed for concentrations below 1.0 wt%, one pit was observed at 1 wt%, and several pits had completely penetrated the specimen at 10 wt% $FeCl_3$. It is emphasized that the interpretation of the results presented in Fig. 7.16 must take into consideration the statistical nature of pitting; namely, what is

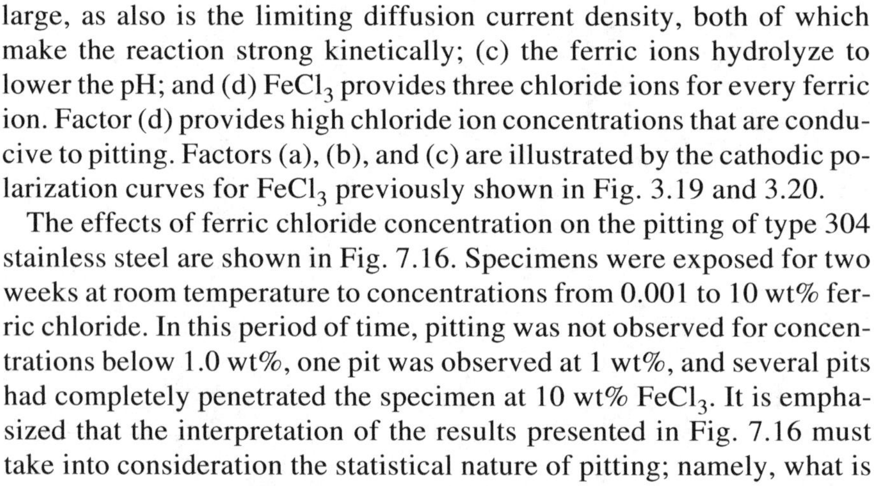


**Fig. 7.16** Effect of ferric chloride concentration in water on pitting of type 304 stainless steel. Two-week immersion at room temperature

the probability that a pit will be produced per unit area? More specifically, more pits per unit area could be found if the sample area had been larger on exposure to the 1.0 wt% solution, and it would be less certain whether pits would be observed at the 0.1 wt% concentration. Thus, in conducting tests of this type, sufficient area or numbers of test specimens must be exposed to show that the probability of pitting in a structure is acceptably low.

The behavior of an aggressive environment, such as one containing ferric chloride, in causing pitting and the observations that are made using this environment for pitting-susceptibility tests, may be understood by reference to Fig. 7.17. For reasons pointed out subsequently, the abscissa is in terms of total current rather than current density, although the values of current correspond to an area of 1 $m^2$. The curve ABCDFG is representative of the anodic polarization of type 304 stainless steel in deaerated 1 N $H_2SO_4$. An anodic peak occurs at B, the passive range is CF, and the transpassive range is FG. The active dissolution range is AB, but in the presence of sufficient chloride ions to prevent passivation, the active curve extends along BHG. The much larger current that would exist for the active state relative to the passive state is evident (i.e., over the potential range 200 to 1100 mV (SHE)). The presence of aggressive anions ($Cl^-$) leading to susceptibility to pitting would result in the polarization curve ABCDHG, with D being the pitting potential at which the current density could increase to H if the passive film were completely removed.

At any state of pitting, the surface is a composite of active and passive areas. The anodic polarization curve for this composite surface is then the sum, at each potential, of the current densities of the passive and active curves weighted by their areas. The dashed curves, $P_1q_1$, $P_2q_2$, and $P_3q_3$, represent the positions of the active curve (initially ABHG) for active surface areas of 0.01, 0.1, and 1.0% of the total area. The polarization curve for the composite surface at any potential is obtained by adding the shifted curve to the passive curve. These composite-surface

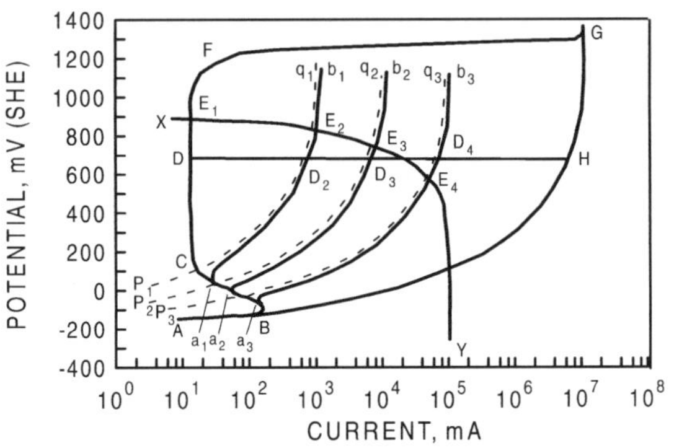


**Fig. 7.17** Schematic representation of shift of polarization curves associated with progressive fractions of pitted surface

curves, $a_1b_1$, $a_2b_2$, and $a_3b_3$, are shown for the respective areas. Thus, the total measured current at any potential is:

$$I_{total} = i_{pass}A_{pass} + i_{pit}A_{pit} \qquad \text{(Eq 7.1)}$$

where $i_{pass}$ and $i_{pit}$ are the current densities associated with the passive and active (pitted) areas, $A_{pass}$ and $A_{pit}$.

The curve XY in Fig. 7.17 is representative of the cathodic polarization of $Fe^{3+}$ (from $FeCl_3$). In the absence of chloride ions, or on immediate contact with the ferric chloride environment, the $Fe^{3+}$ ions will passivate the stainless steel and the corrosion potential will be $E_1$, the intersection of the anodic and cathodic polarization curves. This corrosion potential would be observed in ferric sulfate, which does not induce pitting. However, since $E_1$ is above the pitting potential (D), the alloy will start to pit in a chloride environment with a distribution of now unpassivated areas on the surface. When 0.01% of the surface has pitted, the effective anodic curve is $a_1b_1$, and the corrosion potential will have decreased to $E_2$, the intersection with the cathodic curve. Since this potential is still above the pitting potential, new pits will form and old pits will propagate. The corrosion potential continues to decrease to $E_3$, and then, at potentials below $E_D$, such as $E_4$, new pits should no longer form. However, pits that have already formed may continue to grow because of the aggressive corrosion-product environment in the pits. If there is a protection potential, $E_{prot}$, in the range $E_D - E_C$, below which pits no longer propagate, then when the potential has decreased to $E_{prot}$, pitting should stop. These sequences of pit initiation and propagation allow detection of pitting in a metal/environment system by monitoring the corrosion potential. Initial stages are frequently detected by instability in the corrosion potential with time as the pits form and repassivate. A sustained drop in potential is an indication of established pitting. Figure 7.18 shows the corrosion potential for type 304 stainless

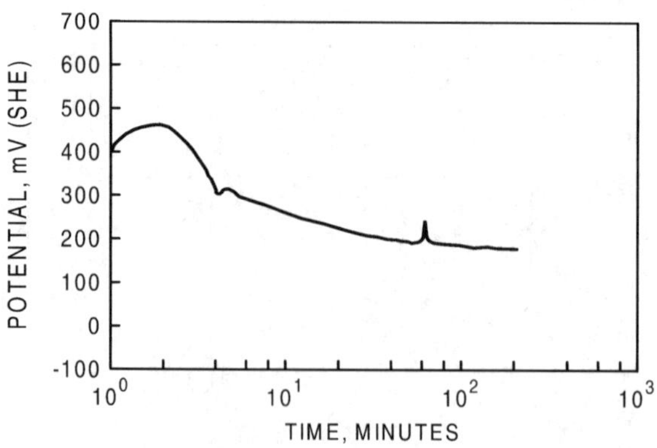


**Fig. 7.18** Change in corrosion potential of type 304 stainless steel with time at 25 °C in 1.5 wt% ferric chloride. Source: Ref 30

steel exposed to 1.5 wt% $FeCl_3$ as a function of time (Ref 30). An initial increase in corrosion potential is observed due to thickening of the passive film prior to pit initiation. A decrease in potential is then observed when pitting is initiated. In more dilute solutions, the corrosion potential does not reach $E_{b,pit}$, and pitting is not immediately initiated, although it may occur in time if the potential remains above $E_{prot}$.

## *Effects of Temperature on Pitting: the Critical Pitting Temperature*

In the presence of aggressive anions such as $Cl^-$, the polarization curves, and, hence, pitting potentials, are sensitive to temperature. This effect is illustrated in Fig. 7.19 in which polarization curves for temperatures from 10 to 90 °C are shown for a modified stainless steel (~18 wt% Cr, ~20 wt% Ni, 5.6 wt% Mo) in a solution of pH = 3 and 3.5 wt% NaCl (Ref 31). Examinations of the surfaces following the scans revealed no pitting at temperatures below 40 °C. At 60 °C, a sharp increase in current density at about 400 mV (SHE) was associated with pitting, identifying this potential as the critical pitting potential at this temperature. The curves show a large change as the temperature is increased from 35 to 60 °C. If a measure of the effect of temperature on the polarization curve, including pitting behavior, is that potential which results in a specified current density (e.g., 100 $mA/m^2$), a plot of this potential as a function of temperature takes the form shown in Fig.

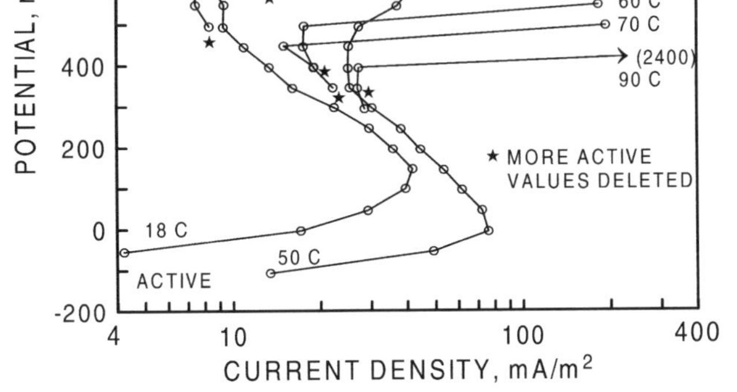


**Fig. 7.19** Effect of temperature on the anodic polarization curves of a modified austenitic stainless steel containing 5.6 wt% Mo in 3.5 wt% NaCl at pH = 3. Redrawn from Ref 31

7.20 (Ref 32). The temperature range of rapid decrease in pitting potential spans the critical pitting temperature. Decreasing the potential scan rate at which the polarization curve is determined frequently leads to a very small temperature range above which pitting is observed but below which pitting is not observed. The center section of the curve of Fig. 7.20 becomes progressively vertical and, in the limit, takes the form of the dashed lines (Ref 32). Two useful interpretations follow. First, for potentials near the center of the passive range, temperatures to the left of the vertical line correspond to conditions of no pitting; temperatures to the right indicate conditions under which pitting will occur. Second, a useful procedure for determining susceptibility to pitting is to hold a specimen potentiostatically in the passive range at a low temperature. The current density will be low corresponding to the passive state. The temperature is slowly increased until a rapid increase in current density indicates initiation of pitting. The temperature at which this occurs is the critical pitting temperature for the alloy in the environment and defines the upper limit for safe exposure.

Determination of the critical pitting temperature is also accomplished by using oxidizing cathodic reactants that establish the potential in the presence of a constant concentration of anions causing pitting. The data in Fig. 7.21 were obtained on exposure of type 317L stainless steel to a constant chloride-ion concentration (Ref 33). Additions of NaOCl, $FeCl_3$ and $K_3Fe(CN_6)$ produced corrosion potentials of about 1100, 900, and 690 mV (SHE). On increasing the temperature in each of these environments, pitting is observed within a few degrees of 30 °C, thus

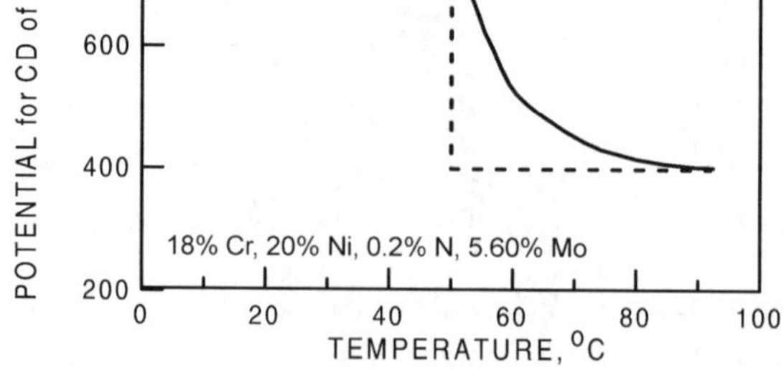


**Fig. 7.20** Temperature dependence of pitting potential defined as potential at which current density reaches 100 mA/m$^2$. Same alloy as Fig. 7.19. Dashed curve approached as potential scan rate used in Fig. 7.19 is decreased. Based on Ref 32

establishing this as the critical pitting temperature, resulting in the division of the graph (Fig. 7.21) into ranges of pitting and no pitting.

In an investigation of several accelerated laboratory tests for determining localized corrosion resistance of high-performance alloys, the procedure just described was called an immersion pitting temperature test and was considered to best simulate and correlate with service performance of the alloys (Ref 34). For a series of 17 alloys, the critical pitting temperatures (determined by increasing the temperature 5 °C at 24 h increments) ranged from 20 to 80 °C in an environment of 4.0 wt% NaCl acidified with 0.01 M HCl to pH = 2 and 0.1% $Fe_2(SO_4)_3$ as an oxidizing agent to increase the corrosion potential. The pitting potential was determined at 70 °C in 4.0 wt% NaCl acidified to pH = 2 with 0.1 M HCl using potentiodynamic scans of 360 mV/h. The correlation between the critical pitting temperature and the critical pitting potential was reasonably good, as shown in Fig. 7.22. A single "potentiostatic" measurement on SANICRO 28 alloy (the potential was increased 24 mV every four days) resulted in the pitting potential identified as 16S. This illustrates, as earlier discussion emphasizes, that the potentiodynamic scan rate can be a significant variable in determining pitting potentials (Ref 34).

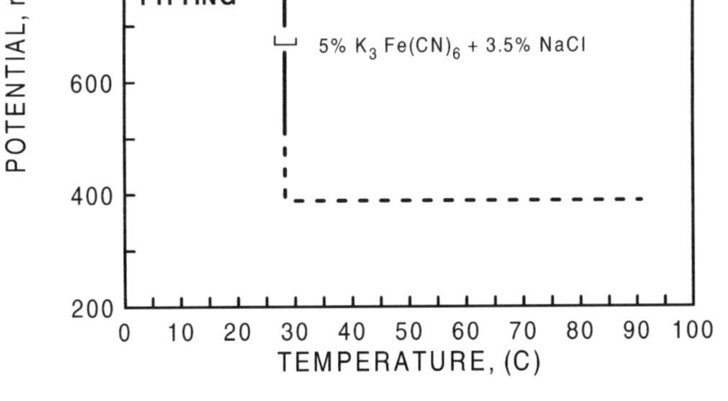


**Fig. 7.21** Pitting temperature range of type 317L stainless steel exposed to chloride solutions of different oxidizing power for 24 and 66 h. Dashed lines are based on potentiodynamic data in Fig. 7.19. Redrawn from Ref 33

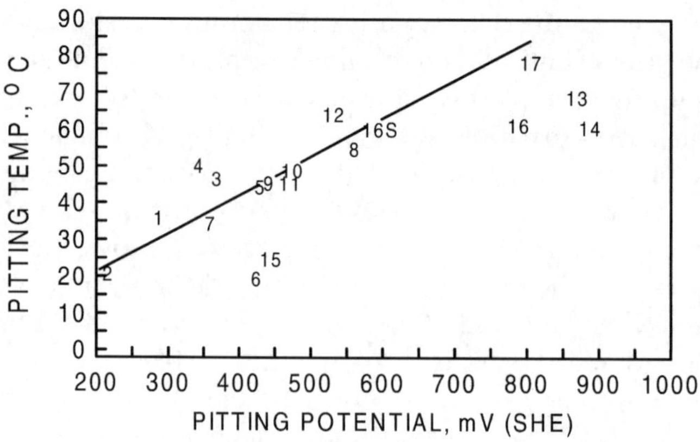


**Fig. 7.22** Correlation between the critical pitting temperature and critical pitting potential of 17 high-performance alloys. The alloys are: (1) 317LM, (2) 3RE60, (3) AF22, (4) 44LN, (5) FERRALIUM ALLOY 255, (6) 20CB-3 Alloy, (7) URANUS 86, (8) 2545LX, (9) JESSOP 700, (10) JESSOP 777, (11) 904L, (12) M-32, (13) AL6X, (14) 1545MO, (15) 825, (16) SANICRO 28 ALLOY, (17) G-3, and (16S) potentiostatic pitting potential, SANICRO 28 ALLOY. (Details can be found in text. Analyses of alloys are given in Ref 34.) (Redrawn from Ref 34)

## *Effect of Alloy Composition on Pitting*

The resistance to pitting corrosion of active-passive-type alloys, particularly those based on iron or nickel, can be increased by selective alloying. The following section is restricted to the stainless steels and the high-performance nickel-base alloys in which the major alloying elements are chromium, iron, molybdenum, tungsten, and nitrogen. Small amounts of titanium and niobium are frequently present but are of greater significance in enhancing resistance to intergranular corrosion rather than pitting. Brief reference is made to sulfur in these alloys since it is detrimental to pitting. A qualitative summary of the effects of alloying elements in austenitic stainless steels on pitting in chloride solutions is given in Fig. 7.23 (Ref 35). Another correlation between composition and tendency for pitting is shown in Fig. 7.24 (Ref 36), in which the pitting potentials are shown to be higher when the critical current densities for passivation ($i_{crit}$) are lower. If low values of this critical current density reflect enhanced structural and compositional integrity of the passive film being formed, increased resistance to pitting should be indicated by higher pitting potentials. The exception in the correlation should be noted by the high pitting potential for the alloy with 0.16% nitrogen. This indicates, as is discussed later, that the mechanism whereby nitrogen influences pitting resistance appears to be unique.

The major alloying element contributing to resistance to pitting corrosion in iron- and nickel-base alloys is chromium. The effect of chromium in reducing both the critical current density and the passivating potential of iron in 1 N $H_2SO_4$ is shown by the polarization curves of

Fig. 5.24. A similar influence of chromium in nickel is shown in Fig. 5.27. As increasing amounts of chromium are added to iron, the relative fraction present in the passive film increases until, at chromium contents of the large-volume commercial stainless steels (18 to 22% Cr),

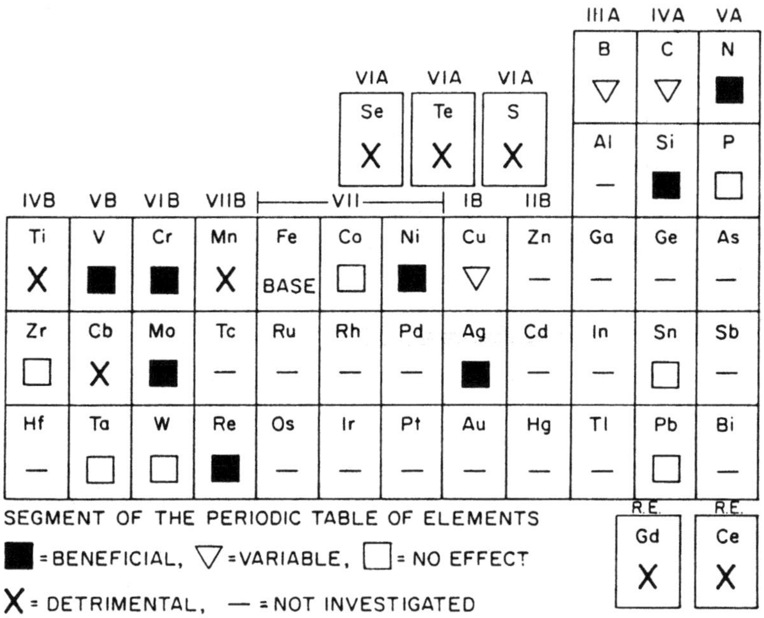


**Fig. 7.23** Effect of elements shown on resistance of stainless steels to pitting in chloride solutions. Source: Ref 35

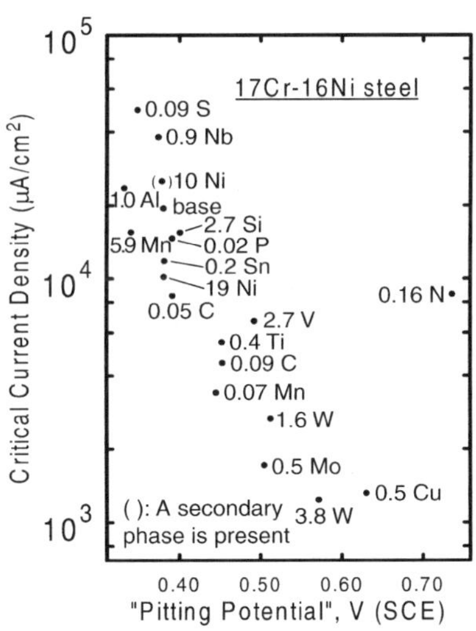


**Fig. 7.24** Relation between the pitting potential of 17 wt% Cr, 16 wt% Ni steels with elements shown in 0.1 N NaCl + 0.25 N $Na_2SO_4$ and the critical current density for passivation in 1 N $H_2SO_4$ + 0.05 N NaCl at 40 °C. Source: Ref 36

the film is >60% chromium. The film is complex in structure and composition, the interpretation of both being complicated by a film thickness that is usually <10 nm and frequently <2 nm. Surface analysis techniques have detected interatomic bonds corresponding predominantly to $Cr_2O_3$ with some $CrO_3$ as an inner or barrier-layer metal/film interface that is largely responsible for the passivity. In an outer, more-porous precipitate layer, bonds have been reported corresponding to $Cr(OH)_3$, $CrO_4^{=}$, and FeOOH. When the environment contains anions such as $Cl^-$ and $SO_4^{=}$, they usually are detected in the passive film. The film progressively becomes more amorphous as the chromium content is increased (Ref 6). The increased resistance to pitting has been attributed to the amorphous structure, assigning the effect to fewer defects capable of initiating pits and to reduced diffusion of aggressive anions such as chloride ions (Ref 37).

The effect of chromium concentration on the pitting potential of iron-chromium alloys in neutral 0.1 N NaCl is shown in Fig. 7.25 (Ref 38). Below 12% chromium, the alloys do not passivate in this environment, and hence, pitting as a breakdown of a passive film does not occur. The influence of chromium in increasing the pitting potential is confined to the range of about 20 to 40% chromium over which the increase is about 700 mV. The effect of chromium in nickel-chromium alloys is shown in Fig. 7.26 (Ref 38). Pure nickel can be passivated in the neutral 0.1 N NaCl environment and exhibits a pitting potential of about 300 mV (SHE). However, additions of chromium are most effective in the range 10 to 20% Cr over which the pitting potential increases by about 500 mV.

Nickel has a very small effect on the anodic polarization behavior of iron, and hence, iron-nickel alloys are of minor significance as corrosion-resistant alloys. However, the addition of nickel to iron-chromium alloys (AISI 200 series) permits conversion of the latter as ferritic alloys to austenitic iron-chromium-nickel alloys (AISI 300 series). In

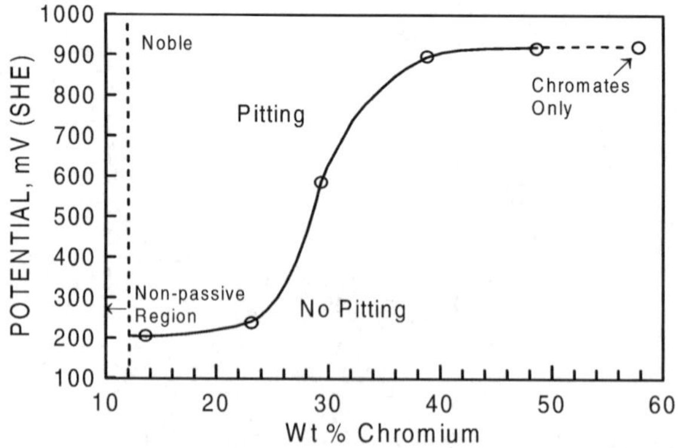


**Fig. 7.25** Critical pitting potentials for Fe-Cr alloys in 0.1 N NaCl at 25 °C. Redrawn from Ref 38

these alloys, nickel provides a relatively small increase in the critical pitting potential as shown in Fig. 7.27 for the neutral 0.1 N NaCl environment (Ref 38).

Molybdenum has a strong influence in increasing the pitting potential of iron-chromium, iron-chromium-nickel, and nickel-chromium alloys (Ref 38). The effect of molybdenum in an Fe-15Cr-13Ni alloy in the same environment of 0.1 N NaCl, as previously reported in Fig. 7.25 to 7.27, is shown in Fig. 7.28 in which additions of 2.5% molybdenum increase the pitting potential by nearly 500 mV at 25 °C. Similar effects of molybdenum in stainless steels are observed in acid chloride environments. In a wide range of environments, the beneficial effects of molybdenum extend to at least 6%. The temperature dependence of the pitting potential should be noted in Fig. 7.28 (Ref 38). At 0% Mo, decreasing the temperature from 25 to 0 °C increases the pitting potential by about 500 mV. Increasing the molybdenum concentration at this lower tem-

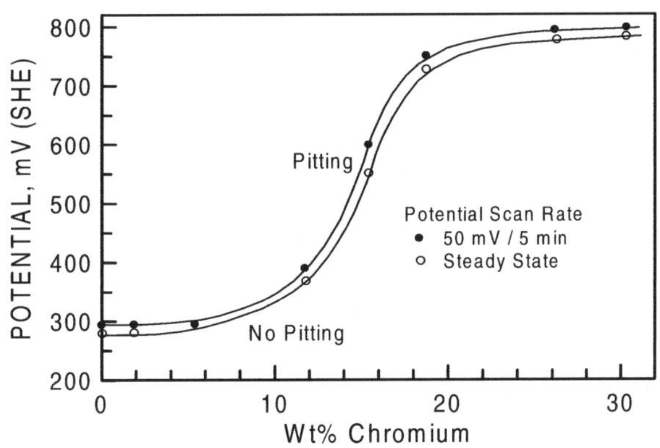


**Fig. 7.26** Critical pitting potentials for Ni-Cr alloys in 0.1 N NaCl at 25 °C. Redrawn from Ref 38

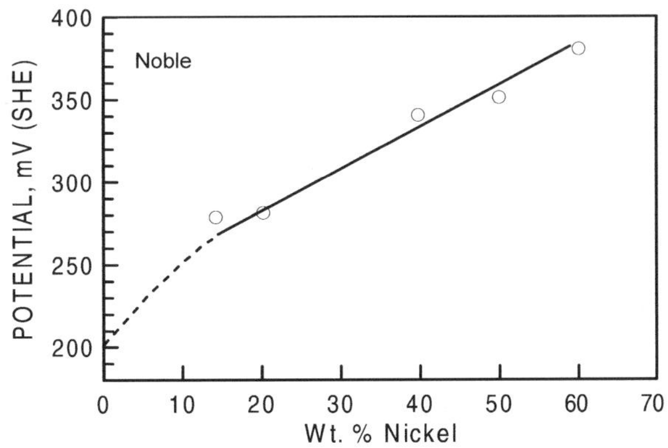


**Fig. 7.27** Critical pitting potentials for 15 wt% Cr-Fe alloys with increasing Ni content in 0.1 N NaCl at 25 °C. Redrawn from Ref 38

perature decreases the pitting potential such that at 2% Mo, the pitting potential is about the same as the molybdenum-free alloy at 25 °C.

The increase in pitting potential and, hence, resistance to pitting in many environments, as a result of molybdenum additions is of considerable economic significance, for example, in the selection of type 316 (2-4% Mo) stainless steel in place of type 304 for pit-inducing environments. As a consequence, a large amount of research has been conducted to understand the mechanism of this influence. Additions of molybdenum to iron do not improve pitting resistance. In fact, the polarization behavior of Fe-Mo alloys in sulfuric acid is similar to that shown in Fig. 5.29 for Ni-Mo alloys. The passivating potential and the current density in the passive potential range are increased, both of which indicate decreasing resistance to corrosion. In acid-chloride environments, such as 1 N HCl, Fe-Mo alloys cannot be passivated. As the chromium content is increased, the contribution of molybdenum to resistance to pitting corrosion increases. For example, in a series of Fe-25Ni-5Mo alloys in 1 N HCl, at 5% Cr addition, the alloy could not be passivated. At 17% Cr, the alloy could be passivated with a pitting potential of about 650 mV (SHE); at 20% Cr, the alloy passivates, and no pitting is observed at any potential up to the transpassive potential. On decreasing the molybdenum content to 3%, complete passivity is not obtained prior to pitting at about 200 mV (SHE). For comparison, 40% Cr is required in an Fe-Cr binary alloy to not show pitting corrosion in the 1 N HCl. The 5% Mo has thus halved the chromium concentration necessary to avoid pitting (Ref 39).

The major question is whether the influence of molybdenum on pitting is related to the structure and composition of the passive film, resulting in restricted pit initiation, or that molybdenum quickly stops or severely retards pit propagation. With respect to pit initiation, the mo-

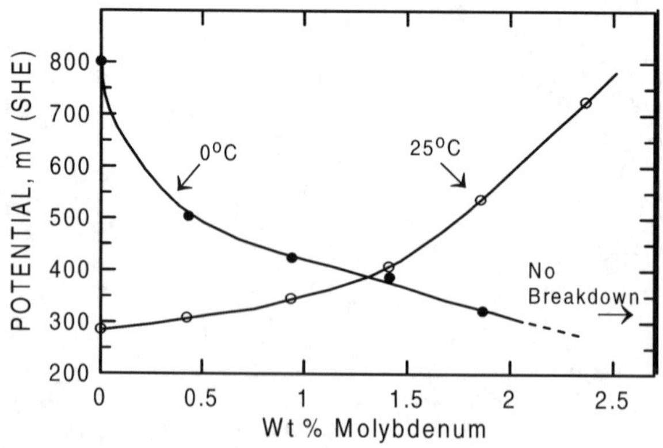


**Fig. 7.28** Critical pitting potentials for 15 wt% Cr, 13 wt% Ni stainless steel with increasing Mo content in 0.1 N NaCl at 25 °C. Redrawn from Ref 38

lybdenum concentration in the passive film increases with increased concentration in the metal substrate, and some increase in film thickness has been reported. Although the amount of molybdenum in the film decreases as the potential is increased, the passive current density is smaller than in the absence of molybdenum. This appears to be inconsistent with the polarization curve of pure molybdenum (Fig. 5.20), which exhibits an abrupt transition to a high dissolution rate in the transpassive region starting near 200 mV (SHE). Explanations for the beneficial effect of molybdenum in enhancing passivity and, hence, pitting resistance include: (a) the hexavalent molybdenum is stable in the presence of chromium in the passive film and is present with $MoO_3$ bonding near the metal interface and $MoO_4^{2-}$ bonding near the solution interface; (b) the thickness and stability of the glassy or amorphous interface oxide is increased; (c) an increase has been observed in the $Cr_2O_3/Cr(OH)_3$ ratio resulting from the decrease in bound water when molybdenum is present; (d) the presence of $MoO_4^{2-}$ enhances the negative charge associated with $CrO_4^{2-}$ in the outer surface of the film, thereby retarding the influx of $OH^-$ and $Cl^-$ ions (Ref 39–42). This interaction between molybdenum and chromium is consistent with the necessity of having chromium present as a major element in the alloy for the molybdenum to be beneficial. There is also evidence that the chloride concentration in the film is decreased as the molybdenum concentration is increased. The interdependence of chromium and molybdenum is also supported by the observation that the passive current density is smaller in Cr-Mo alloys than in pure chromium (Ref 43). Molybdenum may also inhibit pit initiation by changing the composition of the alloy surface in immediate contact with the environment or at the metal/oxide interface. It has been established that on initial contact of a stainless steel with a corrosive environment, the dissolution rate of the iron component of the alloy is largest, followed by that of nickel, leaving an increased surface concentration of chromium. The molybdenum concentration is increased, but the extent decreases as the potential is increased in the passive range. As a consequence, the passive state of the alloy, including pitting resistance, becomes that of this altered surface layer. Nickel-chromium-molybdenum alloys are known to have excellent pitting resistance. Therefore, the formation of surface or interface compositions similar to these alloys would have enhanced pitting resistance.

There is also evidence that the beneficial effect of molybdenum is to interfere with pit propagation. If the mechanism is active at the initiation of localized breakdown of the passive film, then, effectively, pitting will not occur. Based on the low solubility of molybdenum chloride, $MoO_3$, and polymolybdates in acid solutions, one mechanism proposes that molybdenum enhances the formation of salt films of these species within the pit. This can decrease the IR potential drop to the pit

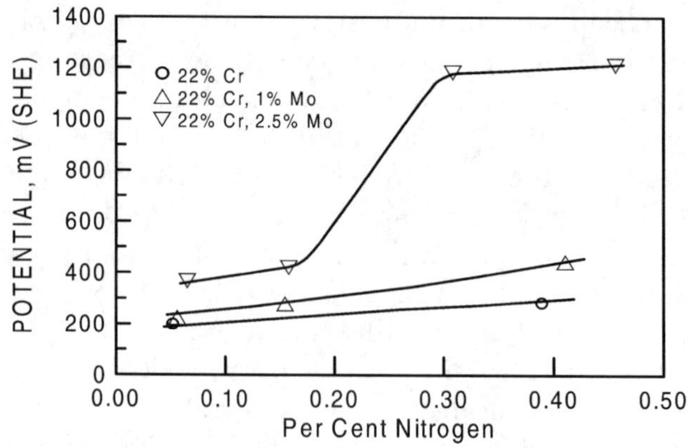


**Fig. 7.29** Nitrogen dependence of pitting potential for austenitic stainless steels containing 22 wt% Cr, 20 wt% Ni, 4 wt% Mn, and 0, 1, or 2.5 wt% Mo. Redrawn from Ref 47

interface, thus increasing the potential at this interface such that repassivation of the pit rapidly occurs, effectively stopping propagation. The effect of molybdenum in promoting an alloy-rich surface of enhanced corrosion resistance, as just discussed, also would contribute to a decreased propagation rate (Ref 44, 45).

Addition of nitrogen to stainless steel as an alloying element increases the resistance to pitting. The beneficial influence requires that the nitrogen be in solid solution and, for this reason, depends on the limit of solubility in the alloy. The limit in pure body-centered cubic (bcc) iron is ~0.01 wt%, and in FCC iron it is ~0.03 wt%, but is increased in the presence of several alloying elements with chromium and molybdenum being the most significant in the commercial stainless steels. Nitrogen-alloyed ferritic steels contain >0.08 wt% N, and the austenitic steels contain up to 0.5 wt% N; higher nitrogen content alloys have been investigated (Ref 46). The effect of alloy composition is illustrated by Fig. 7.29 (Ref 47) in terms of pitting potential as a function of nitrogen content for a 22Cr-20Ni-4Mn alloy with 0, 1, and 2.5% Mo. A factor of the form, %Cr + 3.3(%Mo) + 16(%N), provides a reasonable correlation of composition to the pitting potential as shown in Fig. 7.30 (Ref 47). The beneficial effect of nitrogen in stainless steels also has been demonstrated by exposure to pitting chemical environments such as ferric chloride solutions.

Several mechanisms have been proposed for the beneficial influence of nitrogen. One mechanism is that the reduction of dissolved nitrogen according to the reaction $N + 4H^+ + 3e \rightarrow NH_4^+$ consumes hydrogen ions and prevents acidification by hydrolysis of metal ions from reaching values preventing passive film formation. This mechanism has been discounted on the basis that the beneficial effect of nitrogen is observed at higher potentials than would support this cathodic reaction, and the

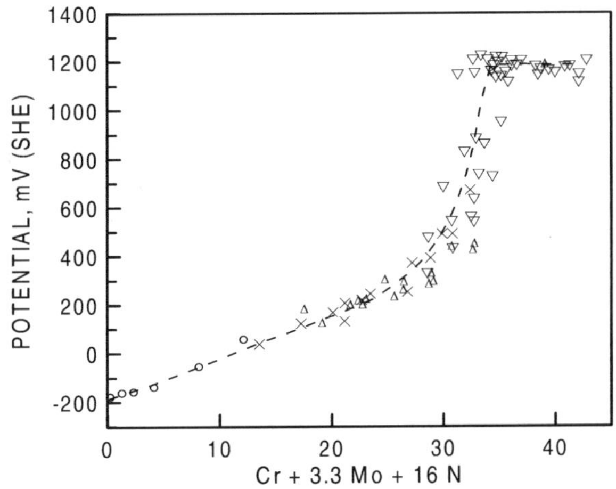


**Fig. 7.30** Pitting potential versus factor, Cr + 3.3 Mo + 16 N. Steels were austenite, martensite, tempered martensite, or ferrite. Composition range: 0–29 wt% Cr, 0–20 wt% Ni, 0.3–4 wt% Mo, 0.01–0.5 wt% N, and 0–0.3 wt% Nb. Redrawn from Ref 47

beneficial effect is observed in high concentrations of HCl in which the neutralizing effect of the reaction would be negligible. A more generally accepted mechanism is that nitrogen is enriched in the metal at the metal/passive-film interface, with greater enrichment in the metal by a factor of about 7 over that in the bulk alloy. The enrichment occurs by selective dissolution of metal ions during initial contact with the aggressive environment or during early stages of film formation. It is proposed that the accumulated nitrogen atoms on the metal surface retard the anodic dissolution rate, thus reducing the current density below values required for pit propagation. Improved resistance to pitting is thereby indicated by an increase in the pitting potential. This mechanism is supported by the observation that nitrogen has little effect on the anodic polarization curve that would be expected if nitrogen were influencing the passive film itself. Also, in the nitrogen bearing steel, small initial pits have been observed that do not propagate, indicating that the effect of nitrogen is to block propagation rather than initiation (Ref 48–50).

## *Effect of Fluid Velocity on Pitting*

The effect of fluid velocity on the corrosion of several commercial materials in seawater is shown in Fig. 7.31 (Ref 51). Three generalized types of materials are indicated by the corrosion behavior. The copper-base alloys, cast iron, and carbon steels tend to progressively increase in corrosion rate with increasing velocity. This is consistent with the schematic representation shown in Fig. 4.10, where the limiting current density for diffusion control of the cathodic reaction increases with

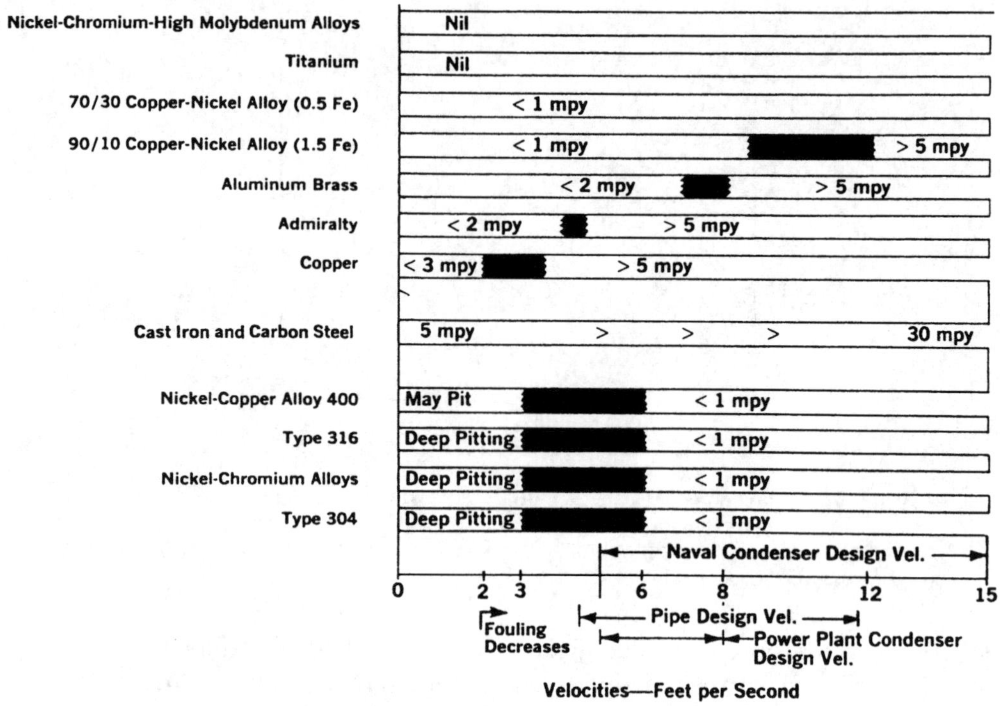


**Fig. 7.31** Effect of seawater velocity on corrosion mode of a range of commercial alloys. Source: Ref 51

increasing velocity. Increased availability of the cathodic reactant (frequently dissolved oxygen) increases the corrosion rate. The second type of behavior is represented by the nickel-chromium-high molybdenum alloys and titanium that form very stable passive films and have nil corrosion rates at all velocities. The third type is represented by the nickel-copper alloy, the nickel-chromium alloys, and the stainless steels, types 316 and 304. The passive films on these latter alloys are less stable such that at low velocities (<3 ft/s) where suspended solids may settle or attach to the surface, localized acidification processes may proceed under the deposits as described in previous sections. The local environment no longer supports passivity, and deep pits may then form under these deposits. At higher velocities, a stable protective passive film forms, pitting does not occur, and the corrosion rate remains small. A particularly velocity-sensitive type of pitting occurs on many active/passive alloys when certain microbial species tend to attach to surface irregularities (Ref 2, 52). Severe pitting can occur under tubercles at these sites as a consequence of the acidic metabolic products of the microbes. With increase in velocity, the probability of developing these localized conditions decreases, and pitting is diminished or does not occur.

### *Effect of Surface Roughness and Oxides on Pitting of Stainless Steels*

A wide range of surface conditions can be encountered on stainless steel products, which influences the tendency for pitting corrosion. Rolling and drawing operations, frequently followed by pickling (nitric/hydrofluoric acid) and passivation (nitric acid) treatments, result in surfaces ranging from smooth and passivated to ones that are rough, scratched, and/or variously coated with oxides residual from annealing. Significant effects on pitting corrosion can be related to surface oxides resulting from heat treatment following fabrication and from welding. These processes result in significant variations in the oxide film structure, composition, and thickness, all of which can influence pit initiation and propagation.

These variables have been investigated by electrochemical methods to determine the influence on the overall polarization behavior and, in particular, the influence on the pitting and protection potentials. Exposure to chemical environments, particularly ferric chloride and acidified sodium chloride solutions, also have been used to evaluate the influence of surface conditions on susceptibility to pitting corrosion. Controlled studies on type 316 stainless steel have shown that the pitting potential is changed by as much as 500 mV using surface preparations ranging from wire brushing and sand blasting to chemical treatments and pickling. Changing the grinding grit size from 36 to 220 to produce a finer surface finish correlated with an increase of 200 mV in the pitting potential in 0.1 M NaCl (Ref 53). Similar effects are observed on type 304 stainless steel by the decrease, with improved surface finish, in the number of unstable current bursts in the passive potential range of the form shown in Fig. 7.13 (Ref 11, 23). The effect of the rougher surfaces is to produce grooves, crevices, and related defects that act as a form of preexisting pit in which the acidification reactions previously discussed can occur, resulting in local conditions causing the propagation of localized corrosion. There is also evidence that an effect of surface roughness is to expose more inclusions and thus increase the number of sites for pit initiation (Ref 54).

## Pitting Corrosion of Carbon Steels

The active-passive behavior of iron as a function of pH is shown in Fig. 5.6. At pH values less than about 9, $i_{crit}$ is sufficiently large that corrosion generally occurs in the active potential range of the polarization curve. Below pH = 3 to 4, the surface remains free of corrosion products, and corrosion is largely due to hydrogen ion reduction. In strong mineral acids, the attack is frequently one of deep irregular pits unless inhibitors (usually pickling inhibitors) are present. At higher pH values (>9), passivity is readily established, and corrosion rates are very small unless passive-film-destroying anions, such as chlorides, are present. In the intermediate pH range, corrosion products form that influence corrosion rates sensitive to the environment, particularly the availability of dissolved oxygen to the metal surface, and to the presence of both aggressive anions that alter the protective character of the corrosion products, and to a wide range of inhibitors that can significantly decrease the corrosion rate. In this intermediate pH range, several conditions can cause localized corrosion, commonly described as pitting, although the surface appearance is generally distinct from that observed for the more strongly passivated alloys such as stainless steels.

The localized pitting-type corrosion of carbon steels can generally be attributed to one or more of the following:

- Selective attack at areas of hot-rolled products where an otherwise protective black oxide has been removed, allowing the exposed area to become anodic with corrosion supported by the cathodic oxide-coated surface
- Partial loss of passive films by insufficient inhibitor concentration in near-neutral environments
- Partial loss of passive films formed at higher pH (>9)
- Partial loss of otherwise protective carbonate or similar mineral deposits
- Localized deposits of inert material from the environment
- Localized microbiological deposits
- Irregular deposits of corrosion products

## *Corrosion Products and Surface Topology*

In the absence of dissolved oxygen or other oxidizing species such as ferric ions, corrosion of iron by reduction of hydrogen ions or by direct reduction of water in near-neutral (pH = 5 to 9) environments results in negligible corrosion rates (<<25 μm/year, or 1 mpy). Oxygen present in the bulk environment (aerated) provides a cathodic reactant that increases the corrosion rate to 50 to 125 μm/year (2 to 5 mpy) under stagnant conditions and to as high as 500 μm/year (20 mpy) at high velocities that reduce the diffusion boundary-layer thickness for oxygen. On an initially clean surface, small uniformly distributed local anodes and cathodes tend to produce uniform corrosion. Even with a bulk pH as low as 4, consumption of hydrogen ions on the surface allows sufficient increase in pH to form a solid corrosion product of black $Fe_3O_4$ and, at sufficiently positive potentials induced by aeration, to the formation of FeOOH and $Fe_2O_3$. The corrosion rate is also controlled by the physical properties of the corrosion product layer. The adherence and porosity of this layer affects the corrosion rate both by its effect on the ohmic resistance to the corrosion current and by restricting the availability of oxygen at the metal/oxide interface. The rates usually decrease significantly with time depending on the adherence and porosity of the corrosion product deposits and the influence of other ions in the environment, particularly chloride ions, which decrease the adherence. Under conditions forming carbonate or other adherent mineral scales, the corrosion rates may become extremely small.

The removal of corrosion products from the walls of carbon steel pipes and tanks, and from carbon steel objects in contact with water and soils for long periods of time, reveals an uneven surface. A frequently encountered attack is so-called "oxygen pitting," which is observed in boilers when the oxygen content of the feedwater has not been controlled to an adequately low value. An example of this type of pitting is shown in Fig. 7.32. In severe cases, the inner wall may have the appearance of Fig. 7.32(a) with a red deposit based on hematite ($Fe_2O_3$) and a black deposit of magnetite ($Fe_3O_4$), the latter always found in the bottom of the pits. The pits usually have a rounded geometry as shown to the right in Fig. 7.32(a), where the deposits have been removed, and in Fig. 7.32(b), which shows the cross section of the pipe wall.

Various degrees of localized attack, including severe pitting, may occur on hot fabricated carbon steel products containing missing oxide scale. These areas may result from physical damage during manufacture, shipping, application, service, or from spalling of the oxide from the surface on cooling. The oxide is usually black $Fe_3O_4$, which is relatively adherent and a sufficient electron conductor to support cathodic reactions such as oxygen reduction. Hence, oxide-free areas tend to become anodic in aqueous environments, and the large cathode-to-anode area ratio induces large current densities and correspondingly high local corrosion rates.

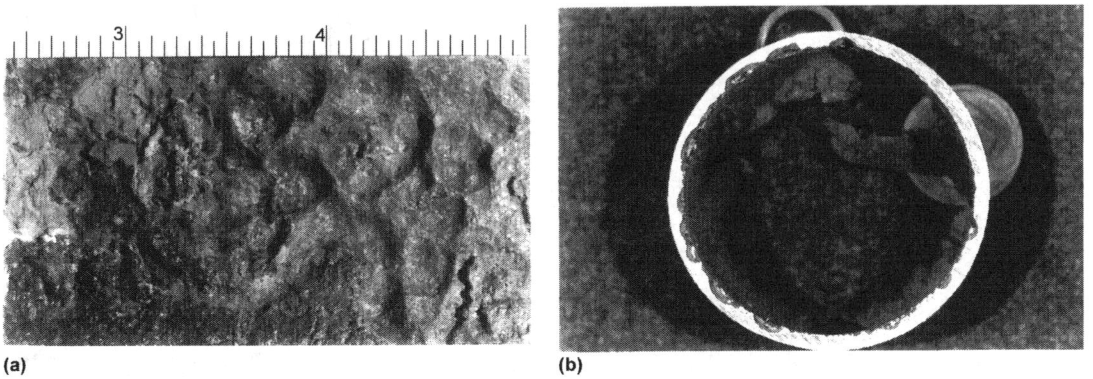


(a) (b)

**Fig. 7.32** (a) Pitting corrosion of inner wall of boiler tube due to insufficient deaeration of feedwater. Corrosion products brush removed from right side of section. (b) Cross section of pipe wall showing distribution of pits

Pitting of plain carbon steels also is associated with small deposits of inert materials. At near-neutral and higher bulk pH values, oxides readily form on an initially clean surface as previously described. These surfaces can support cathodic reactions, particularly the reduction of dissolved oxygen. Due to an oxygen deficiency under the inert deposit, protective films are destroyed, and dissolution of the metal leads to the buildup of $Fe^{2+}$ ions. These ions hydrolyze according to the reaction:

$$Fe^{2+} + H_2O \rightarrow Fe(OH)^+ + H^+ \qquad \text{(Eq 7.2)}$$

leading to a decrease in pH. In addition, as discussed earlier, the local buildup of positive charge and the potential gradient into the region under the deposit cause migration of $Cl^-$ ions from the environment in amounts related to their concentration in the bulk environment. It has been demonstrated that these ions shift the equilibrium in Eq 7.2 to the right, which further decreases the pH. The result is that the occluded region can be very acidic relative to the bulk environment, which may be neutral or even alkaline. As a consequence, increased corrosion occurs within the localized region at a rate related to the potential and environment in the pit.

## *Analysis of Pitting of Carbon Steels: Electrochemical Behavior*

The operation of an occluded cell resulting from an inert deposit on iron is analyzed by referring to polarization curves representative of conditions under the deposit and over the deposit-free surface. Since areas having these respective conditions are in electrical contact through the iron substrate, the surface consists of coupled cells and acts as a mixed electrode. The analysis is based on summing these curves weighted by the areas of the respective surfaces. It is similar to that developed previously for active-passive alloys such as stainless steels but takes into account the lower stability of the passive film on carbon steels in the local environment. Polarization curves are shown in Fig. 7.33(a) that are representative of conditions allowing formation of a passivated surface (curve A) and a nonpassivated surface (curve B) due to chloride ions. Curve C represents the cathodic polarization curve for dissolved oxygen on the passive surface. All curves are in units of current density. The two anodic curves are representative of what would be expected for iron passivated in an essentially neutral environment and iron remaining active in a chloride environment. For analysis, the polarization curves are again shown in Fig. 7.33(b), but in terms of current for a mixed electrode surface of 1 $m^2$. The total anodic current is given by:

$$I_a = A_{pit} i_{pit} + A_{pass} i_{pass} \qquad \text{(Eq 7.3)}$$

In this example, it is assumed that 1% of the surface (0.01 $m^2$) is covered by the occluded (pitted) area. To establish the steady-state corrosion condition, it is necessary to equate total anodic current to total cathodic current for conservation of charge. Since the active surface of the pits is 0.01 $m^2$, the polarization curve representative of these active areas is given by curve D, established by displacing curve B to lower currents by a factor of 0.01. The total anodic current at any potential is the sum of the current from curve A, for the passivated surface, and curve D, for the active surface within the pits. Curve E is the resulting total anodic polarization curve. In the limit of negligible solution resistance between the pitted and passivated regions, steady-state corrosion is represented by the intersection of the total anodic curve and the cathodic curve. This establishes the corrosion potential, $E_{corr}$, and corrosion current, $I_{corr}$. At this potential, the corrosion rate in the pit is proportional to the current density at this potential, $i_{corr,pit} = I_{corr,pit}/0.01\ m^2$, which in this case also is given by the unit area (1 $m^2$) polarization curve for the environment in the pit. This current density is quite high and represents an upper limit since the ohmic resistance of the inert deposit plus the corrosion-product deposits in the pit would reduce the current. An IR potential decrease in the pit due to either deposits or geometrical effects of large-depth-to-cross-sectional-area ratio will further depress the potential at the bottom of the pit, with a corresponding decrease in current density (and penetration rate). This relates to following curve B to lower potentials.

Plain carbon steels in acid environments, up to pH = 4, corrode predominantly by hydrogen-ion reduction, although at the upper limit of this range, dissolved oxygen becomes a significant contributor. Although corrosion rates in this pH range are usually prohibitively high, carbon steels may be used for short periods of time and may encounter

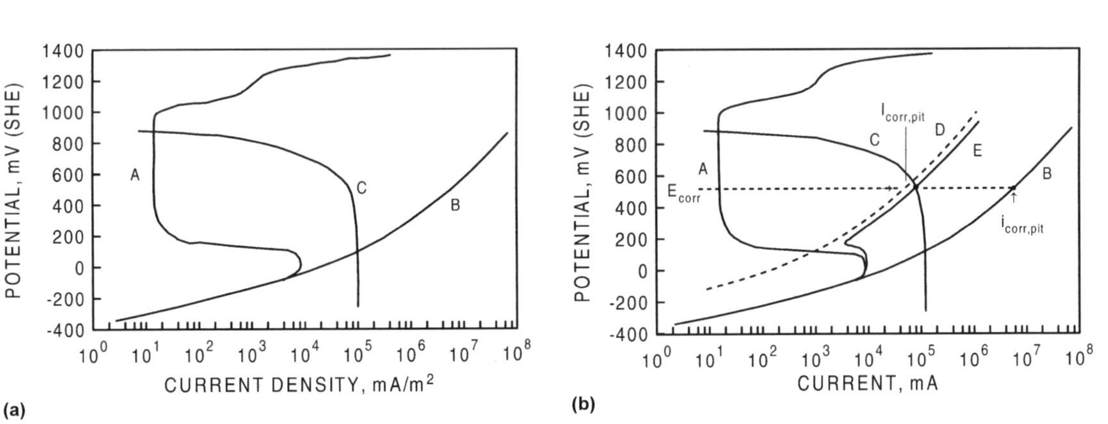


**Fig. 7.33** (a) Schematic polarization curve for iron showing passivity (curve A), active corrosion (curve B), and for oxygen reduction (curve C). (b) Effective polarization curve (curve E) when pitting has activated 1% of the surface (Details can be found in text.)

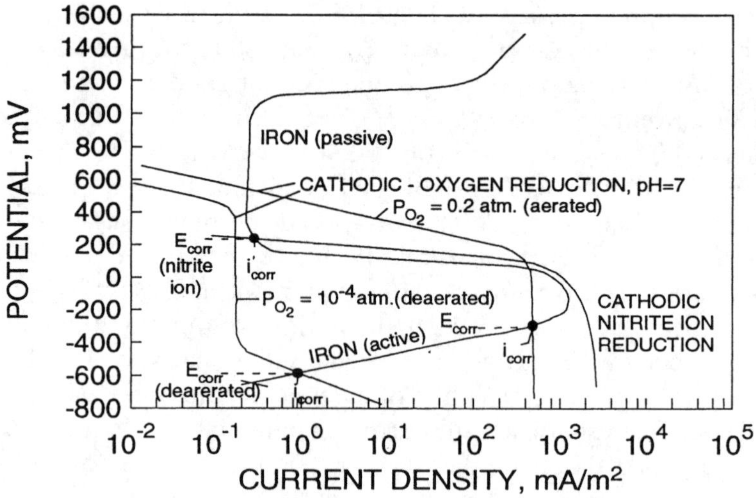


**Fig. 7.34** Idealized polarization curve for iron in neutral solution. Idealized cathodic polarization curves for sodium nitrite and oxygen under aerated and deaerated conditions. $E_{corr}$, $i_{corr}$ indicated for each environment

acidic environments in cleaning operations. Under these conditions, the attack occurs, unless inhibitors are present, as deep pits with sharp edges, and is frequently nucleated by chemical attack on inclusions in the metal, resulting in cavities that then propagate. Acid-soluble inclusions are more effective in pit nucleation than the less-soluble oxide and silicate inclusions. These variables can be important when surface appearance following acid cleaning is an important consideration. Also, there will be a difference of attack on surfaces parallel and perpendicular to the rolling direction.

Pitting of plain carbon steels can result from improper selection and/or control of inhibitors. Particular care is required with passivating inhibitors that contain sodium nitrite, $NaNO_2$, or sodium dichromate, $Na_2Cr_2O_7$. Cathodic polarization curves for near-neutral solutions of these materials will intersect the anodic polarization curves for steels in the passive range and provide very low corrosion rates. This behavior is shown in Fig. 7.34 for sodium nitrite. However, the passive film can be destroyed if the pH becomes too low, if concentrations of anions such as $Cl^-$ are too high, or if the concentration of inhibitor decreases below a critical value. Other anions can also affect the concentration of inhibitor required to provide protection (Ref 55). As a consequence, different domestic waters may require more inhibitor than others and appreciably greater amounts than when the environment is distilled water. These conditions lead to local destruction of the film with formation of local anodic sites at which the current density becomes very high due to the large cathodic area provided by the unpitted passive film. An example of pitting in bright cold-rolled, low-carbon steel sheet in water due to

None <Critical % >Critical %

Sodium nitrite concentration

**Fig. 7.35** Effect of $NaNO_2$ concentration as an inhibitor for the corrosion of low carbon steel in water. Exposure was two months at room temperature.

depletion of adequate $NaNO_2$ to maintain a passive film is shown in Fig. 7.35. With adequate inhibitor, the sheet remains bright for the two months of exposure. Two types of attack are observed as the inhibitor concentration decreases below the critical value and the passive film is penetrated. At some sites, pitting immediately penetrates the sheet. At other sites, following film penetration, corrosion tunnels under the passive film for some distance and then penetrates the sheet. The latter is a form of filiform corrosion that usually occurs as tunneling under polymeric coatings on steel.

## Pitting Corrosion of Copper

### *Analysis of Pitting of Copper with Reference to the Pourbaix Diagram*

According to the Pourbaix diagram (Fig. 7.36) (Ref 7), copper is thermodynamically stable with respect to corrosion by hydrogen-ion reduction or the direct reduction of water at any pH (line a is below lines 14, 7, and 17). Exceptions to this stability may occur in the presence of strong complexing agents for copper (cyanide and ammonium ions). In the absence of these agents, oxidizing agents in the environment that raise the potential of copper above the region of immunity (Cu area in

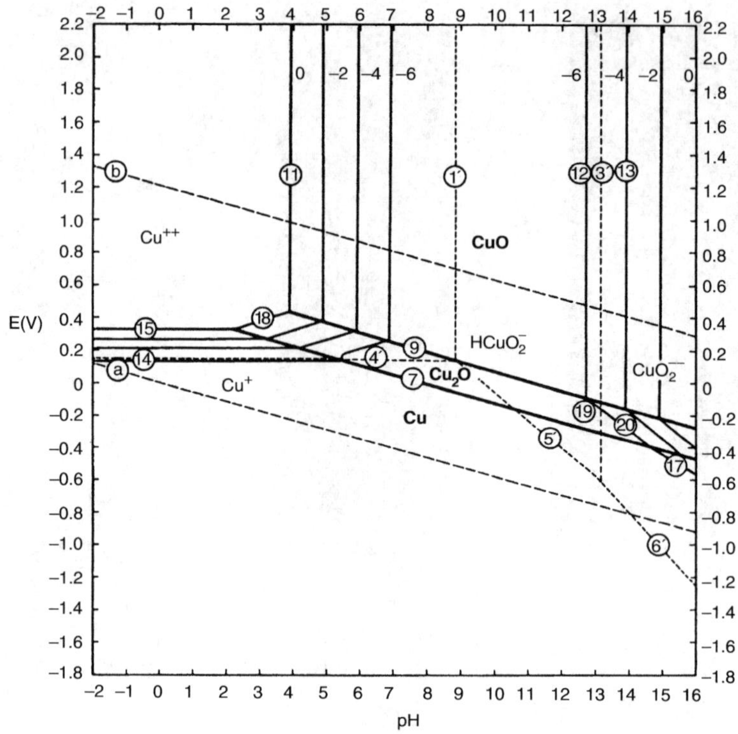


**Fig. 7.36** Pourbaix diagram for the system copper-water. Source: Ref 7

the Pourbaix diagram) lead to active corrosion or to possible passivation, depending on the pH as can be determined by Fig. 7.36. It is evident that for a $Cu^{2+}$ ion activity of $10^{-6}$ (line 14), the range of possible passivation extends from slightly acid (pH = 5) to strongly alkaline. In the absence of chloride ions, the oxide film formed on copper ($Cu_2O$, possibly overlaid by CuO) is reasonably protective (i.e., a state of actual passivity exists), although it is not as protective as the passive films that form on the more strongly passive metals, including iron, nickel, chromium, and related alloys. In view of the generally small concentrations of $Cu^{2+}$ ions ($<10^{-6}$) found in most environments, passive film formation would be expected over the pH range of about 6 to 12. However, it is emphasized that since copper is stable with respect to hydrogen-ion reduction, corrosion must relate to dissolved oxygen (aerated environment) or other oxidants in the environment.

## *Variables in the Pitting of Copper*

There are several major factors contributing to the tendency for copper to undergo pitting-type corrosion; these include:

- The tendency to form carbonate-related scale (water hardness). This can be a major factor in the absence of naturally occurring negatively-charged colloidal substances that act as inhibitors. These organic inhibitors are present in many surface waters but absent in deep water supplies. As a consequence, copper has a greater tendency to pit in the latter environment.

- Chloride ion concentration. Although this is an important variable, the effect is greatest in the range 0 to 20 ppm.
- The amount of dissolved carbon dioxide that influences the pH and the tendency to form carbonate-related scales
- The amount of dissolved oxygen that controls the corrosion potential
- The presence of surface films that can support the cathodic reduction of oxygen. In particular, carbonaceous films remaining after bright-annealing treatments in reducing atmospheres. These annealing treatments decompose the drawing and rolling lubricants but do not "burn away" the carbon residue.
- Other factors controlling the electrochemical potential. Increasing potential increases the tendency to pit, and decreasing potential enhances protection. The shift in potential may be due to either externally applied currents or currents due to galvanic coupling.

Fundamentally, pitting depends on the presence of sufficient oxygen or other oxidants to raise the potential above the immunity-potential range. At low pH values, as stated previously, active, rapid, and reasonably uniform corrosion will occur if the potential is maintained above line 14 on the Pourbaix diagram. At higher pH, where oxide films form, there is general agreement on the distribution of species present at the pit, but there is some controversy over the mechanism of pit propagation, and considerable uncertainty over the initiation mechanism.

## *Mechanisms of Pitting of Copper*

With reference to Fig. 7.37 (Ref 56), the following mechanism has been proposed for pit initiation of copper in near-neutral environments containing chloride and carbonate ions (Ref 56, 57). The site of the initiation is an otherwise protective $Cu_2O$ oxide film. To obtain copper-ion buildup in the defect requires that the potential be above line 7 on the Pourbaix diagram for copper (Fig. 7.36), which could be accomplished by sufficient dissolved oxygen in the environment. A critical contributor to initiation and propagation is formation of CuCl, a very insoluble salt. It is stable relative to $Cu_2O$ only in acid solution, as shown by the potential/pH diagram in Fig. 7.38 (Ref 56). Therefore, it should not form relative to $Cu_2O$ in near-neutral environments. It appears, however, that kinetically CuCl will form in preference to or in association with $Cu_2O$. Once formed, the CuCl hydrolyzes according to the reaction:

$$2CuCl + H_2O \rightarrow Cu_2O + 2Cl^- + 2H^+ \quad \text{(Eq 7.4)}$$

The acidity so produced and the coexistence of CuCl and $Cu_2O$ in the pit (Fig. 7.37) indicates that the conditions within the pit correspond to the intersection of lines 12, 51, and 55 in Fig. 7.38 (Ref 56), that is, at a pH = 3.5 and a potential of +270 mV (SHE). These conditions are repre-

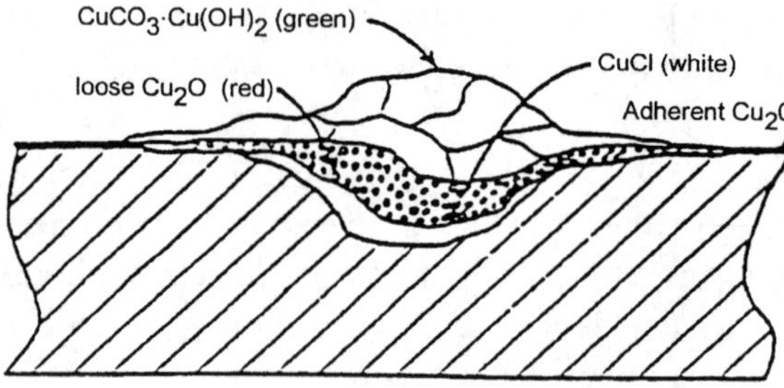


**Fig. 7.37** Pit in copper in the presence of Brussels water. Cross section would show the presence of red $Cu_2O$ and white CuCl beneath a mushroom of green malachite. Source: Ref 56

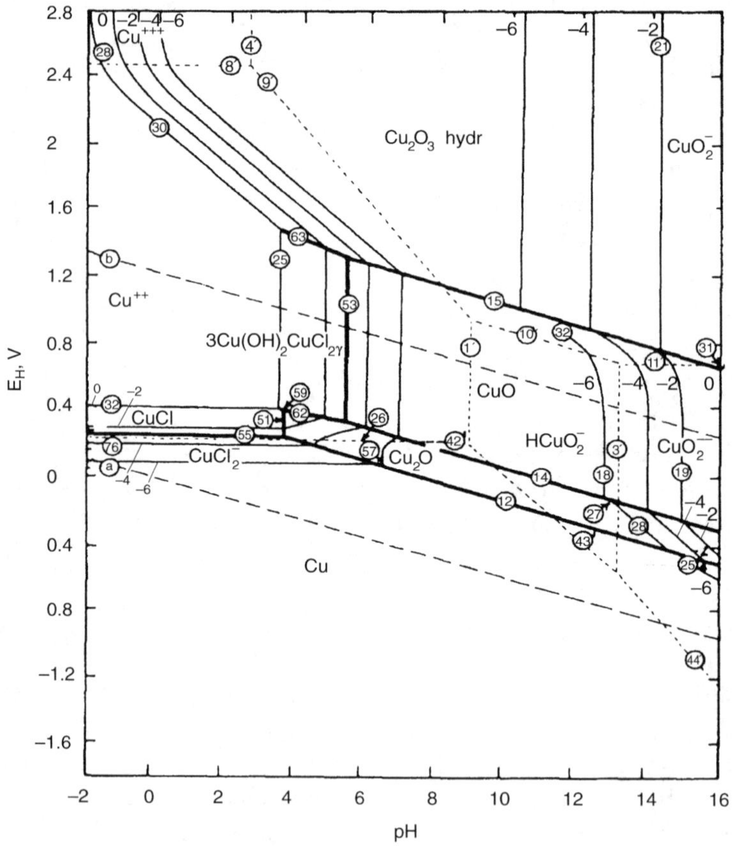


**Fig. 7.38** Potential-pH diagram for the ternary system Cu-Cl-$H_2O$ at 25 °C (355 ppm $Cl^-$). Source: Ref 56

sented by the small circle in Fig. 7.39; the vertical bar represents the range of potentials that might exist depending on degree of aeration for copper in near-neutral solutions, here, pH = 8. Shaded regions show pH-potential conditions for general corrosion. The critical factor governing growth of the pit is whether the potential in the bottom of the pit is greater than +270 mV (SHE) (Ref 56). If greater, the pit propagates by dissolution of copper as $Cu^{2+}$; if less, $Cu^{2+}$ deposits as metallic copper. Since the corrosion potential is measured above the pitted surface, the potential in the bottom of the pit will be lower by the IR potential decrease into the pit. This potential decrease is estimated to be 100 to 150 mV. Hence, if the measured corrosion potential is above about 420 mV (SHE), the pit is activated and propagation occurs. At the potential of the lower limit of the vertical bar, a protective passive film with no pitting is expected. Increased aeration, for example, which causes the potential to rise above 270 mV (SHE) in the pit will support pit propagation. Conversely, if the potential is held (or maintained by the environment) below a measured potential of 420 mV (SHE), propagation should not occur; this potential may be identified as the protection potential. The approximate potential range of this protection potential at pH = 8 is identified by the dashed lines in Fig. 7.39. An important aspect of this model is the large passive surface that supports the cathodic reaction and thereby controls the potential. Table 7.3 gives a comparison of the concentrations of the several species in the bulk environment and in the pit (Ref 56).

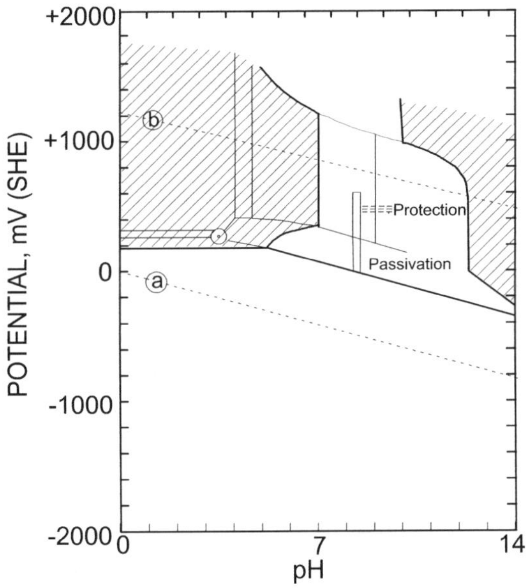


**Fig. 7.39** pH-potential diagram for copper used in the analysis of corrosion in Brussels water. Shaded regions indicate pH-potential conditions for corrosion. Vertical bar defines corrosion potential limits for pitting at pH = 8. Source: Ref 56

Another model of pitting in copper in hard waters (calcium bearing), including the pit geometry and species distribution, is shown in Fig. 7.40 (Ref 57, 58). As just discussed, small variations in surface conditions can determine whether the corrosion product is cuprous chloride (CuCl) or cuprous oxide ($Cu_2O$). The figure represents a local region of entrapment of cuprous chloride under the cuprous oxide film, which is protective over the bulk of the surface. The $Cu_2O$ film acts as a bipolar-membrane electrode across which cuprous ions, $Cu^+$, from CuCl formed by the corrosion reaction at the Cu/CuCl interface, are oxidized to cupric ions, $Cu^{2+}$, at the underside of the membrane. The $Cu^{2+}$ ions are reduced to $Cu^+$, as ions or as cuprous chloride, the reduction resulting from electrons produced by oxidation of copper and the diffusion of electrons through the cuprous chloride film. Thus, the corrosion step is that of oxidation of copper to $Cu^+$ and formation of CuCl at the Cu/CuCl interface. The net process is a form of reductive dissolution in which a higher valence species ($Cu^{2+}$) reacts with the lower valence state ($Cu^0$) to produce the intermediate valence state ($Cu^+$). $Cu^+$ is available at the upperside of the $Cu_2O$ film either by diffusion through the film or by

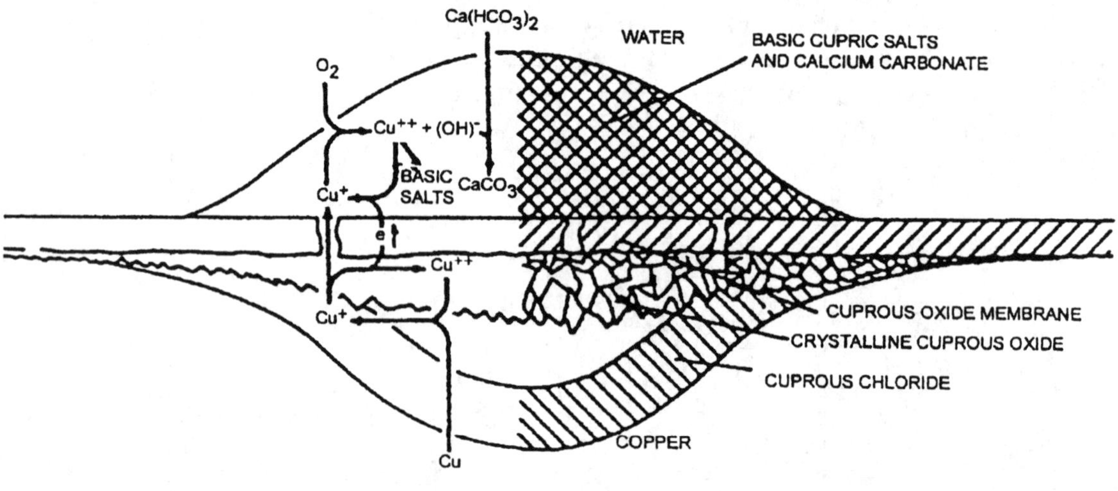


**Fig. 7.40** Model of positions of corrosion products and reaction paths in pitting of copper in hard water (details can be found in text). Source: ref 57, 58

**Table 7.3 For copper, concentrations of species in the bulk environment and in the pit**

| Species | Concentration in environment, ppm | Concentration in pit, ppm |
|---|---|---|
| $CO_2$ | 229 | ... |
| $SO_3$ | 46 | ... |
| $Cl^-$ | 22 | 273 |
| $Cu^{2+}$ | <1 | 250 |
| pH | 7.9 | 3.5 |

Source: Ref 56

transfer through defects in the film as shown in the figure. In the region above the film, the $Cu^+$ is oxidized by oxygen diffusing from the environment to $Cu^{2+}$, which forms solid $CuCO_3$-$Cu(OH)$. Even though these are reasonably insoluble salts, some $Cu^+$ and $Cu^{2+}$ are present in solution, and the diffusion of these ions provides the transport mechanism. The oxidized $Cu^{2+}$ then picks up an electron at the upper surface of the $Cu_2O$ and is reduced to $Cu^+$. Since the net effect of the processes in Fig. 6.40 is corrosion of copper as $Cu^+$ at the metal interface and the reduction of oxygen at the outer surface of the mound, the mechanism does not invoke the usual argument that pitting is driven by a large cathode-to-anode area ratio. Rather, it involves oxidation and reduction on the upper and lower sides of the $Cu_2O$ membrane with species at these positions diffusing from the Cu/CuCl interface and from the environment through a deposit of carbonates and salts. Also, the production of $OH^-$ from the reduction of oxygen increases the local pH and accounts for the observed increased carbonate deposit above the pit site. Although this mechanism has not been widely considered, it is reasonable and could apply to any system in which two or more valence states exist for the corroding species, which then permits the reductive dissolution processes previously described.

## Pitting Corrosion of Aluminum

### *The Passive Film on Aluminum*

As with other active-passive-type metals and alloys, the pitting corrosion of aluminum and its alloys results from the local penetration of a passive oxide film in the presence of environments containing specific anions, particularly chloride ions. The oxide film is $\gamma$-$Al_2O_3$ with a partially crystalline to amorphous structure (Ref 13, 59). The film forms rapidly on exposure to air and, therefore, is always present on initial contact with an aqueous environment. Continued contact with water causes the film to become partially hydrated with an increase in thickness, and it may become partially colloidal in character. It is uncertain as to whether the initial air-formed film essentially remains and the hydrated part of the film is a consequence of precipitated hydroxide or that the initial film is also altered. Since the oxide film has a high ohmic resistance, the rate of reduction of dissolved oxygen or hydrogen ions on the passive film is very small (Ref 60).

It is generally accepted that the passive film contains flaws that are the favored sites for pit initiation (Ref 13, 14, 60). The flaws occur predominately at sites of intermetallic phase particles in the substrate aluminum, particularly copper and iron-bearing intermetallics. Correlations have been made of flaw shape and distribution with these intermetallic particles. The effect of the flaw is the inability to form passive films over the particles that are as protective as those formed on the aluminum substrate. This correlation supports the observation that the number of flaws per unit area is related to the purity of the material, and to its thermal history when the solubility in the alloy of the intermetallic phase is temperature dependent. Pit initiation sites also have been attributed to the surface condition of the substrate aluminum, for example, scratches, and to stresses in the oxide film.

The high specific resistivity of the passive film over nonflawed areas limits the availability of electrons for support of cathodic reactions such as hydrogen-ion, water, and dissolved-oxygen reduction (Ref 13, 60). As a consequence, both anodic dissolution of the aluminum and the cathodic reaction are associated largely with the flawed regions of the film. The film over a flaw (such as a particle of $CuAl_2$ or $FeAl_3$) supports the cathodic reactions by being thinner and having a greater electronic conductivity. Even within a flawed region, the surface appears to be predominantly cathodic, estimated to be as high as 99% of the internal area (Ref 14). As a consequence, the effective cathodic polarization curve, the sum of the hydrogen-ion, water, and oxygen-reduction curves, will be sensitive to the number of flaws and hence purity, alloy content, and thermal treatment of the aluminum. This leads to considerable variation in the potential at the intersection of the anodic and cathodic polarization curves and would account for the wide scatter in reported corrosion potentials for aluminum in a given environment.

## *Polarization Behavior of Aluminum*

A representative anodic polarization curve for 99.99% Al in deaerated 0.1 M NaCl is shown in Fig. 7.41 (Ref 61). The initial corrosion potential is about –750 mV (SHE) from which the potential and

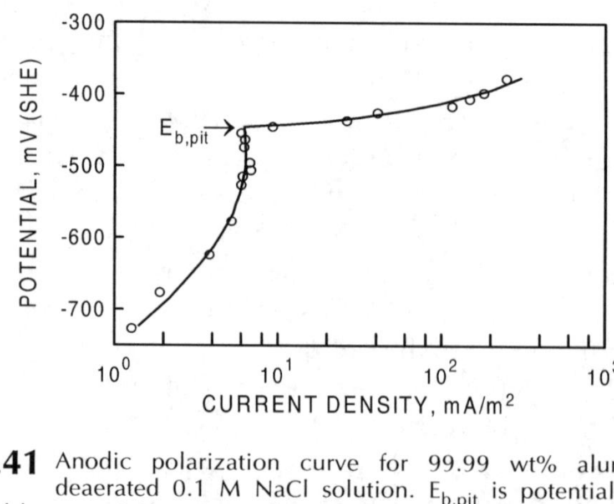


**Fig. 7.41** Anodic polarization curve for 99.99 wt% aluminum in deaerated 0.1 M NaCl solution. $E_{b,pit}$ is potential at which upscan of the potential, starting at the corrosion potential, results in sudden increase in current density. Redrawn from Ref 61

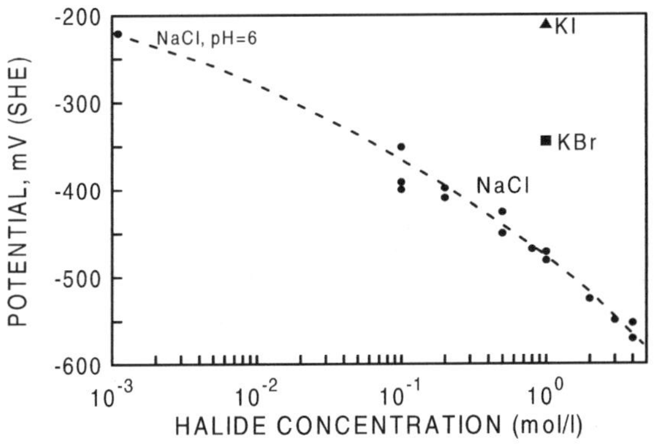


**Fig. 7.42** Pitting potential of 99.99 wt% aluminum in several halide environments. All environments are pH = 11 except as indicated for pH = 6. Redrawn from Ref 60

current density continuously increase characteristic of a preexisting passive film. The corrosion potential results from the intersection of the passive region of the aluminum anodic polarization curve and the combined polarization of hydrogen-ion, water, and oxygen reduction largely at the flawed areas. The sharp increase in current density at –450 mV (SHE) is associated with the onset of pitting and identifies the critical pitting potential, $E_{b,pit}$, for this chloride concentration. The dependence of the pitting potential on chloride ion concentration is shown in Fig. 7.42 (Ref 60, 62). Although the curve implies a limiting $E_{b,pit}$ = –220 mV (SHE) in the chloride-free environment, pitting occurs only at very high potentials in the absence of pit-inducing anions.

## *Mechanisms of Pitting Corrosion of Aluminum*

As the potential is increased from $E_{corr}$, the anodic current increases by migration of aluminum ions through the less-protective film at the flawed areas. Chloride ions are attracted into these areas by electrolytic migration and are incorporated into the film, further decreasing its specific resistivity. The pH in the flaw, and now developing pit, is determined by the balance between the hydrolysis of the aluminum ions, which lowers the pH, and an increase in pH resulting from hydrogen evolution as a cathodic reaction and by outward migration of hydrogen ions. The pH in the pit tends to stabilize near 3.5, independent of the bulk environment pH, which is the value for equilibrium of $Al^{3+}$ ions with $Al(OH)_3$, indicating the presence of $Al(OH)_3$ in a nonprotective form in the pit (Ref 63). At the critical pitting potential, the balance of these processes is to produce a pH at which the local oxide film is no longer stable. The pit then propagates, supported largely by cathodic hydrogen-ion reduction within the pit. A modification of this mechanism is based on the proposal that in the limit, $AlCl_3$ precipitates such that an acid chloride environment exists with properties (pH and specific resistivity) of a saturated solution of $AlCl_3$. Further, if the precipitation of this salt as a film decreases the aluminum dissolution rate, the degree of hydrolysis decreases and the pH increases, favoring reestablishment of the oxide. Thus, the critical pitting potential becomes, with addition of any IR potential drop into the pit, that potential for coexistence of the salt, $AlCl_3$, and the oxide. The mechanism is supported by observations of hydrogen evolution from pits and that pitting potentials can be related to the corrosion potential of aluminum in saturated $AlCl_3$ when IR potential drops into the pit are also considered (Ref 14, 64).

The transition from nonpitting to pitting is very potential sensitive. For example, in 1 M NaCl the current density has been observed to change from $10^{-6}$ to $10^{-3}$ A/cm$^2$ on increasing the potential from –530 to –520 mV (SHE) (Ref 61). It has been shown that pitting is always observed at potentials > –510 mV (SHE) and never observed at potentials < –520 mV (SHE) in 3% NaCl solutions (Ref 13). It also follows from this mechanism that the pitting potential should decrease with an increase in the chloride-ion concentration as shown in Fig. 7.42. At the higher concentrations, a lower potential is sufficient to provide the driving force to increase the chloride-ion concentration to the critical value resulting in pit propagation.

The pitting potential of aluminum in chloride solutions is essentially independent of temperature, which differs significantly from the response of stainless steels, the latter exhibiting a critical pitting temperature as discussed previously. The pitting potential for a specific chloride concentration is also relatively independent of the bulk environment pH, indicating that the controlling factor in conversion of a flaw to a propagating pit is the rapid development of the critical chloride concentration and pH in the flaw that dissolves the local oxide film. It also is observed that in aerated environments, the corrosion potential is the same as the pitting potential, indicating that the cathodic oxygen reduction, largely within the flawed regions, is sufficient to raise the corrosion potential to the pitting potential.

# Crevice Corrosion

## *General Description*

Although much of the previous discussion is applicable to pitting and crevice-type corrosion in that both involve occluded cells, crevice corrosion exhibits several distinguishing features. A significant difference is that a crevice has the geometry of a preexisting site for the occluded cell. As a consequence, the initiation stages for the two modes differ. Crevice geometries conducive to crevice corrosion include the following:

- Overlapping metal/metal or metal/nonmetal surfaces
- Bolts, nuts, and washers
- Flanged joints
- Irregular surfaces associated with scratches and welds
- Poorly adhering surface coatings
- Inert surface deposits

Major parameters affecting crevice corrosion are summarized in Fig. 7.43 (Ref 65). Although most of these are related to the analysis of pitting corrosion, modeling of crevice corrosion considers the crevice geometry described in terms of a crevice gap, or width, and a crevice depth. These factors govern how rapidly and to what extent changes occur in the crevice leading to localized corrosion.

The mechanism of crevice corrosion involves the following sequence. Immediately following access of the environment into the crevice, the metal-ion concentration will increase and the oxygen concentration will decrease, with both processes occurring slowly from dissolution of a passive film on the crevice walls and more rapidly from an active crevice surface. The metal ions will hydrolyze, but the effect in the lowering of the pH is initially countered by the $OH^-$ ions resulting from the oxygen reduction. After the occluded oxygen is consumed, the cathodic reaction is predominantly on the outer surfaces, with continued dissolution of the metal in the crevice decreasing the local pH by hydrolysis. The result is a corrosion cell driven by a large cathode-to-anode area ratio. At a sufficiently low pH in the crevice, any passive film that might have been present will be destroyed, and active corrosion becomes the mechanism of propagation of corrosive attack. The rate at any part of the crevice will depend on the local interface potential

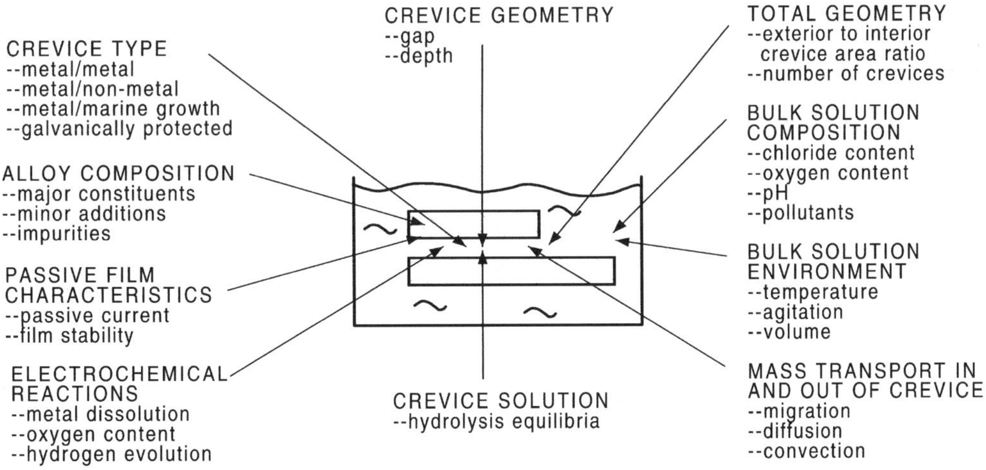


**Fig. 7.43** Parameters and variables influencing crevice corrosion. Source: Ref 65

as related to the effective polarization curve at the position. This potential may become strongly dependent on the IR potential drop into the crevice and, hence, on the crevice geometry. The current flow from the crevice to the outer surface is partially supported by the inward migration of anions such as $Cl^-$, which further lowers the pH. In these respects, the mechanisms of crevice and pitting corrosion are similar.

In view of this sequence, the crevice geometry parameters of gap width and depth become important. If the gap is sufficiently wide and shallow, oxygen depletion and chloride-ion influx will decrease and metal-ion buildup will be less due to increased diffusion of corrosion products from the crevice. The pH decrease due to hydrolysis of cations will be less, the passive film may be preserved, and if so, crevice corrosion will not occur. These factors are reversed for deep, narrow crevices, and at some critical geometry, crevice corrosion will occur. As with pitting, increased concentration of chloride ions in the environment will increase chloride-ion concentration in the crevice and increase the probability of initiating crevice corrosion.

## *The Critical Potential for Crevice Corrosion*

The mechanisms of crevice corrosion described indicate that for a given metal, environment, and crevice geometry, a critical potential should exist below which the crevice corrosion will not occur. A critical potential is observed for creviced specimens when subjected to polarization measurements or when placed in chemical environments of varying oxidizing power providing a range of potentials spanning the critical potential. Figure 7.44 is representative of polarization measurements of an active-passive-type alloy susceptible to pitting and crevice corrosion. In the absence of a crevice, a breakdown potential, $E_{b,pit}$, is measured as the potential at which pitting is initiated. Continuing and then reversing the scan results in an anodic loop that terminates at a protection potential, $E_{prop,pit}$. In the presence of a crevice, conditions for localized corrosion preexist and the equivalent of a breakdown potential, $E_{b,crevice}$, is observed at a lower potential than for pitting. On cyclic scanning, an anodic loop also is observed and may result in a protection potential, $E_{prot,crevice}$, on the downscan. Because of the sensitivity of crevice-corrosion initiation to the geometry of the crevice, reproducibility and significance of a critical crevice potential is limited. However, critical crevice potentials are always lower than pitting potentials. For alloys very susceptible to crevice attack, it may be almost impossible to measure a pitting potential because of the initiation of crevice attack at interfaces between the test specimen and imbedding or masking materials used to hold and provide electrical contact to the specimen in making electrochemical measurements. Figure 7.45(a) shows crevice attack at the metal/epoxy interface of a polarization-measurement specimen; Fig. 7.45(b) shows crevice attack at the edge of an enamel coating used to mask a defined area on the metal surface.

For alloys showing high susceptibility to crevice corrosion, measurements of the pitting potentials are of limited value since failure in service by crevice corrosion would predominate. Polarization measurements can be useful in showing relative susceptibility of alloys to crevice corrosion. Figure 7.46 shows results from cyclic polarization measurements on specimens of three alloys containing O-rings to produce crevices (Ref 66). The environment was aerated water with 3.5

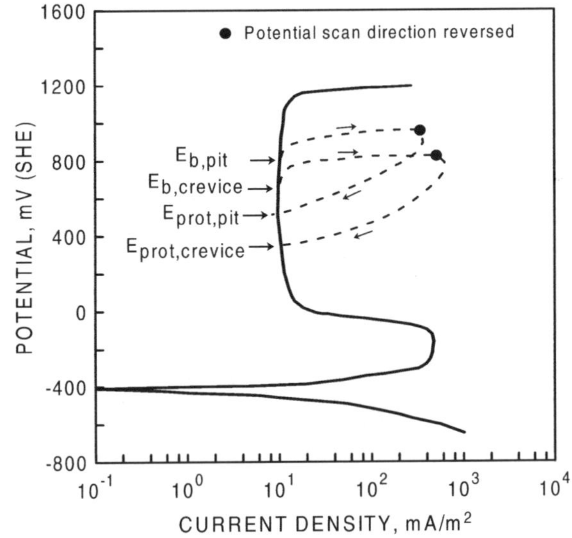


**Fig. 7.44** Schematic polarization curve for an alloy susceptible to localized corrosion. Pitting is initiated at $E_{b,pit}$ and stops at $E_{prot,pit}$. Crevice corrosion starts at $E_{b,crevice}$ and stops at $E_{prot,crevice}$.

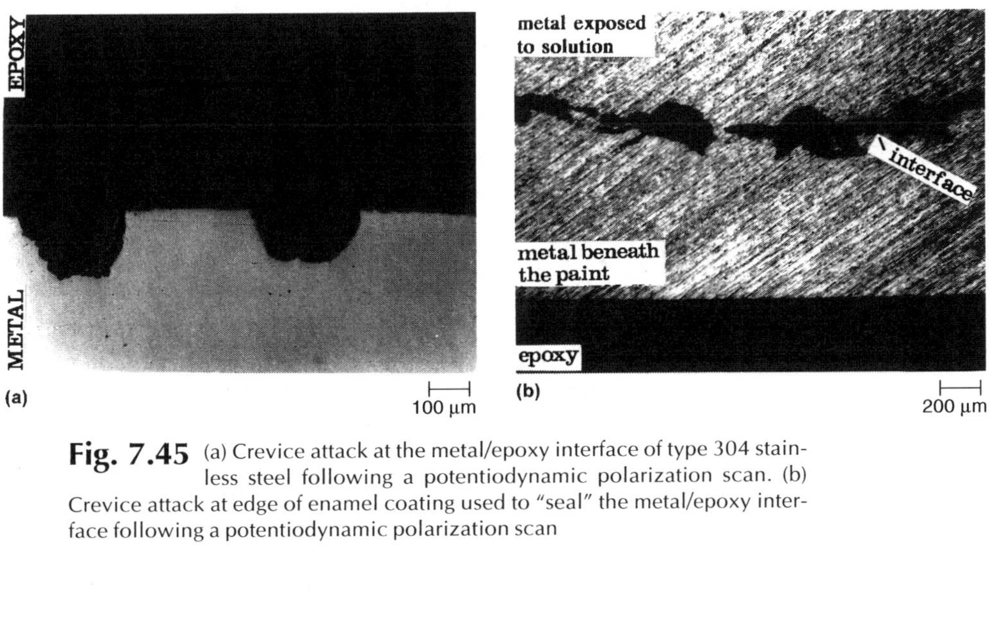


**Fig. 7.45** (a) Crevice attack at the metal/epoxy interface of type 304 stainless steel following a potentiodynamic polarization scan. (b) Crevice attack at edge of enamel coating used to "seal" the metal/epoxy interface following a potentiodynamic polarization scan

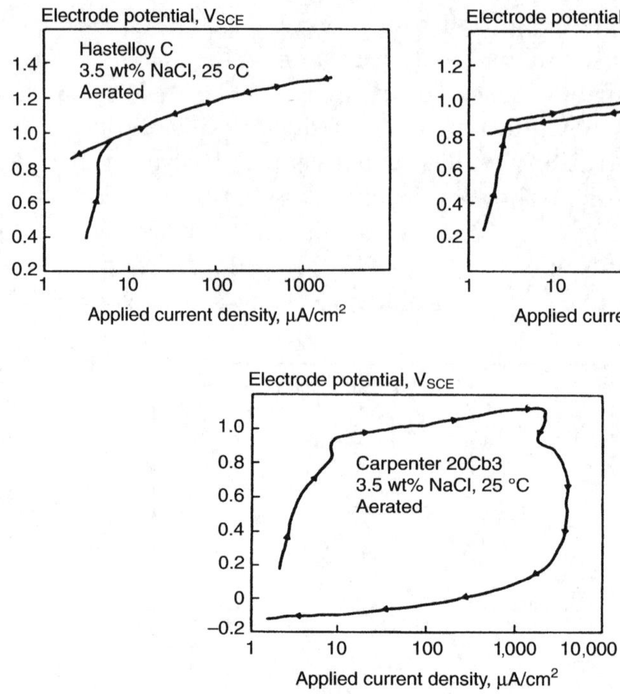


**Fig. 7.46** Cyclic potentiodynamic polarization curves for specimens with synthetic crevices in aerated 3.5 wt% NaCl solution at 25 °C. (a) Hastelloy C. (b) Incoloy 825. (c) Carpenter 20Cb3. Source: Ref 66

wt% NaCl at 25 °C. The value of the breakdown potential and size of the loop are measures of susceptibility to crevice corrosion in the environment. The relative behavior represented by these three curves substantiated the behavior of these alloys when creviced samples were exposed to seawater for two years. Using samples with identical geometries in the seawater, the weight loss for the alloy of Fig. 7.46(a) was 0.16 mg/cm$^2$; for Fig. 7.46(b), 4.1 mg/cm$^2$; and for Fig. 7.46(c), 26.1 mg/cm$^2$.

## *Evaluation of Crevice Corrosion*

A recommended configuration for investigating the susceptibility of an alloy to crevice corrosion is shown in Fig. 7.47 (Ref 67). A polymeric, grooved washer is held in place with either a polymeric bolt or insulated metal bolt. The relative responses of several alloys to a similar type of corrosion testing are shown in Fig. 7.48 for a range of $FeCl_3$ concentrations. For comparison, the square specimens were exposed to show relative resistance to pitting corrosion. It is evident that resistance to both crevice and pitting attack decreases with increasing $FeCl_3$ concentration (increasing corrosion potential and chloride concentration, and decreasing pH) and with decreasing alloy content. For repro-

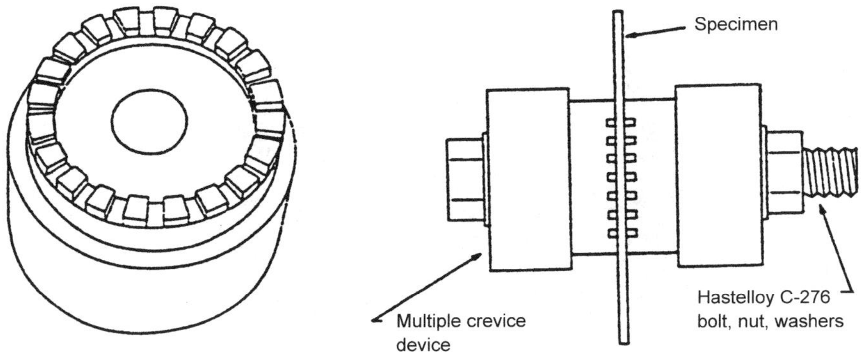


**Fig. 7.47** ASTM crevice corrosion test assembly. Source: Ref 67

ducibility of results, the washer geometry must be carefully controlled as well as the torque exerted on the bolt to control the crevice width. In an extensive series of crevice corrosion tests using the washer crevice assembly on 46 stainless alloys in seawater, crevice width as determined by the torque applied to the bolt was shown to have an effect on the extent of crevice corrosion (Ref 68).

## Microbiologically Influenced Corrosion

Microbiologically influenced corrosion (MIC) is an area concerned with the effects of microorganisms (bacteria, fungi, and algae) in natural and industrial water systems on the corrosion of structural materials. The effects can be highly detrimental, resulting in surprisingly shorter lifetimes for the structural components than expected. Most of the time, the mode of corrosion is localized (i.e., pitting or crevice corrosion). The scenario is often such that, based on the general environmental conditions (pH, $O_2$ content, $Cl^-$ concentration, etc.) localized corrosion is not expected, but due to MIC, the material prematurely fails by pitting or crevice corrosion. When addressing the mechanisms associated with MIC, it must be recognized that microorganisms often attach to the material surface, and through their metabolic processes, modify the local solution chemistry at the material surface relative to the bulk solution chemistry. Thus, the solution at the material surface may become much more corrosive than the bulk solution.

### *Biofilms (Ref 69–71)*

Microorganisms may either be freely suspended within the bulk solution (planktonic existence) or attached to a surface (sessile existence). When a material is first immersed in an aqueous solution, a thin layer of organic matter (referred to as the conditioning film) is adsorbed onto

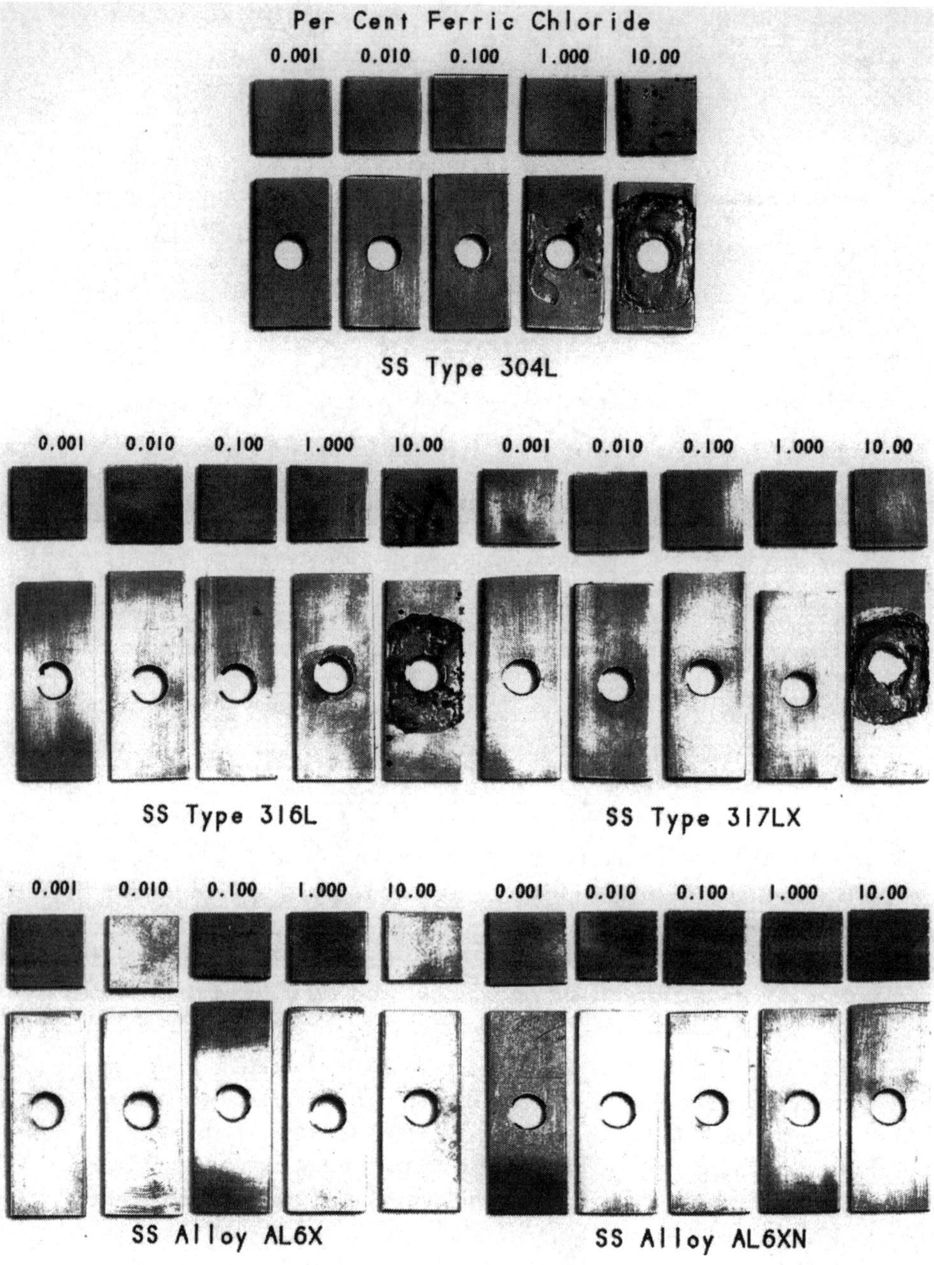


**Fig. 7.48** Response of five austenitic stainless steels to pitting and crevice corrosion. Alloys exposed 1 month at room temperature in indicated concentrations of $FeCl_3$ solutions. (Numbers represent weight percent.)

the surface. Planktonic microbes attach to this nutrient source, thereby becoming sessile microbes. As the attached microorganisms replicate and secrete adhesive extracellular polymeric substances (exopolymer material), a "biofilm" is created at the material surface. Within a few days to several weeks, a mature biofilm is established. The characteristics of the biofilm, including the thickness, morphology, and degree of chemical heterogeneity, depend on the types of microorganisms present, which often vary from site-to-site within the biofilm. For example, aerobic microbes (those requiring oxygen) may exist in regions containing dissolved oxygen, whereas anaerobic microbes (those that can only function in the absence of oxygen) may exist in other regions that have been previously depleted of oxygen by aerobic microbes. Thus, the biofilm is complex, dynamic, three-dimensionally heterogeneous, and often involves synergistic life-support processes involving different types of microorganisms.

### *Microorganisms and Effects on Solution Chemistry within Regions of the Biofilm (Ref 72, 73)*

The microorganisms that have been commonly associated with MIC are:

- Sulfate-reducing bacteria (SRB)
- Sulfur/sulfide-oxidizing bacteria
- Iron/manganese-oxidizing bacteria
- Aerobic slime formers
- Organic acid-producing bacteria
- Organic acid-producing fungi

For each, the generally required environmental condition (aerobic or anaerobic), the primary metabolic processes related to MIC, and the resultant chemical species that can increase corrosion rates are summarized in Table 7.4. These characteristics are discussed in the following paragraphs.

Sulfate-reducing bacteria (SRB) (e.g., *Desulfovibrio*) are anaerobic and reduce sulfate to sulfide according to the reaction, $SO_4^{2-} + 8H \rightarrow S_{2-} + 4H_2O$ (or $SO_4^{2-} + 8H \rightarrow H_2S + 2H_2O + 2OH^-$) (Ref 74), where the necessary hydrogen may be supplied as the cathodic-reaction product of the corrosion process in the anaerobic (deaerated) solution (Ref 75), that is, by reduction of hydrogen ions in an acidic solution ($H^+ + e \rightarrow H$) or by direct reduction of water in a neutral or basic solution ($H_2O + e \rightarrow H + OH^-$). The sulfide usually shows up as hydrogen sulfide ($H_2S$) or, if ferrous ions are available (e.g., through the corrosion of iron-based alloys, $Fe \rightarrow Fe^{2+} + 2e$), as black ferrous sulfide, FeS. From a corrosion point of view, sulfide ions can be very damaging through acceleration of the anodic dissolution process (e.g., they can significantly lower the pitting potential of passive alloys by degrading the passive film, as discussed in the section "Effects of Sulfide and Thiocyanate Ions on Polarization of Type 304 Stainless Steel" in Chapter 5). SRB are often found in the region of the biofilm nearest to the metal surface because this region has the highest probability of being deaerated when the bulk solution is aerated.

Sulfur/sulfide-oxidizing bacteria are aerobic and oxidize sulfide to elemental sulfur ($S^{2-} \rightarrow S + 2e$), sulfide to sulfate ($S^{2-} + 4H_2O \rightarrow SO_4^{2-} + 8H^+ + 8e$), or sulfur to sulfate ($S + 4H_2O \rightarrow SO_4^{2-} + 8H^+ + 6e$). The bacteria in this family that produce elemental sulfur probably do not contribute directly to corrosion (elemental sulfur is not corrosive at near-ambient temperatures) but can form bulky deposits with anaerobic zones beneath, suitable for growth of SRB. On the other hand, the bacteria that oxidize sulfide or sulfur to sulfate (e.g., *Thiobacillus*) produce sulfuric acid ($H^+$ and $SO_4^{2-}$ ions), with pH values as low as 1.0 reported. The lower pH environment can accelerate corrosion for two reasons. First, corrosion-product or passive films tend to be less stable and, therefore, less protective at low pH values (they can completely dissolve at a critically low pH, as shown in Chapter 2 on Pourbaix diagrams); thus, the acidic environment can detrimentally influence the anodic behavior of the alloy. The second reason is that the equilibrium half-cell potential for the $O_2 + 4H^+ + 4e = 2H_2O$ reaction increases as the pH decreases (as described in the section "Half-Cell Reactions and Nernst-Equation Calculations" in Chapter 2), thereby influencing the cathodic polarization behavior, which, in turn, can produce a higher corrosion potential, $E_{corr}$, and, consequently, a higher corrosion rate.

**Table 7.4 Microorganisms commonly associated with microbiologically influenced corrosion (MIC), generally required environmental conditions, metabolic processes related to MIC, and resultant chemical species that can increase corrosion rates**

| Microorganism | Generally required environmental condition | Primary metabolic process related to MIC | Resultant chemical species that can increase corrosion rate |
|---|---|---|---|
| Sulfate-reducing bacteria (SRB) (e.g., *Desulfovibrio*) | Anaerobic | Reduce sulfate ($SO_4^{2-}$) to sulfide ($S^{2-}$), which usually shows up as hydrogen sulfide ($H_2S$) or, if Fe is available, as black iron sulfide (FeS) | $S^{2-}$ |
| Sulfur/sulfide oxidizing bacteria (e.g., *Thiobacillus*) | Aerobic | Oxidize sulfide ($S^{2-}$) to elemental sulfur (S), sulfide to sulfate ($SO_4^{2-}$), or sulfur to sulfate. Strains that produce sulfate (e.g., Thiobacillus) create sulfuric acid ($H_2SO_4$) | $H^+$ (lower pH) |
| Iron oxidizing bacteria (e.g., *Gallionella*) | Aerobic | Oxidize ferrous ions ($Fe^{2+}$) to ferric ions ($Fe^{3+}$) | $Fe^{3+}$ |
| Manganese oxidizing bacteria | Aerobic | Oxidize manganous ions ($Mn^{2+}$) to manganic ions ($Mn^{3+}$) | $Mn^{3+}$ |
| Aerobic slime formers (e.g., *Pseudomonas*) | Aerobic | Produce extracellular polymers referred to as "slime." Slime can prevent oxygen from reaching the material surface. | $O_2$ (lower) |
| Organic-acid producing bacteria (e.g., *Clostridium*) | Anaerobic | Produce organic acids | $H^+$ (lower pH) |
| Organic acid-producing fungi | Aerobic | Produce organic acids and may produce anaerobic sites for SRB | $H^+$ (lower pH) |

As just noted, aerobic sulfide-oxidizing bacteria use sulfides to produce sulfates, whereas anaerobic SRB utilize sulfates to produce sulfides. The two types of bacteria often are found in the same biofilm, at different locations (aerobic versus anaerobic), with both participating in this synergistic sulfur cycle.

Iron-oxidizing bacteria (e.g., *Gallionella*) are aerobic and oxidize ferrous ions (which may be produced by the corrosion of iron-base alloys) to ferric ions ($Fe^{2+} \rightarrow Fe^{3+} + e$). Similarly, manganese-oxidizing bacteria are aerobic and oxidize manganous ions to manganic ions ($Mn^{2+} \rightarrow Mn^{3+} + e$). Ferric and manganic ions are powerful cathodic-reaction species. The standard equilibrium half-cell potentials for the $Fe^{3+} + e = Fe^{2+}$ and $Mn^{3+} + e = Mn^{2+}$ reactions are quite high (+771 and +1,541 mV [SHE], respectively), and it is known that the exchange-current density for the $Fe^{3+} + e = Fe^{2+}$ reaction is high. These characteristics result in higher cathodic reaction rates ($Fe^{3+} + e \rightarrow Fe^{2+}$ and/or $Mn^{3+} + e \rightarrow Mn^{2+}$) at a given potential, increasing the value of $E_{corr}$, and, thereby, possibly increasing the corrosion rate (especially if $E_{corr}$ exceeds the pitting potential of a passive alloy). It is noted that ferric ion-base solutions often are used for accelerated corrosion tests, for example, ferric chloride ($FeCl_3$) solutions, where both the ferric ions and the chloride ions contribute to the corrosion.

Aerobic slime formers (e.g., *Pseudomonas* and *Siderocapsa*) produce extracellular polymers (exopolymers) consisting of sticky strands that bind the bacterial cells and various particulates to the material surface. The resulting "slime" can prevent the oxygen in the bulk environment from reaching the surface, thereby creating, locally, a deaerated or anaerobic condition. This condition per se can lead to accelerated corrosion through the classic pitting and crevice corrosion mechanisms; that is, the local oxygen depletion is followed by hydrogen-ion production, producing a lower pH (by metal-ion hydrolysis) and increased chloride concentration (by chloride migration to maintain charge balance) (explained in the sections "Pit Propagation" and "Crevice Corrosion" in this chapter). Furthermore, the local anaerobic condition provides an ideal site for SRB growth.

Various anaerobic bacteria (e.g., *Clostridium*) are capable of producing organic acids (e.g., acetic, formic, or propionic acid). Thus, their actions can lower the pH at the material surface, thereby creating a more corrosive local environment. Certain fungi, normally aerobic, also are capable of producing organic acids as well as producing anaerobic sites for SRB growth.

## *Ennoblement*

It is generally observed that the corrosion potential, $E_{corr}$, for a given metal/alloy is higher in a natural biotic aqueous environment (one containing a natural distribution of microorganisms) than in essentially the same environment but without the microorganisms (a reference, or control, abiotic environment). Thus, $E_{corr}$ is "ennobled" by the presence and actions of the microorganisms. A possible mechanism by which this ennoblement occurs is associated with the cathodic component of the overall corrosion process. If the microorganisms produce additional oxidizers (species that can undergo reduction, or cathodic, reactions), then, as illustrated in Fig. 7.49, not only can $E_{corr}$ be increased, but also pitting corrosion can be initiated. In Fig. 7.49, two hypothetical alloys are shown, A and B, with B having a lower breakdown potential for pitting corrosion, $E_{b,pit}$. Both alloys remain passivated in the absence of MIC (i.e., when only the oxygen-reduction cathodic reaction is occurring). However, in the presence of MIC, when an additional cathodic reaction(s) is occurring, alloy A remains passivated, but alloy B undergoes pitting corrosion. It is noted that the current density of the cathodic curve for the MIC oxidizer(s), over a certain potential range, is shown to decrease with decreasing potential rather than approach a diffusion limit. This unusual behavior, which is often, but not always, observed, is believed to be caused by the depletion of the MIC oxidizer(s), since it only exists in the biofilm and, therefore, cannot be replenished from the bulk solution. The general types of polarization behaviors depicted in Fig. 7.49 have been reported by several investigators (Ref 76–78).

As seen in Table 7.4, specific microorganisms can produce $Fe^{3+}$, $Mn^{3+}$, or $H^+$ ions within the biofilm at the metal surface, all of which are oxidizers (i.e., $Fe^{3+} + e \rightarrow Fe^{2+}$, $Mn^{3+} + e \rightarrow Mn^{2+}$, $2H^+ + 2e \rightarrow H_2$). The resultant additional, or enhanced, cathodic reaction(s), relative to

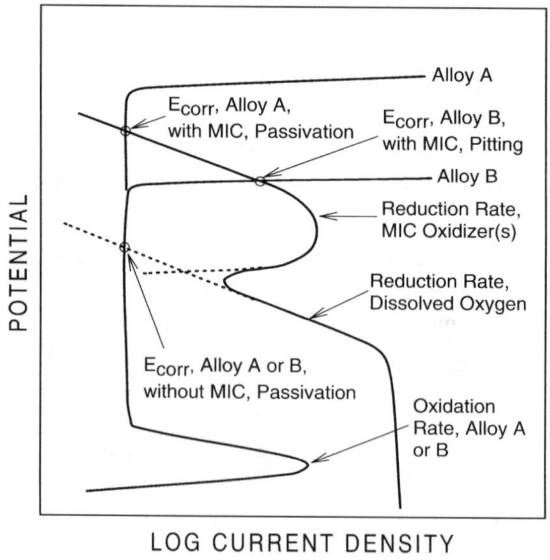


**Fig. 7.49** A mechanism for ennoblement by microbiologically influenced corrosion

the reduction of dissolved oxygen in a near-neutral aerated bulk solution ($O_2 + 2H_2O + 4e \rightarrow 4OH^-$), could increase the observed corrosion potential. Other oxidizers produced by microorganisms also have been reported. For example, hydrogen peroxide has been observed in marine biofilms, and the additional cathodic reaction for the peroxide under acidic, low-oxygen conditions ($H_2O_2 + 2H^+ + 2e \rightarrow 2H_2O$) was shown to be capable of producing the observed ennoblement (Ref 77). Another example involves $MnO_2$, produced by manganese-oxidizing bacteria (e.g., *Siderocapsa*), which establishes an additional cathodic reaction ($\gamma MnO_2 + H_2O + e \rightarrow \gamma MnOOH + OH^-$) that has been shown to account for the ennoblement observed in a fresh river-water system (Ref 78).

## *Biocides (Ref 73, 79, 80)*

In the context of MIC, the purpose of a biocide is to kill the microorganism(s) responsible for the increased corrosion rates. First, it is important to establish that the corrosion problem is due to the presence and actions of microorganisms and not due to other corrosion mechanisms; this may not be a simple, straightforward task. An analysis for MIC should include a number of considerations (Ref 79). For example, a given type of microorganism generally has a temperature range for optimum growth and function of only about 10 to 20 °C. Therefore, as a diagnostic test, if the system temperature is increased, and the resultant corrosion rate decreases, the problem is probably due to MIC. In other types of corrosion, not associated with microorganisms, the corrosion rate almost always increases with increasing temperature. Identifications of the microorganisms present should be accomplished, especially the sessile bacteria in the biofilm since they are directly responsible for the MIC. Consideration of the form of corrosion is also necessary; with few exceptions, MIC is localized, producing pitting or crevice corrosion. Mounds, or tubercles, are often produced above the pits and consist of corrosion products, cells, and extracellular products (Fig. 7.1d). Analyses of the chemical species within the biofilms or tubercles can provide evidence of MIC in light of the metabolic processes of the various microorganisms, as previously described. Sometimes, a proof that the problem is due to MIC can only be accomplished by adding a biocide to a test loop and observing that the corrosion problem is mitigated.

Typical biocides include hypochlorous acid, chlorine dioxide, hypobromus acid, hydrogen peroxide, ozone, ultraviolet-light treatment, phenolics, aldehydes, and quaternary ammonium compounds (Ref 73, 80). A brief description of each follows (Ref 73, 80). Hypochlorous acid is probably the most commonly used biocide and also one of the most powerful oxidizing agents. The sources of hypochlorous acid are chlorine gas and sodium hypochlorite. In aqueous solutions, sodium hypochlorite hydrolyzes to hypochlorous acid (HOCl) and sodium hydroxide; then, depending on pH, the hypochlorous acid dissociates to hydrogen ions and hypochlorite ions ($OCl^-$). The hypochlorite-ion concentration determines the biological killing capacity. Hypochlorous acid becomes ineffective at pH values greater than 9. Chlorine dioxide ($ClO_2$) has more oxidizing power than hypochlorous acid and is also more costly. Hypobromus acid (HOBr) has recently been replacing hypochlorous acid as a biocide in water with pH > 8. Hydrogen peroxide ($H_2O_2$) is a relatively strong biocide, approximately equivalent to hypochlorous acid. It is nonpolluting but requires large doses with long contact times to be effective. Ozone is the strongest oxidant among the biocides but is highly toxic and must be generated and stored on site. High-intensity ultraviolet lamps are sometimes used for biocidal treatment, especially in low-flow systems and demineralized-water systems. Phenolics and more complex compounds (e.g., pentachlorophenols) can be effective against both bacteria and fungi. However, many are long-term environmental pollutants and, therefore, are restricted in use. Of the aldehydes (e.g., formaldehyde and glutaraldehyde), glutaraldehyde is the most widely used. The quaternary ammonium compounds (derivatives of ammonium salts, such as alkyldimethyl benzyl ammonium chlorides) kill microorganism cells by damaging the cell membranes. The membranes' effectiveness as a permeability barrier is reduced; therefore, the cells are not able to maintain their chemical balance with the extracellular medium, and eventually die.

## Intergranular Corrosion

### *Relationship of Alloy Microstructure to Susceptibility to Intergranular Corrosion*

Precipitation from unstable solid solutions is generally initiated at grain boundaries since these provide short-circuit diffusion paths and nucleation sites for precipitation of phases. Accompanying these processes are compositional gradients associated with solute-depletion zones extending from the grain boundaries and precipitated-phase interfaces into adjacent grains. The result is a lower composition of the diffusing species in solid solution at, and adjacent to, these interfaces, which may be sufficient to permit corrosion in these localized regions. An important variable is the extent of continuity of the precipitate phase in the grain boundary. If the precipitate is continuous around the grains as shown in Fig. 7.50(a), all grain boundaries will have a continuous region around them that has been depleted in the diffusing solute atoms. If the precipitate is discontinuous, as shown in Fig. 7.50(b), regions exist

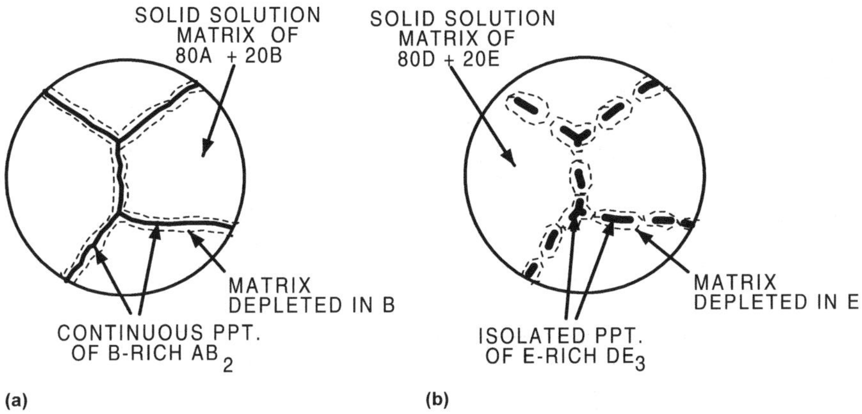


**Fig. 7.50** Schematic representation of microstructures susceptible to intergranular corrosion. (a) Continuous precipitation of B-rich $AB_2$. (b) Discontinuous precipitation of E-rich $DE_3$

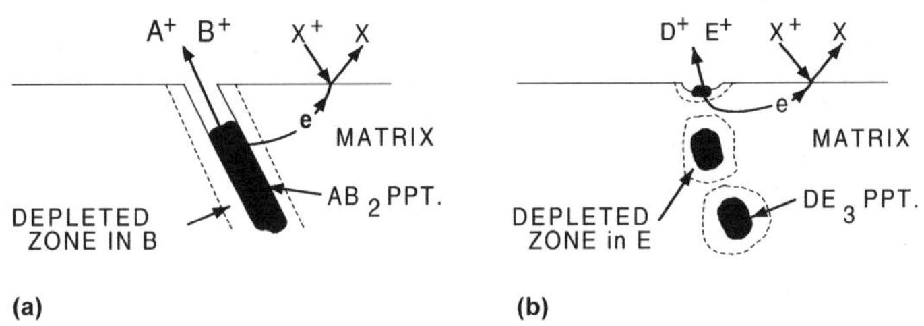


**Fig. 7.51** Interface profile of intergranular corrosion when the precipitate phase is anodic to the matrix phase. (a) Preferential corrosion of continuous $AB_2$ phase. (b) Preferential corrosion of discontinuous $DE_3$ phase

in the grain boundaries between precipitate particles that are unaltered or less altered in chemical composition.

The corrosive attack will depend on the relative extents to which the unaltered alloy within the grains, the depleted zones adjacent to grain boundaries, and the precipitate phase tend to act as cathodic or anodic surfaces. The latter surfaces, of course, undergo corrosive attack, which is supported by the cathodic reaction occurring on cathodic surfaces. If the precipitated second phase is continuous and anodic to both the solute-depleted and more-remote undepleted solid solution surrounding it, the precipitate will corrode, leaving a continuous crevice that tends to propagate around the grains as shown in Fig. 7.51(a). If the second phase is discontinuous and anodic, the precipitate will corrode, leaving isolated pits along the grain boundaries as shown in Fig. 7.51(b). The former is obviously the more severe condition leading to complete separation of grains along the boundaries and eventual disintegration of the alloy.

If the solute-depleted solid solution adjacent to the grain-boundary precipitate is anodic, and the cathodic reaction is supported by the pre-

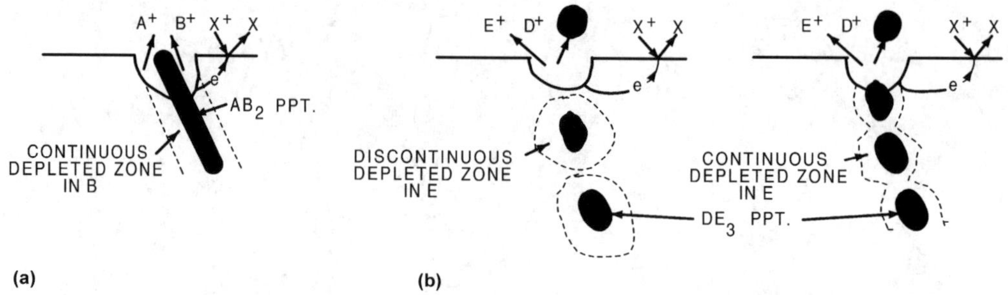


**Fig. 7.52** Interface profile of intergranular corrosion when solute-depleted zone is anodic to precipitate and undepleted matrix. (a) Intergranular attack when precipitate and solute-depleted zone is continuous. (b) Intergranular attack when precipitate and depleted zones are discontinuous

cipitate particle and/or the unaltered solid solution within the grain, the corrosive attack is localized in the region near the precipitate as shown in Fig. 7.52. The corrosive attack will be discontinuous or continuous depending on the distribution of the precipitate in the boundary. Alternatively, it is rare to observe unaltered solid solution within the grain as anodic to the depleted zone in the grain boundary. In the cases described, large local corrosion rates usually occur due to the large cathode-to-anode area ratios. As a result, the localized corrosion rate may be several orders of magnitude greater than that of a homogeneous alloy.

## *Intergranular Corrosion of Austenitic Stainless Steels*

Austenitic stainless steels are the most significant class of corrosion-resistant alloys for which intergranular corrosion can be a major problem in their satisfactory use. The problem is most often encountered as a result of welding but also may result from stress-relief annealing or incorrect heat treatments. Intergranular corrosion also can occur in ferritic stainless steels and in nickel- and aluminum-base alloys.

**The Fe-Cr-C Equilibrium Relationships in Stainless Steels.** The metallurgical processes occurring in austenitic stainless steels causing susceptibility to intergranular corrosion (sensitization) and methods to either prevent or remove susceptibility, are illustrated by the physical metallurgy of the selected alloys in Table 7.5. These are all austenitic stainless steels, and after quenching from elevated temperatures are es-

**Table 7.5 Compositions of selected austenitic stainless steels**

| Type | wt% carbon (max) | wt% Cr | wt% Ni | Other (max) |
|---|---|---|---|---|
| 302 | 0.15 | 18 | 9 | ... |
| 304 | 0.08 | 19 | 10 | ... |
| 304L | 0.03 | 19 | 10 | ... |
| 321 | 0.08 | 18 | 11 | (a) |
| 347 | 0.08 | 18 | 11 | (b) |

(a) wt% Ti = 10 × wt% C (min); (b) wt% Ta + wt% Cb (Nb) = 10 × wt% C (min)

sentially solid solutions of the alloying elements in face-centered cubic (fcc) iron. The most important alloying element influencing susceptibility to intergranular corrosion is carbon since it can react with chromium to form a carbide in the grain boundaries of the steel. The carbide is based on $Cr_{23}C_6$, but in the presence of iron, an iron-containing carbide, $(Cr,Fe)_{23}C_6$, which contains 60 to 70 wt% chromium, is observed. As a consequence of this high chromium content, the precipitation is accompanied by removal of chromium from a narrow volume of the austenite matrix on both sides of the grain boundary precipitate. If the chromium content in this volume is reduced to values that no longer support passive-film formation, then environments normally creating passivity and low corrosion rates will produce very high localized intergranular corrosion rates.

The solubility of carbon in an 18Cr-8Ni wt% stainless steel is shown in Fig. 7.53 (Ref 81). The region marked austenite is the single-phase, fcc solid solution of carbon in interstitial lattice sites. Below this region, the chromium carbide containing a small amount of iron, $(Cr,Fe)_{23}C_6$, is in equilibrium with the austenite. Dashed vertical lines have been added at 0.03, 0.08, and 0.15 wt% carbon corresponding to the maximum carbon contents of types 304L, 304 and 302 stainless steels, respectively. The austenite region (all chromium and carbon dissolved) is reached by heating type 304L to 950 °C, type 304 to 1075 °C and type 302 to 1200 °C. The recommended practice for heat treatment of these steels is to heat to 1000 to 1100 °C and cool rapidly enough to prevent precipitation of the chromium-rich carbide on cooling; in most cases, water quenching is used. It is noted that at these temperatures, some undissolved car-

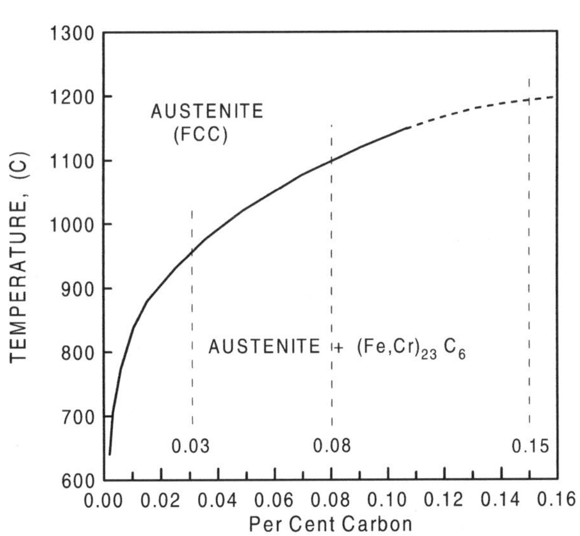


**Fig. 7.53** Solubility of carbon in 18 wt% Cr-8 wt% Ni stainless steel. Maximum carbon contents of types 304L (0.03%), 304 (0.08%), and 302 (0.15%) are shown. Based on Ref 81

bide will exist at temperature in the type 302, but this will not cause corrosion problems in the as-quenched steel since sufficient chromium is uniformly distributed throughout the structure. In the 18 wt% chromium alloy, the extent of solubility of carbon in austenite decreases with increasing nickel concentration (Ref 82). This effect can be a factor in increasing the susceptibility to intergranular corrosion of stainless steels with >10 wt% nickel.

These stainless steels are unstable when the high-temperature, single-phase austenite is cooled below the solubility curve. The significant temperature range extends down to about 500 °C. If cooling rates are fast enough down to this temperature to prevent precipitation, then for most practical purposes, precipitation will not occur at lower temperatures because the diffusion rate of chromium to the grain boundaries is too slow. Since this temperature is above that for processes involving aqueous environments, properly quenched stainless steels do not become susceptible to intergranular corrosion in industrial practice. The problem is, therefore, to avoid circumstances during fabrication of these alloys that will precipitate carbide and make them susceptible when placed in service.

**Effect of Thermal History of Austenitic Stainless Steels on Susceptibility to Intergranular Corrosion.** The time dependence for the local depletion of chromium sufficient to cause susceptibility to intergranular corrosion as functions of temperature and carbon content is of the form represented in Fig. 7.54 (Ref 83). The curves are typical of type 3*xx* alloys with nominal chromium concentrations of 17 to 25 wt% and, since they represent times for initiation of intergranular corrosion,

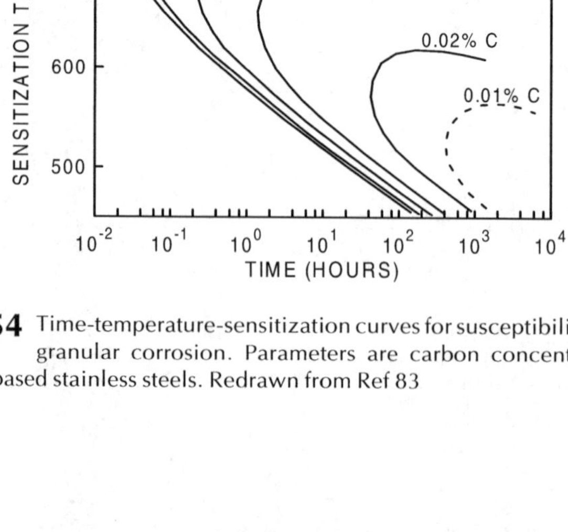


**Fig. 7.54** Time-temperature-sensitization curves for susceptibility to intergranular corrosion. Parameters are carbon concentrations in type 304-based stainless steels. Redrawn from Ref 83

their position also will depend on the aggressiveness of the corrosive environment. Specific data are shown in Fig. 7.55 for 18 wt% Cr, 10 wt% Ni stainless steels containing 0.05 and 0.027 wt% C; these are representative of types 304 and 304L stainless steels (Ref 84). The "C"-type curves define the time at each temperature beyond which carbides precipitate and lead to excessive intergranular corrosion in boiling acidified copper sulfate solution. The maximum allowable time at the most rapid precipitation temperature for the 0.050 wt% C alloy is about 40 s and for the 0.027 wt% C alloy, about 400 s, a factor of about ten times. Figures 7.54 and 7.55 represent the isothermal time dependence for initiation of susceptibility (Ref 83, 84). More generally, it is important to know the cooling history below the solubility limit (Fig. 7.53) and down to about 500 °C, which is required to prevent damaging carbide precipitation. Only limited information of this type is available, but rough estimates of the time-temperature relationships for initiation of susceptibility during continuous cooling can be made by shifting the isothermal curves to slightly longer times and slightly lower temperatures.

Grain-boundary chromium depletion by carbide precipitation can be significantly retarded or effectively eliminated by adding strong carbide-forming elements such as titanium, niobium, and tantalum, which form highly insoluble carbides and lower the carbon available to form chromium-rich carbides. Typical compositions are represented by types 321 and 347 in Table 7.5. These steels are used most effectively to reduce susceptibility to intergranular corrosion resulting from welding as discussed subsequently. However, problems can develop in the use of these steels in the fabrication of large pieces of equipment when forming and fabrication operations (including welding) introduce internal stresses that must be removed by heat treatments called stress-relief

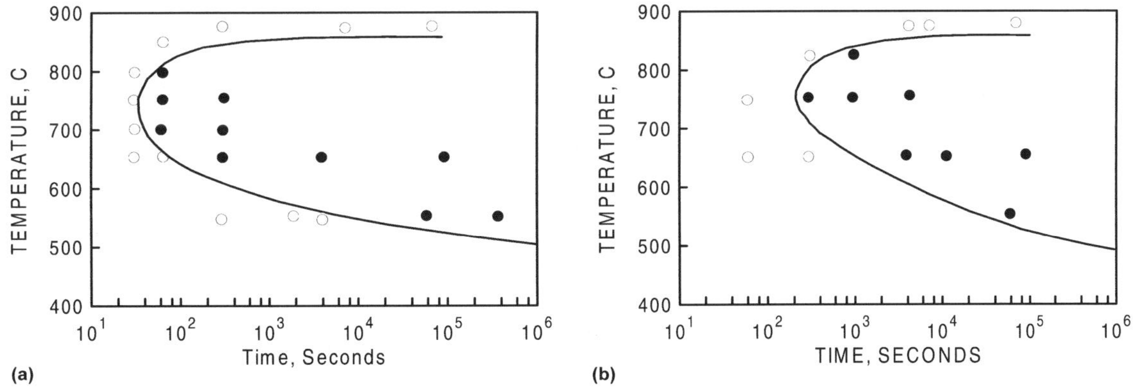


**Fig. 7.55** Effect of carbon content on susceptibility to intergranular corrosion of 18 wt% Cr-10 wt% Ni stainless steels in boiling acidified copper sulfate. Open circle, no corrosion; solid circle, intergranular corrosion. (a) 0.050% C, 18.22% Cr, 10.95% Ni, 0.049% N. (b) 0.027% C, 18.35% Cr, 10.75% Ni, 0.043% N. Redrawn from Ref 84

anneals. These treatments require heating for long periods of time into the same temperature range that causes carbide precipitation and, therefore, can result in susceptibility to intergranular corrosion. The effects of long-time annealing on the susceptibility of type 347 stainless steel to attack by boiling nitric acid are represented in Fig. 7.56 as contour lines between which a range of corrosion rates is given (Ref 85). Two comparisons should be made between these curves and those in Fig. 7.55. First, the time of exposure is much longer for the type 347, hours rather than seconds. For example, the time for the initial identified corrosion is about 40 s for the 0.05 wt% carbon type 304 and about 0.3 h (1100 s) for the type 347. The slower rate of sensitization in the type 347 shows the effect of the tantalum plus niobium in this steel in reducing the carbon available for precipitation of chromium carbide. Also, the shapes of the contours in Fig. 7.56 indicate that with increasing time at a specific temperature, the corrosion rate first increases and then decreases. This shape is also shown by the curves of Fig. 7.54. The rate initially increases with time as chromium is depleted adjacent to the grain boundaries due to chromium-rich carbide precipitation at the grain boundaries. In time, the carbon content of the austenite is reduced to a low value, and chromium, which diffuses much more slowly than carbon, diffuses from the grains into the chromium-depleted grain-boundary regions, again restoring the ability to form a corrosion-resistant passive film.

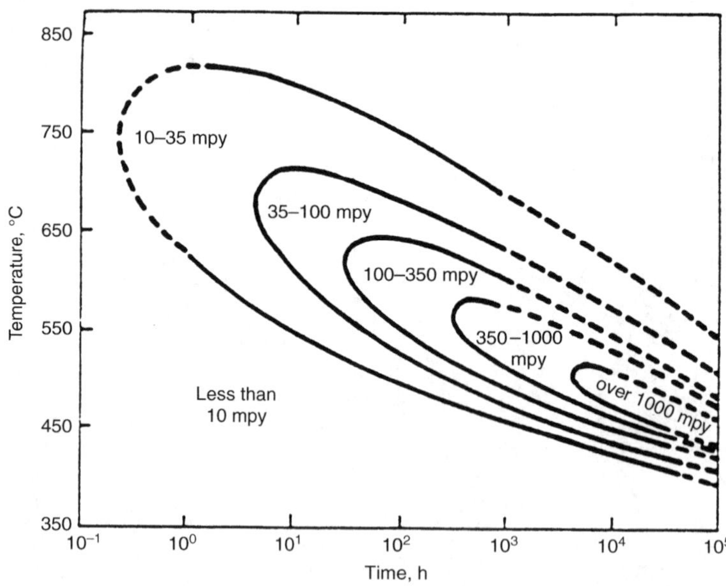


**Fig. 7.56** Time-temperature-sensitization curves for intergranular corrosion of type 347 stainless steel in boiling 65% nitric acid. mpy, mils per year. Source: Ref 85

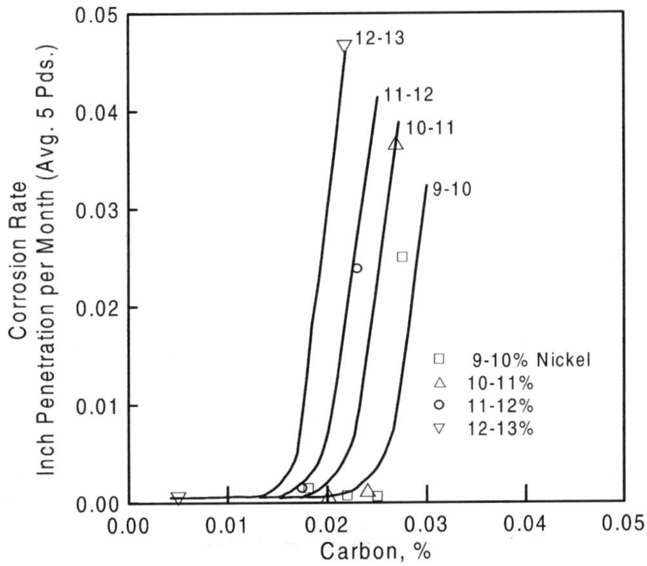


**Fig. 7.57** Effect of carbon and nickel content on intergranular corrosion penetration rate of 18 wt% Cr-base stainless steels. Alloys sensitized for 100 h at 550 °C. Immersion in boiling 65% nitric acid. Pds., periods (48 h) of exposure. Redrawn from Ref 84

An example of the interrelationship between carbon and nickel contents on susceptibility to intergranular corrosion of type 304 stainless steels is shown in Fig. 7.57 (Ref 84). The corrosion rate in boiling 65% nitric acid is plotted as a function of carbon content for 18 wt% chromium stainless steels with four ranges of nickel content. The time of heat treatment is 100 h at 550 °C. It is evident that the corrosion rate increases rapidly beyond 0.02 wt% carbon, which emphasizes that the 0.03 wt% carbon maximum for type 304L stainless steel is on the border line of holding the corrosion rate to reasonable values. The curves also show that increasing the nickel concentration at a given carbon concentration, particularly in the range of 0.01 to 0.03 wt% carbon, increases the amount of intergranular attack. This effect is due to the decrease in solubility of carbon in the austenite with increasing nickel concentration as previously mentioned.

## *Intergranular Corrosion of Ferritic Stainless Steels*

Susceptibility to intergranular corrosion also can occur in ferritic stainless steels (Ref 86–90). As with the austenitic stainless steels, the extent of the susceptibility is a function of the chemical composition and the thermal history of the steel. Also, the mechanism of intergranular attack is essentially the same for both classes of stainless steels, specifically, attack of lowered-chromium-content regions adjacent to precipitated chromium-rich carbides and nitrides. However, there are distinct differences in the conditions under which susceptibility is developed. This contrast in behavior of the ferritic versus austenitic stainless steel results from the much greater rate of precipitation from the body-centered cubic (bcc) structure of the ferritic alloy. The greater rate is attributed to the lower solubility of carbon and nitrogen, and to the more rapid decrease in solubility with decrease in temperature, in the ferritic alloy. Also, the diffusion rates of carbon and nitrogen are greater in the bcc ferrite structure.

Three groups of ferritic stainless steels can be identified, each being characterized by chemical compositions and thermal treatments leading to susceptibility to intergranular corrosion. The first group is the series of AISI type 4*xx* alloys with the approximate compositions of 0.08, 0.12, and 0.2 wt% carbon maximum with 13, 17, and 27 wt% chromium, respectively. The second group, frequently classed as intermediate-purity alloys, contains 0.02 wt% carbon maximum, 0.025 wt% nitrogen maximum, 25 to 27 wt% chromium, 0.5 wt% titanium, and 1 to 4 wt% molybdenum. A third group, referred to as ultrahigh-purity alloys, contains <0.005 wt% carbon, <0.01 wt% nitrogen, 26 to 30 wt% chromium, and 1 to 4 wt% molybdenum plus small amounts of niobium and/or titanium (Ref 87, 89).

In contrast to the austenitic stainless steels, the type 4*xx* ferritic steels become susceptible to intergranular corrosion when either water quenched or air cooled from 925 to 1100 °C (Ref 86). Because of the use of higher-carbon contents and the lower solubility of carbon and nitrogen in these alloys, the supersaturation on cooling is large, which, with the rapid rate of decrease in solubility and rapid rate of diffusion of carbon and nitrogen, results in precipitation of chromium-rich carbides and nitrides in the grain boundaries even on water quenching. The accompanying decrease in chromium concentration in the ferrite adjacent to the precipitates leads to susceptibility to localized corrosion. Thus, the basic condition allowing susceptibility is the same as with the austenitic stainless steels, but occurs much more rapidly in the ferritic bcc crystal structure as compared with the fcc austenitic structure. On reheating the water-quenched or air-cooled 4*xx* ferritic alloys into the temperature range 425 to 925 °C, chromium from within the grains diffuses rapidly into the chromium-depleted regions, and the susceptibility is progressively decreased with increasing time at temperature. The times for removal of susceptibility are of the order of minutes at temperatures down to 600 to 700 °C, but are significantly longer at lower temperatures. This decrease in susceptibility on reheating below 925 °C is consistent with the observation that very slowly cooled alloys from the higher temperature are not susceptible, having spent sufficient time in the temperature range below 925 °C to diffuse chromium into the chromium-depleted regions adjacent to the carbide or nitride precipitates immediately following their formation.

The ultrahigh-purity group of ferritic stainless steels containing molybdenum are not susceptible to intergranular corrosion on water quenching. Three factors contribute to this behavior: (a) the very low carbon and nitrogen concentrations of these alloys reduce the tendency for these alloys to precipitate chromium-rich carbides and nitrides on quenching; (b) small amounts of niobium combine with carbon and nitrogen to reduce their effective concentrations; and (c) 1 to 4 wt% molybdenum decreases the rate of diffusion of nitrogen and, hence, the nitride precipitation rate. These factors decrease or prevent precipitation of chromium-rich carbides and nitrides such that chromium-depleted regions subject to localized corrosion are not formed. On reheating, these alloys may become susceptible with a time-temperature dependence similar to the austenitic alloys. The difference in rate of sensitization of ferritic versus austenitic alloys is illustrated in Fig. 7.58 (Ref 91). The significantly faster kinetics of carbide precipitation in the ferritic alloys supports the requirement for very rapid cooling to prevent sensitization. The dotted curve also shows that recovery from the sensitized condition by diffusion of chromium into the depleted grain boundaries can be accomplished in relatively short times. This recovery is more rapid in the ferritic alloys since the diffusion rate of chromium is greater in the bcc structure than in the fcc structure. These curves must be taken as representative of the behavior; their position may be very sensitive to carbon, nitrogen, and molybdenum concentrations.

Because of the greater carbon and nitrogen contents of the intermediate-purity ferritic stainless steels, prevention of susceptibility to intergranular corrosion is more difficult than with the ultrahigh-purity alloys. Small amounts of niobium and/or titanium are added to combine

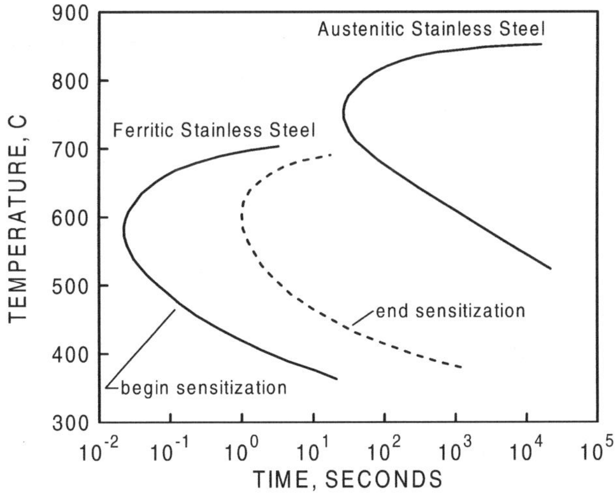


**Fig. 7.58** Time-temperature-sensitization curves for austenitic and ferritic stainless steels of equivalent chromium content. Redrawn from Ref 91

with carbon and nitrogen to reduce the amounts available for chromium carbide and nitride formation. Molybdenum retards the rate of precipitation, presumably by reducing the diffusion rate of nitrogen as it does in the ultrahigh-purity alloys. These alloys require very high cooling rates to suppress sensitization and, for this reason, may be restricted to thin sections if intergranular corrosion is to be avoided.

### *Intergranular Corrosion of Welded, Cast, and Duplex Stainless Steels*

Stainless steels containing both austenite and ferrite phases are encountered in welds made with type 308 stainless steel filler metal, in most stainless steel castings, and in alloys referred to as duplex stainless steels with approximately equal amounts of the two phases. The latter steels span the composition ranges of 20 to 28 wt% Cr, 2.5 to 6 wt% Ni, 1 to 4.5 wt% Mo, and 0.03 to 0.08 wt% carbon; controlled amounts of copper and nitrogen also may be present. In the as-quenched condition, these steels are quite resistant to intergranular and pitting-type corrosion. On holding in the temperature range 500 to 800 °C, the duplex steels precipitate chromium-bearing carbides, nitrides, and other intermetallic phases in austenite/ferrite interfaces, causing chromium depletion adjacent to the interfaces and subsequent susceptibility to intergranular corrosion (Ref 92).

The susceptibility of welded and cast stainless steels to intergranular corrosion following thermal histories in the temperature range 500 to 800 °C depends on the amount and distribution of the ferrite phase. The microstructure consists of austenite/austenite and austenite/ferrite grain boundaries, the relative amounts depending on composition and thermal treatment (Ref 93). Since the chromium-rich carbides tend to precipitate preferentially at austenite/ferrite phase boundaries in preference to austenite/austenite boundaries, the continuity of the precipitate-containing boundaries subject to intergranular corrosive attack will depend on the relative amount and distribution of the two phases. If the amount of ferrite is too low (<3 to 5%), sufficient carbon is not drained from the austenite/austenite boundaries to the austenite/ferrite interfaces to prevent sensitization of the former boundaries; therefore, destructive susceptibility to intergranular corrosion may occur. In contrast, if the amount of ferrite is too high, it can form a continuous network, and sensitizing thermal treatments will result in a continuous chromium-depleted path and susceptibility to destructive intergranular corrosion.

### *Intergranular Corrosion of Nickel-Base Alloys*

Nickel-base alloys under certain conditions of composition, thermal history, and environment are susceptible to intergranular corrosion.

These materials are frequently referred to as high-performance alloys for use in aggressive environments. They have a wide range of compositions based on 40 to 70 wt% Ni; representative ranges of alloy additions include 15 to 50 wt% Cr, 15 to 30 wt% Mo, <20 wt% Fe, and important but controlled amounts of Al, Ti, Nb, Ta, Co, and W. The carbon content is <0.10 wt% and frequently limited to 0.02 wt%. Because of the fcc crystal structure of nickel, the alloys are classed as austenitic. The major alloying element contributing to the highly protective passive film is chromium, which forms, as with stainless steel, a chromium-rich oxide film based on $Cr_2O_3$. In general, these alloys can develop susceptibility to intergranular corrosion when thermal histories result in grain-boundary precipitates that alter the local composition of the austenite below values capable of maintaining a protective passive film (Ref 94, 95).

Because of the broad variation in composition and response to thermal treatment of the nickel-base alloys, it is not possible to generalize mechanisms responsible for developing susceptibility to intergranular corrosion. Therefore, the following discussion of the behavior of a Ni-Mo-Cr alloy is used to illustrate the complexity of an interrelationship between alloy composition, heat treatment, corrosion environment and corrosion rate. The alloy has the nominal composition in weight percent of 14.5 to 16.5 Cr, 15 to 17 Mo, 3 to 4.5 W, and 4 to 7 Fe with maximum limits on carbon and silicon. The alloys for which the corrosion data are shown in Fig. 7.59 contained 0.045 to 0.06 wt% carbon and 0.53 to 0.80 wt% silicon and were initially quenched from 1225 ± 15 °C (2235 ± 25 °F), which produced a dispersion of $M_6C$ type carbides (M = Mo, W, Si) in austenite (Ref 94). These carbides were not involved in the subsequent corrosion behavior or heat treatments. Heat

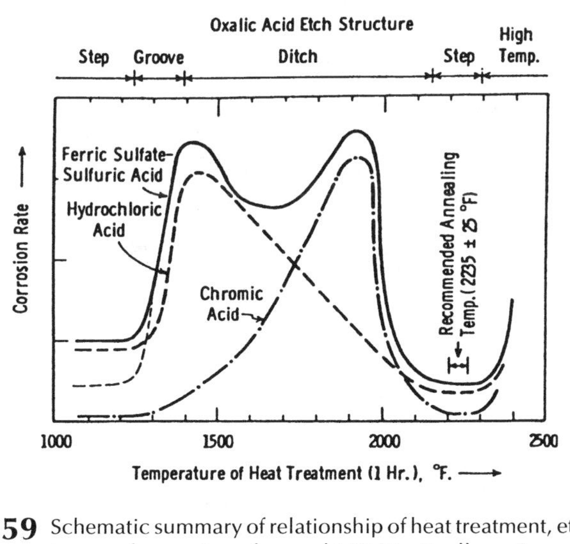


**Fig. 7.59** Schematic summary of relationship of heat treatment, etch structure, and corrosion of wrought Ni-Mo-Cr alloys. Representative analyses of alloys investigated to establish this summary. 15% Cr, 16% Mo, 5% Fe, 4% W, 0.045–0.065% C. Source: Ref 94

treatments were conducted for 1h in the temperature range 425 to 1300 °C (800 to 2375 °F), followed by 24 h corrosion tests in boiling hydrochloric acid, sulfuric-acid/ferric-sulfate, and chromic acid. The oxidizing power, and corrosion potentials, of these environments increase in the respective order with hydrochloric acid also contributing chloride ions to the environment.

It is evident in Fig. 7.59 that maximum attack occurs following holding near 760 °C (1400 °F) and near 1050 °C (1920 °F) but is selective with respect to the corrosive environment. The lower heat-treating temperature develops susceptibility to the boiling hydrochloric acid, the higher temperature develops susceptibility to boiling chromic acid, and the boiling sulfuric-acid/ferric-sulfate environment causes attack following heat treatment over the entire temperature range. The mode of attack relative to temperature range is indicated at the top of the figure. The modes are grooves or deep ditches into the grain boundaries and steps in the surface due to differences in attack depending on grain orientation. The grain-boundary precipitate at the lower temperatures (~760 °C) (~1400 °F) occurs as a thin, continuous molybdenum-rich phase related to $Ni_7Mo_6$ that depletes the adjacent austenite in molybdenum. Molybdenum in solid solution in nickel-base alloys imparts corrosion resistance in nonoxidizing chloride-bearing environments, and its depletion near the grain boundaries by precipitation of the molybdenum-rich phase would be conducive to intergranular attack in hydrochloric acid. Precipitation of a chromium-rich, sigma-type phase has been proposed as responsible for the high corrosion rate in the oxidizing environments containing $Fe^{3+}$ and $Cr^{6+}$ ions following heat treatments near 1050 °C (1920 °F). Here, the mechanism is similar to that occurring in stainless steels in which precipitates leading to localized depletion of chromium result in intergranular corrosion in highly oxidizing environments.

The corrosion behavior of the aforementioned Ni-Mo-Cr alloy is significantly changed by reducing the carbon and silicon contents to a maximum of 0.01 wt% C and 0.08 wt% Si (Ref 95). The effect of heat treatment following quenching from 1150 °C is to produce a single maximum in the corrosion rate on exposure to the nonoxidizing boiling hydrochloric acid environment following heat treatments near 760 °C, and a maximum in the corrosion rate in the oxidizing boiling sulfuric-acid/ferric-sulfate environment following heat treatments to 870 °C. The major precipitate forming on reheating into the temperature range 650 to 1100 °C is the molybdenum-rich $Ni_7Mo_6$. It forms predominantly in the grain boundaries as a thin, continuous precipitate at the lower temperatures and as a discontinuous precipitate at the higher temperatures. The shift to a peak corrosion rate at the higher temperature in the oxidizing environment is attributed to an unidentified chromium-rich phase, possibly the sigma phase, in the grain boundaries.

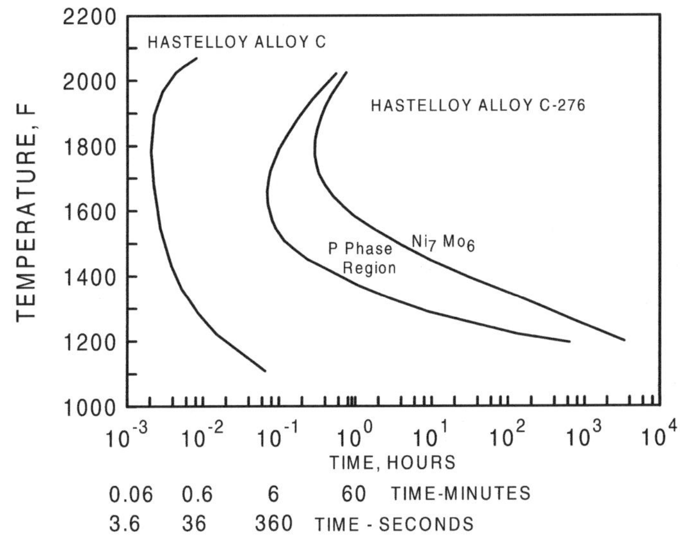


**Fig. 7.60** Comparison of the time-temperature-transformation curves of Hastelloy alloys C and C-276. The latter contains less carbon and silicon. Redrawn from Ref 95

The more significant result of the lower carbon and silicon content alloy is the reduced susceptibility to intergranular corrosion. This has been correlated with the much slower rate of precipitation in this alloy when compared with the higher carbon and silicon concentrations. The difference is evident in Fig. 7.60, where the time of appearance of precipitates is some 30 times longer for alloys with the lower concentration of these elements. Consistent with this difference in precipitation rate is the ability to satisfactorily weld the latter alloys without introducing susceptibility to intergranular corrosion.

## *Intergranular Corrosion of Aluminum-Base Alloys*

Intergranular corrosion can occur in aluminum-base alloys, the extent depending on the environment, and the composition and thermal treatment of the alloy. Development of susceptibility to intergranular corrosion is more closely related to galvanic effects between precipitated phases in grain boundaries and the immediately adjacent matrix than with stainless steels. In the latter, precipitation of chromium-rich carbides in the grain boundaries results in an adjacent chromium-depleted matrix that cannot maintain a protective passive film. The local corrosion is caused by the large area of the grain exposed to the environment and the ability of the passive film on this surface to conduct electrons, which supports the cathodic reaction. This large cathode-to-anode area ratio concentrates the corrosion at the grain-boundary, chromium-depleted matrix. With aluminum alloys, the poor conductivity of the passive film limits its influence in supporting the cathodic reaction. As a result, the relative electrochemical potentials (anodic or cathodic) of a grain-boundary precipitate and the immediately-adjacent matrix generally govern the mechanism of intergranular corrosion (Ref 59).

Aluminum alloys in which the magnesium/silicon ratio is controlled to the stoichiometric ratio for the intermetallic compound $Mg_2Si$ have minor tendency toward intergranular corrosion since the electrochemical potentials of this compound and the matrix are similar. When this ratio is lower, silicon may precipitate in the grain boundaries, where it supports the cathodic reaction and induces corrosion in the adjacent matrix. Alloys in which $Mg_5Al_8$ or $MgZn_2$ can precipitate in grain boundaries following certain thermal treatments may be susceptible to intergranular corrosion since these phases are anodic to the adjacent matrix and will be preferentially attacked. A variable in the severity of attack is the extent to which the phases are continuous or discontinuous in the boundary (Ref 96–98).

The susceptibility to intergranular corrosion of aluminum-copper alloys has been investigated extensively, and although the precipitated phase, $CuAl_2$, provides a surface supporting the cathodic reaction, the mechanism appears to involve more than a simple galvanic interaction. As the solid-solution copper concentration in aluminum is increased in the range 0 to 5 wt%, the pitting potential increases in chloride solutions from –520 to –340 mV (SHE). Grain-boundary precipitation of $CuAl_2$ creates a zone in the adjacent matrix, which is depleted in solid-solution copper concentration (Ref 60, 99). As a consequence, if the environment leads to a corrosion potential between –520 and –340 mV (SHE), which is reasonable, then severe pitting will occur in the depleted zone, and intergranular corrosion is initiated.

## *Susceptibility of Stainless Steels to Intergranular Corrosion due to Welding*

Each position at and near a weld undergoes a specific time-temperature history as the welding electrode passes. Figure 7.61 shows representative temperature profiles at the indicated positions near a weld as the welding electrode passes (Ref 100). The temperature band from 1200 to 1600 °F (650 to 980 °C) is the temperature range in which susceptibility to intergranular corrosion due to grain-boundary chromium depletion develops. Curve B is the most important time-temperature history since at this distance from the weld line, the alloy remains in the critical temperature range for the longest period of time during which $(Cr,Fe)_{23}C_6$ can precipitate. Nearer the weld (curve A), the temperature increases above and then decreases rapidly through the range, and farther from the weld, the maximum temperature attained is below that for damaging precipitation to occur. As a consequence, the region of susceptibility to corrosion when placed in a corrosive environment is a band extending from "B" out to "C," as shown in Fig. 7.61(b).

When welding conditions and the corrosive environment lead to intergranular corrosion in an austenitic stainless steel such as type 304, the following alternatives may be considered as possible solutions to the problem:

- Reheat the welded component to 1000 to 1100 °C to redissolve the $(Cr,Fe)_{23}C_6$, and water quench to retain the homogeneous austenite phase. It is generally impractical to use this alternative since the components may be too large, may distort due to rapid cooling, and more often, will be part of a system of components welded in place and cannot be removed.
- Substitute a low-carbon stainless steel such as type 304L that can usually be welded without developing intergranular precipitation to the extent that it becomes susceptible to corrosion. The low-carbon stainless steels are somewhat more expensive, and care must be exercised that there is no carbon pickup during welding.
- Substitute type 321 or 347 for 304. Type 321 contains Ti and 347 contains Ta and Nb. These elements have more negative free energies of formation of their carbides than Cr and, therefore, tend to more readily combine with the carbon, thus leaving the Cr in solid solution. In some cases, these carbides will dissolve very near the fusion line and allow chromium carbide to still precipitate in the heat-affected zone during welding. This is due to the much greater concentration of chromium in the alloy favoring precipitation kinetically even though the Ti- and Nb-bearing carbides are more stable thermodynamically. The band in which carbide precipitation is observed after welding types 321 and 347 is very narrow, which is responsible for referring to the localized corrosion as knife-line attack (Ref 101).

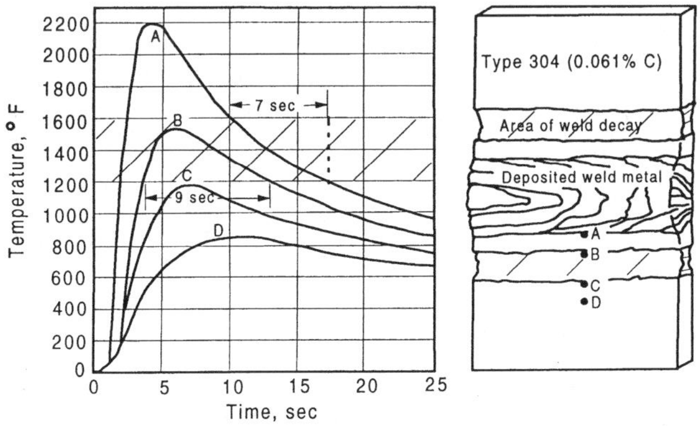


**Fig. 7.61** Temperature-time histories at indicated positions during electric arc welding of a type 304 stainless steel. Source: Ref 100

There is a significant difference in the appearance of welded sections of austenitic versus ferritic stainless steels following exposure to environments causing intergranular attack. As just described, in austenitic stainless steels the maximum attack occurs in the heat-affected zone at some distance from the weld fusion zone associated with a reheat temperature of about 760 °C. In contrast, the type 4*xx* ferritic stainless steels become susceptible to intergranular corrosion following exposures to temperatures above 925 °C (see the section "Intergranular Corrosion of Ferritic Stainless Steels" in this chapter). The region of maximum attack is therefore nearer to the weld fusion zone and even in the weld deposit itself (Ref 86).

## *Measurement of Susceptibility of Stainless Steels to Intergranular Corrosion*

**ASTM Chemical Environment Test Standards.** Since intergranular corrosion is one of the most serious problems in the satisfactory application of stainless steels, several procedures are available for the measurement of the susceptibility of these steels to this type of corrosion. The procedures have been formalized as standardized tests, designated as ASTM A 262 (Ref 102), and are widely accepted as a basis for certifying that a specific stainless steel meets specifications. A limitation of these tests is that they specify specific environments rather than the environment of the actual application. In many applications, however, reasonable correlations have been established between acceptable response to the tests and successful service performance.

Most applications of stainless steels, particularly in the chemical process industry, are for oxidizing environments, extending from dissolved oxygen to nitric acid, and depend on these conditions to produce and maintain a protective passive film. The test environments are therefore oxidizing and have been selected to provide a range of positive half-cell potentials. Information is given in Table 7.6 on the tests, including procedures, corrosion potentials generated, and the selectivity of the attack on the surface if the steel is susceptible (Ref 91). A schematic polarization curve for a stainless steel heat treated to give maximum corrosion resistance is shown in Fig. 7.62 (Ref 91). Listed to the right of the figure are the major A 262 test environments placed at the corrosion potentials that they tend to produce at the surface of the stainless steel. It should be noted that the tests do not require a potentiostat to produce these potentials, but rather depend on the equilibrium half-cell potentials and polarization parameters ($i_o$, $\beta$, and $i_D$) of the cathodic reactions to electrochemically produce the corrosion potentials indicated by the arrows. The exception is the position labeled "oxalic acid electrolytic etch." This is a screening test conducted in the very positive potential range using either a galvanostat or potentiostat. Steels showing no attack fol-

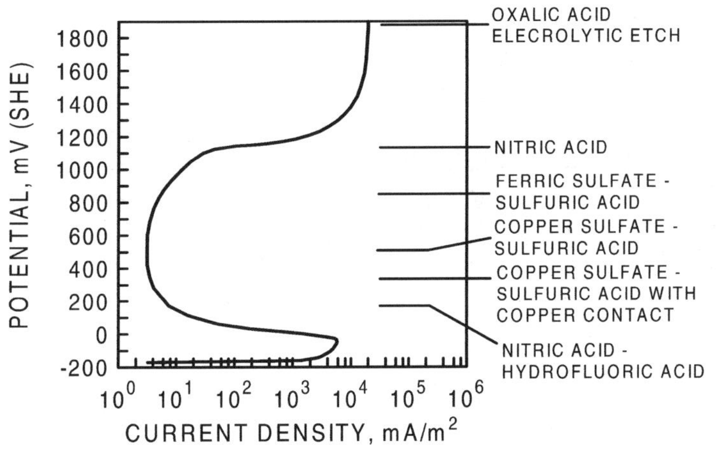


**Fig. 7.62** Approximate potentials developed on stainless steels in the indicated ASTM standard test environments. The polarization curve is representative of type 304 stainless steel in 1 N $H_2SO_4$. Based on Ref 91

**Table 7.6 Summary of chemical tests used for the determination of susceptibility to intergranular corrosion of iron-nickel-chromium alloys**

| Test name | Usual solution composition | Test procedure | Quantitative measure | Potential range, V (SHE) | Species selectively attacked |
|---|---|---|---|---|---|
| Nitric acid test | 65 wt% $HNO_3$ | Five 48 h exposures to solution; solution refreshed each period | Average weight loss per unit area of five testing periods | +0.99 to +1.20 | 1. Chromium-depleted areas 2. Sigma phase 3. Chromium carbide |
| Acid ferric sulfate test (Streicher test) | 50 wt% $H_2SO_4$ + 25 g/L ferric sulfate | 120 h exposure to boiling solution | Weight loss per unit area | +0.7 to +0.9 | 1. Chromium-depleted areas 2. Sigma phase in some alloys |
| Acid copper sulfate test | 16 wt% $H_2SO_4$ + 100 g/L $CuSO_4$ (+metallic copper) | 72 h exposure to boiling solutions | 1. Appearance of sample on bending 2. Electrical resistivity change 3. Change in tensile properties | +0.30 to +0.5 | Chromium-depeleted areas |
| Oxalic acid test | 100 g $H_2C_2O_4$–$H_2O$ + 900 mL $H_2O$ | Anodically etched at 1 A/cm$^2$ for 1.5 min | Geometry of attack on polished surface at 250× or 500× | +1.70 to 2.0 or greater | Various carbides |
| Nitric-hydrofluoric acid test | 10% $HNO_3$ + 3% HF | 4 h exposures at 70 °C solution | Comparison of ratio of weight loss of laboratory annealed and as-received samples of same material | Corrosion potential of 304 = +0.14 to +0.54 | 1. Chromium-depleted areas 2. Not for sigma phase 3. Used only for Mo-bearing steels |
| Hydrochloric acid test | 10% HCl | 24 h in boiling solution | 1. Appearance of sample after bending around mandril 2. Weight loss per unit area | (a) Redox potential = +0.32 (b) Corrosion potential = –0.2 ± 0.1 | 1. Alloy-depleted areas 2. Not for sigma phase |
| Nitric acid, $Cr^{4+}$ test | 5 N $H_2SO_4$ + 0.5 N $KCr_2O_7$ | Boiling with solution renewed every 2–4 h for up to 100 h | 1. Weight loss per unit area 2. Electrical resistivity 3. Metallographic examination | (a) Redox potential = +1.37 (b) Corrosion potential of 304 = +1.21 | Solute segregation to grain boundaries |

(Based on Ref 91)

lowing the oxalic acid etch on a polished surface will not be attacked in any of the chemical environments, and, hence, no additional testing is required. Some attack following the oxalic acid test, however, does indicate that one or more of the subsequent tests could cause attack. The subsequent test that most nearly correlates to the actual environmental conditions to which the stainless steel will be exposed in service should therefore be selected.

Shown in Fig. 7.63 are the effects of chromium additions on the anodic behavior of an iron-nickel alloy containing 8.3 to 9.8 wt% nickel, the nickel content of stainless steels such as type 304 (Ref 91). In Fig. 7.64, the effects on the anodic polarization curve of heating type 304 stainless steel for the indicated number of hours at 650 °C are shown (Ref 103). The tests in Table 7.6 can be interpreted in terms of these figures. Assume that precipitation of $(Cr,Fe)_{23}C_6$ in the grain boundaries reduces the chromium content of the matrix to 3.54 wt% Cr in the narrow region adjacent to the precipitates. According to Fig. 7.63, this composition will corrode at the rate $i_{corr,1}$ (~$10^7$ mA/m$^2$) in the copper-sulfate/sulfuric-acid/copper-contact test environment. The properly heat treated steel containing 19.2 wt% Cr in solid solution will corrode at the rate $i_{corr,2}$ (~$10^{-2}$ mA/m$^2$). Thus, the corrosion rate at the chromium-depleted grain boundaries is about $10^9$ times faster and ac-

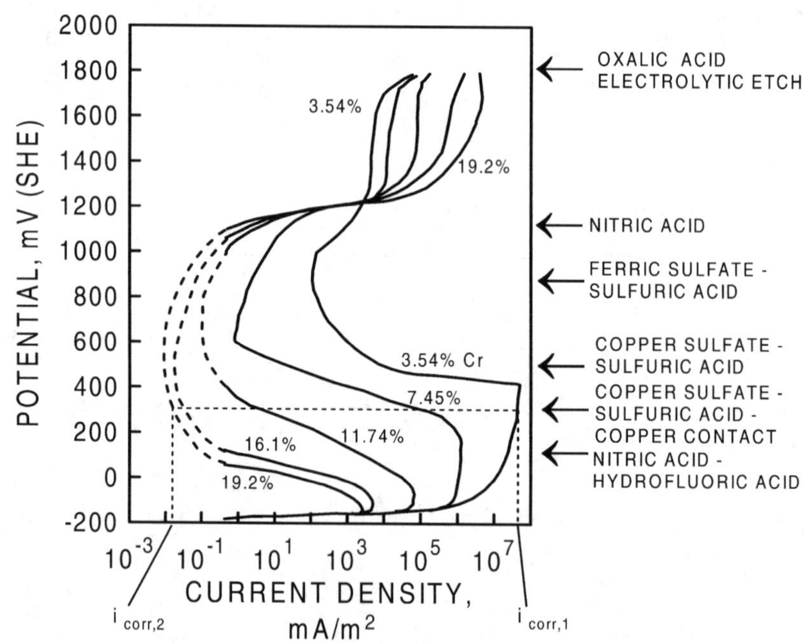


**Fig. 7.63** Effect of indicated Cr contents on the anodic polarization of stainless steels with 8.3–9.8 wt% Ni. 1 N $H_2SO_4$. Arrows indicate potentials developed in the corresponding ASTM standard test environments. Based on Ref 91 with dashed sections added as estimates of passive regions

counts for the selective intergranular corrosion. With reference to Fig. 7.64, when a sensitized stainless steel is potentiostatically polarized, the measured current, for example, for the alloy after 1000 h at 650 °C, is the current from the rapidly corroding grain boundaries with a local current density of $i_{corr,1}$ plus the small current contribution from the passive surface of the grains with a current density of $i_{corr,2}$ (Fig. 7.63). It is evident that the influence of sensitization will be different at different potentials, and hence, the results of a corrosion test will depend on the test solution used. Also, from Table 7.6, each solution selectively attacks different parts of the alloy microstructure. As stated previously, in practice, the test environment should be selected that produces a potential closest to the potential that will be produced on the steel by the environment that it contacts in service.

**Electrochemical Evaluation of Susceptibility to Intergranular Corrosion.** The determination of the susceptibility of stainless steels to intergranular corrosion using electrochemical measurements relates to the sensitivity of the polarization curve to the amount of chromium in solid solution. This influence for homogeneous alloys is shown in Fig. 7.63. The effect of holding a type 304 stainless steel at 650 °C for increasing times on the polarization curve is shown in Fig. 7.64. The shift in the polarization curve to larger current densities for the alloy held at longer times at temperature is related to the increasing contribution to the measured current density by the progressively greater amount of chromium depletion in the grain boundaries. That is, the surface is a changing composite of passivated grain surface with low current den-

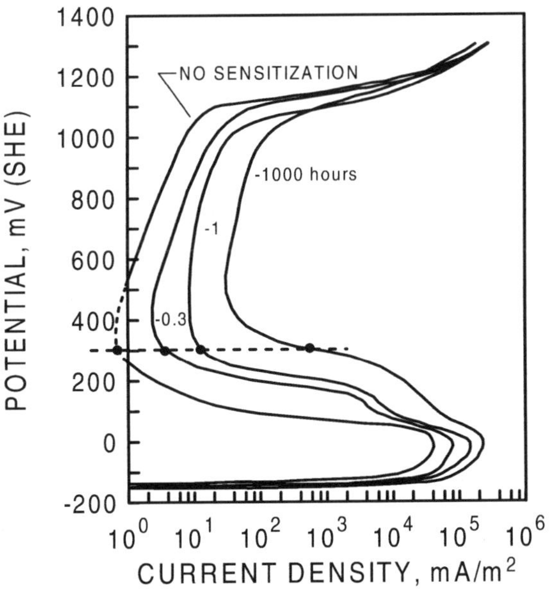


**Fig. 7.64** Effect of sensitization time at 650 °C on anodic polarization of type 304 stainless steel in 2 N $H_2SO_4$. Redrawn from Ref 103

sity and poorly passivated or active surface adjacent to the grain boundaries with high current density.

A more sensitive and quantitative electrochemical evaluation than represented by Fig. 7.64 is to conduct an electrochemical potentiodynamic reactivation (EPR) scan under carefully prescribed conditions (Ref 104). The environment is 1 N $H_2SO_4$ with 0.01 M KSCN, which, in the potential range of the current density maximum, accelerates the chemical removal of the passive film at a rate depending on the film composition as controlled by the chromium content of the underlying alloy. The rate of attack is greater the lower the chromium content, particularly below ~13% Cr and, hence, is more aggressive toward the film over the chromium-depleted grain boundary areas than to the more protective film over the grain surfaces. The procedure is to establish the corrosion potential of a polished specimen (polished with 1.0 μm diamond compound) in the deaerated solution at 30 °C. The corrosion potential is usually in the range of –210 to –110 mV (SHE); if not, the specimen is briefly cathodically cleaned at –360 mV (SHE) to allow the corrosion potential to be in this range. The steel is immediately passivated by holding at +440 mV (SHE) for 2 min followed by measurement of the downscan curve at a rate of 6000 mV/h, terminating the scan at the corrosion potential.

A schematic representation of downscan polarization curves using the EPR procedure is shown in Fig. 7.65 (Ref 93). A sensitized stainless steel will result in an "anodic loop" with size depending on the degree of sensitization. With the specified rapid downscan rate, the passive film

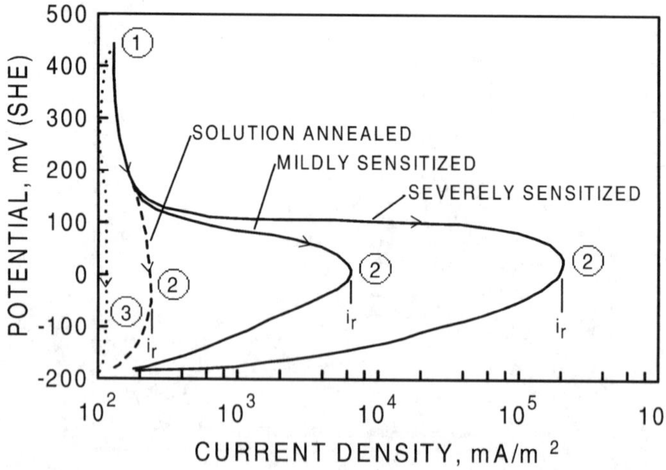


**Fig. 7.65** Schematic EPR (electrochemical potentiokinetic reactivation) curves for three amounts of sensitization of an austenitic stainless steel. Passive film formed at (1). Downscans pass through maximum attack at (2). Environment: 1 N $H_2SO_4$ + 0.01 M KSCN at 30 °C. Curve (3) is observed if passive film continues to form on downscan. Source: Ref 93

formed on holding a nonsensitized steel at +440 mV (SHE) remains during the downscan, and the curve is nearly vertical as shown by the dashed curve in Fig. 7.65. In some cases (the dotted curve), the passive film may continue to form during the downscan, resulting in a curve associated with a decreasing current density. Also shown in the figure are downscan curves showing increasing current density in the potential range of ±200 mV (SHE). These curves result from thermal histories leading to mildly and severely sensitized conditions. Since the current density over the passivated grain surfaces is very low, the higher current densities observed for the sensitized steel are predominantly due to dissolution of the chromium-depleted zones at the grain boundaries resulting from selective loss of the passive film. This local current density is sufficiently high (particularly in the presence of the KSCN (see Fig. 5.34) to make a large contribution to the measured current even though the area of the sensitized zone is a small fraction of the total area.

Two quantities that are used to evaluate the degree of sensitization are the maximum current density during reactivation, $i_r$, and the area enclosed by the "anodic loop." Since the downscan rate is constant, the potential axis can be converted to a time axis, and integration between the curves for sensitized and nonsensitized conditions gives the total charge density, Q (coulombs/cm$^2$), transferred due to grain-boundary attack. The sensitized area depends on the grain size and the width of the chromium-depletion zone along the grain boundary undergoing dissolution. Since it is not practical to measure this width for each evaluation, a value of 0.5 μm depletion into the grain (1.0 μm total width) is used based on scanning electron microscopy images and profiles of the chromium composition across the boundary. It is recognized that the width varies depending on time and temperature of sensitization and the relative grain orientation. Also, the area undergoing dissolution increases as the attack progresses. The degree of sensitization is expressed as the normalized integral charge density, or $P_a = Q/GBA$, where GBA is the exposed chromium-depleted grain-boundary area per unit specimen area. For convenience, $P_a$ may be expressed as $P_a = A_s$ $(5.095 \times 10^{-3} \exp(0.347\ X))$ where $A_s$ is the area of the specimen used in the polarization measurement, and X is the ASTM grain size number (Ref 104).

The results of EPR measurements on type 316 stainless steel quenched and reheated for 2, 4, 5, 20, and 40 h at 600 °C are shown in Fig. 7.66 (Ref 105). Areas within the anodic peaks increase with heat treatment time; the associated values of $P_a$ are 0.05, 0.29, 0.77, 3.90, and 7.36 C/cm$^2$. The time-temperature dependence of EPR values for this steel are shown in Fig. 7.67, in which the C-curve represents the time limit beyond which the sensitized steel fails the ASTM A 262E test (boiling $H_2SO_4 + CuSO_4$) (Ref 105). For this correlation, heat treat-

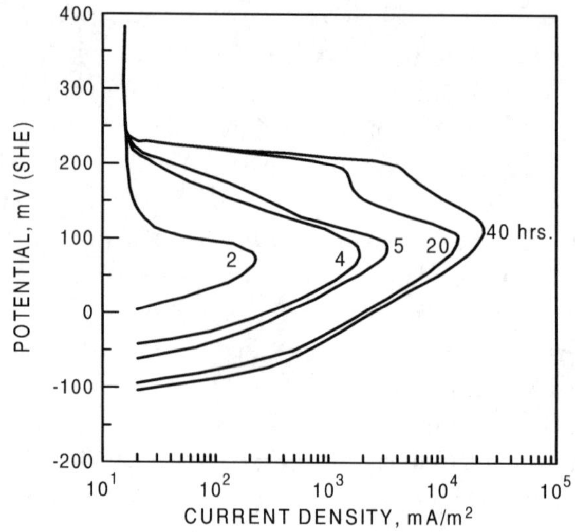


**Fig. 7.66** EPR curves for type 316 stainless steel sensitized to intergranular corrosion by heating at 600 °C for 2, 4, 5, 20, and 40 hours. Redrawn from Ref 105

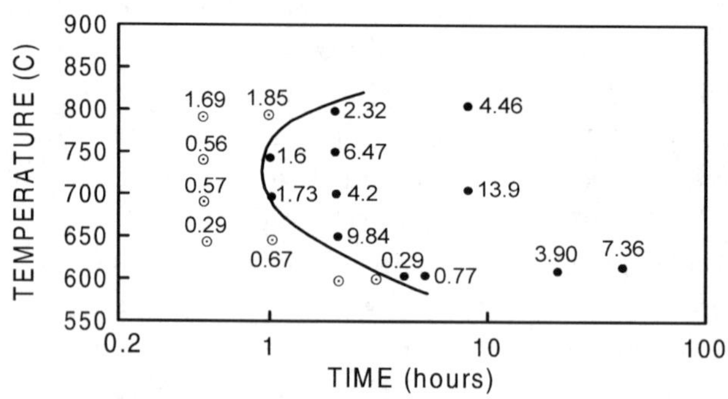


**Fig. 7.67** Correlation of EPR test values on type 316 stainless steel with ASTM A 262E test for susceptibility to intergranular corrosion. Circles indicate time-temperature treatments prior to test. Numbers at points are EPR values ($P_a$) of the same steel. C-curve defines times beyond which steel does not pass ASTM A 262E. Redrawn from Ref 105

ments resulting in EPR values less than approximately $P_a = 2.0$ C/cm$^2$ will result in passing ASTM A 262E.

EPR measurements can be used to show the distribution of sensitization in the heat-affected zone of welded stainless steels. $P_a$ values for test specimens cut parallel to and progressively away from the weld fusion zone of a type 304 stainless steel are shown in Fig. 7.68 (Ref 104). The shape of this curve is consistent with the time-temperature thermal histories shown in Fig. 7.61 and indicates that the steel was in the critical temperature zone for sensitization for the longest time at positions 100 mils (2.54 mm) from the fusion line.

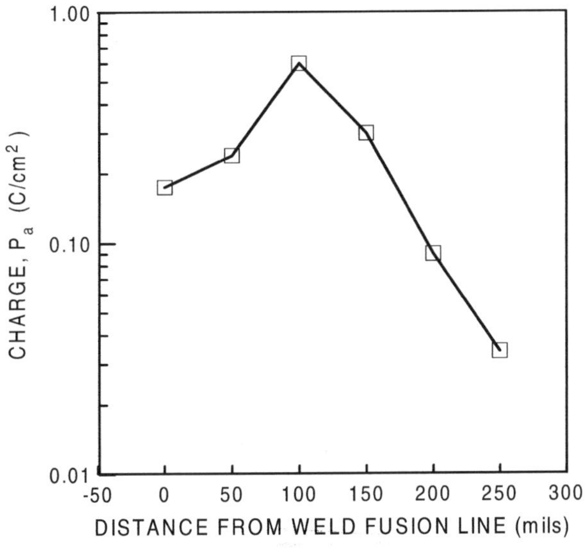


**Fig. 7.68** Effect of welding on sensitization as a function of distance from the weld fusion line of type 304 stainless steel as determined by the EPR test. Redrawn from Ref 104

As with all standardized tests (e.g., the ASTM A 262 procedures previously discussed), correlations must be established between the EPR $P_a$ values and service performance. For example, a criterion of $P_a < 2$ C/cm$^2$ has been proposed for adequate resistance to intergranular corrosion leading to intergranular stress-corrosion cracking (IGSCC) of type 304 and 304L pipe and welds. Other limits would be set depending on the material, application, and environment (Ref 105, 106).

## Environment-Sensitive Fracture

From studies of service behavior and from extensive laboratory investigations, the well-established terms *stress-corrosion cracking* (SCC) and *corrosion fatigue* have been shown to relate to a continuum of failure modes classified as environment-sensitive fracture. In many environments, the addition of stress, with associated strains, introduces a variable that can result in brittle failure in the sense of very limited plastic flow in otherwise ductile materials such as the stainless steels. Environment-sensitive fractures propagate at an advancing crack tip at which, simultaneously, the local stresses can influence the corrosion processes, and the corrosion can influence the crack-opening processes. Since these processes proceed by kinetic mechanisms, they are time and stress dependent with the result that the crack propagation rate can become very sensitive to the stress application rates. Conventional SCC usually has been associated with static stress, but this is seldom realized in service due to variations in operating conditions including start-up and shut-down cycles. Furthermore, observations of changes in ductility as a function of strain rate during slow strain-rate testing in corrosive environments has provided useful information on conditions under which SCC is probable. Higher cyclic stress/strain rates merge into the ranges of conventional fatigue behavior, but crack propagation rates are now also influenced by a corrosive environment. That is, modes of failure are then classed as conventional corrosion fatigue.

## *Characteristics of Environment-Sensitive Cracking*

Table 7.7 is an overview of alloy/environment systems for which SCC has been reported (Ref 107). Most of these systems have been investigated extensively to establish the variables influencing the cracking phenomena, including alloy composition and microstructure as established by thermal and mechanical treatment, environment composition, state of stress (static and cyclic), time, temperature, and electrochemical potential. From these investigations, several generalizations can be made relating environment-sensitive cracking to the state of stress (both magnitude and time dependence), to the material and to the environment (based on Ref 108).

- Crack propagation occurs only as a result of tensile stress regardless of the source of the stress. Stresses may result from service such as structural loads and internal pressures, or the stresses may be residual as a result of fabrication, including welding, bending, and in press fits as observed in heat exchanger tubes and tube sheets.
- Cracking generally is restricted to metal/environment conditions, which, in the absence of stress, show negligible corrosive attack. In particular, metals and alloys whose corrosion resistance depends on maintaining a stable passive film undergo stress cracking in environments that cause, or when environments change to cause, local instabilities in the film. Thus, stress concentrations associated with pitting or stress rupture of the passive film can lead to crack propagation. Stable films will not allow crack initiation or will immediately repair when local film rupture occurs. Stripping of the film will lead to general corrosion and to the absence of critical stress concentration sites.
- It follows, then, that small changes in the environment may initiate cracking. These include small changes in concentrations of species such as chloride ions, which initiate pitting and/or prevent film repair. Changes in concentration of cathodic reactants such as dissolved oxygen can shift the corrosion potential to values at which the passive film is not stable and cracking occurs. It also follows

that small concentrations of critical species may control the cracking.

- In general, alloys are much more susceptible than pure metals to environmental stress cracking. Both alloy composition and microstructure are significant variables, and hence, thermal treatments, including welding can affect response to SCC. While cracking can be frequently related to unique alloy/environment combinations, most alloys are susceptible to cracking in the presence of a number of environmental species.
- Even in conventionally ductile materials, environment-sensitive cracking results in macroscopically brittle failure. The leading edge of the crack usually advances in steps even under static loading, and small, but variable, amounts of plastic deformation may occur at the advancing edge of the crack. Although stress corrosion cracks may propagate by branching, this is not always observed, and for this reason, scanning electron microscopy of fracture surfaces is generally required to differentiate between statically and cyclically

**Table 7.7 Some environment-alloy combinations known to result in stress-corrosion cracking**

| | Alloy system | | | | | | | | |
|---|---|---|---|---|---|---|---|---|---|
| | | | | | Stainless steels | | | | |
| Environment | Aluminum alloys | Carbon steels | Copper alloys | Nickel alloys | Austenitic | Duplex | Martensitic | Titanium alloys | Zirconium alloys |
| Amines, aqueous | … | • | • | … | … | … | … | … | … |
| Ammonia, anhydrous | … | • | … | … | … | … | … | … | … |
| Ammonia, aqueous | … | … | • | … | … | … | … | … | … |
| Bromine | … | … | … | … | … | … | … | … | • |
| Carbonates, aqueous | … | • | … | … | … | … | … | … | … |
| Carbon monoxide, carbon dioxide, water mixture | … | • | … | … | … | … | … | … | … |
| Chlorides, aqueous | • | … | … | • | • | • | … | … | • |
| Chlorides, concentrated, boiling | … | … | … | • | • | • | … | … | … |
| Chlorides, dry, hot | … | … | … | • | … | … | … | • | … |
| Chlorinated solvents | … | … | … | … | … | … | … | • | • |
| Cyanides, aqueous, acidified | … | • | … | … | … | … | … | … | … |
| Fluorides, aqueous | … | … | … | • | … | … | … | … | … |
| Hydrochloric acid | … | … | … | … | … | … | … | • | … |
| Hydrofluoric acid | … | … | … | • | … | … | … | … | … |
| Hydroxides, aqueous | … | • | … | … | • | • | • | … | … |
| Hydroxides, concentrated, hot | … | … | … | • | • | • | • | … | … |
| Methanol, plus halides | … | … | … | … | … | … | … | • | • |
| Nitrates, aqueous | … | • | • | … | … | … | • | … | … |
| Nitric acid, concentrated | … | … | … | … | … | … | … | … | • |
| Nitric acid, fuming | … | … | … | … | … | … | … | • | … |
| Nitrites, aqueous | … | … | • | … | … | … | … | … | … |
| Nitrogen tetroxide | … | … | … | … | … | … | … | • | … |
| Polythionic acids | … | … | … | • | • | … | … | … | … |
| Steam | … | … | • | … | … | … | … | … | … |
| Sulfides plus chlorides, aqueous | … | … | … | … | • | • | • | … | … |
| Sulfurous acid | … | … | … | … | • | … | … | … | … |
| Water, high-purity, hot | • | … | … | • | … | … | … | … | … |

Source: Ref 107

stressed metals in corrosive environments and brittle fracture in the absence of corrosion.

- Environmental stress cracking may follow transgranular or intergranular paths depending on the metal/environment combination and frequently on the microstructure of an alloy. Intergranular stress-corrosion cracking (IGSCC) is frequently observed under conditions exhibiting intergranular corrosion in the absence of stress. The localized corrosion may then be referred to as stress-assisted intergranular corrosion. Intergranular corrosion also may occur by the penetration of corrosion products along grain boundaries. If these products are sufficiently brittle to crack under stress, allowing access of the environment to the crack front, then repetition of these steps provides the mechanism for intergranular cracking.

Thus, environment-sensitive cracking is related to conditions that are on the borderline between low corrosion rates in the absence of stress and extremely localized attack associated with surface tensile stresses and progressing in the form of cracks. These conditions are most frequently met with active-passive type alloys such as stainless steels, nickel-base alloys, aluminum-base alloys, and plain-carbon and low-alloy steels in higher pH environments capable of forming passive films. Cracking is also observed in copper-base alloys in environments that form "tarnish" films susceptible to cracking under stress, particularly when the tarnish has penetrated grain boundaries. In general, slight changes in environment (frequently associated with changes in oxidation potential of the environment, and hence, its tendency to support cathodic reactions) can lead to either very high stability in the passive potential region of the polarization curve or to destruction of passive films and the establishment of general corrosion. Either of these changes will decrease the susceptibility to localized corrosion. These factors and the problems in defining the position, magnitude, and variations of stress over the metal surface, particularly as a function of time, complicate the prediction of conditions under which stress-sensitive environmental cracking will occur, the design of standardized tests and establishment of research procedures for study of the phenomena, and the development of theories that could act as guides in its control.

## *Evaluation of Susceptibility to Environment-Sensitive Cracking*

Evaluation of susceptibility to environment-sensitive cracking encounters the usual problems of attempting to simulate in the laboratory conditions that reflect service performance. Two approaches are usually taken, both of which must provide a final consistent prediction of service behavior. One of these approaches is to duplicate as closely as possible the physical shape and state of stress (for example, static or cyclic loading) of the material in service and to use environments that are representative as closely as possible to those encountered in service. The other approach is to set up standardized tests, preferably designed on the applicable fundamental theory (the electrochemistry of corrosion and the mechanics of materials), and from such tests develop reliable correlations to service behavior. An advantage of the more fundamental investigations is that they provide greater insight into the phenomena and may lead to very useful short-time screening tests from which the most probable satisfactory materials may be selected.

A wide variety of test conditions, and particularly specimen geometries, have been used in establishing standard tests and in research on environment-sensitive cracking. Representative examples of test specimens are shown in Fig. 7.69 (Ref 109). A number of considerations enter into the choice of test specimen:

- The form and availability of the metal (e.g., availability as sheet, tubes, pipes, etc.)

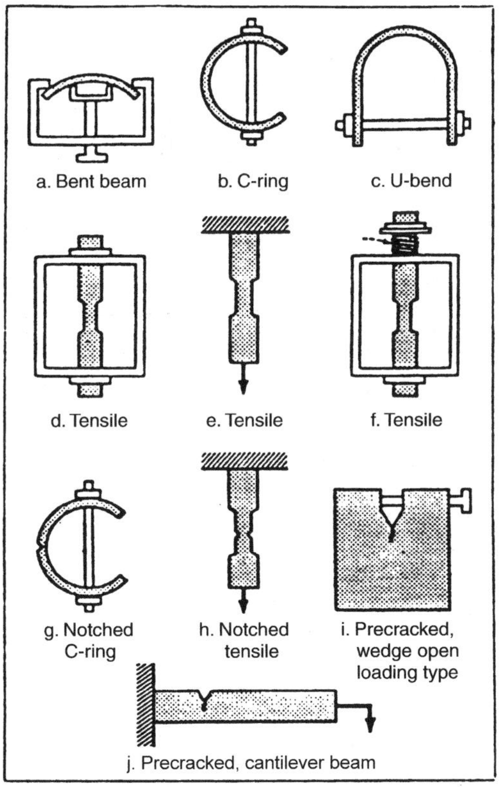


**Fig. 7.69** Types of specimens for investigating stress-corrosion cracking (SCC) and corrosion fatigue. Source: Ref 109

- The type of loading. Represented in the figure are conditions of simple tension, with and without notches, tension in the outer surface of bent specimens, and precracked specimens such as i and j.
- The method of applying the load to give the desired stress. Specimens e, h, and j require external supports and means of application of the load. The other specimens are self stressed, which is an advantage where large numbers of specimens are to be exposed, where tests are to be conducted over long periods of time, and where conditions of temperature and pressure require enclosed systems. Specimens e and h are typically used for slow-strain-rate tests to reveal susceptibility to SCC. Specimens e, h, and j, and rotating bent-beam geometries, are used for cyclical stressing to evaluate susceptibility to corrosion fatigue.
- The method of calculating the stress, both initially and over time. The analytical expressions of strength of materials are used to calculate the maximum stress for specimens a to h. Specimens i and j are two of a number of geometries used for tests conforming to the requirements of fracture mechanics. It is important to establish that the calculated stresses do actually exist and that they do not change with time or that changes with time are known. For example, after cracking starts, the conditions at the propagating edge of the crack (shape and environment) will change, and hence, the local stress will change.
- Corrosion of the bolts and frames must be considered for those geometries using self loading, such as by bolts in b, c, g, and i or by frames in a, d, and f. Corrosion products from the supports can affect the corrosion of the test specimens, and particularly, galvanic coupling between dissimilar metals could shift the corrosion potential of the test specimen from the value that it would have under uncoupled conditions.

## *Scope of Environment-Sensitive Fracture*

The scope of environment-sensitive fracture can be represented by the modified Venn diagram of Fig. 7.70, in which the modes leading to fracture are identified as stress corrosion, hydrogen embrittlement, and corrosion fatigue (Ref 110). All of these modes include a synergism between stress (static and cyclic) and electrochemical reactions at crack tips that provide mechanisms for crack growth. In the limit, stress corrosion refers to static stress at a crack tip that induces environmentally influenced crack opening mechanisms leading to crack growth. The corrosion-related mechanisms include the successive cracking of passive films or salt films with exposure of the substrate metal, which then undergoes active corrosion. Growth also can occur by crack-opening

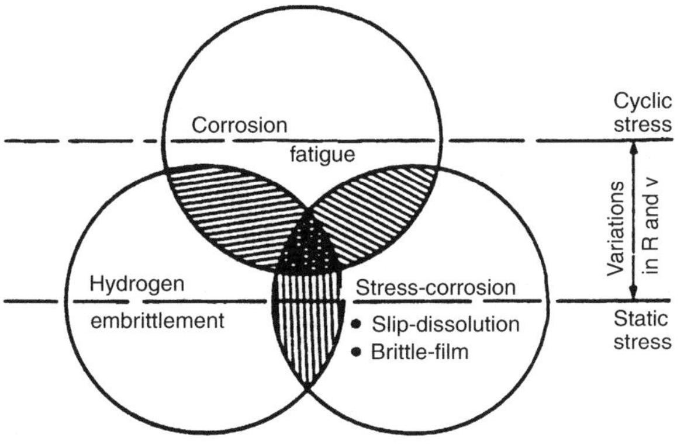


**Fig. 7.70** Venn diagram illustrating the interrelationship between stress corrosion, corrosion fatigue, and hydrogen embrittlement. R, stress ratio; v, strain rate. Source: Ref 110

mechanisms influenced by environments, resulting in a reduction in metal-to-metal bonds at the crack tip.

A number of materials are susceptible to hydrogen embrittlement when the crack tip environment is sufficiently low in pH, and the potential is sufficiently negative to allow reduction of hydrogen ions or water to hydrogen. As discussed later, both adsorbed and absorbed hydrogen atoms at the crack tip are involved in mechanisms for crack growth under stress and ultimate failure. Under repeated loading, fatigue crack growth rates are enhanced by mechanisms ranging from adsorption of species from the environment to contributions directly related to static stress-corrosion cracking and hydrogen embrittlement.

The dominant mode of failure depends upon the environment and material, and with some materials, both the composition and thermal treatment may be critical variables. As a result, a spectrum of modes is observed in which at one end the dominant factor is corrosion and at the other end, the state of stress (Ref 111). Corrosion appears to be dominant, for example, with carbon steels in nitrate solutions and certain aluminum alloys in the presence of chlorides. In these cases, preexisting active paths such as pits and grain-boundary attack initiate cracking, and crack growth is associated with active corrosion. Examples of strain-generated active corrosion sites, related to film rupture at crack tips causing crack growth, include brass in ammonia and austenitic stainless steels in chlorides. In contrast, the state of stress is the dominant factor for cracking mechanisms associated with hydrogen embrittlement. Although a number of material/environment systems are susceptible to hydrogen embrittlement, the dominant examples are associated with high-strength steels in water with and without chlorides.

The major concern in environment-sensitive cracking is crack-propagation rate since this is almost always the controlling factor in time to failure. Therefore, it is important to identify mechanisms involving both corrosion and state of stress at the crack tip that control its growth rate. Because of the restricted geometry of the crack tip, direct experimental evidence is difficult to obtain as to how these two factors, corrosion and stress, interact at the crack tip. Contributions to understanding the corrosion component have come from two types of investigations. Studies of pitting and crevice corrosion, in which variables establishing the electrolyte composition, for example, pH and electrochemical potential in these occluded regions, have an obvious relationship to the occluded region of a stress corrosion crack. A contribution has been made to the influence of stress (and associated strain) on these corrosion mechanisms by subjecting smooth surfaces to various stress and strain histories to determine the influence of environment, electrochemical potential, and strain rate on ductility as a measure of susceptibility to environmental embrittlement. It is then inferred that similar effects apply at a crack tip during environment-sensitive cracking. Results of these types of investigations are covered in the next section.

A significant contribution to understanding the stress component of environment-sensitive cracking has come from the application of the concepts of fracture mechanics. Specifically, fracture mechanics provides information on the state of stress at a crack tip in terms of variables, including the geometry of the crack, its size and position relative to the structure in which it occurs, the magnitude of stress, the maximum and minimum stress when cyclic, and the time profile of application of stress (constant, increasing to produce constant strain rate, or cyclic). Some of these variables are identified in Fig. 7.70. A brief review of fracture mechanics and the significance of these variables is discussed in a following section. Fracture mechanics has contributed significantly to the understanding of environment-sensitive cracking by providing insight into cracking mechanisms and into the design of components where environmental influences are a factor in performance, in particular, in the determination of whether existing surface defects in stressed structures will grow due to corrosion, eventually resulting in failure by brittle crack propagation. In piping and tank systems, fracture mechanics has contributed to the prediction of failure by leaking rather than by rupture where surface cracks are growing under corroding conditions.

## *Material/Environment Variables Affecting Crack Initiation and Growth*

**Relationship of Potential to Environment-Sensitive Cracking.** Figure 7.71 is a schematic representation of the potentiodynamic polarization curve of a metal with potential regions of active corrosion, passivity, and pitting identified, all related to the anodic polarization behavior. The lower or cathodic section is representative of conditions under which hydrogen would be produced by reduction of hydrogen ions or water. The hatched regions are representative of those potential ranges generally associated with environment cracking. The lower (I) of these is obviously associated with potential ranges of hydrogen formation where cracking is predicted to occur by mechanisms of hydrogen embrittlement. The two upper ranges of susceptibility span potentials associated with instability of protective passive films. This occurs in the vicinity of and just above the anodic-peak current density of the anodic polarization curve, potential region II. In this potential range, stress-induced crystallographic slip can produce surface offsets that crack the passive film and expose the substrate surface, which then undergoes rapid local active corrosion (crack propagation). In the potential region III just below the critical pitting potential, stress-initiated cracks in the passive film lead to rapid local corrosion at rates related to those observed in pitting.

A limitation to associating stress-cracking tendencies with potential regions as represented by Fig. 7.71 is that the ranges indicated relate to potentials measured at the surface under either freely corroding conditions or potentials established by a potentiostat or other external

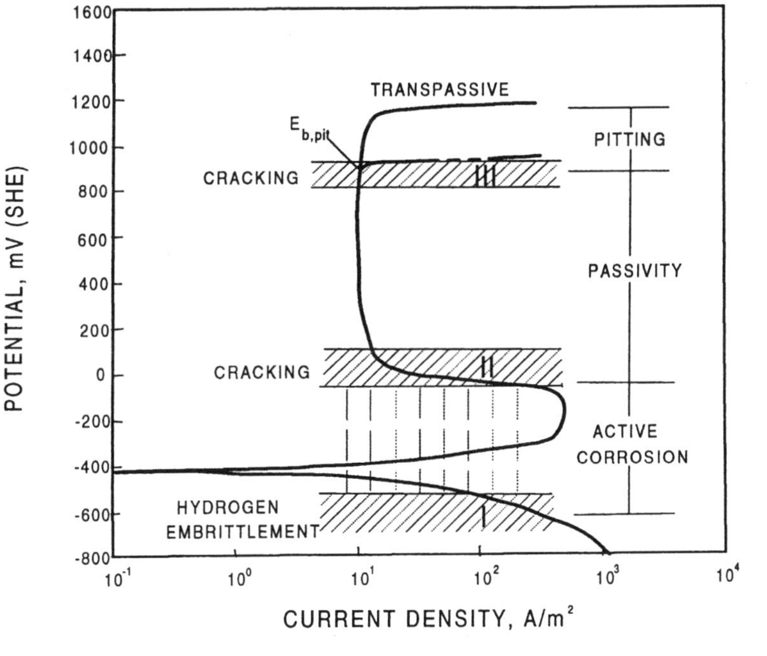


**Fig. 7.71** Potential ranges of stress-corrosion cracking by (I) hydrogen embrittlement, (II) cracking of unstable passive film, and (III) cracking initiated by pits near the pitting potential. Vertical dashed lines define potential range over which nonpassivating type films may crack under stress.

sources. Crack propagation, however, is influenced by the potential at the leading edge of the crack, and this may differ significantly from the externally measured potential due to the IR potential drop into the crack. This is particularly true when the potential drop leads to conditions for hydrogen embrittlement. The factors are similar to those discussed in the section "An Analysis of Pitting Corrosion in Terms of IR Potential Changes in Occluded Regions and Relationship to Polarization Curves," relating potential drops to pit initiation and propagation.

A schematic representation of surface profiles corresponding to the mode of attack at increasing potentials in relationship to environmental cracking is shown in Fig. 7.72. At the lowest potential (Fig. 7.72a), hydrogen embrittlement is associated with crack propagation from surfaces undergoing little or no general corrosion because of the low potential. These potentials are in the range of cathodic protection, and in fact, hydrogen embrittlement may occur while systems are under cathodic protection. At slightly higher potentials (Fig. 7.72b), but below the anodic-peak potential of active-passive type alloys, active general corrosion occurs, resulting in an uneven surface but without cracking. At potentials just above the anodic peak potential (Fig. 7.72c) (identified as potential region II in Fig. 7.71), and in some cases, in the extended lower potential ranges, deformation by slip produces in the surface an offset that cracks the passive film and exposes clean surface as illustrated in Fig. 7.73. It is evident that whether the exposed region actively corrodes and initiates a crevice, which then propagates as a crack, or repassivates blocking propagation depends on the relative rates of the two processes. In particular, which actually occurs depends on the po-

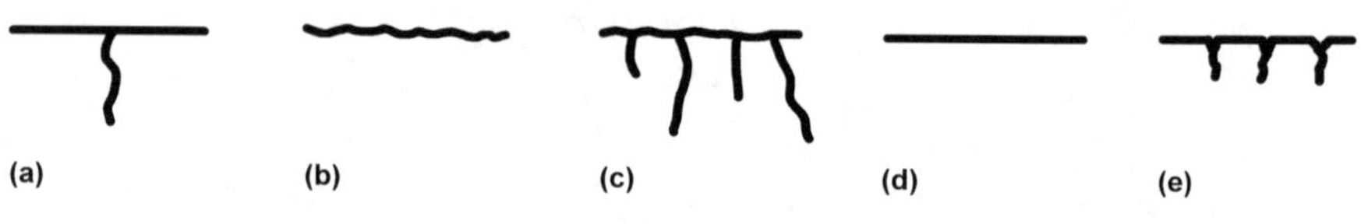


**Fig. 7.72** Schematic representation of stress induced surface profiles representative of the potential ranges identified in Fig. 7.71. (a) Hydrogen embrittlement. (b) Active corrosion. (c) Passive film cracking. (d) Passivity. (e) Pit-initiated cracking

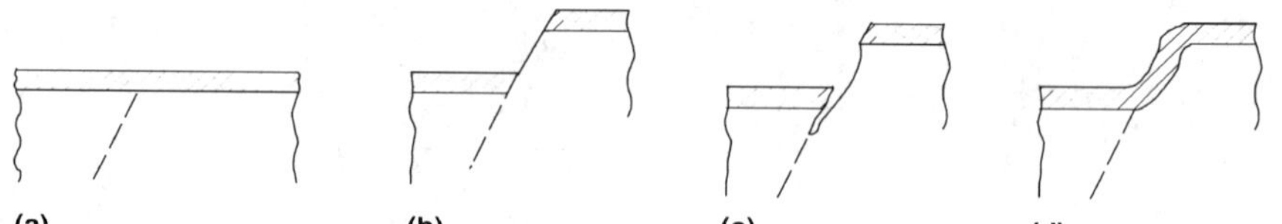


**Fig. 7.73** Schematic representation of (a) passive film, (b) passive film rupture by stress-induced slip resulting in exposure of bare substrate, (c) crack initiation by anodic dissolution initiating crevice corrosion conditions before repassivation of exposed substrate, and (d) repassivation of exposed substrate before crack initiation.

tential, with tendency to repassivation increasing with increasing potential. As discussed subsequently, it also depends on the strain rate, which governs the rate at which the exposed regions are produced and the time allowed for repassivation. On further increasing the potential within the passive-potential range (Fig. 7.72d), the patterns of SCC may change, and cracking may not exist when stable highly protective passive films can rapidly form. At higher potentials (Fig. 7.72e), pitting may occur with initiation of stress corrosion cracks from the bases of the pits.

The potential scan rate at which a potentiodynamic polarization curve has been determined may be a significant variable in the identification of potential ranges over which SCC can be anticipated. The effect of scan rate on the anodic polarization curve of an active-passive type alloy is shown schematically in Fig. 7.74. With a slow scan rate, the time is sufficient to form a stable passive film at the lowest potential in the passive range; thus, the anodic peak occurs over a relatively small potential range. At a fast scan rate, there is less time for formation of the film, and complete passivity is attained at a higher potential; thus, the anodic peak occurs over a wider potential range. As a consequence, if a tensile strain rate is applied that can crack a preexisting passive film, and the potential is in the range identified by SCC in Fig. 7.74, then the exposed substrate will corrode at the high current density indicated for the fast-scan curve, repassivation may never occur locally, and a stress corrosion crack propagates. An example of this scan-rate effect is shown for a carbon steel in boiling 35% NaOH in Fig. 7.75 (Ref 112). The predicted potential range for SCC is shown. There is a larger effect of scan rate for carbon steel in boiling 4 N $NaNO_3$ as shown in Fig. 7.76, in which the relatively narrow potential range of the anodic peak becomes a range of about 1800 mV, and susceptibility to SCC exists over this very wide range of potentials (Ref 112).

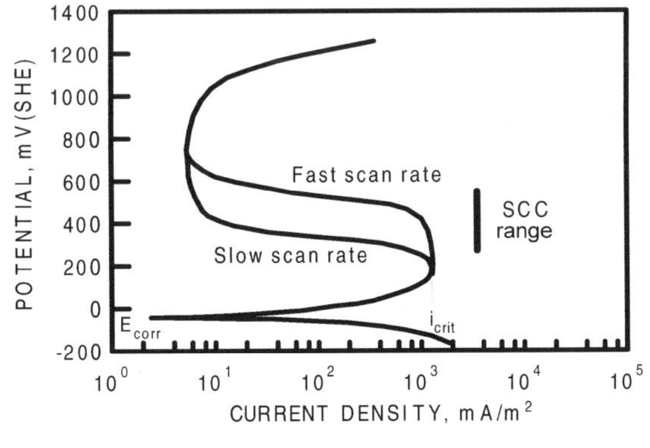


**Fig. 7.74** Schematic representation of the effect of scan rate on potentiodynamic polarization curve of an active-passive type alloy and the range of potentials of predicted stress corrosion cracking

Potential ranges of susceptibility to SCC also have been identified by determining the polarization curve during rapid straining (e.g., at strain rates of the order of $10^{-2}$ $s^{-1}$). A shift of the polarization curve to larger

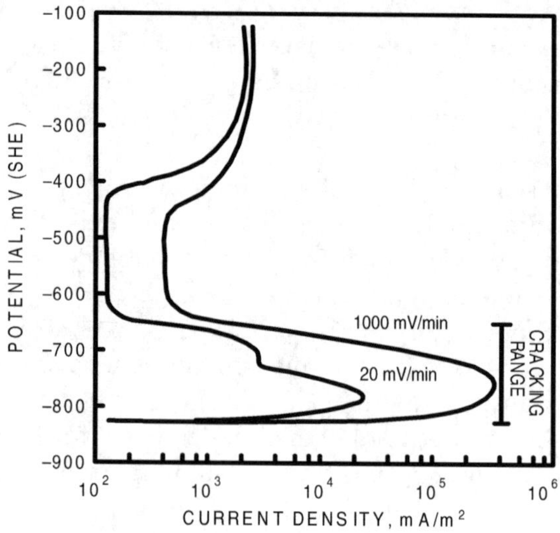


**Fig. 7.75** Potentiodynamic polarization curves at two scan rates for carbon steel in boiling 35% NaOH and potential range of cracking. Redrawn from Ref 112

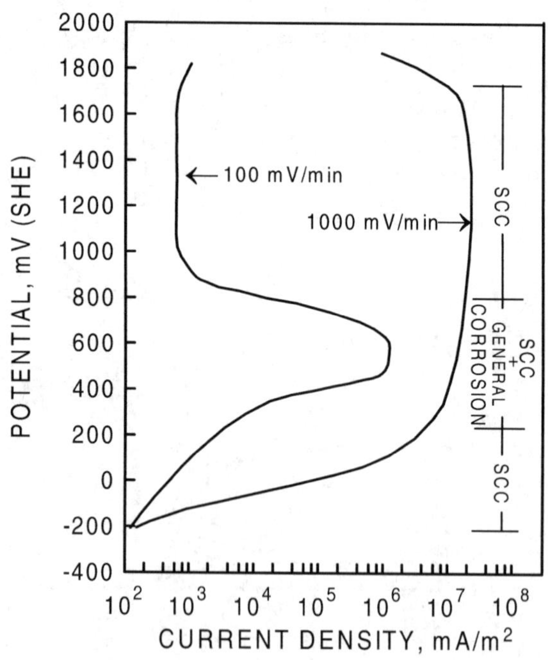


**Fig. 7.76** Potentiodynamic polarization curves at two scan rates for carbon steel in boiling 4 N $NaNO_3$. Form of corrosion in different potential ranges identified. Redrawn from Ref 112

current densities is attributed to exposure of bare metal at strain-induced cracks in surface films. Criteria to identify the potential ranges of susceptibility are based on comparing the current densities with and without straining. One criterion depends on a procedure to calculate, based on the amount of strain, the fraction of the surface that is bare. The current density on the bare metal is then calculated from which an average propagation rate is determined. The ratio of the current density on the bare surface to that on the filmed surface is then calculated as a function of potential. Potentials at which this ratio is greater than ten and the calculated growth rate is greater than $10^{-10}$ to $10^{-9}$ $ms^{-1}$ indicate susceptibility to SCC due to rapid preferential corrosion at the bare sites (Ref 113). Other criteria include exceeding a critical ratio of the change in current density on straining to the current density in the absence of straining, and to observing potential ranges in which this ratio is rapidly changing (Ref 114). (Crack growth rates $<10^{-10}$ $ms^{-1}$ are considered to be insignificant from a practical standpoint.)

Ranges of pH and potential associated with SCC of carbon steels in several environments are shown in Fig. 7.77 in relationship to the Pourbaix diagram for the iron/water system (Ref 115). It is noted that each of the regions of susceptibility span conditions for existence of a stable oxide ($Fe_2O_3$ and $Fe_3O_4$) and aqueous environments containing corrosion product ions ($Fe^{2+}$, $Fe^{3+}$, and $HFeO_2^-$). Susceptibility to SCC is, therefore, a consequence of stresses cracking passive films and exposing the substrate to active corrosion, accompanied by changes in the

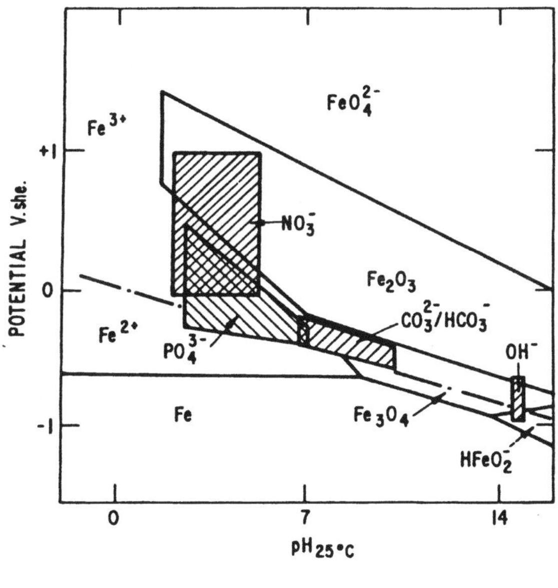


**Fig. 7.77** Relationship between pH/potential conditions for severe cracking susceptibility of mild steel in various environments and the stability region for solid and dissolved species on the potential-pH diagram. Source: Ref 115

environment in the crack that sustains crack propagation at the crack tip. The sharpness of the crack is enhanced by formation of passive films on the crack walls restricting widening by active corrosion.

An example of the relationship between environment and potential on SCC of a pipeline steel as it influences the time to failure ratio is shown in Fig. 7.78 (Ref 68). The environments are solutions of hydroxide, carbonate-bicarbonate, and nitrate; the effect of SCC is represented as the ratio of failure time in the environment to failure time in inert oil; the strain rate was constant at $2.5 \times 10^{-6}$ s$^{-1}$. The cracking range for the hydroxide environment is about 200 mV with the greatest effect at –700 mV (SCE); the cracking range for the carbonate-bicarbonate is 150 mV with shortest cracking time at –400 mV (SHE). In nitrate solutions, the cracking range is much broader, extending from –50 to 1300 mV (SHE). It also is evident that the effect of the nitrate environment is much greater than for the other two, the time for failure being reduced to about 1% of that in inert oil near 550 mV (SHE). The corrosion potential for the pipeline steel in each of the environments is indicated and allows the failure-time ratio to be determined for these freely corroding conditions. Since environmental variables such as dissolved oxygen and traces of other oxidizing agents can influence the corrosion potential, reference to curves of the form shown in Fig. 7.78 allows prediction of how changes in the corrosion potential will affect failure time.

Figure 7.79 illustrates the influence of alloy composition on the potential dependence of the failure-time ratio of carbon steel in a carbonate-bicarbonate environment (Ref 116, 117). The failure-time ratio is

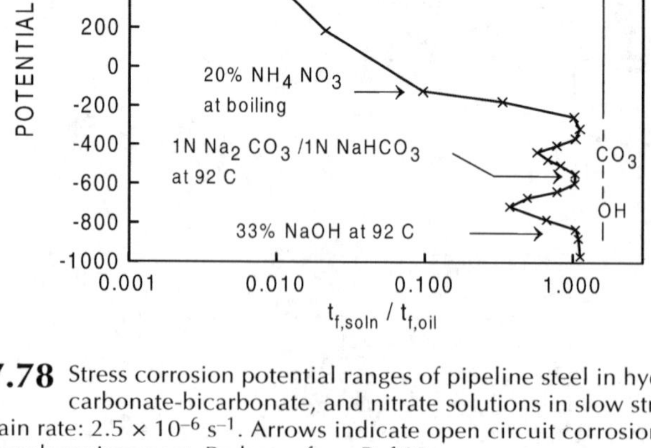


**Fig. 7.78** Stress corrosion potential ranges of pipeline steel in hydroxide, carbonate-bicarbonate, and nitrate solutions in slow strain-rate test. Strain rate: $2.5 \times 10^{-6}$ s$^{-1}$. Arrows indicate open circuit corrosion potentials for each environment. Redrawn from Ref 68

increased by addition of chromium, nickel, or molybdenum with the latter steel showing the least susceptibility to SCC. However, the relative influences of these alloy additions can be sensitive to the environment as shown in Fig. 7.80, in which the molybdenum-containing steel has the lowest failure-time ratio, and the susceptibility of this steel occurs over a much wider potential range (Ref 116, 117). Figures 7.79 and 7.80 also illustrate the effect of environment on the potential for minimum failure-time ratio. The minima occur near –440 mV (SHE) in the carbonate-bicarbonate environment and near –760 mV (SHE) in the NaOH environment. This shift is consistent with the shift in the potential of the

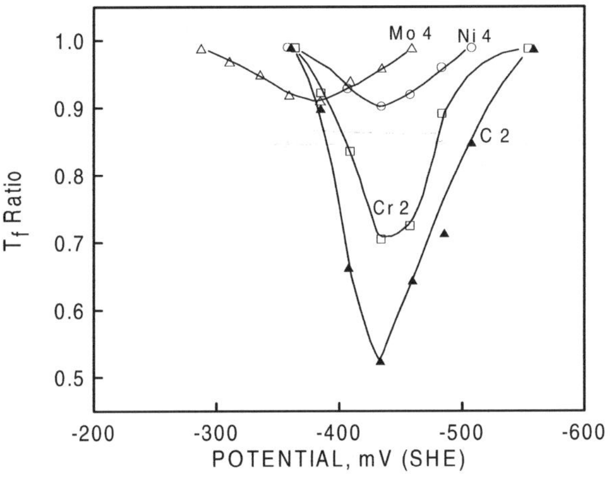


**Fig. 7.79** Effects of applied potential upon time-to-failure ratio in slow strain rate tests of low-alloy ferritic steels in 1 N $Na_2CO_3$ + 1 N $NaHCO_3$ at 75 °C. C 2 (0.27% C carbon steel), Cr 2 (0.09% C, 1.75% Cr), Ni 4 (0.09% C, 6.05% Ni), and Mo 4 (0.10% C, 5.00% Mo). Redrawn from Ref 116, 117

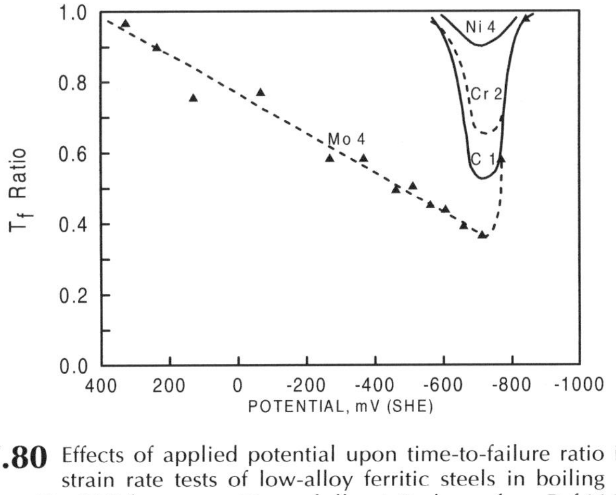


**Fig. 7.80** Effects of applied potential upon time-to-failure ratio in slow strain rate tests of low-alloy ferritic steels in boiling 8.75 N NaOH (see Fig. 7.79 for compositions of alloys). Redrawn from Ref 116, 117

anodic peak in the polarization curve when the pH is higher (i.e., the anodic peak is at a lower potential in the NaOH environment).

**Relationship of Strain Rate to Environment-Sensitive Cracking.** Measurements of ductility under tensile loading over a wide range of strain rates can provide significant information on a material's tendency for stress cracking. The ductility can be evaluated in terms of elongation at fracture, reduction in area at fracture, or time to failure. The ductility is expressed either in absolute values or as a ratio of the value in a corrosive environment to that in an inert environment. Additional variables are the environment, electrochemical potential, variations in material composition and treatment, and temperature. The form of the strain-rate dependence of the ductility is different for stress-corrosion versus hydrogen-embrittlement cracking as illustrated in Fig. 7.81 (Ref 118). When the failure mechanism is SCC, there is a range of strain rates over which there is a decrease in ductility. In this range, it is proposed that the crack advances by a critical sequence of successive steps of passive-film rupture by emerging dislocations, local active corrosion at the exposed offset, and repassivation at the offset. The ductility remains high at slower strain rates where there is sufficient time for the exposed substrate to repassivate, thus blocking crack growth. At higher strain rates, the ductility remains high because deformation is occurring so rapidly that the corrosive environment does not have time to influence the deformation process.

The relationship between ductility and strain rate under conditions conducive to hydrogen embrittlement is also shown schematically in Fig. 7.81. Under these conditions, the controlling factor is the absorption of hydrogen resulting from the reduction of hydrogen ions. The slower the strain rate, the longer the time for absorption of hydrogen,

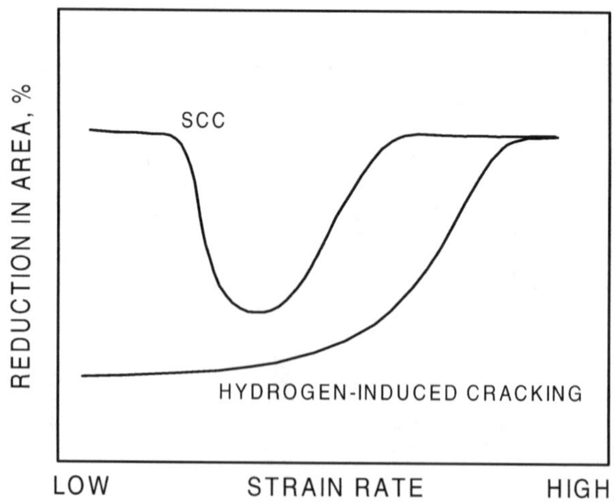


**Fig. 7.81** Schematic representation of the effect of strain rate on SCC and hydrogen-induced cracking. Redrawn from Ref 118

and therefore, the lowest ductility occurs at the slowest strain rate. As the strain rate increases, the ductility progressively increases due to the decreased time for hydrogen absorption until the time for hydrogen embrittlement is negligible, beyond which the ductility is high and not influenced by the environment.

Strain-rate dependence of ductility of the form shown in Fig. 7.81 is presented in Fig. 7.82 for a carbon steel in a carbonate-bicarbonate environment (Ref 119). The ductility is represented as the ratio of the reduction in area (RA) in the environment relative to the value in inert oil. The tests were conducted at the indicated constant potentials and illustrate that the strain-rate dependence can be sensitive to the potential, particularly the minimum ductility and the strain rate at which the minimum occurs. It follows, as an illustration, that if small changes in the environment, such as dissolved oxygen, shift the potential from –720 to –680 mV (SHE), significant changes in susceptibility to SCC would be predicted.

High-strength AISI 4340 steel is representative of a material susceptible to hydrogen embrittlement (lower schematic curve in Fig. 7.81). Results of tests on this alloy in artificial seawater are shown in Fig. 7.83 (Ref 120). The ductility, expressed as reduction in area at fracture, increases progressively with increased strain rate until values are reached equal to those observed in air. The strain-rate dependence, however, depends on the electrochemical potential maintained during the straining. At the lower potential (–1000 mV (SCE) or –760 mV (SHE)), the embrittlement is greater (10 versus 16% RA) and persists to higher strain rates due to the greater rate of hydrogen evolution at the lower potential.

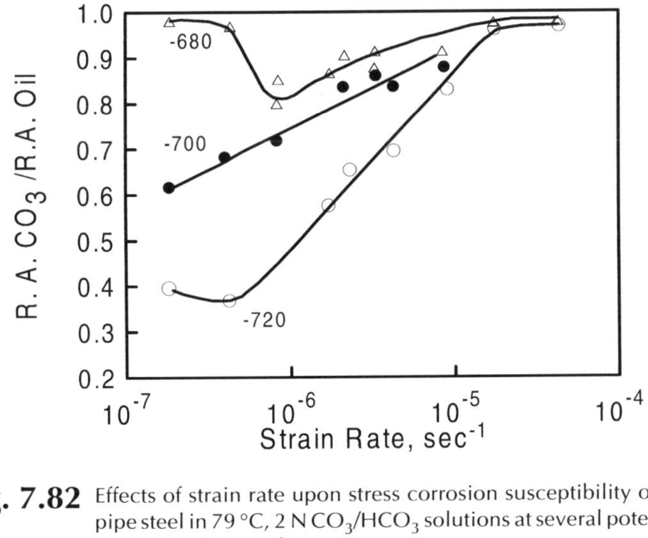


**Fig. 7.82** Effects of strain rate upon stress corrosion susceptibility of line pipe steel in 79 °C, 2 N $CO_3/HCO_3$ solutions at several potentials relative to SHE. Redrawn from Ref 119

Aluminum alloys, particularly the high-strength compositions, are susceptible to environmental cracking, both in aqueous environments and in air as a function of relative humidity. This susceptibility is particularly sensitive to alloy composition and thermal treatment, which is shown by differences in the dependence of ductility on strain rate. Understanding these differences can contribute to identification of mechanisms of the strain-rate sensitivity. A summary of the influence of strain rate on the ductility of 2000-, 5000-, and 7000-series aluminum alloys in environments represented by 3% NaCl + 0.3% $H_2O_2$ is shown in Fig. 7.84 (Ref 121). The 7000 series shows susceptibility to hydrogen embrittlement at strain rates below $10^{-5}$ to $10^{-6}$ $s^{-1}$. Although there is

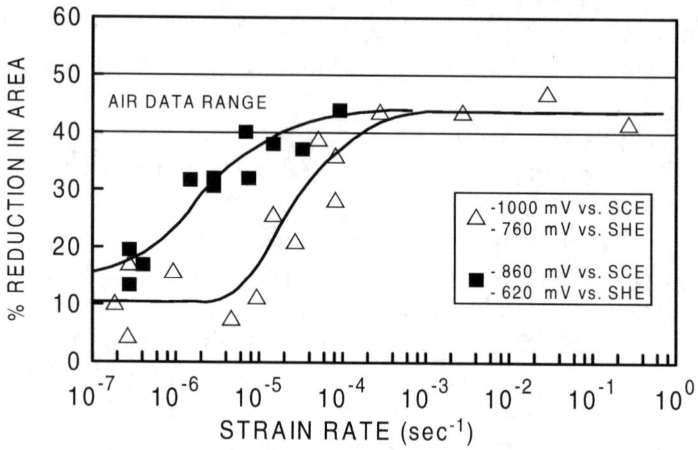


**Fig. 7.83** Relationship between strain rate and ductility for AISI 4340 steel in ASTM artificial ocean water at two cathodic polarization potentials. Redrawn from Ref 120

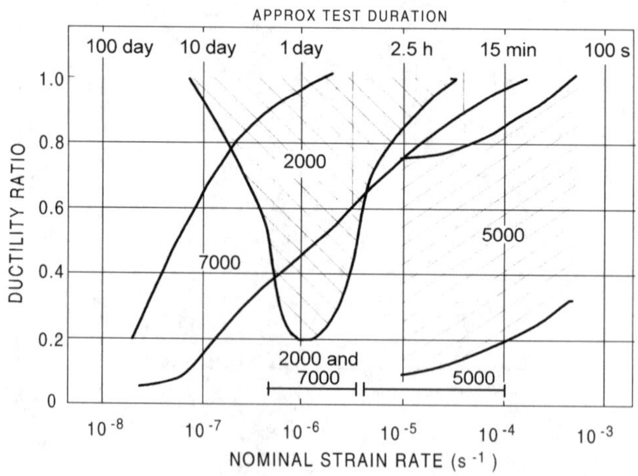


**Fig. 7.84** Strain-rate regimes for studying SCC of 2000-, 5000-, and 7000-series aluminum alloys. Source: Ref 121

some uncertainty about the embrittling mechanism in the 2000 series, the results summarized in the figure indicate that crack propagation is controlled by rates of strain-induced passive-film fracture allowing rapid corrosion of the exposed substrate metal relative to rates of repassivation. The results of the 5000 series indicate susceptibility to loss in ductility by cracking at relatively high strain rates. The mechanism of cracking is uncertain since measurements were not reported for lower strain rates to establish whether a minimum occurs in the scatter band or that the band continues to decrease, indicating a predominant hydrogen-embrittlement mechanism. Test-duration times identified at the top of the figure illustrate that very long periods are required to investigate the strain-rate dependence of the ductility at very low strain rates.

**Relationship of Composition and Heat Treatment to Environment-Sensitive Cracking of Low-Alloy and High-Strength Steels.** This section is an overview of the environment-sensitive stress cracking of nonstainless types of steels. These include the carbon and low-alloy steels that are not heat treated by quenching and tempering, the frequently called high-strength steels, which consist of both low- and higher-alloy quenched and tempered steels, and the higher-alloyed precipitation-hardenable and maraging steels. Several of these steels are also strengthened by cold working, which may have an effect on susceptibility to environment cracking. These steels range in yield point from <50 ksi to >350 ksi.

Representative environments for which SCC has been reported in carbon steels are included in Table 7.7. The sensitivity of these steels to changes in composition and environment are illustrated by the effects of potential in Fig. 7.78 to 7.80 and by the slow strain-rate data of Fig. 7.82 and 7.83. These data support the conclusion that environment cracking is related to the susceptibility of the passive films to crack under stress, to the subsequent crack growth due to anodic dissolution and/or hydrogen embrittlement during the period of exposure of the alloy substrate, and to rates of repassivation of the exposed areas. Actual crack-front growth mechanisms are discussed in some detail in a later section.

Stress-corrosion cracking of steels tends to be intergranular at the lower-strength levels, with crack growth primarily dominated by corrosion processes of anodic dissolution at the crack tip (Ref 122). Cracking tends to be transgranular at the higher-strength levels with growth dominated by stress accompanied by hydrogen-embrittlement mechanisms. There is a gradual transition from one mechanism to the other as summarized in Table 7.8 (Ref 122). It is significant to find that potential, strain rate, and in particular, yield strength, are generally more important variables than composition, thermal treatment, or microstructure for these steels. This distinction is not too clear because the latter three variables determine the yield strength. Nevertheless, useful correlations have been developed between environment cracking tendencies and yield strength.

Many investigations of SCC in terms of time-to-failure for a large number of carbon and alloy steels in chloride solutions have indicated that the susceptibility is very low in steels with yield strengths below 160 ksi; rather, general corrosion occurs. Susceptibility may be observed up to 180 ksi, and then increases rapidly in the range 180 to 210 ksi as shown in Fig. 7.85 (Ref 123). In the latter strength range, failure time becomes sensitive to the particular steel and its heat treatment. At yield strengths above approximately 200 ksi (Fig. 7.86) (Ref 123), the

**Table 7.8 Gradual transition from one mechanism of failure to another, intergranular corrosion to brittle fracture**

| | | | |
|---|---|---|---|
| | | **Intergranular corrosion** | |
| Corrosion dominated | Stress-assisted intergranular corrosion | Dissolution controlled intergranular fracture along preexisting active paths | Steel in $NH_4NO_3$<br>Steel in $NaNO_3$<br>Steel in NaOH<br>Steel in $Na_2CO_3$ + $NaHCO_3$ |
| (Solution specificity)<br>↑<br>↓ | Slip-step dissolution | Transgranular fracture along strain-generated active paths | C steel in $CO_3$-$HCO_3$ (higher strain rates)<br>Ni steels in $MgCl_2$, C steel in CO-$CO_2$-$H_2O$<br>C and low alloy steels in liquid $NH_3$<br>Ti steel in $CO_3$-$HCO_3$—high stresses, slow strain rate tests |
| (Solution not specific)<br>Stress dominated | Surface energy lowering | Mixed crack path by H adsorption at subcritically stressed sites | C steel in OH or $CO_3$-$HCO_3$, low strain rate tests, low potential<br>Medium-strength steel in OH, $CO_3$, acetates, etc., low potential<br>High-strength steel in $H_2O$,$Cl^-$ |
| | | **Brittle fracture** | |

Source: Ref 122

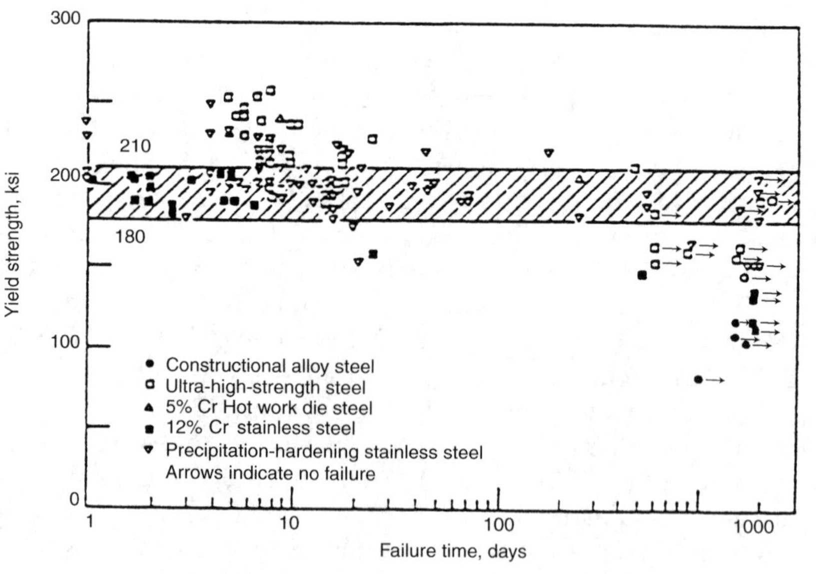


**Fig. 7.85** Stress-corrosion behavior of steels exposed to marine atmospheres at 75% of the yield strength. Source: Ref 123

failure time becomes very sensitive to the steel; in each case, the failure time decreases rapidly with increase in yield strength. It should be noted that the data in these figures relate to a wide range of steels in terms of composition and heat treatment. Representative effects of specific alloying elements on stress-corrosion resistance of alloy steels that can be heat treated to yield strengths up to about 200 ksi are shown in Table 7.9 (Ref 124).

The microstructures associated with the steels in Fig. 7.85 and 7.86 (Ref 123) vary from tempered martensite and bainite for the low-alloy steel to various dispersions of precipitated phases in the other alloys. The respective strength levels are given in Fig. 7.85 and 7.86. In general, failure time correlates to the strength of the steel with the major role of the microstructure being to control the strength rather than to influence the cracking mechanism. Although crack propagation in steels of lower strength in chloride environments may occur by active-path

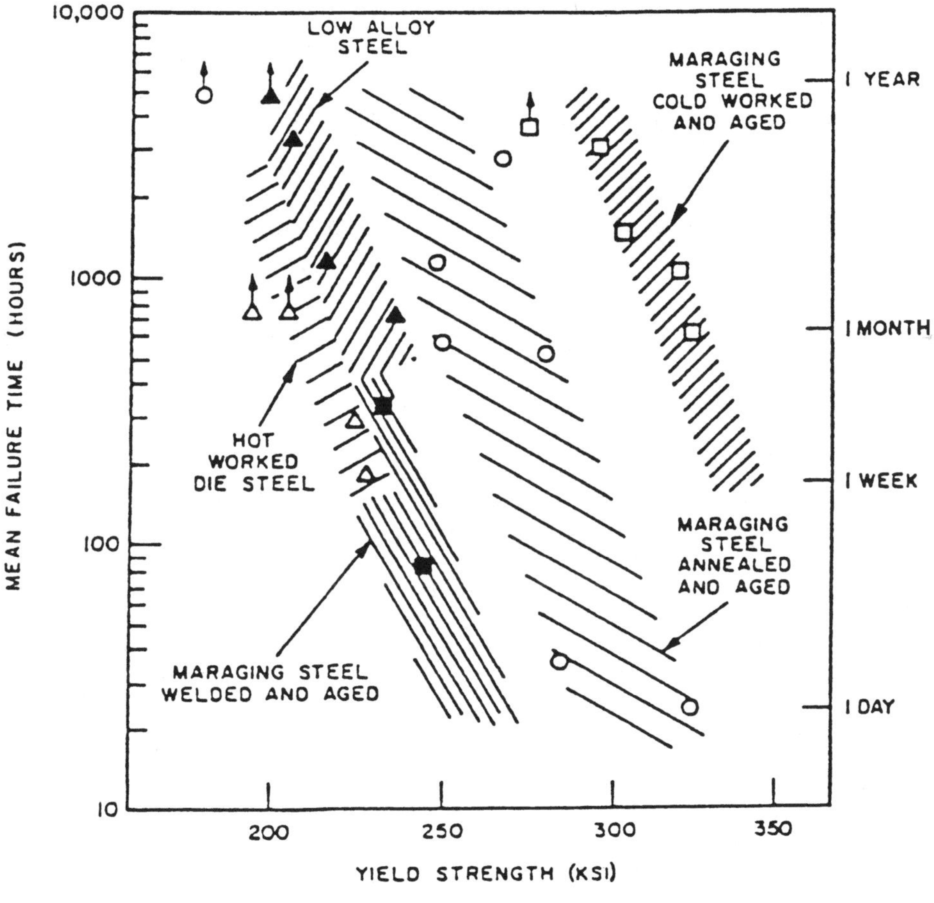


**Fig. 7.86** Relationship between yield strength and mean failure time for high-strength steels exposed as bent-beam tests in distilled water. Specimens were exposed at stress of 75% of the yield strength. Source: Ref 123

anodic dissolution, there is general agreement that the cracking mechanism of the high-strength steels, produced by heat treatment or cold working, is hydrogen embrittlement. This mechanism, particularly under cyclic loading, is enhanced by the presence of sulfide ions, which tend to inhibit hydrogen-atom recombination on the metal surface and thereby increase hydrogen absorption at a crack tip. As a consequence, carbon and low-alloy steels with yield points below 100 ksi have experienced SCC in hydrogen sulfide environments (Ref 122, 124). In chloride environments, the cracking susceptibility is essentially insensitive to pH in the range pH = 2 to 9; as would be expected, the susceptibility increases at lower pH and decreases in the range pH = 9 to 13.5.

In contrast to SCC of carbon and low-alloy steels in chloride, sulfide, and sulfuric acid environments by hydrogen-embrittlement mechanisms, cracking in several environments is attributed to passive-film cracking and/or active-corrosion-path anodic-dissolution penetration mechanisms (Ref 124). These environments include nitrates, hydroxides, ammonia, carbon-dioxide/carbonate solutions, and aqueous carbon-monoxide/carbon-dioxide. Nitrate-bearing solutions are encountered in coal distillation and fertilizer plants; hydroxide solutions in the production of NaOH and in crevices of steam boilers; and ammonia cracking has occurred in tanks and distribution systems for agricultural ammonia applications.

In nitrates, cracking of low-carbon steels occurs along preexisting active corrosion paths associated with ferrite grain boundaries (Ref 125). Although several impurities are known to segregate in these boundaries, correlations have been made with essentially continuous films of iron carbide or segregated carbon. Maximum susceptibility occurs in the range of 0.005% C; it is proposed that lower carbon contents do not

**Table 7.9 Effect of alloying elements on stress-corrosion resistance**

| Element | Base alloy: AISI 4340 0.4C-1.7Ni-0.7 Cr | AISI 4120 0.2C-1Cr-0.3 Mo | HY 150 0.12C-5Ni-0.5Cr-0.6Mo-0.25Mn-0.1V |
|---|---|---|---|
| C | Decrease (0.2–0.4) | Decrease | |
| Mn | Decrease (0–5) | No effect | No effect (0.25–0.75) |
| Ni | No effect (0–9) | Increase | Slight effect (4.5–6.5) |
| Cr | No effect (0–12) | Increase | Decrease (0.6–2.0) |
| Mo | No effect (0–2) | Increase | Decrease (0.6–1.0) |
| V | | Increase | Slight effect (0.007–0.014) |
| Nb | ... | Increase | ... |
| Ti | ... | Increase | ... |
| Zr | ... | Increase | ... |
| B | ... | No effect | ... |
| Cu | ... | No effect | ... |
| Si | ... | No effect | ... |
| S | No effect (0.004–0.024) | Beneficial | ... |
| P | No effect (0.002–0.027) | Decrease | ... |
| O | ... | Decrease | ... |
| N | ... | Decrease | No effect (0.007–0.015) |

Note: Ranges of alloy contents (wt%) evaluated are shown in parentheses. Values were not quoted for the 4140 steel but were reported to be within ranges conventionally used for low-alloy steels. Source: Ref 124

provide sufficient carbon to create an active path, and at higher carbon contents the carbon is present in pearlite. Cracking occurs predominately by an electrochemical mechanism at the carbide/ferrite interface with the carbide supporting the cathodic reaction, although direct attack on the carbide has been reported. Prior cold working reduces susceptibility to cracking presumably by mechanically breaking and redistributing the grain-boundary segregation. Figure 7.76 indicates that passive film formation is associated with cracking. How this is related to crack initiation and propagation at ferrite grain boundaries is uncertain. It is accepted that cracking is largely intergranular, which is consistent with an active path mechanism, although transgranular cracking has been reported. Cracking tendency increases with increasing nitrate concentration and is greater for ammonium nitrate and least for sodium nitrate solutions. This difference correlates with the lower pH of the ammonium nitrate solution. Time to failure decreases with increase in temperature (Ref 124).

**Relationship of Composition and Heat Treatment to Environment-Sensitive Cracking of Stainless Steels.** Depending largely on composition, the stainless steels are classed as austenitic (AISI 300 series, fcc), ferritic (AISI 400 series, bcc), duplex (austenite plus ferrite), martensitic, or precipitation hardening. The approximate composition ranges of each of these classes of stainless steels are given in Table 7.10. A representative list of environments in which the austenitic stainless steels have been observed to crack under stress is included in Table 7.7. As with pitting and crevice corrosion, environments containing chloride ions are the most frequent contributors to stress-environment cracking, although the susceptibility may be greater in other environments (e.g., the austenitic stainless steels in polythionic acids) (Ref 126–128).

Two generalizations are frequently made with respect to the cracking response of the stainless steels in chloride-bearing environments. One is that the ferritic stainless steels are immune to cracking relative to the austenitic alloys. Although the cracking tendency is much lower, cracking of ferritic stainless steels has been encountered when chlorides are present. This tendency has been reduced with the development of ferrit-

**Table 7.10 Approximate composition ranges of major classes of stainless steels**

| Type | %Cr | %Ni | %C | Other |
|---|---|---|---|---|
| Austenitic(a) | 16–26 | 6–37 | 0.03–0.25 | Mn, Si, Mo, Ti, Nb, N |
| Ferritic | 11–29 | | 0.005–0.20 | Mn, Si, Mo, Ti, Nb, N |
| Duplex | 18–28 | 4–6 | 0.02–0.10 | Mn, Si, Mo, Ti |
| Martensitic | 11–18 | | 0.15–1.20 | Mn, Si, Mo, W, V |
| Precipitation hardening | 10–18 | 4–25 | 0.03–0.30 | Mn, Si, Mo, Cu, Ti, N, Al, Ta |

There is also a large group of austenitic alloys with compositions ranging to 100% Ni, 50% Cr, 16% Mo, and controlled amounts of Nb, Cu, Ti, and W.

ic alloys having very low concentrations of carbon and small, controlled amounts of Mo, Ni, and Cu (Ref 129).

A second generalization relating to chloride cracking of austenitic alloys is that it does not occur at stresses below one-half the yield strength or below 60 °C and rarely is observed below 80 °C (Ref 82, 130). Above these temperatures, the time for failure decreases rapidly. The magnitude of the cracking response, however, is sensitive to alloy composition, thermal history (particularly heat treatments resulting in sensitivity to intergranular corrosion), and the environment. The latter is illustrated by the data in Fig. 7.87 (Ref 131), which relates the concentration ranges for dissolved oxygen and chloride ions at 250 to 300 °C to SCC in type 304 stainless steel, depending on the presence of a microstructure showing grain-boundary carbide precipitation (sensitization). It should be noted that the compositions are represented on a logarithmic scale, and hence to prevent cracking, very low chloride concentrations are required at high-oxygen concentrations and conversely for high-chloride concentrations. The data indicate that there are critical concentrations of chloride and oxygen, which, if exceeded, result in cracking. The inverse form of this interrelationship is consistent with an increase in the corrosion potential with increased oxygen concentration (i.e., the chloride concentration must decrease as the oxygen concentration increases to prevent cracking). Alternatively, the passive film formed in the higher-oxygen environment may result in a thicker passive film, which, however, on cracking, results in more severe localized corrosion, which is then associated with crack propagation. In the cracking range, the mode is intergranular when the steel is sensitized; otherwise, the mode is transgranular.

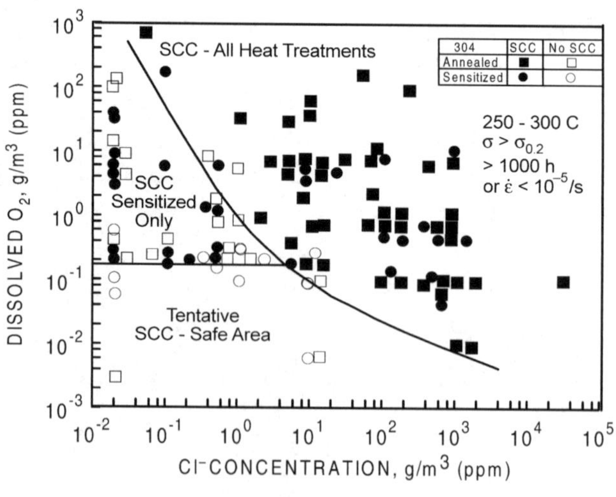


**Fig. 7.87** Synergistic effect of chlorides and oxygen on the SCC of 304 stainless steel. Source: Ref 131

The effect of alloying elements on tendency for austenitic stainless steels to stress-corrosion crack in chloride solutions is summarized in Fig. 7.88 (Ref 35). Recognizing that a chromium concentration in the range of 18 to 20 wt% is needed for passivity and that detrimental elements are held to low concentrations, the nickel concentration has a significant influence on SCC as shown in Fig. 7.89 (Ref 132). The figure shows the effect of nickel content on the susceptibility to SCC of stainless steel wires containing 18 to 20 wt% chromium in $MgCl_2$ boiling at 154 °C. At very low concentrations, the alloys are ferritic and show the

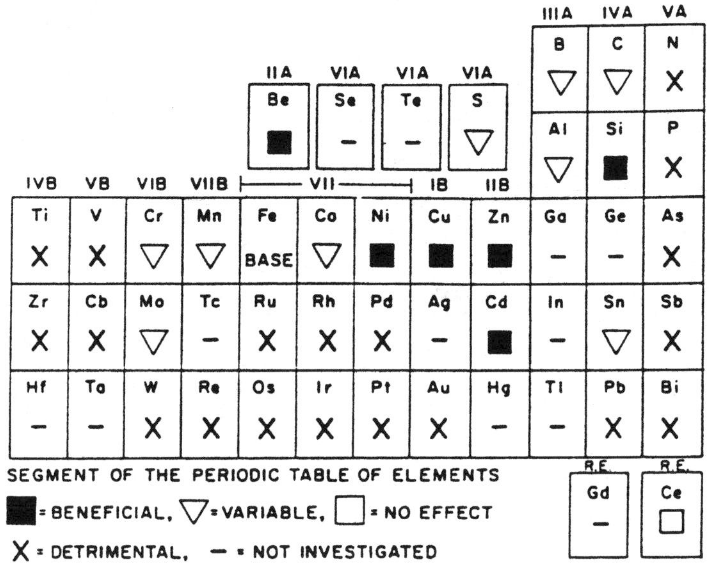


**Fig. 7.88** Effect of elements on resistance of stainless steels to SCC in chloride solutions. Source: Ref 35

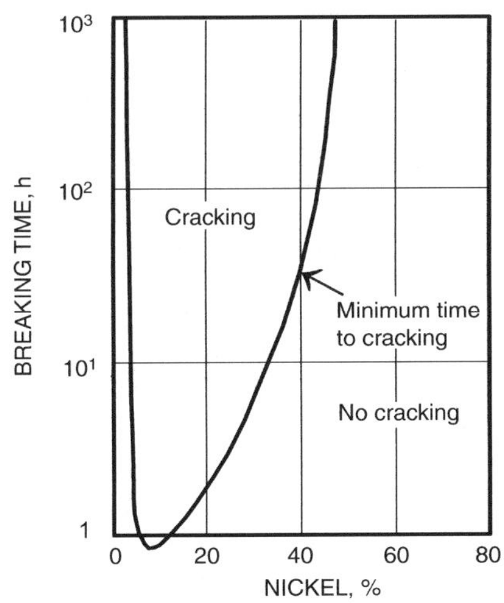


**Fig. 7.89** Stress-corrosion cracking of iron-chromium-nickel wires in boiling 42% magnesium chloride. Redrawn from Ref 132

resistance to cracking characteristic of ferritic stainless steels. There is a minimum in the resistance near 10 wt% Ni, with approximately 40 wt% Ni required to regain failure times approaching that of the nickel-free alloy. In general, for severe chloride environments at elevated temperatures, the high-nickel stainless steels or nickel-base alloys (>40 wt% Ni) are required to ensure protection against SCC (Ref 133).

In the properly heat treated condition, the standard ferritic stainless steels such as AISI 430, 434, and 436 are more resistant to SCC in chloride environments than the austenitic stainless steels. Improper heat treatment, and in particular, welding, results in a material with poor ductility and susceptibility to SCC. These limitations are significantly reduced by increasing the chromium content to 25 to 30 wt% and using careful melting procedures to reduce the carbon (0.002 to 0.02%), nitrogen (0.005 to 0.02%), oxygen, and hydrogen contents. Titanium and/or niobium also may be added to stabilize the carbon as insoluble phases. These alloys are essentially immune to SCC. However, because of the requirement to maintain the very low concentration of interstitial impurities, precaution must be used to avoid contamination in welding with subsequent susceptibility to SCC (Ref 35, 134).

The compositions of duplex stainless steels allow microstructures of approximately equal amounts of austenite and ferrite, and with proper welding methods, this ratio can be maintained. The alloys, relative to the austenitic alloys, are somewhat more resistant to SCC in chloride and chloride/hydrogen-sulfide environments than the single-phase austenitic alloys. One contributing factor to the better resistance is the blocking effect of the ferrite phase in the microstructure to the propagation of cracks through the austenite phase. Susceptibility to intergranular SCC is reduced in duplex stainless steels because the sensitization associated with carbide precipitation occurs predominately at the austenite/ferrite phase interfaces rather than at austenite/austenite grain boundaries. Hence, continuous chromium-depleted paths do not exist along which stress-assisted intergranular corrosion will propagate (see the section "Intergranular Corrosion of Ferritic Stainless Steels" in this chapter). High temperature, higher hydrogen-sulfide concentrations, and lower pH decrease the more favorable behavior of the duplex alloys (Ref 135).

**Relationship of Composition and Heat Treatment to Environment-Sensitive Cracking of Aluminum Alloys.** Those aluminum alloys strengthened by cold working only, particularly the 1000-series alloys, do not develop susceptibility to SCC. The so-called high-strength alloys are strengthened by thermal/mechanical treatments, which result in solid-state precipitation of one or more intermetallic phases that restrict dislocation motion and, hence, increase strength. Their susceptibility to SCC varies extensively with alloy composition and the thermal/mechanical treatment. While susceptibility tends to increase with increase in strength level, the stress-corrosion mechanisms related to the microstructure resulting from processing are more important in governing susceptibility.

Selected characteristics of the several series of aluminum alloys whose composition and heat treatment influence stress-corrosion susceptibility are presented in the following sections (ranges of alloy content also are given) (Ref 96–98, 136–138):

- *2xxx-series (Al-Cu(2.6 to 6.3 wt%)-Mg(0.5 to 1.6 wt%)):* Copper and magnesium are in solid solution at elevated temperatures. Following quenching, as functions of time and temperature, these elements separate progressively from the solid-solution matrix as coherent Cu-rich zones in the aluminum-rich crystal matrix. These zones grow to semicoherent precipitates and finally to the stable phases, $CuAl_2$ in the Mg-poor alloys, and $CuMgAl_2$ in the Mg-rich alloys. At ambient temperatures, the strength is increased by formation of the coherent zones; stable phase precipitates are not observed in the grain boundaries. At elevated temperatures (e.g., 175 °C), the semicoherent and, in time, the stable phases, form. This artificial aging is accompanied by an initial increase and then decrease in strength. Of particular importance to SCC susceptibility is formation of the stable phase in the grain boundaries, with regions adjacent to the grain boundaries denuded of both solute elements and coherent precipitates. In the 2*xxx* series of alloys, the denuded matrix along the grain boundary is anodic to both the stable precipitates and to the incompletely alloy-depleted matrix within the grains. As overaging progresses, the matrix is uniformly depleted in copper and magnesium, and the potential difference between exposed grain boundaries and matrix grains becomes small, thereby decreasing susceptibility to SCC. It should be noted that susceptibility also can be sensitive to the cooling rate from the initial solid-solution state. In a critical cooling-rate range, $CuAl_2$ and/or $CuMgAl_2$ can form in the grain boundaries in association with denuded adjacent solid solution, thereby creating susceptibility to SCC. Sufficiently rapid cooling avoids this condition and very slow cooling results in a condition, equivalent to severe overaging.
- *5xxx-series (Al-Mg(0.8 to 5.1 wt%)):* Although the solid solubility of magnesium in aluminum is large (17.4 wt% at 450 °C) and magnesium can be retained in solution on quenching, subsequent thermal treatments do not result in useful increases in strength. Unlike the 2*xxx*-series alloys, coherent Mg-rich zones do not form that impede dislocation motion and usefully increase strength. These alloys are strengthened by cold working. However, long times (months to years) at ambient temperatures and shorter times at elevated temperatures result in grain-boundary precipitation of

$Mg_5Al_8$. This phase is very anodic to the matrix solid solution and leads to intergranular corrosion in the absence of stress and to SCC in the presence of stress. Alloys with less than 3 wt% Mg are generally free of susceptibility to SCC. Susceptibility of alloys with higher magnesium concentrations depends on composition, time/temperature thermal histories, and cold working. Cold working enhances precipitation of $Mg_5Al_8$, with small amounts of cold working preferentially increasing grain-boundary precipitation. Larger amounts result in uniform precipitation throughout the matrix and a decrease in continuous precipitation in the grain boundaries. As a consequence, local anodes and cathodes are more uniformly distributed and susceptibility to SCC is decreased.

- *6xxx-series (Al-Mg(0.5 to 1.1 wt%)-Si(0.4 to 1.4 wt%)):* The Mg/Si ratio is usually adjusted such that the equilibrium phase that separates from the high-temperature solid solution is $Mg_2Si$. On quenching, these elements are retained in solid solution. Subsequent time/temperature treatments allow strengthening through the stages of Mg- and Si-rich coherent zones, a semicoherent $Mg_2Si$ precipitate and the stable $Mg_2Si$. In general, these alloys are not susceptible to SCC. The exact reason is not clear since the potential of the $Mg_2Si$ in chloride environments is very anodic to the solid-solution matrix. This anodic potential, however, rapidly increases (becomes less anodic) with time and approaches that of the matrix solid solution. The local galvanic coupling and, hence, susceptibility to SCC is reduced. Since $Mg_2Si$ reacts with water to form $SiO_2$ and MgO, these oxides may quickly coat exposed $Mg_2Si$ particles and reduce, if not prevent, their galvanic coupling with the matrix.
- *7xxx-series (Al-Zn(1.0 to 7.6 wt%)-Mg(2.5 to 2.7 wt%)-Cu(0.1 to 2.8 wt%)):* In these alloys, the Al-Zn-Mg solid solution formed at elevated temperatures is retained on quenching with subsequent time/temperature/mechanical treatments increasing strength to the highest levels of the commercial alloys. The precipitation sequence can be summarized as solid solution → spherical coherent Zn- and Mg-rich zones → ordered zones → semicoherent precipitate → $MgZn_2 + Mg_3Zn_3Al_2$ (Ref 97). The intermediate stages are associated with maximum precipitation strengthening. Susceptibility to SCC is extremely sensitive to the thermal/mechanical history of the alloy but correlation of resulting microstructures with SCC has been only partially successful. At critical stages of precipitation and at critical cooling rates from the initial solid-solution-treatment temperature, $MgZn_2$ and $Mg_3Zn_3Al_2$ precipitate in the grain boundaries. A precipitate-free zone, whose composition varies depending on the thermal history of the alloy, can form around and between the

> stable grain boundary precipitates. It has been proposed that critical boundary compositions form that are sufficiently anodic to the bulk grains that local attack under stress results in intergranular SCC. Underaging or overaging is associated with smaller differences in potential and resistance to SCC is greater.

All aluminum alloys contain controlled concentration limits of Cr, Mn, Zr, Ti, and Fe. These elements form high-melting intermetallic compounds with aluminum that influence grain size on solidification. Their insolubility in the solid-solution alloys results in particle distributions that restrict grain growth following mechanical working. Of particular importance is the stringering or banding of these intermetallic phases in the direction of plastic flow during hot and cold working. For example, in rolled sheet and plate, the grains, even after annealing, are elongated between the bands by the restricted growth across the bands by insoluble particles. As a consequence, these products usually have significantly different mechanical properties in the longitudinal (rolling), long-transverse, and short-transverse (normal to rolling plane) directions; the properties in the latter direction are poorest, including resistance to SCC. Since environmental cracking in high-strength alloys is almost always intercrystalline due to factors just discussed, and develops preferentially along grain boundaries perpendicular to the stress, susceptibility to SCC varies significantly with direction of the stress in the sheet. The anisotropy of grain shape is illustrated in Fig. 7.90 along with boundaries (dark lines) along which cracking occurs under stress (Ref 97). Note that grain boundaries extend predominantly in planes whose normal is in the short-transverse direction; in contrast, the smallest grain-boundary area occurs perpendicular to the longitudinal direction. The effect of loading direction is shown qualitatively in Fig. 7.91 (Ref 98) in which the time dependence for failure under sustained tensile stress is shown for the three directions in a rolled plate. For each loading direction, a threshold stress exists below which failure does not occur, with this value being significantly lower for the short-transverse direction.

A schematic representation of the simultaneous influence of aging (precipitation from solid solution) on strength and resistance to SCC for 7*xxx*-series aluminum alloys is shown in Fig. 7.92 (Ref 97). The stages identified as I, II, and III correspond to stages of aging. In stage I, both strength and stress-corrosion resistance change rapidly; coherent zones of precipitate are forming within the grains, and grain-boundary precipitation accompanied by an adjacent denuded region is generally observed. In stage II, coherent zones are progressively replaced by semicoherent precipitates within the grains, further precipitation occurs in the grain boundaries, the rate of strengthening decreases, and the resistance to SCC increases. Overaging is occurring in stage III; the stable phases progressively form both within the grains and in the grain

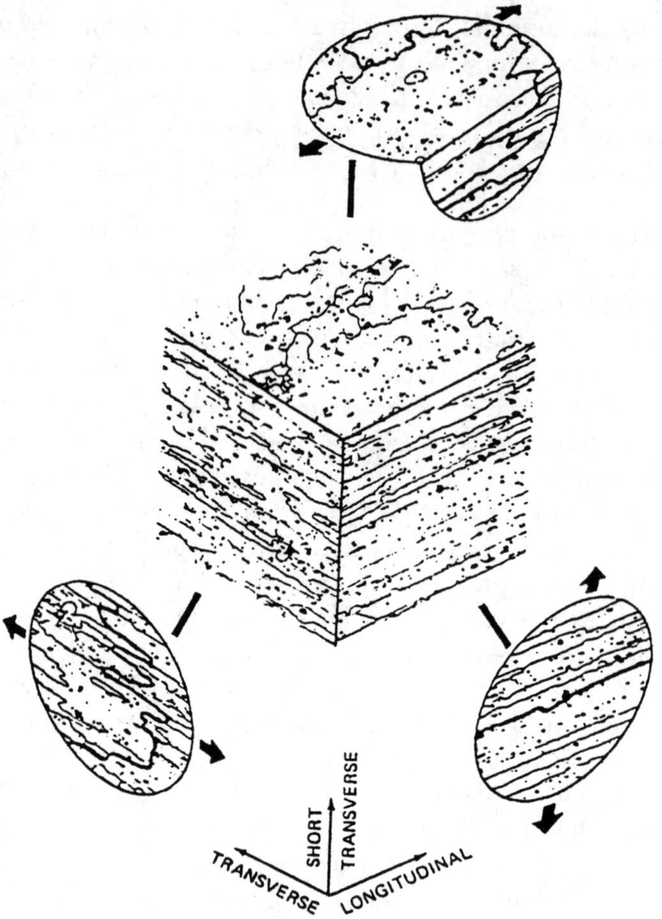


**Fig. 7.90** Effect of stressing direction on the intergranular stress-corrosion crack path in susceptible high-strength aluminum alloy. Dark boundaries are representative of ones favored for cracking for indicated direction of applied stress. Source: Ref 97

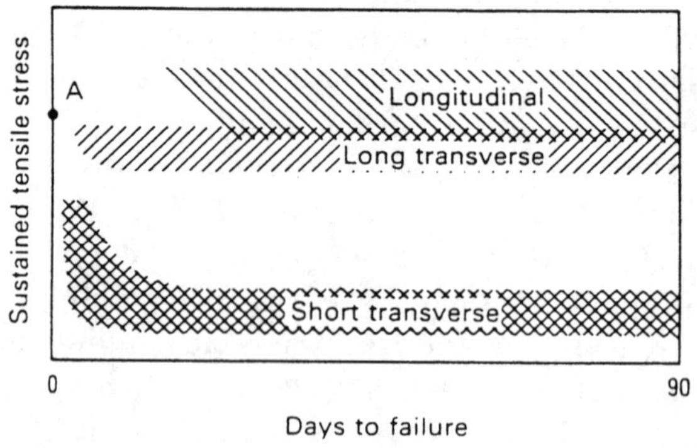


**Fig. 7.91** Sustained tensile-stress failure time for 76 mm (3 in.) plate of 7075-T651 aluminum alloy. Shaded bands indicate combinations of stress and time known to produce SCC in specimens intermittently immersed in 3.5% NaCl solution. Point A is the minimum yield strength in the long transverse direction for plate 76 mm (3 in.) thick. Source: Ref 98

boundaries, and the composition of the matrix becomes progressively depleted in alloying elements both within the grains and at the grain boundaries. The behavior exhibited by these curves is reasonably representative of most of the aluminum-alloy series discussed briefly earlier. The shapes of the curves shift with respect to magnitude and the relative positions of the maximum and minimum values as a function of aging conditions. In particular, for the 7*xxx*-series alloys, it is significant that the minimum in the resistance to SCC occurs before the maximum strength. It also should be noted that the resistance to SCC rapidly increases in stage III, and that appreciable resistance can be attained with relatively small decrease in strength. This is illustrated quantitatively in Fig. 7.93 for the alloys 7075 and 7178 (Ref 97). First, it should be noted how severe susceptibility to SCC can depress the strength in the short-transverse grain direction (7 ksi) relative to the nonenvironmentally affected yield strength in the longitudinal grain direction (≈85 ksi). Second, it is noted that after 25 h aging, the stress-corrosion threshold has increased to 45 ksi, while the yield strength is still greater that 70 ksi. These data illustrate the general necessity to overage these alloys in order to have acceptable resistance to SCC.

**Relationship of Composition to Environment-Sensitive Cracking of Copper Alloys.** Stress-corrosion cracking of copper alloys is observed to be intergranular or transgranular, depending on the specific alloy, the environment, the potential, and, in some cases, the stress level. Small changes in any of these may result in a change in the cracking mode.

Although a number of copper alloys undergo SCC, the Cu-Zn alloys (brasses) have received the greatest attention with respect to both their service performance and in research. Of particular significance is the

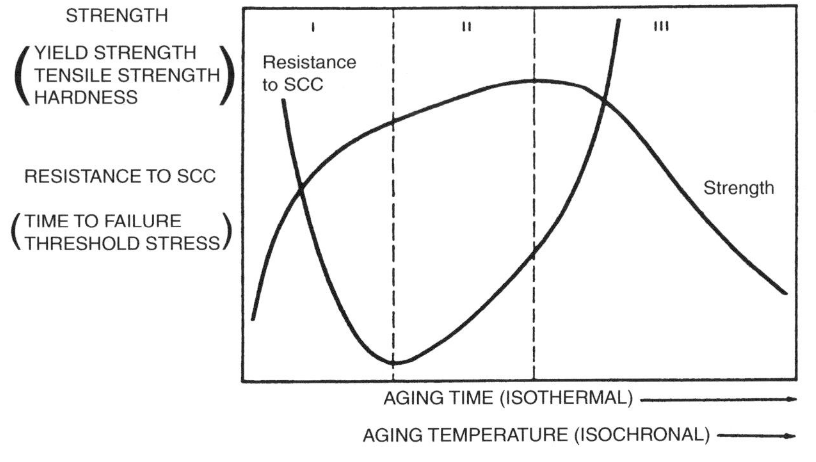


**Fig. 7.92** Relationship between strength and SCC resistance during aging of high-strength 7*xxx*-series aluminum alloys. Source: Ref 97

cracking of these alloys in moist ammonia atmospheres and in aqueous solutions containing ammonia under either externally applied stresses or from residual stresses following mechanical working. Cracking associated with residual stresses is commonly referred to as season cracking, the term having originated from failures observed to occur in cartridge brass (Cu-30 wt% Zn) after extended exposure to moist atmospheres containing small amounts of ammonia (Ref 139). The susceptibility to cracking increases with zinc content, and the mode of cracking is intergranular when a relatively thick tarnish film of cuprous oxide, $Cu_2O$, is present. Since under most service conditions, tarnish films form on the copper-zinc alloys, intergranular cracking is the failure mode most commonly encountered (Ref 140).

In the presence of ammonia in solution and at pH > ~5 to 7, the anodic dissolutions of copper and zinc are immediately associated with formation of the complex ions, $Cu(NH_3)_n^+$ (n = 2 to 5) and $Zn(NH_3)_4^{2+}$ as

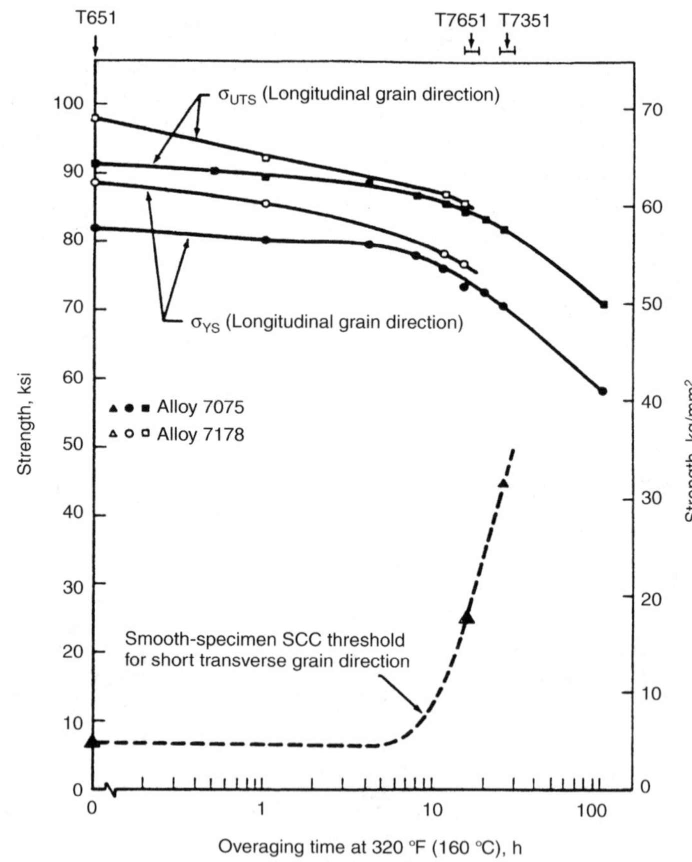


**Fig. 7.93** Effect of artificial aging at 320 °F on the strength and smooth-specimen SCC threshold stress of 7075-T651 and 7178-T651 aluminum alloys. Source: Ref 97

soluble corrosion products (Ref 141, 142). At lower pH, the concentrations of the complex ions decrease, and the $Cu^{2+}$ ion becomes dominant as the soluble corrosion product. In the presence of dissolved oxygen, the cuprous ammonium complex, $Cu(NH_3)_n^+$, is oxidized to the cupric complex, $Cu(NH_3)_n^{2+}$, and the reduction of this species on the surface of the alloy provides the cathodic reaction supporting the continued anodic dissolution of the alloy. The reactions are (Ref 143):

$$Cu + nNH_3 \rightarrow Cu(NH_3)_n^+ + e \quad \text{(Eq 7.6)}$$

$$2Cu(NH_3)_n^+ + 1/2\,O_2 + H_2O \rightarrow 2Cu(NH_3)_n^{2+} + 2OH^- \quad \text{(Eq 7.7)}$$

$$Cu(NH_3)_n^{2+} + e \rightarrow Cu(NH_3)_n^+ \quad \text{(Eq 7.8)}$$

Thus, in the presence of dissolved oxygen, the mechanism is autocatalytic in that the corrosion product of the anodic reaction, Eq 7.6, through the reaction of Eq 7.7, progressively supports the cathodic reaction, Eq 7.8. Not only is the cathodic reactant regenerated, but also the concentration increases such that the corrosion rate tends to increase with time.

The conditions under which, and the mechanisms whereby, the tarnish film forms are complex. The tarnish films differ from passive films in that they are less protective and much thicker, up to 10 μm (Ref 144). There is evidence that the films contain micropores which are filled with liquid and that ion transport through these solution-containing paths supports formation of tarnish at the metal/tarnish interface. The tarnish consists of platelets of cuprous oxide, $Cu_2O$, having crystal lattice orientations related to the lattice orientation of the substrate brass grains. The film grows into the substrate with a tendency for preferential penetration along grain boundaries, presumably due to segregation of zinc atoms along these interfaces. When tarnish is present, intergranular cracking occurs by the repeated sequence of cracking of the grain boundary oxide, access of solution to the unreacted grain boundary at the depth of the crack, followed by further formation of oxide, which then cracks at a critical increment of penetration (Ref 143, 145).

The rate of tarnish formation in ammoniacal solutions at the free-corrosion potential is greatest in the pH ranges of ~ 6.5 to 7.5 and >11. Proposed reactions include:

$$2Cu^{2+} + H_2O + 2e \rightarrow Cu_2O + 2H \quad \text{(Eq 7.9)}$$

or, assuming oxidation of the copper to immediately form adsorbed cuprous complexes at the brass/tarnish interface (Ref 143):

$$2Cu(NH_3)_2^+ \text{ (adsorbed)} + 2OH^- \rightarrow Cu_2O + 4NH_3 + H_2O \quad \text{(Eq 7.10)}$$

It is assumed that electrons move through the tarnish to the tarnish/solution interface and support the cathodic reactions of oxygen and/or cupric-complex reduction.

Tarnishing is enhanced if the electrochemical potential is increased either by oxidizing species other than oxygen in the solution or by an externally applied potential; the rate is decreased if the potential is decreased. The rate of formation and final thickness of the tarnish increases with an increase in the zinc content of the brass, an increase in the complex-ion concentration in the solution and an increase in temperature. The effect of the zinc concentration in the alloy has been attributed to the selective depletion of zinc at the surface, creating a surface with enhanced reactivity for the formation of $Cu_2O$. The selective depletion of zinc is readily related to the greater electrochemical activity of zinc relative to copper, but is confined to a few atom layers at the surface because of the slow rate of solid-state diffusion at practical service temperatures (Ref 146). The effect of the zinc content of the brass also has been attributed to the reaction (Ref 142, 147):

$$Zn + 2Cu(NH_3)_2^{2+} \rightarrow Zn^{2+} + 2Cu(NH_3)_2^{+} \quad \text{(Eq 7.11)}$$

The cuprous complex ion then reacts to form tarnish (Eq 7.10). The zinc concentration of the tarnish is very small due to the greater stability of the zinc ammonium complex in solution relative to zinc oxide that would coexist with the $Cu_2O$ in the tarnish. These effects of increasing zinc concentration of the brass are consistent with the fact that tarnishing occurs less readily on copper. It is also consistent with the greater susceptibility of the high-zinc brasses (20 to 30 wt% Zn), which more readily tarnish, to IGSCC.

In fresh ammoniacal solutions, free of the complex ions, tarnish does not form and intergranular cracking does not occur. However, if the brass corrodes into a restricted volume of ammoniacal solution such that the concentration of the cupric-ammonium-complex corrosion product increases due to anodic dissolution to $Cu^{2+}$ and $Zn^{2+}$ ions, a critical concentration is rapidly reached at which tarnish forms, and failure by intergranular corrosion in a short time is observed. Hence, the solution-volume to alloy-surface-area ratio and the contact time of the solution would be influential in determining the onset of intergranular cracking. These observations are consistent with the experience that brass with 20 to 40 wt% Zn and containing residual manufacturing stresses readily cracks in moist air containing ammonia. A thin, condensed layer of water readily absorbs both oxygen and ammonia. Because of the small volume of liquid in the film, anodic dissolution quickly increases the copper-complex concentration above that for tarnish formation, and cracking occurs.

If, in ammoniacal solutions, the pH is not in the pH range previously indicated, ~6.5 to 7.5 and >11, or the complex-ion concentration is maintained very low by a large solution-volume to surface-area ratio, or under very low concentrations of oxygen, the tarnish film does not tend to form. In these cases, the mode of cracking is intergranular for alloys having <~18 wt% Zn and transgranular for greater zinc concentrations (Ref 148). This change in mode has been attributed to the effect of zinc in changing the dislocation structure from cells of entangled dislocations at the lower zinc contents to planar arrays of dislocations at the higher compositions. It is significant that pure copper has been observed to crack intergranularly under nontarnishing conditions but does not stress corrosion crack when a tarnish film is present. The tarnish film does not preferentially penetrate the grain boundaries of the copper, presumably due to lack of zinc atoms in these interfaces. As a consequence, the mechanism of alternate cracking and incremental growth of the oxide along the grain boundary does not occur.

The effects of alloying elements, other than zinc, with copper on SCC in ammoniacal solutions have been investigated (Ref 144, 148–150). In solid-solution alloys of 0 to 6 wt% Al and 0 to 25 wt% Ni, IGSCC is observed in tarnishing solutions with time to failure increasing in the order Cu-Ni > Cu-Al > Cu-Zn. In nontarnishing solutions, the Cu-Ni alloys failed intergranularly at all compositions. In contrast, the Cu-Al alloys up to 3 wt% Al failed intergranularly, but failed transgranularly at 6 wt% Al (Ref 148). This behavior is consistent with observations on the Cu-Zn brasses where, at low zinc concentrations, the dislocation structure is cellular but planar at high concentrations. A similar transition in dislocation structure occurs with the Cu-Al solid solutions, but with the Cu-Ni alloys, the dislocation structure remains cellular with increasing nickel concentration, and a change from intergranular to transgranular mode of cracking is not observed (Ref 148). In an investigation of stress cracking of a number of copper-base alloys in ammoniacal solutions, failure time under tarnishing conditions was observed to decrease as the initial corrosion potential was increasingly lower than that of pure copper (Ref 149). For example, the corrosion potential becomes progressively lower relative to copper with increasing zinc content, being about 120 mV lower at 30 wt% Zn. On the basis that the tarnish film is a large cathodic surface, the difference in potential driving localized corrosion at a break in the film (i.e., crack initiation) would be greater the higher the zinc content.

Stress–corrosion cracking of copper-zinc alloys can occur in environments other than ammoniacal solutions (Ref 114, 147, 151, 152). Included are nitrogen-bearing compounds such as amines and aniline, as well as sulfates, nitrates, nitrites, acetates, formates, and tartrates. These environments can produce tarnish films of $Cu_2O$ similar to the films formed in ammoniacal solutions. Both the rate of formation and the thickness of the tarnish films tend to be significantly smaller than found in the ammoniacal environments, but when the tarnish is present, cracking is also predominantly intergranular. There is evidence that citrates and tartrates form complex ions with copper, but the role of these in the mechanism of tarnish formation and cracking has not been investigated to the extent that it has in the ammoniacal solutions (Ref 147).

Although the corrosion potential at which cracking occurs was not reported for many of these environments, it has been established that the electrochemical potential is a significant variable. For example, the polarization behavior of 70Cu-30Zn (wt%) brass has been used to determine conditions of pH and potential at which SCC would be predicted (Ref 114). Stress-corrosion cracking was observed consistent with the predicted conditions. As a consequence, if the corrosion potential is below the potentials at which cracking could be produced, failure would not be expected unless other species were present that would increase the corrosion potential into the cracking potential range.

Slow-strain-rate tests have been used to evaluate the stress-corrosion tendency of Admiralty metal, 71Cu-28Zn-1Sn (wt%), in solutions of a number of oxyanions (Ref 153). The solutions were adjusted to pH = 8 and the potential was controlled at 300 mV (SHE) during straining. The order of decreasing promotion of susceptibility to SCC was $NO_2^- > NO_3^- > ClO_3^- > SO_4^= > MoO_4^= > Cl^-$. In all environments, $Cu_2O$ films were observed to form and the cracking was intergranular. However, the highest corrosion potential was 210 mV (SHE), and since this was substantially lower than the test potential, the tendency of the environments to produce cracking under open-circuit conditions was not reported. Again, the results would be directly applicable if species were present that raised the corrosion potential to 300 mV (SHE).

## *Mechanisms of Environment-Sensitive Crack Growth*

As with pitting and crevice corrosion, identification of mechanisms of stress-related environment-sensitive cracking is complicated by establishing, either experimentally or theoretically, the environmental conditions at a crack tip. In addition to the factors considered previously relating to pitting and crevice corrosion (i.e., local acidification due to metal-ion hydrolysis), passive film formation and IR potential drops causing the potential at the crack tip to differ from that of the surface, the major additional variable in environmental cracking is the state of stress surrounding the crack tip. Depending on the alloy composition, the microstructure as established by thermal and mechanical treatments, and the environment, cracks follow transgranular or intergranular paths. Observations of the morphology and mechanisms for the propagation of these two modes of environment-sensitive cracking are discussed in the following sections. Reviews can be found in references 110, 115, 145, 154, and 155.

**Mechanisms of Transgranular SCC.** Transgranular SCC occurs predominantly with alloys and environmental conditions forming passive films. On smooth surfaces, cracks may be initiated by stress-induced glide of dislocations to the surface resulting in offsets (as shown in Fig. 7.73), which are larger than the passive film thickness and thereby expose the substrate to dissolution. If the new surface immediately repassivates, cracking is not initiated; otherwise, a crevice is created that subsequently propagates as a crack under the control of mechanisms involving the environment and stress state at the crack tip. Since the width of the cracks is very small relative to the depth, the tip growth rate must be very much larger than the lateral rate of corrosive attack. This requires that, as the crack progresses, the sides of the crack must very quickly repassivate, resulting in a lateral growth rate restricted by the low passive-current density.

A mode of stress-corrosion crack propagation of stainless steels is multiple parallel penetration (cracking), frequently initiated at grain boundaries along specific crystallographic planes as shown in Fig. 7.94 (Ref 156). The orientation of these planes relative to the fracture plane (the plane of the photograph) is governed by the orientation of the crystal lattice of the grain in which the penetration occurs. Parallel cracks tend to merge or coalesce, and sheets of material between the cracks are

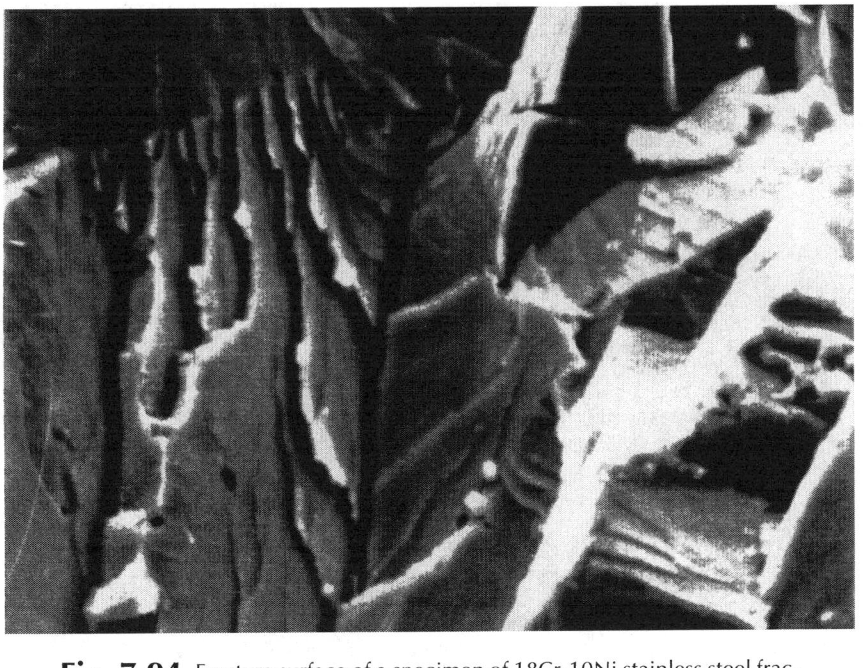

**Fig. 7.94** Fracture surface of a specimen of 18Cr-10Ni stainless steel fractured in $MgCl_2$ solution boiling at 154 °C. Multiple fractures coalescing by plastic tearing between adjacent cracks. 500×. Source: Ref 156

ruptured as fracture progresses. The extent to which the penetration or cracking is mechanical cleavage or stress-assisted local corrosion is uncertain. If it is mechanical cleavage, then an influence of the environment on the cleavage strength must exist, since the stress to cause fracture is much lower in the presence of a corrosive environment.

Stress-corrosion crack growth also has been associated with localized multiple tunnels penetrating into the material along a crack tip resulting in fracture surfaces of the form shown in Fig. 7.95 (Ref 156). The appearance of the fracture surface is observed to be sensitive to the stress level at which the crack propagates. Schematic representations of crack mechanisms that proceed by tunnel formation at low and high stress levels are also shown in Fig. 7.96. At low stress across the plane of tunnel formation, radial growth of the tunnels proceeds until the wall between the tunnels is very thin. These then fracture, resulting in grooved surfaces. At high stresses, fracture of the between-tunnel wall occurs while these walls are relatively thick. The appearance of selective attack at emergent slip planes intersecting the tunnels (Fig. 7.95 and shown schematically in Fig. 7.96) indicates that stress-corrosion crack propagation can be associated with plastic deformation in the material near the crack interface.

Observations of the growth of transgranular stress-corrosion cracks at free surfaces, and examinations of fracture surfaces, have established that for several metal/environment systems, cracks propagate

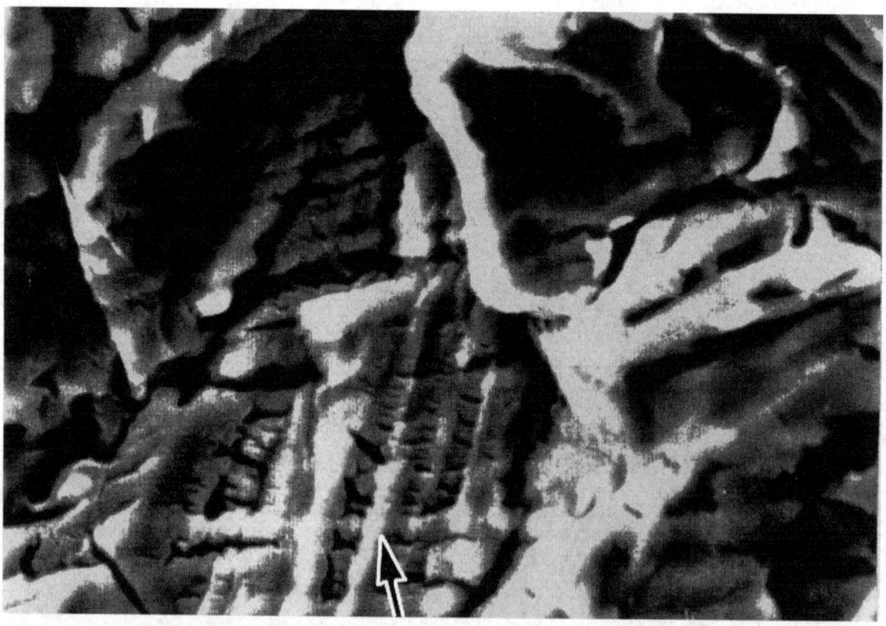

**Fig. 7.95** Transgranular fracture surface of a specimen of 18Cr-10Ni steel illustrating the effect of emergent slip planes upon the lines of parallel tunnels indicated by the arrow. 5 N $H_2SO_4$ + 0.5 N NaCl. 2000×. Source: Ref 156

intermittently (Ref 154). Increments of growth involve periods of stagnation followed by cleavage along specific crystallographic planes, which stops after propagating a characteristic distance. Successive markings on the fracture surface perpendicular to the growth direction are associated with the periods of stagnation of the crack front. The intermittent character of the growth is also supported by periodic acoustical emissions and by fluctuations in the corrosion potential associated with the opening of bare surface during the cleavage step. The cleavage increments occur in times of the order of microseconds and the stagnation step lasts from milliseconds to seconds.

One mechanism proposed for the intermittent crack growth is embrittlement of the alloy ahead of the crack as a result of the corrosion processes at the edge of the crack during the stagnation period (Ref 154). Causes of embrittlement have included injection of lattice vacancies associated with anodic dissolution at the crack tip, preferential dealloying, pinning of dislocations, and absorption of hydrogen. The latter, of course, is not applicable where hydrogen-ion or water reduction is not possible. At a critical stage of embrittlement, a cleavage

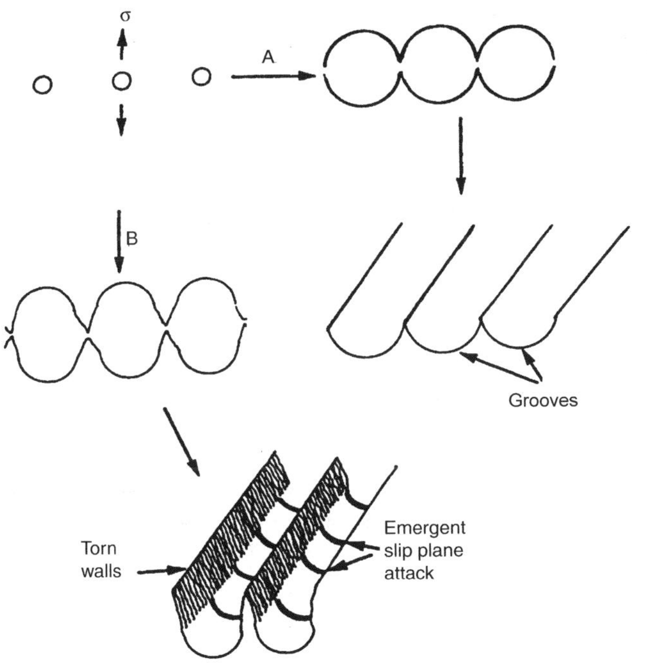


**Fig. 7.96** Schematic drawing of a crack mechanism that proceeds by tunnel formation. Two different situations are described: (A) A low stress across the plane of tunnel formation. Radial growth of the tunnel proceeds until the walls are very thin. These then fracture resulting in grooved surfaces. (B) A high stress acts across the plane of tunnel formation. Fracture of the tunnel walls occurs while they are relatively thick. In addition, glide processes are initiated on the grain under the action of the stress, and selective attack occurs where the emergent slip planes intersect the tunnels. Source: Ref 156

crack is initiated and propagates an incremental distance of the order of 5 μm (Ref 154, 157). The environment again has access to the crack tip where the corrosion process is reestablished and the sequence of steps is repeated. Each period of stagnation appears to be associated with blunting of the crack tip by either anodic dissolution or plastic deformation, or by both. Several mechanisms have been proposed to account for the termination of each cleavage increment. One is that the cracking proceeds to the depth of the embrittled region ahead of the crack, where it is arrested by its inability to proceed by plastic deformation rather than cleavage. This mechanism, however, can be questioned because the slow rate of solid-state diffusion, except for hydrogen, precludes formation of a brittle zone ahead of the crack front equal to the observed increment of cleavage (Ref 154).

Other mechanisms of stress corrosion attribute crack growth to processes that are restricted to the immediate vicinity of the crack front; they do not consider discontinuous cleavage events of the type just discussed. Also, the crack tip is modeled as blunted by dissolution and/or plastic deformation, or both. The maximum rate of crack advance, if the controlling condition is anodic dissolution at a bare crack tip, is obtained by application of Faraday's law. For a crack of depth "a," the growth rate is $da/dt = iM/nF\rho$ ($i$ = average current density along the crack front, $M$ = atomic weight of the metal atom, $n$ = valence of the metal ion, $F$ = Faraday's constant, and $\rho$ = density). The anodic dissolution rate may be greater than that of a stress-free surface due to the strained lattice at the crack tip.

Since the crack-tip growth rate is generally less than that accounted for by a clean, actively corroding surface, the lower observed growth rate has been attributed to passive-film or salt-film formation at the crack tip. It is proposed that the stress field at the tip maintains the successive processes of film rupture by slip offset of the surface, active dissolution at the offset causing an increment of advancement, and that the accompanying current densities cause repassivation. The processes are repeated along the crack front as stress-induced dislocation movement cracks these films. This mechanism assigns crack growth to the dissolution of the exposed substrate immediately following passive-film rupture. To be consistent with steady-state crack growth, the mechanism requires a critical balance between film cracking and repassivation, which is consistent with the fact that the conditions for stress corrosion are generally very specific. If the repassivation rate is slow, then cracking is to be expected in the potential range II in Fig. 7.71 (i.e., just above the anodic current maximum of the polarization curve—the potential region of initial passive film formation). In contrast, if the repassivation rate is fast, SCC is expected in potential range III in Fig. 7.71, which is just below the pitting potential. Here, exposed substrate tends to immediately repassivate at a slip offset but is restricted in doing so by the presence of aggressive anions associated with the tendency to initiate pitting.

Crack-tip growth mechanisms have been proposed that do not involve dislocation movement explicitly, but rather, in response to the stress field at the crack tip, interstitial atoms diffuse to the region of the stress field to reduce the stress; substitutional atoms also will diffuse to the tip if the local stress is thereby reduced. Crack-tip growth would be increased if this local change in alloy composition enhances dissolution during slip displacement or alters the passive film such that it is more easily ruptured by dislocations emerging to the surface. That is, there is continuously produced at the crack tip a film that is more easily ruptured than the more stable passive film on the sides of the crack (Ref 158).

Crack-opening mechanisms have been proposed that simply relate to the effect of environment and local alloy composition on the atom-to-atom bond strength at the crack tip. Reduction in this bond strength has been attributed to stress-induced changes in alloy composition as just described and to adsorption of atoms from the environment. Since dislocation movement is not considered in the mechanism, breaking bonds in the plane of the crack propagation leads to a cleavage-type rupture (Ref 159).

A strongly stress-dependent mechanism for crack growth has been proposed based on the argument that there is a constant driving force to reduce the stress by surface migration of atoms from the crack tip along the surface leading away from the tip (Ref 160). This migration of atoms from the tip is equivalent to migration of surface vacancies to the tip, thereby producing an opening of one lattice spacing per vacancy. To be consistent with observed crack-growth rates, significantly larger rates of surface migration must exist than expected for clean surfaces. These enhanced rates have been attributed to the decrease in bonding of atoms at the surface as a consequence of the environment, including the presence of an overlying salt film in the vicinity of the crack tip, in which case diffusion is enhanced at the metal/salt interface.

A contributing, if not controlling, mechanism for crack growth rate may be transport of corrosive reactants to the crack tip and/or corrosion products from the tip. This transport may be bulk flow of the environment into the crack as it advances or it may be diffusion of species such as $Cl^-$, $H^+$, and $O_2$.

**Mechanisms of IGSCC.** An example of transition from transgranular to intergranular SCC in a stainless steel is shown in Fig. 7.97 (Ref 156). Transgranular cracking has occurred by processes of multiple crack nucleation followed by coalescence as described in the previous section. The fracture surface associated with IGSCC is characterized by facets of the individual grains, several of which are shown in the top part of the figure. Intergranular SCC is usually, but not exclusively, associated with those alloys that are susceptible to intergranular corrosion. The correlation is not always observed since alloy/environment systems are known that exhibit susceptibility to intergranular corrosion but not intergranular stress-corrosion cracking and conversely. When the penetration rate is greater when tensile stresses exist across the grain boundary, the mode of cracking has appropriately been called stress-assisted intergranular corrosion. As with intergranular corrosion, intergranular stress-corrosion cracking is generally related to one or more of the following conditions: (a) preferential penetration of a corrosion product, usually an oxide, along grain boundaries; (b) presence of second phases distributed along the grain boundary; (c) presence of regions along the grain boundaries that have been depleted with alloying elements as a result of precipitation of second phases; and (d) segregation of alloying elements in the boundary. These conditions are discussed in the following paragraphs.

Those alloy/environment systems that form relatively thick corrosion-product layers (e.g., brass in ammonia environments) frequently exhibit preferential penetration of the corrosion products along grain boundaries. These are generally brittle products such that cracking will occur on reaching a critical depth in the presence of tensile stresses across the grain boundary. The environment again has access to the

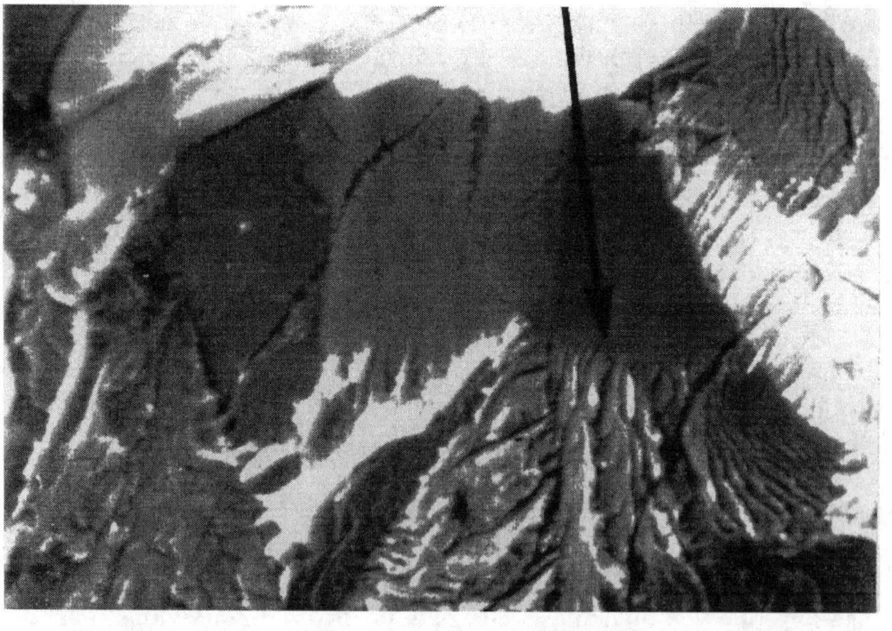

**Fig. 7.97** Transition from transgranular to intergranular cracking that has occurred by a process of multiple crack nucleation followed by coalescence. 18Cr-10Ni steel in $MgCl_2$ solution boiling at 154 °C. Source: Ref 156

crack tip, an increment of corrosion product forms, the increment cracks, and the process continues to be repeated.

Intergranular corrosion associated with the presence of second phases in grain boundaries is discussed in the section "Intergranular Corrosion." These phases may occur following slow cooling from elevated temperatures, or on reheating supersaturated solid solutions retained by quenching from elevated temperatures at which the precipitating phase is soluble in the matrix phase. The precipitated phase, the adjacent solid-solution matrix denuded of solute by the precipitation, and the bulk grains usually exhibit different corrosion potentials, and hence, the more anodic of these locations will preferentially corrode. In most cases, the frequently continuous denuded region along the grain boundary is anodic and is responsible for intergranular corrosion. Intergranular stress corrosion cracking occurs if stress enhances the intergranular corrosion penetration rate. Critical stages of precipitation, for example, in 7*xxx*-series aluminum alloys, lead to minimum resistance to SCC as shown in Fig. 7.92 (Ref 97). However, the actual cracking mechanism for these alloys is probably hydrogen embrittlement due to hydrogen atoms produced by the cathodic reaction supporting the anodic dissolution. The hydrogen embrittlement mechanisms are described briefly in the following section.

**Mechanisms of SCC due to Hydrogen Embrittlement.** When crack tip conditions of pH and potential cause hydrogen-ion or water reduction, the resulting hydrogen atoms are adsorbed to the surface then transported into the substrate by lattice diffusion and by migration along dislocations. Two mechanisms have been proposed to account for an increment of crack growth. One is that the expanded lattice of the high-triaxial-stress state near the advancing edge of the plastic/elastic boundary in advance of the crack tip (explained in the subsequent section "Overview of Fracture Mechanics") enhances the hydrogen concentration. Dislocation mobility is thereby reduced such that relief of stress by plastic flow is less favorable than by local cleavage. An increment of cleavage related to the depth of hydrogen transport occurs, which again allows access of the environment to the crack tip and the process is repeated. An alternate mechanism is based on observations that hydrogen atoms will diffuse to voids where they form hydrogen gas under pressure. This process is enhanced by the triaxial-stress field at the plastic/elastic boundary, resulting in void growth with subsequent joining of voids in the form of local microcracks. Since both of these mechanisms take place ahead of the crack tip, internal cracks form slightly in advance of the actual crack tip and propagate back to the tip, resulting in an increment of crack-tip opening. The cracking morphology has been observed to be both intergranular and transgranular. Repetition of these processes accounts for the hydrogen-embrittlement mode of environment cracking (Ref 145).

## *Application of Fracture Mechanics to the Evaluation of Environment-Sensitive Fracture*

**Background.** Application of forces to materials containing discontinuities such as holes, pits, notches, and cracks results in the concentration of stress in the vicinity of these discontinuities. In the absence of discontinuities, increasing uniaxial stress, for example, results in elastic followed by plastic strain (initiated at the yield-strength stress), both associated with lateral contraction (i.e., normal to the axis of the stress). Ultimately, failure occurs by ductile rupture on either a microscopic or macroscopic scale, or it occurs by cleavage related to bond breakage along selected crystallographic planes with little or no plastic deformation. The amount of strain at failure depends on the properties of the material; the material is macroscopically ductile if the strain is large and macroscopically brittle if the strain is small. At the leading edge of a discontinuity such as a notch or crack, lateral contraction is restricted by material just above and below the discontinuity creating a local state of triaxial stress confined to a small volume of material at the leading edge. The important consequence is that the induced triaxial stress state allows higher stresses in the volume before plastic flow occurs and hence increases the probability that microscale ductile rupture or cleavage become favored modes of failure. It should be noted that in the presence of stress concentrators, both ductile rupture and cleavage may appear macroscopically as brittle fracture in that little net strain is observed in the object, but at the microscopic level the fracture processes are very different.

The relevance of the foregoing discussion to environment-sensitive cracking (SCC and corrosion fatigue) is (a) the corrosive environment can initiate discontinuities that become stress concentrators; (b) corrosion at the leading edge of the crack increases the crack depth until failure occurs either by penetration through a pipe or tank wall by plastic collapse or by macroscopic brittle fracture; and (c) the state of stress at the crack tip influences the corrosion mechanisms responsible for crack tip growth. The latter include active dissolution, passive film fracture, hydrogen embrittlement, and the mechanisms discussed previously for penetration into the metal at the crack tip. Since fracture mechanics has contributed significantly to current understanding of the interrelationship between these aspects of environment-sensitive cracking, a brief overview of fracture mechanics is given in the following section.

**Overview of Fracture Mechanics.** The objective of fracture mechanics is to establish the maximum section stress that can be applied to a material containing a sharp crack of defined geometry without propagating the crack and, in particular, result in partial or complete fracture

(Ref 161– 164). Under static loading, a stress less than the critical value neither extends the crack nor causes fracture. Dynamic loading can cause subcritical crack growth to above the critical size, resulting in fracture. When crack growth occurs under repeated stress application (fatigue loading), the stress-time history is the significant variable. Since environmental conditions at the crack tip can influence the crack growth rate, fracture mechanics concepts contribute to both theoretical and applied aspects of SCC. In particular, the fracture mechanics approach to fatigue failure, when combined with the effect of the environment, contributes to a better understanding of corrosion fatigue.

For purposes of discussions of the application of fracture mechanics to environment-sensitive cracking, the three crack geometries shown in Fig. 7.98 form the basis of analysis. Figure 7.98(a) represents a through crack of width 2a in a section of plate B thick and W wide. Figure 7.98(b) is representative of a through edge crack and also of each edge of the through crack of Fig. 7.98(a). The more frequently encountered geometry is the surface crack shown in Fig. 7.98(c), which would be of particular significance where environmental factors can affect crack initiation and propagation. The discontinuities are variously referred to as cracks or notches, the former usually developing in service, and the latter (Fig. 7.98b) artificially introduced in test specimens, although repeated loading may be applied to initiate a sharp crack from the base of the notch. In any case, the sharpness of the crack expressed as a radius of curvature is another variable. All three cases in Fig. 7.98 conform to a fracture mechanics mode I opening configuration, which is the only one considered here and the one most commonly analyzed.

The nominal or macroscopic stress on the section (here, uniaxial) is $\sigma = P/BW$, where P is the load applied to the component, and B and W are shown in Fig. 7.98. In the limit of distances sufficiently removed from a notch or crack to no longer be influenced by it, $\sigma$ is the uniform cross-section stress in the material. However, in the vicinity of the

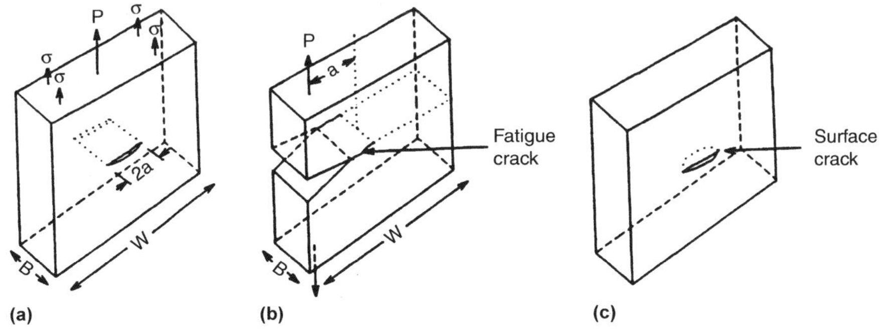


**Fig. 7.98** Three types of cracks analyzed by fracture mechanics methods. (a) Through crack of width 2a. (b) Through edge crack of depth a. (c) Partial surface crack of width 2a

notch or crack tip, a state of stress exists that is described with reference to a coordinate system with origin at the notch tip as shown in Fig. 7.99 (Ref 163). The y-axis is parallel to the load (P) direction; the x-axis is the direction of crack propagation; and the z-axis is along the section thickness, B. Another variable is the notch or crack-tip radius, which will be designated as $\rho$ and approaches zero for an infinitely sharp crack.

In the plane of the crack (dotted plane in Fig. 7.98b), the x- and y-axis stresses for an elastic material are given by:

$$\sigma_y = K/\sqrt{2\pi x} + Cx^0 + Dx^{1/2} + \ldots \quad \text{(Eq 7.12)}$$

$$\sigma_x = K/\sqrt{2\pi x} \quad \text{(Eq 7.13)}$$

$$K = \beta\sigma\sqrt{\pi a} \quad \text{(Eq 7.14)}$$

where K is called the stress-intensity factor; $\sigma$ is the nominal cross-section stress, P/BW; and "a" is half the length of a through crack (Fig. 7.98a) or the depth of an edge crack. The terms following the first for $\sigma_y$ form a series to account for $\sigma_y = \sigma$ at values of x beyond which the effect of the crack becomes negligible. Near the crack ($x \rightarrow 0$), the first term dominates the y-direction stress. $\beta$ is a geometry factor whose value depends on the specimen shape and crack depth (Ref 163). For the through crack, Fig. 7.98(a), in an infinitely wide plate ($W = \infty$), $\beta = 1$; for a small through edge notch, Fig. 7.98(b), without the crack at the base of the notch, $\beta \approx 1.12$; for a small crack at the base of a notch, Fig. 7.98(b), $\beta \approx 3$ and "a" is equal to the length of the crack; for a deep crack

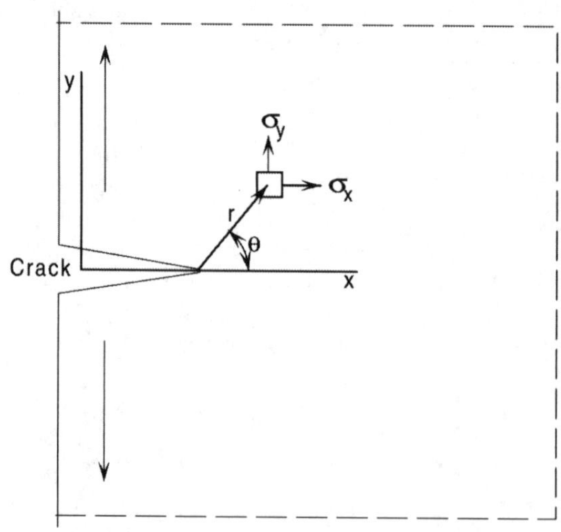


**Fig. 7.99** Polar coordinates used to locate element under stress in the stress field surrounding the tip of a surface crack

at the base of a notch, Fig. 7.98(b), $\beta = 1.12$ and "a" is equal to the depth of the notch plus the depth of the crack. The latter two cases approximate the situation at the leading edge of a surface crack, Fig. 7.98(c), initiated by and growing from a corrosion pit or by a notch created by intergranular corrosion (Ref 163).

An expression for $\sigma_z$ was not included in Eq 7.12 to 7.14 because its value depends on position along the crack line. This is one of the most significant factors in fracture mechanics analysis. Consider first that the stresses are elastic. As P is increased, the thickness, B, tends to decrease. However, the material just above and below the crack surface is unloaded, since this is a free surface, and hence does not tend to contract as shown schematically in Fig. 7.100 (Ref 163). This material restricts a volume of material just beyond the crack line from contracting and, in doing so, generates tensile stresses, $\sigma_z$, in the z-direction extending into this volume along the crack line. Since $\sigma_y$ and $\sigma_x$ are also tensile stresses, a distribution of three-dimensional (triaxial) states of tension exists within the material parallel to the crack line; the triaxial stress will have a maximum value at the midpoint (B/2) of the crack line. When the constraint is sufficient (B large enough) to prevent contraction at a position in the z-direction, then strains are confined to the x-y plane and a state of *plane strain* is said to exist. At the plate surface intersected by the crack line, the value of $\sigma_z$ must be zero, leaving only $\sigma_x$

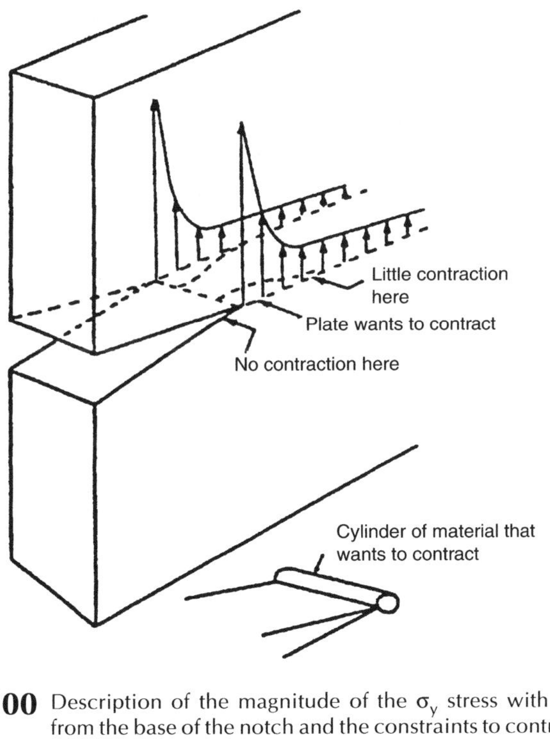


**Fig. 7.100** Description of the magnitude of the $\sigma_y$ stress with distance from the base of the notch and the constraints to contraction of a small cylinder of material at the leading edge. Source: Ref 163

and $\sigma_y$, which are in the x-y plane, and a state of *plane stress* exists within a small zone at the crack tip.

The forms of Eq 7.12 to 7.14 indicate that the stresses tend to infinity as $x \rightarrow 0$ (see Fig. 7.99 with $r = x$ in the plane of the advancing crack) (Ref 163). Two factors limit this stress. First, the equations apply to a crack tip with zero radius, which is not physically possible. Second, in real materials the stresses cannot be increased indefinitely without yielding by plastic flow, and hence there exists a small, approximately cylindrical volume of material undergoing plastic deformation along the crack line as illustrated schematically in Fig. 7.101 (Ref 164). The size of the cross section of this volume will depend on the yield strength of the material, being larger the lower the yield strength, and on the stress intensity factor, K. A plastic volume created under plane stress starts at the plate surface and decreases in cross section away from the surface as conditions change from plane stress to plane strain. Thus, the larger the value of B is, the larger the fraction of material along the crack front that will be in a state of plane strain. This change in the dominance of plane strain relative to plane stress as B increases is an important factor in governing the transition from failure dominated by plastic flow to that dominated by macrobrittle fracture or, in the limit, by cleavage.

In plane stress, plastic flow starts at the yield strength, $\sigma_{YS}$. In plane strain, however, the presence of $\sigma_z$ associated with the restriction of strain in the z-direction decreases the local shear stress such that the tensile stress in the y-direction must be increased to $\sigma_y = 3\sigma_{YS}$ before plastic flow starts. This is a limiting condition and would not be reached if failure by brittle cleavage occurred at a lower stress. Also, the state of triaxial stress within the localized volume along the notch front cannot extend to the leading edge of the notch since this is also a free surface at which $\sigma_x = 0$. At this surface, the state of stress must be plane stress but changes very quickly to plane strain with increasing distance into the

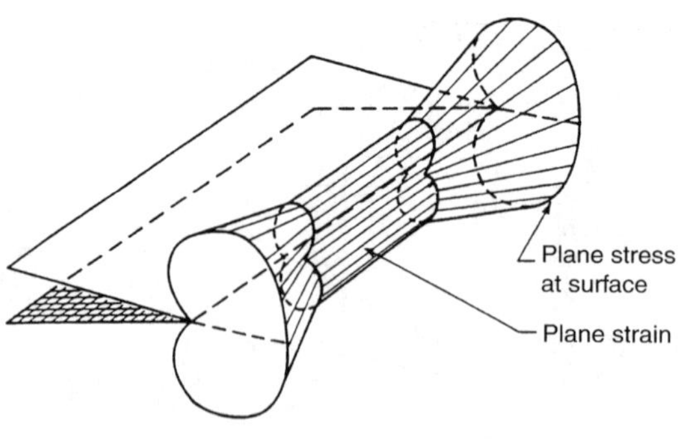


**Fig. 7.101** Through-thickness plastic zone in a plate of intermediate thickness. Larger plane stress volume starting at the surface tapers into the smaller plane strain volume with distance into the material parallel to the leading edge of the notch. Source: Ref 164

material from the leading edge of the notch. Therefore, on increasing P (and, therefore, $\sigma_y$), plastic flow at the notch edge starts at the yield strength of the material. On further increase in P, the size of the cylinder of plastically deformed material increases. At the advancing edge of the cylinder, the stress state approaches plane strain, and $\sigma_y$, if deformation is to continue, approaches $3\sigma_{YS}$. Beyond the advancing edge of the cylinder, an elastic stress field extends into the material with maximum value of $\sigma_y = 3\sigma_{YS}$ at the plastic/elastic stress-state interface and then decreases as $1/\sqrt{x}$ with increasing x. In most cases, the materials properties are such that the plastic zone is small compared to the elastic stress field, and therefore, Eq 7.12 to 7.14 are reasonably applicable, although they were based on elastic stresses only.

If B is small enough, the plane-strain condition may never be reached in the material along the crack line, and failure occurs by plastic flow, initially in the plane containing the crack line but then shifting to planes at 45° leading to a decrease in cross section similar to that observed on deformation of a tensile test bar of a ductile material. Thus, a given material may fail by local ductile flow if B is small but by nonductile fracture if B is large, in both cases the failure being macroscopically brittle because of the small amount of strain at fracture in each case. Conversely, for a given B, a material of high yield strength may exhibit macrobrittle behavior, but a material of lower yield strength would appear ductile.

The stress state of a crack, whether in a component or in a test specimen, is given by Eq 7.12 to 7.14. The effect of increasing P is contained in $K = \beta\sigma\sqrt{\pi a}$ ($\sigma = P/BW$) such that an increase in P results in an increase in K. The value of K at which failure occurs will depend on the thickness, B, of the material. As B decreases, the conditions for stress triaxiality decrease, and in the vicinity of the notch, plane stress predominates. Response to increasing P is then to produce plastic flow, and relatively large values of P (and hence K) are reached before failure. As B is increased, the triaxiality increases until a condition of plane strain predominates. Plastic flow is restricted, and values of $\sigma_y$ leading to brittle fracture are reached at lower P than for plastic flow. A schematic representation of the dependence of K on B is shown in Fig. 7.102 (Ref 163). The value of K for failure , $K_c$, decreases asymptotically to a limiting value corresponding to a stress causing fracture under the plane strain conditions that now exist. This limiting value of K is called the plane-strain fracture toughness, $K_{Ic}$. At small B, failure is associated with large amounts of plastic strain, which decreases as B increases. At larger values of B when $K_c = K_{Ic}$, the strain at fracture is very small. However, the actual fracture may occur as very localized ductile rupture or as brittle cleavage, depending on the material, its microstructure, temperature, etc.

Figure 7.102 shows that there is a value of B beyond which the fracture response in terms of $K = K_{Ic}$ is independent of the thickness and depends on the material only. Hence, $K_{Ic}$ is a material constant, with the significance that if a crack of length "a" exists at any place in the material and the constraints are such as to produce plane-strain conditions, then forces resulting in a $K = K_{Ic}$ will cause macrobrittle fracture. Forces resulting in lower values of K will not produce failure even though the defect exists. For B values not meeting the plane-strain condition, the value of K causing fracture depends on B, and while useful in failure analysis for components of the material having the same thickness as used in determining the K versus B curve, these values of K are not characteristic materials properties independent of component geometry. It is important to emphasize that this analysis holds for conditions under which the crack does not grow (otherwise K changes). Thus, under cyclic loading, fatigue-crack growth may occur. Corrosive environments also may cause growth under both static and repeated loading. Under any of these conditions, crack growth may increase until the now increased value of K becomes equal to $K_{Ic}$, at which time brittle failure will occur.

**Fracture Mechanics Investigations of Stress Corrosion under Static Loading.** These investigations of SCC incorporate the concepts of fracture mechanics using precracked test specimens. For a given environment, the crack growth rate is determined as a function of the applied stress intensity factor, K. Since K is proportional to the nominal stress, σ, the latter is the test variable governing the state of stress at the crack tip. The test specimens are of the general form of Fig. 7.69(i) with a sharp crack produced at the tip of the notch by repeated (fatigue) loading before exposure to the corrosive environment. Because of the importance of relating time-to-failure to crack-growth rate, stresses are se-

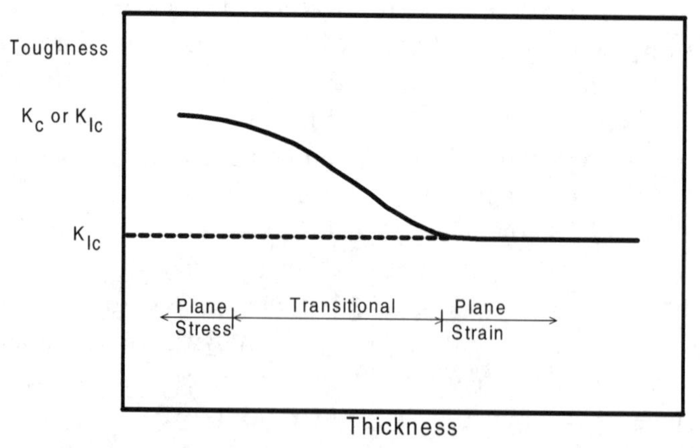


**Fig. 7.102** Dependence of toughness upon thickness showing the transition from plane stress to plane strain and asymptotic approach of $K_c$ to $K_{Ic}$. Redrawn from Ref 163

lected to give growth rates spanning several orders of magnitude. It is then practical to plot growth rate on a logarithmic scale as a function of K. A relationship of the form shown schematically in Fig. 7.103 is generally observed from which three stages of crack growth are noted (Ref 115). In stage I, the growth rate is very sensitive to increases in K and may approach an almost vertical slope; in stage II, the growth rate is almost constant for a range of increasing values of K; and in stage III, the growth rate increases rapidly with increasing K, the curve becoming asymptotic to the value of K corresponding to propagation of fracture at the critical stress intensity, $K_{Ic}$. That is, application of a load initiating this value of K causes immediate fracture propagation independent of the corrosive environment. For this reason, the stage III portion of the curve usually is not measured in stress-corrosion investigations. A linear relationship for stage I, as shown, implies that stress-corrosion cracks have a finite growth rate regardless of how low the applied stress may be. That is, there is no threshold value of K, $K_{TH}$, below which a stress-corrosion crack will not grow. Some alloy/environment systems exhibit an almost vertical stage I, in which case a threshold value for K designated $K_{ISCC}$ can be established; in other systems the curve bends to a limiting K value and a $K_{ISCC}$ can be assigned. Because of the long time required to measure very low growth rates with reasonable accuracy, a $K_{ISCC}$ may be designated as the value of K at a specified low growth rate as shown in Fig. 7.103. At least for many aluminum alloys, crack-growth

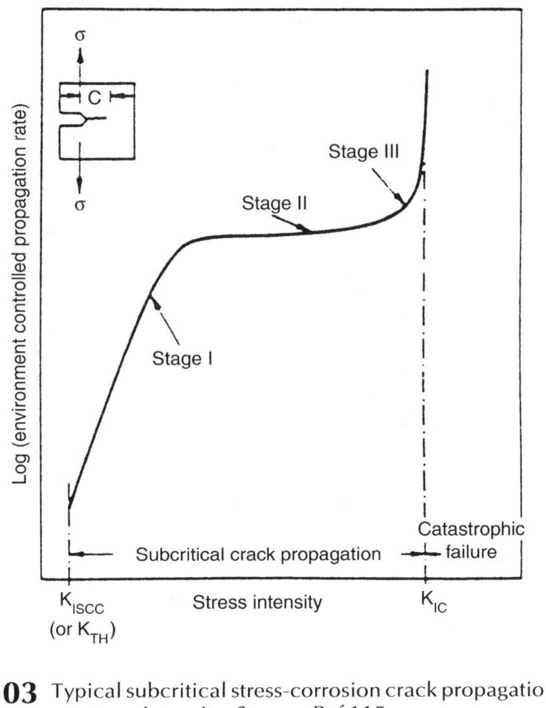


**Fig. 7.103** Typical subcritical stress-corrosion crack propagation rate versus stress intensity. Source: Ref 115

rates as low as $10^{-11}$ m/s, requiring extremely long observation times, have been measured (Ref 159). As a consequence, uncertainty may exist when data taken over shorter times at higher K values are extrapolated to estimate allowable stress intensities for safe times-to-failure.

It is important to emphasize the significance of the shape of the crack-growth rate versus stress intensity of the form shown in Fig. 7.103. Because of the steep slope of the relationship in the vicinity of a $K_{TH}$ (e.g., Fig. 7.104), crack growth behavior is frequently divided into the two ranges of K, stage I and stage II. Hence, in the presence of preexisting cracks, such cracks, on exposure to an environment, effectively either do not propagate at all ($K < K_{TH}$) or do so at a rate relatively independent of K ($K > K_{TH}$) (i.e., at the rates of stage II).

Models to account for stage I in Fig. 7.103 require a stress-dependent, environment-sensitive crack-opening mechanism at the leading edge of the crack accompanied by a very small corrosion rate on the crack sides. All of the mechanisms presented in the section "Mechanisms of Environment-Sensitive Crack Growth" that relate to events at the crack tip have been considered as controlling the crack-growth rate in stage I.

The small slope of the stage II section of the crack-growth rate versus K curve is attributed to corrosion-related, diffusion-controlled processes in the crack. Steady-state diffusion mechanisms are required to account for the fact that the crack growth rate is essentially constant

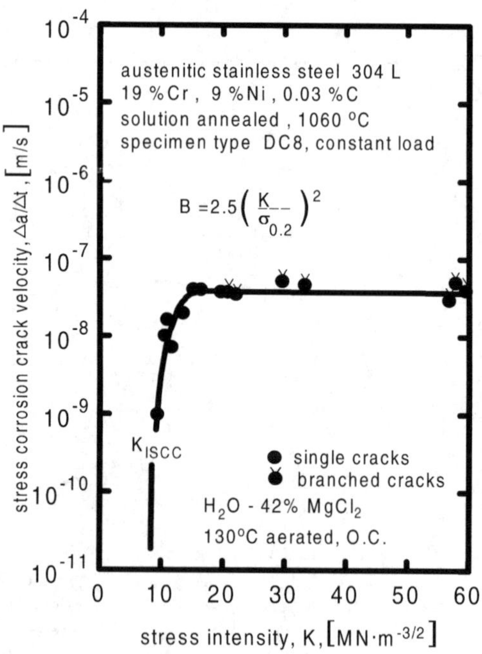


**Fig. 7.104** Effect of stress intensity on stress-corrosion crack growth rate for type 304L stainless steel in aerated $MgCl_2$ at 130 °C. Symbols indicate whether propagation occurs as a single or branched crack. Source: Ref 165

over the stage II range of crack-tip stress intensities. Both transport of the liquid into the crack and rates of diffusion of reactant and/or product species in the liquid in the crack may be rate determining. These species may include aggressive anions such as chlorides, hydrogen ions diffusing in directions governed by the bulk pH and the pH at the crack tip due to metal ion hydrolysis, and metal ions diffusing from the crack tip. Diffusion of cations through salt films, if they form, and possibly through thin passive films at the crack tip also may be rate controlling, making the stage II growth rate essentially independent of the stress intensity.

The dependence of stress-corrosion-crack velocity on stress intensity for a type 304L stainless steel in 42 wt% $MgCl_2$ in water at 130 °C is shown in Fig. 7.104 (Ref 165). The stage I section is nearly vertical and extrapolates to $K_{ISCC}$ = 8 MN/m$^{3/2}$. The stage II section is essentially independent of K. Figure 7.105 shows that in stage I the cracks are single, straight, and transgranular with only microscopic branches (Ref 165). This is in contrast to stage II, in which there are multiple macroscopic branches as shown in Fig. 7.106. Type 304 stainless steel, with higher carbon content than type 304L, is more susceptible to sensitization and, hence, to intergranular corrosion than type 304L. The crack velocity versus K relationship reflects the degree of sensitization as shown in Fig. 7.107 (Ref 166). Although this steel is usually water quenched, air cooling from 1060 °C results in mild sensitization as compared to severe sensitization resulting from reheating to and holding for 50 h at 630 °C (see the section "Intergranular Corrosion"). The more highly sensitized alloy has a $K_{ISCC}$ of about 8 MN/m$^{3/2}$ in the 22 wt%

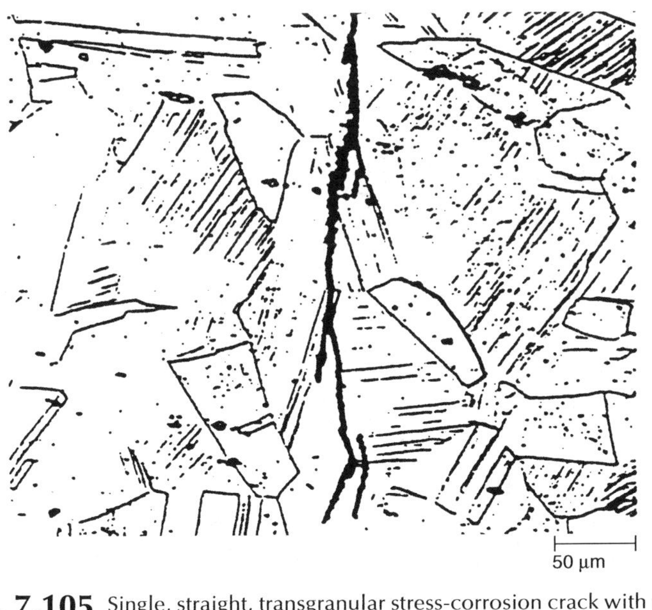


**Fig. 7.105** Single, straight, transgranular stress-corrosion crack with only microscopic branches. Conditions can be found in Fig. 7.104.
Source: Ref 165

NaCl solution at 105 °C, which compares to about 35 $MN/m^{3/2}$ for $K_{ISCC}$ for the less-severely sensitized alloy. The air-cooled alloy exhibited transgranular stress corrosion and the severely-sensitized material cracked intergranularly as shown in Fig. 7.108.

The effect of stress intensity on the stress corrosion crack-growth rates of seven austenitic stainless steels in aerated 22% NaCl solution at 105 °C is shown in Fig. 7.109 (Ref 166). It is evident that the relationship is sensitive to the particular austenitic alloy and that the composition of the austenite has a much greater effect on the threshold stress intensity, $K_{TH} = K_{ISCC}$, than on the maximum growth rate, which is relatively independent of the stress intensity (the plateau region). The effect of nickel content on the stress-corrosion threshold-stress-intensity of 17 alloys (including those shown in Fig. 7.109) with approximately 18 wt% Cr is shown in Fig. 7.110. The shape of this curve is similar to that of Fig. 7.89, both showing a minimum resistance to SCC in the vicinity of 10 to 20 wt% Ni. Corresponding to the nickel content for this minimum in the threshold stress was a maximum in the crack-growth rate, being over 20 times greater for nickel concentrations corresponding to the minimum in Fig. 7.110 compared with alloys with

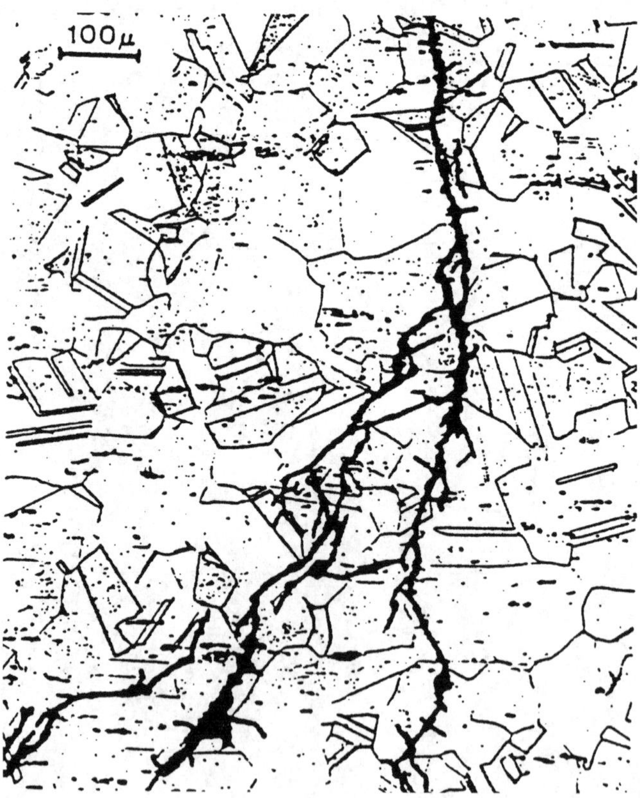


**Fig. 7.106** Stress-corrosion crack with three macroscopic branches. Conditions can be found in Fig. 7.104. Source: Ref 165

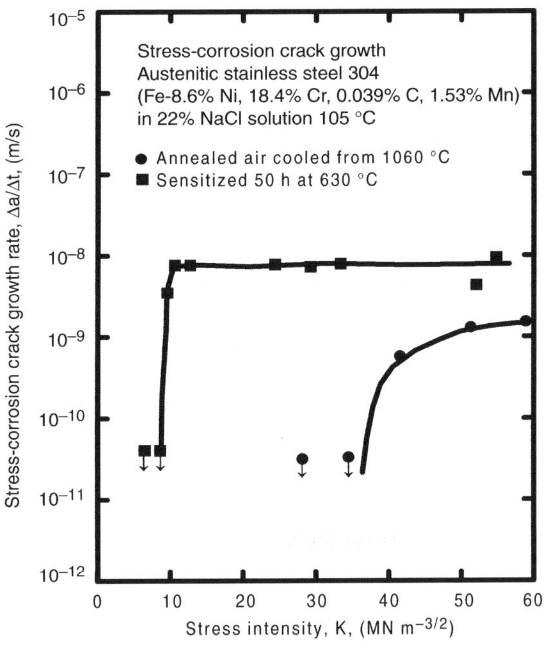


**Fig. 7.107** Effect of sensitization heat treatment on stress-corrosion crack-growth rate of type 304 stainless steel in 22% NaCl solution at 105 °C. Sensitized 50 h at 630 °C. Increased chromium depletion at grain boundaries results in increased growth rate and lower threshold K. Source: Ref 166

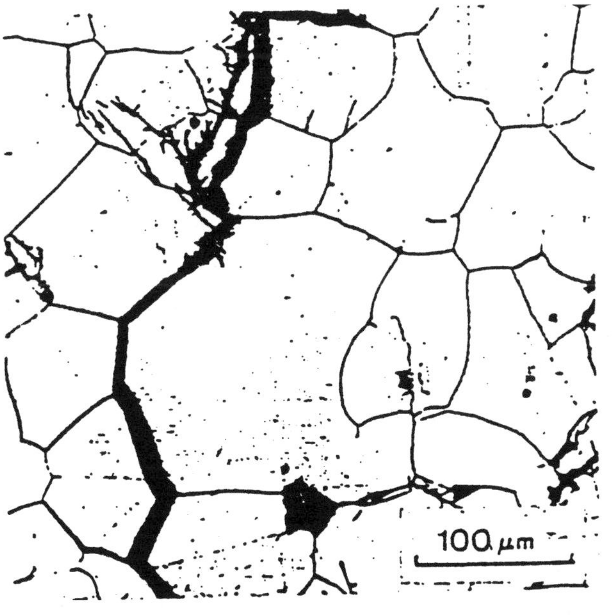


**Fig. 7.108** Intergranular corrosion in type 304 stainless steel resulting from sensitization heat treatment of 50 h at 630 °C. Conditions can be found in Fig. 7.107. Source: Ref 166

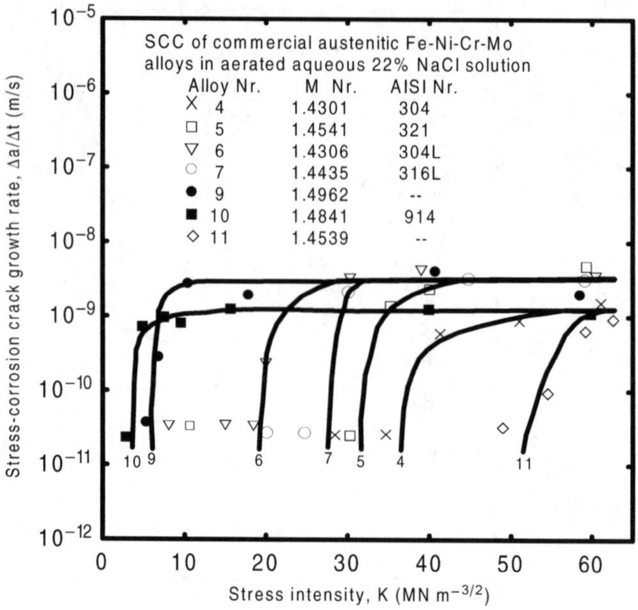


**Fig. 7.109** Effect of stress intensity on the growth rate of stress-corrosion cracks in several austenitic stainless steels. Alloy compositions can be found in Ref 166. Redrawn from Ref 166

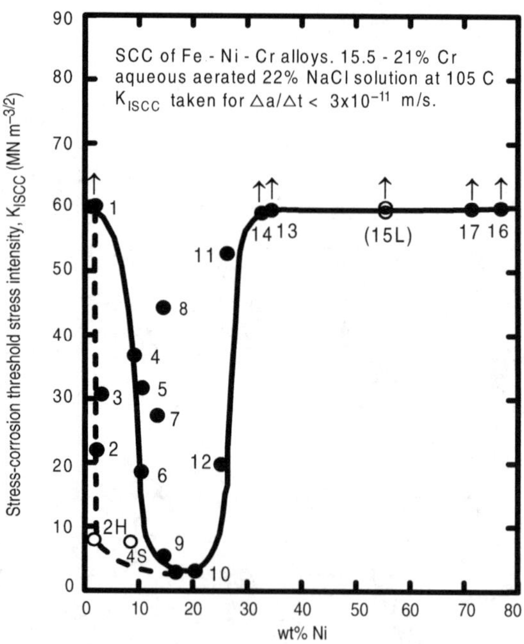


**Fig. 7.110** Effect of nickel content on stress-corrosion threshold stress intensity of Fe-Ni-Cr alloys with about 18% Cr. Alloy compositions can be found in Ref 166. 2H = AISI 431, untempered martensite. 4S = AISI 431, sensitized. Source: Ref 166

>32 wt% Ni. The very low nickel alloy (No. 1) is ferritic and did not stress-corrosion crack in the 22% NaCl solution at 105 °C; also, the alloys containing >32 wt% Ni did not crack. The latter are stable austenitic alloys. Alloys designated by the data points 4 to 11 are heat treated to result in a metastable austenitic structure. Whether the susceptibility to stress cracking over this composition range is associated with this metastability or is related to stress-corrosion mechanisms sensitive to the nickel concentration has not been established. The stress-corrosion threshold stress for the austenitic alloys in Fig. 7.110 (15.5 to 21.0 wt% Cr, 13.0 to 24.5 wt% Ni) increases with molybdenum concentration as shown in Fig. 7.111. It is evident that molybdenum has a strong favorable influence in increasing resistance to SCC as it has for reducing tendency for pitting corrosion.

Examples of the influence of several environmental and compositional variables on the effect of stress intensity on the stress-corrosion crack velocity in 7000-series aluminum alloys are shown in Fig. 7.112 to 7.115 (Ref 159). Figure 7.112 shows the reasonably close scatter of the data in multiple tests on two high-strength alloys exposed to saturated aqueous NaCl at 23 °C. It should be noted that, although both of these alloys are in the 7000 series, there is a large difference in crack-growth rate at a given K value. In general, fracture-mechanics-type investigations of aluminum alloys in a given medium have shown that the crack-growth rate versus K relationship is very sensitive to alloy composition and heat treatment, both variables governing the microstructure. Figure 7.113 is representative of the effect of various

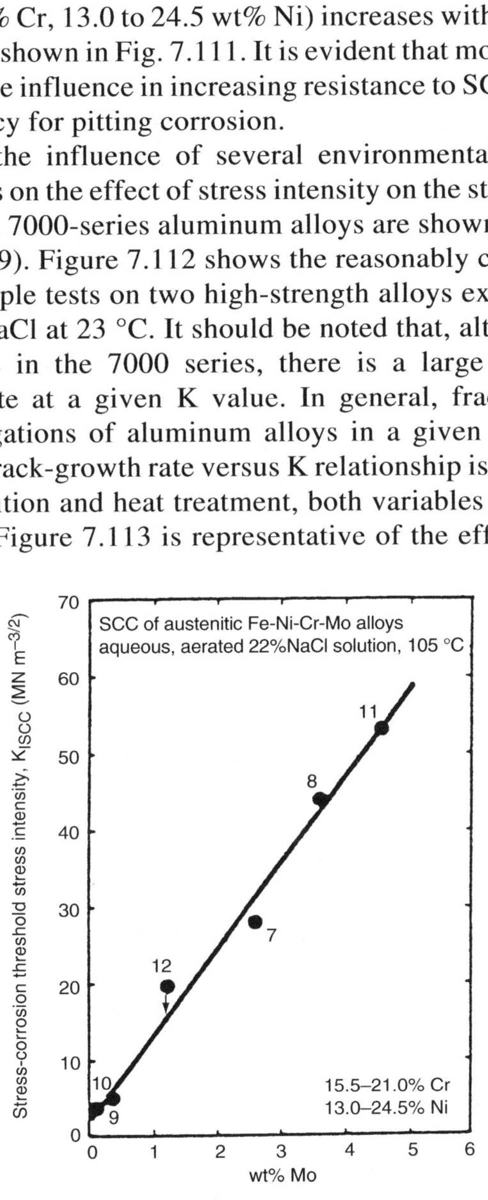


**Fig. 7.111** Effect of molybdenum content on stress-corrosion threshold stress intensity of austenitic stainless steels. Source: Ref 166

anions on the K versus crack-velocity curve for one of the high-strength aluminum alloys. The halide anions ($Cl^-$, $I^-$, and $Br^-$) increase the growth rate in the stage II section of the curve by a factor greater than $10^2$ over the other environments listed. In fact, the crack velocity is relatively insensitive to the latter rather large group of anions in water and to distilled water itself, the stage II velocity ranging to somewhat greater than $10^{-8}$ m/s. Although not shown in the figure, the stage II velocity decreases progressively with dilution of the halide-ion concentration. As would be expected for the effect of temperature on corrosion rate, both stage I and II sections of the curve of crack velocity versus stress intensity shift to higher crack-growth rates with increase in temperature (Fig. 7.114). The effect is greater in stage I, indicating that the influence of temperature is greater on crack-tip growth mechanisms than when the growth rate is diffusion controlled as proposed for stage II.

Fracture mechanics investigations have shown that the crack-growth rates of a number of alloys are sensitive to the relative humidity of the

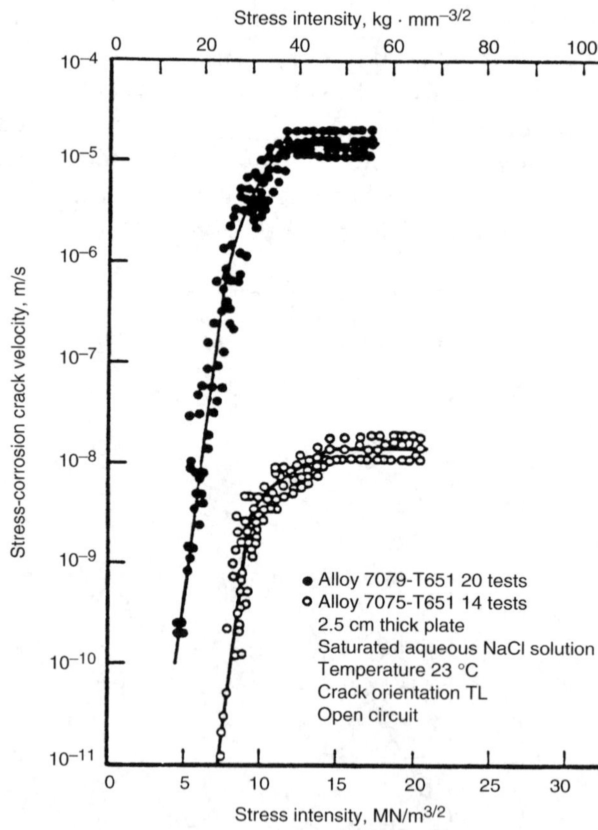


**Fig. 7.112** Dependence of corrosion-crack-growth rate on stress intensity for two high-strength aluminum alloys in saturated NaCl solution at 23 °C. Crack orientation TL (stress in transverse direction; crack propagation in longitudinal direction). Source: Ref 159

air during stressing. High-strength aluminum alloys are particularly sensitive as shown in Fig. 7.115 where both the stage I and stage II rates are decreased as the humidity is decreased (Ref 159). The effect is to be expected in view of the reactivity of aluminum with air and the enhanced effect of adsorbed water in producing oxide films. Crack-tip, stress-induced dislocation movement results in film cracking and establishes the mechanism of passive film cracking, substrate exposure, and repassivation. As discussed previously, depending on the alloy, the rate-controlling mechanism may be anodic dissolution or hydrogen embrittlement.

It should be noted that in Fig. 7.112 to 7.115, the slope in stage I is not as steep as for the stainless steel in Fig. 7.104. A vertical stage I implies a threshold value of K ($K_{TH}$ or true $K_{ISCC}$) below which SCC is not initiated as discussed in relationship to Fig. 7.103. At least for some environments, it appears that a true $K_{ISCC}$ value can be assigned for stainless steels and titanium alloys. It is uncertain that this can be done for

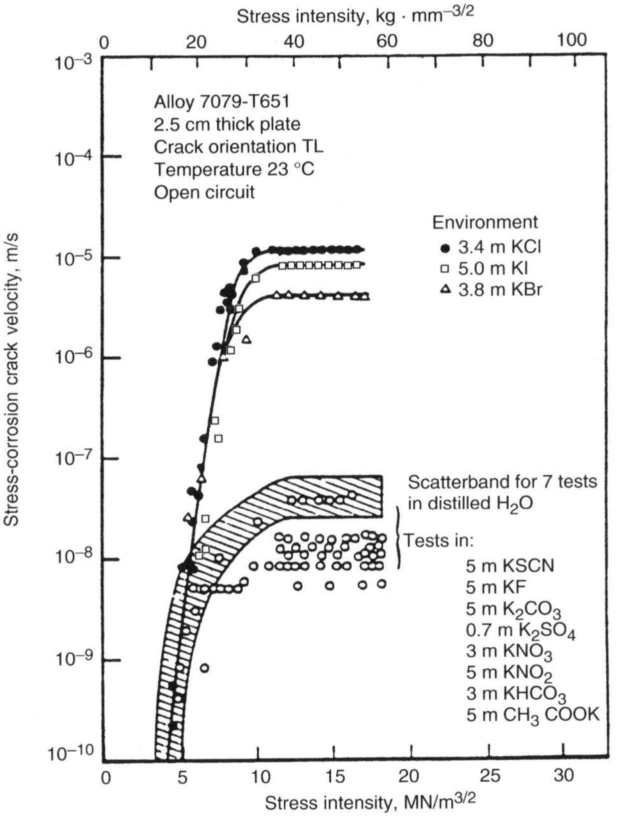


**Fig. 7.113** Dependence of stress-corrosion-crack-growth rate on stress intensity of a high-strength aluminum alloy in several aqueous environments. Crack orientation TL (stress in transverse direction; crack propagation in longitudinal direction). Source: Ref 159

aluminum alloys, and perhaps others, in which case crack growth occurs at even the lowest practical stress level. For these alloys, the design criterion to avoid failure by stress corrosion is based on the stress intensity, K, and the time for the crack to grow to an unacceptable extent rather than on holding the stress intensity below some threshold value below which no growth will occur.

For those metal/environment systems with a reasonably well-defined $K_{ISCC}$, a procedure resulting in a graphical representation illustrated in Fig. 7.116 may be used (Ref 161). Constant nominal stresses, $\sigma$, are applied to a set of fracture-mechanics-type specimens to provide initial stress intensities, K, estimated to be between $K_{ISCC}$ and $K_{Ic}$ (solid circles along the ordinate at 1 min in Fig. 7.116). These specimens are placed in the environment, and the time at which fracture occurs is observed. During exposure, stress-corrosion cracks will propagate, increasing the crack depth, "a", and thus causing the stress intensity to increase ($K = \beta\sigma\sqrt{\pi a}$). When the crack has propagated to a critical depth at which $K = K_{Ic}$, the specimen fails by rapid fracture by mechanical

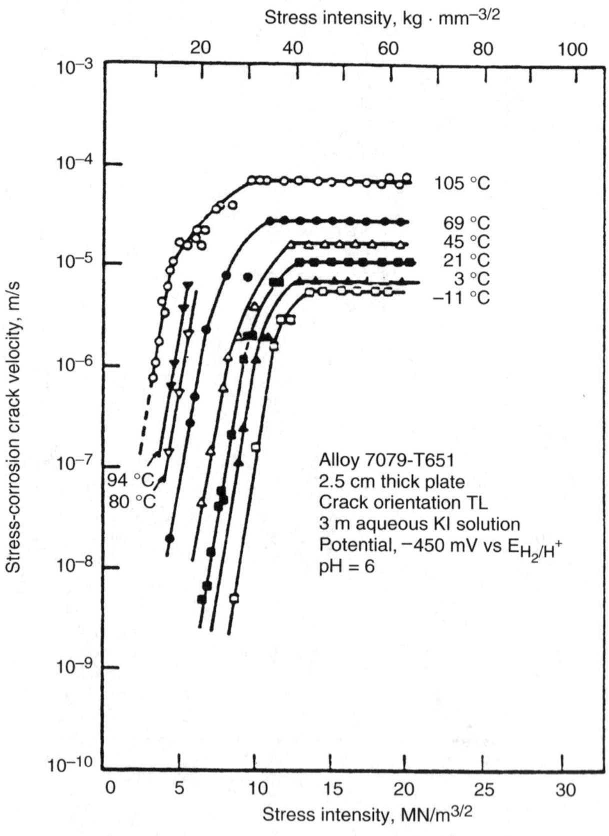


**Fig. 7.114** Dependence of stress-corrosion-crack-growth rate on stress intensity for a high-strength aluminum alloy at various temperatures. Source: Ref 159

crack-opening mechanisms independent of the environment. The $K_{Ic}$ at fracture is calculated from the value of "a" at fracture, determined by examining the fracture surface and measuring the position of the transition from the surface discolored by corrosion products to the rapidly propagating mechanical crack. The times of failure and the calculated values of $K_{Ic}$ are designated by the crosses, and the dashed lines connecting to the initial values of K are estimates of the increases in K with time due to stress-corrosion crack propagation. The fact that the calculated $K_{Ic}$ values are reasonably constant, independent of exposure time, supports the conclusion that the final failure was related to the fracture behavior of the material and not to the environment. It is evident that there is a value of K that results in a dashed curve which converges to $K_{ISCC}$ and will be associated with an indefinitely long time to failure. It may be possible to estimate $K_{ISCC}$ from the initial set of specimens, and if not, the initial results provide a guide to the selection of loads providing K values spanning $K_{ISCC}$.

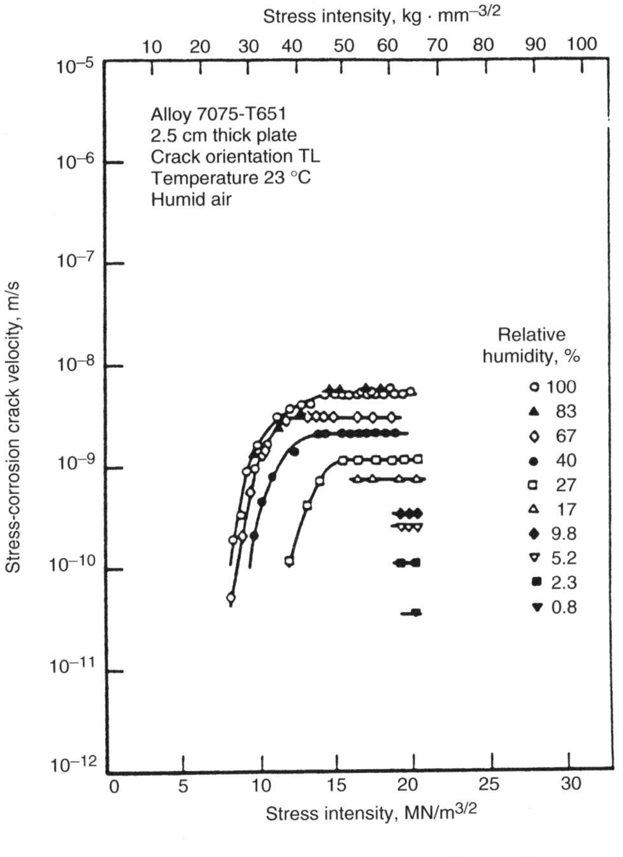


**Fig. 7.115** Dependence of stress-corrosion-crack-growth rate on stress intensity for a high-strength aluminum alloy at several relative humidities. Crack orientation TL (stress in transverse direction, crack propagation in longitudinal direction). Source: Ref 159

**Fracture Mechanics Investigations of SCC under Repeated Loading: Corrosion Fatigue.** The simplest representation of the response of a material to repeated loading is a plot of the cyclic stress amplitude versus the number of cycles to failure—the S-N curve. A representative example of the influence of environment and surface condition is shown in Fig. 7.117(a) and (b) for a 13 wt% Cr ferritic steel cyclically loaded as a rotating bent beam in air, distilled water and 1 wt% aqueous NaCl solution (Ref 167). Figure 7.117(a) shows the response for smooth bars and, Fig. 7.117(b), for notched bars. Both of the curves for tests in air show the characteristic limiting stress below which the failure does not occur regardless of the number of cycles. This behavior is generally exhibited by steels in nonreactive environments and defines a material property known as the endurance limit. For some other materials, such as aluminum alloys, the fatigue curve continues to slowly decrease, and a fatigue-resistance property cannot be assigned without defining an endurance limit as the stress leading to failure at a specified number of cycles. It is evident that the reactive environments of distilled water and 1 wt% NaCl solution lower the curves and make the definition of an endurance limit less certain since the curves tend to continue to decrease with increasing numbers of cycles. Also, in the presence of a notch, the curves converge to a smaller range of stress levels than for the smooth bars. These results are presented as representative of conventional fatigue behavior. The positions of the curves, of course, will be sensitive to the environment, notch geometry, and stress history, particularly the ratio of maximum-to-minimum stress per cycle.

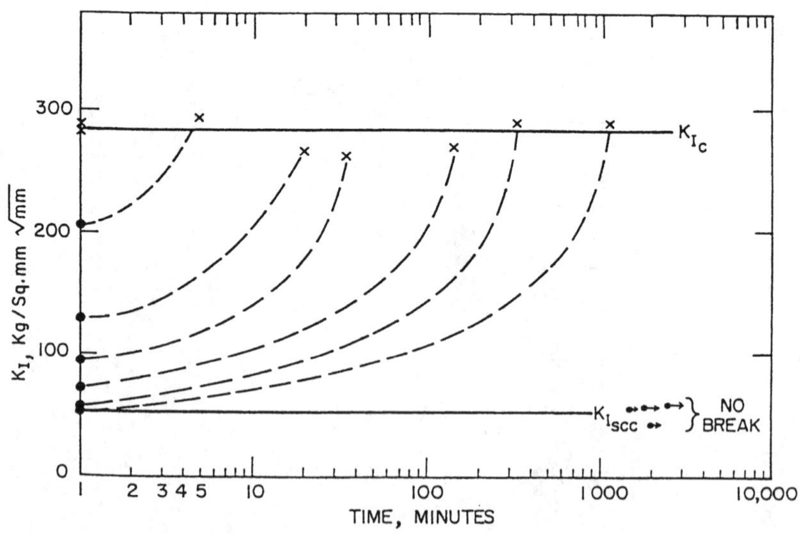


**Fig. 7.116** Effect of stress intensity on time for cracking of AISI 4340 steel in 3.5% NaCl solution in cantilever tests under dead load. Dashed lines show estimated change in stress intensity from an initial value (solid circles) to final fracture (crosses). Source: Ref 161

Fatigue cracks are initiated at stresses below the conventional yield stress at which bulk plastic flow occurs. In this stress range, surface inhomogeneities and favorably oriented grains allow slip by movement of dislocations, which produce surface offsets that, due to localized work hardening (slip interference), are not reversed as the stress is reversed. As a result, surface intrusions and extrusions of the form proposed in Fig. 7.118 are produced (Ref 162). The intrusions and extrusions,

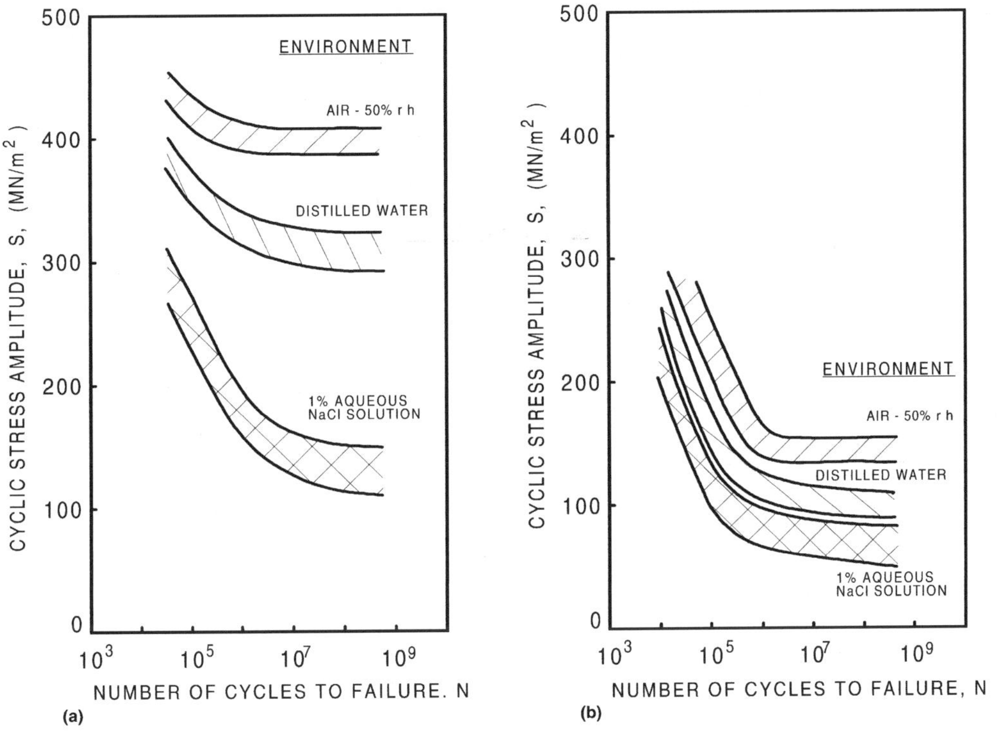


**Fig. 7.117** Conventional S-N fatigue curves for 13% Cr steel (X 20 Cr 13) determined in indicated environments. (a) Smooth rotating bend specimen. (b) Notched rotating bend specimens. ($\sigma_{0.2}$ = 610–650 MN/m$^2$; $\sigma_{UTS}$ = 760–830 MN/m$^2$; mean load = 0; frequency = 50 Hz; temperature = 23 °C.) Source: Ref 167

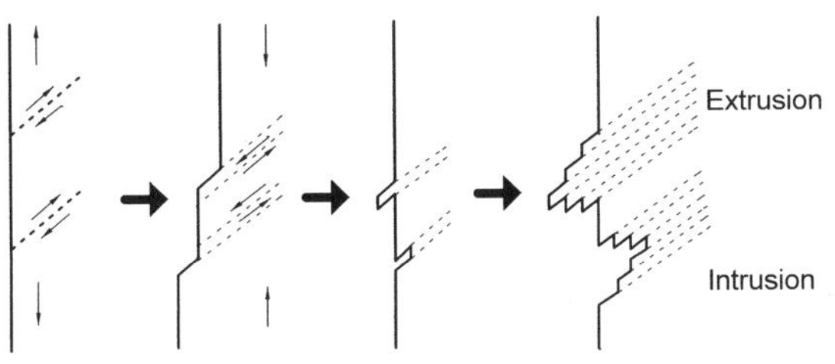


**Fig. 7.118** Model of local plastic deformation by dislocation glide producing surface extrusions and intrusions that initiate fatigue cracks. Source: Ref 162

cracks along slip bands, and cracked or debonded nonmetallic inclusions, initiate fatigue cracks that grow incrementally with each stress cycle. In the presence of aggressive environments, the stress-corrosion mechanisms discussed in the section "Mechanisms of Environment-Sensitive Crack Growth" become repetitively active because of the cyclic stress.

Environment-sensitive fracture under cyclic loading has been investigated extensively by use of fracture-mechanics-specimen configurations with interpretation of test results in terms of the principles of fracture mechanics. From these investigations, the important variables in corrosion fatigue have been more clearly defined, and progress has been made in relating these variables to mechanisms for crack growth under cyclic loading in aggressive environments. The investigations have led to procedures for using cyclic-stress crack-growth rate data to predict the time for a surface flaw to grow to failure. Further, in components such as pipes and tanks, conclusions can be made as to whether failure will be by relatively benign leaking, ductile rupture fracture, or catastrophic brittle cleavage fracture.

The major variables in fracture-mechanics investigations of environment-sensitive cracking under cyclic loading include (Ref 163, 168):

- The maximum ($\sigma_{max}$) and minimum ($\sigma_{min}$) stresses in the cycle, and through $K = \sigma\sqrt{\pi a}$, the maximum and minimum stress intensities, $K_{max}$ and $K_{min}$. The associated stress-intensity-factor range, $\Delta K = K_{max} - K_{min}$, also is used.
- The stress ratio, $R = \sigma_{min}/\sigma_{max} = K_{min}/K_{max}$. $R = 0$ for a cyclic stress from $\sigma_{min} = 0$ to any $\sigma_{max}$, and $R < 0$ if $\sigma_{min}$ is compressive (negative). Since $K_{max} = \Delta K/(1 - R)$, only two of these variables are independent.
- The frequency of the cyclic stress and the stress application profile (i.e., sinusoidal wave, triangular wave with differences in rise and fall time, and square wave)
- The crack-growth rate, usually expressed as da/dN, determined from the slope of the curve of crack depth, a, versus the number of stress cycles, N

The response of a material/environment system to cyclic loading of a fracture mechanics specimen usually is expressed graphically as log crack-growth rate, log da/dN, versus log crack-tip stress-intensity range, log $\Delta K$, imposed by the cyclic loading. Less frequently, the growth rate is plotted versus the maximum stress intensity factor, $K_{max}$, during the stress cycle. Use of $K_{max}$ as the independent variable has an advantage when comparisons are being made between cracking response under sustained stress (zero frequency), relating to SCC as discussed in the section "Fracture Mechanics Investigations of Stress Corrosion under Static Loading," and cracking under cyclic loads relating to corrosion fatigue. The principal parameters are the thermal or mechanical treatment of the material, the frequency (f), the stress-intensity-factor range ($\Delta K$), the ratio (R) of the cycle, and the cycle profile.

Since the interrelationship of these variables in establishing the relationship between fatigue crack growth rate and stress intensity is sensitive to each specific material/environment system, only limited generalizations of corrosion-fatigue growth-rate phenomena have been possible. Because of these complexities, the following discussion is restricted to the types of relationships of crack-growth rate versus stress intensity that are represented by Fig. 7.119 to 7.125. These examples illustrate the types of influences associated with the several variables and form the basis for discussion of mechanisms proposed to account for the shape of the crack-growth rate versus stress-intensity range relationship. The materials are a low-carbon and a high-strength steel, an aluminum-base alloy and a titanium-base alloy.

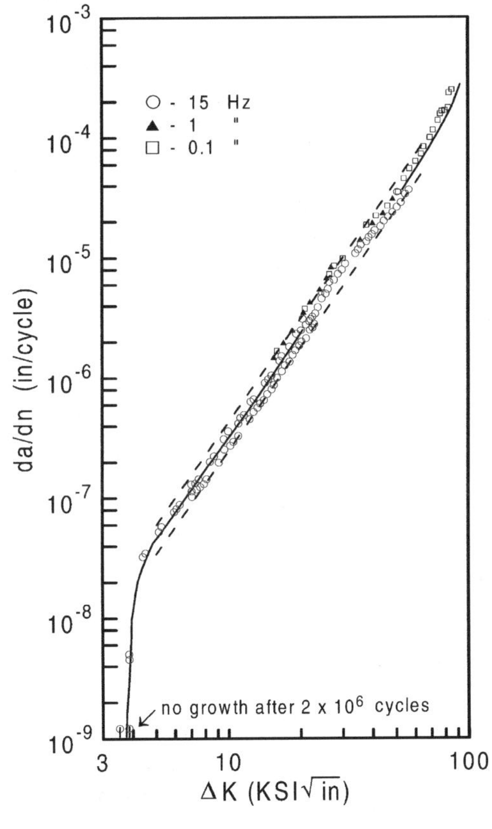


**Fig. 7.119** Fatigue-crack-growth rates as a function of stress-intensity amplitude for X-65 line pipe steel in air. Frequency 0.1–15 Hz, at R = 0.2. Redrawn from Ref 169

Figure 7.119 shows the fatigue crack growth rates for carbon steel tested in air at frequencies of 0.1, 1, and 15 Hz (Ref 169). It is evident that all values follow a linear log-log relationship over an intermediate range of $\Delta K$ but curve downward at lower stress-intensity ranges approaching a threshold value, $\Delta K_{TH}$, below which fatigue crack growth does not occur. At higher stress-intensity ranges, the curve bends upward toward a value of $\Delta K$ having a $K_{max} = K_{Ic}$, where fracture occurs due to a single overload at $K_{max}$. It is evident that the growth rate is not sensitive to load frequency in the inert environment; other measurements have shown that the crack-growth rate is essentially independent of the cycle profile (i.e., sinusoidal, sawtooth, or square wave). A similar behavior is observed for all of the materials to be discussed and this type of curve, based on inert environment tests, is used as a reference to which curves showing the effects of environment and other test variables are compared. For reactive materials, particularly aluminum-base alloys, very low-humidity environments are required to establish the reference curve. Electron microscopy examinations of the fracture surfaces reveal successive striations corresponding to the cyclic loading, the mechanism of crack opening being brittle cleavage or ductile depending on the material, $\Delta K$ and R.

In many metal/environment systems, the corrosion-fatigue behavior can depend on whether the stress-intensity range, $\Delta K = K_{max} - K_{min}$, in cyclic loading extends to a $K_{max} > K_{ISCC}$ determined under sustained load. If this occurs, and the time during the cyclic loading for which this condition exists is long enough, a contribution due to SCC is added to the growth rate associated with mechanical crack-opening mechanisms which may also be influenced by the environment. As a consequence, corrosion-fatigue behavior is frequently differentiated between crack growth behavior below and above a $\Delta K$, designated as $\Delta K_{ISCC}$. Corrosion fatigue below this $\Delta K_{ISCC}$ has been designated as "true corrosion fatigue," whereas above $\Delta K_{ISCC}$, stress-corrosion-enhanced corrosion fatigue occurs. Establishing a value for $\Delta K_{ISCC}$ is complicated by the fact that $K_{ISCC}$ is determined under sustained loads but is being related to crack growth rate under cyclic loading. Setting $\Delta K_{ISCC} \cong K_{ISCC}$ is justified under the conditions that when R ($=K_{min}/K_{max}$) is zero, then $K_{min} = 0$ and values along the abscissa, $\Delta K$, are also values of $K_{max}$. Thus, when $R = 0$ and if $\Delta K = K_{max} > K_{ISCC}$, then during part of the cycle the stress intensity will be large enough to cause an increment of crack growth by SCC mechanisms. In assuming this relationship, it is important to recognize that cyclic loading may affect the sustained SCC mechanisms. In this case, the $\Delta K$ below which true corrosion fatigue occurs is designated as $\Delta K_{SCC}$ to distinguish it from $\Delta K_{ISCC}$ (Ref 170). It is generally found that $\Delta K_{SCC} < \Delta K_{ISCC}$. It should be noted that in relating sustained-load cracking data to that obtained by cyclic loading, the stress-corrosion crack-growth data are presented as da/dt, whereas the corrosion-fatigue data are presented as da/dN, which must be multiplied by frequency to establish a time rate.

Corrosion fatigue crack growth data for a high-strength maraging steel in air and in 3% NaCl solution at frequencies of 6, 60, and 600 cpm are shown in Fig. 7.120 (Ref 171). The behavior is representative of true corrosion fatigue since the stress-intensity-factor range of these data is below $K_{ISCC}$ (identified along the abscissa). The log crack-growth rate versus log $\Delta K$ curves are all linear and parallel to the reference curve of the inert air environment. Similar behavior is observed for some aluminum and titanium alloy/environment systems. Crack-growth rate is observed to increase with decreasing cycle frequency. Measurements were not made at sufficiently low stress-intensity-factor ranges to determine whether negative deviations from linearity occur for each curve and thereby establish a threshold $\Delta K_{TH}$ as observed in Fig. 7.119. The data do suggest, however, that if negative deviations terminate at a $\Delta K_{TH}$ for each curve at lower $\Delta K$s, the threshold stress-intensity range would be lower in the 3% NaCl solution than in air and would be progressively lower with decreasing frequency. The requirement that the growth rate versus $\Delta K$ be essentially linear under environmental effects below $\Delta K_{SCC}$ appears to be an accepted requirement for true corrosion

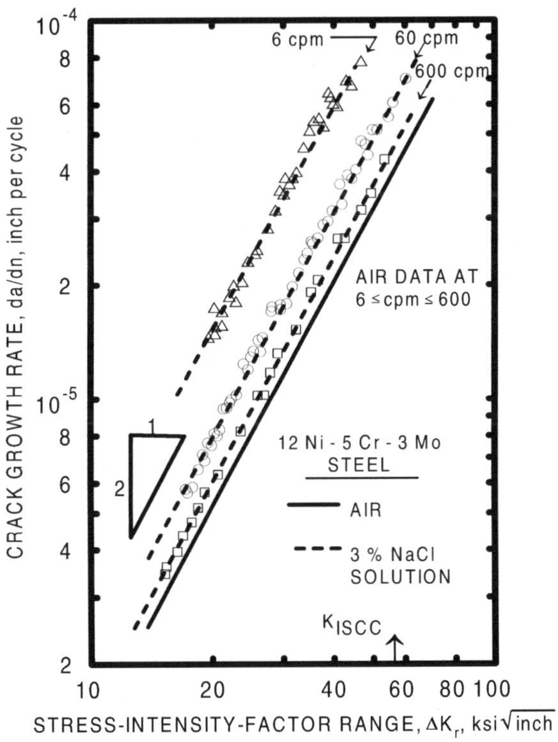


**Fig. 7.120** Corrosion-fatigue-crack-growth rate as a function of stress-intensity range for a maraging steel in air and 3% NaCl solution.
Source: Ref 171

fatigue. A less certain requirement is that the environmental data lie parallel to the inert data and that there should be a systematic shift, if any, with frequency.

The increase in fatigue crack growth rate of the high-strength steel on exposure to the 3% NaCl solution has been attributed to hydrogen embrittlement. Results of changing frequency and cycle profile indicate that the hydrogen-producing step occurs only during increasing strain at the crack tip associated with the increasing stress part of the cycle (Ref 171). During this part of the cycle, new metal surface is being exposed to the environment at which reduction of water or hydrogen ions occurs to produce adsorbed H-atoms; these diffuse into the plastically-deformed region just in advance of the crack tip and migrate toward the plastic/elastic interface where the triaxial stress state is greatest, and the expanded lattice supports a larger hydrogen concentration. Voids are initiated in this region and propagate as cracks back to the fatigue-crack interface. This mechanism of hydrogen-embrittlement cracking accounts for the fact that the crack-growth rate is faster than is estimated for anodic dissolution. Hence, hydrogen embrittlement is proposed as the crack-growth, rate-controlling mechanism. Since lower cyclic frequencies increase the time-per-cycle for exposure of new surface with the environment, more hydrogen is absorbed, and the increment of growth by cracking is increased. Further, it has been shown that the increase in crack-growth rate is the same for sinusoidal and symmetrical sawtooth cycles at the same frequency, both having essentially the same strain rate during the cycle rise time (Ref 171). For unsymmetrical sawtooth-cycle profiles, the longer the stress rise time is, the greater the increase in crack-growth rate will be. In contrast, for square waves of very steep rise time, the effect of the environment is very small. In this case, most of the cycle is at constant stress, and the enhancement of the growth rate over the purely mechanical rate is small. A hydrogen-embrittlement crack-growth mechanism also is supported by observations using electron microscopy. At high frequencies, ductile striations are associated with growth increments per cycle, but at low frequencies, in stage II, the striations are characteristic of brittle fracture. This is consistent with the longer time available for hydrogen embrittlement per cycle at the lower frequency.

The fatigue crack growth rate as a function of stress-intensity range is more complex for those metal/environment systems that undergo sustained-load SCC and exhibit a $K_{ISCC}$ as shown in Fig. 7.103. This more complex behavior also is observed for those systems, such as some high-strength steels in seawater and Ti-base alloys, that do not stress-corrosion crack under sustained loads but will crack under cyclic loads. The corrosion-fatigue crack growth rate as a function of stress-intensity range for the high-strength steel 4340M is shown in Fig. 7.121 (Ref 172). The reference curve is that obtained with cyclic loading in vacuum at a frequency of 4 Hz. The data follow an approximately linear log-log relationship for ΔK from 10 to 40 $MN/m^{3/2}$ and extrapolate at the lower limit to a $\Delta K_{TH} \cong 6$ $MN/m^{3/2}$ and to an upper limit of $K_{Ic} \cong 100$ $NM/m^{3/2}$. Although data are not shown, the crack-growth rate in vacuum as a function of ΔK was essentially independent of frequency.

Sustained SCC measurements on the 4340M steel in distilled water established that $K_{ISCC} = 13$ $MN/m^{3/2}$. The effect of this environment on the corrosion fatigue behavior for this steel for frequencies from $10^{-3}$ to 4 Hz is included in Fig. 7.121. Examination of the data leads to the following observations:

- The environment increases the fatigue crack growth rate at all frequencies.
- Below ΔK = 10 $NM/m^{3/2}$, the growth rate increases rapidly with increase in ΔK. The relationship is linear, independent of frequency, and indicates a threshold $\Delta K_{TH} \cong 6$ $NM/m^{3/2}$. The threshold

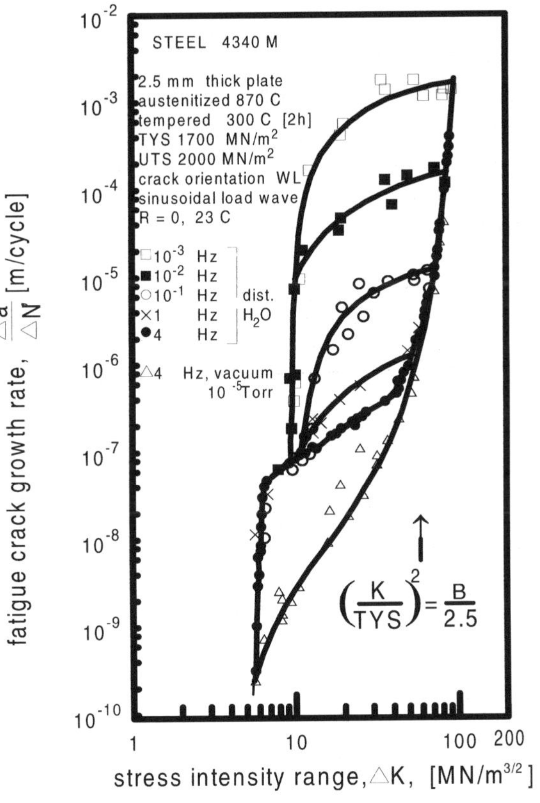


**Fig. 7.121** Corrosion-fatigue-crack-growth rate as a function of stress-intensity range for high-strength 4340M steel in vacuum and distilled water at 23 °C. Data for vacuum and indicated frequencies and R = 0. Source: Ref 172

stress-intensity-range values are approximately the same for the two environments.

- For $\Delta K > 10$, the crack growth rate increases rapidly with increase in $\Delta K$, the increase starting at higher $\Delta K$ the higher the frequency. The slope of the data, which are initially approximately linear, decreases toward a "plateau" similar to that observed for sustained-stress SCC.
- The crack growth rate at which the plateau ranges of the data occur increases with decreasing frequency.
- Crack growth rates for all frequencies merge with the growth rate in vacuum as $\Delta K$ increases.
- The effect of environment and frequency on fatigue crack growth rate can be very large. At $K = 11$ to $12$ MN/m$^{3/2}$, the growth rate at 4 Hz in distilled water is about $10^2$ greater than in vacuum and the rate at $10^{-3}$ Hz is greater by an additional factor of about $10^3$.

The linear relationship between log fatigue crack growth rate and log $\Delta K$ for $\Delta K < 10$ MN/m$^{3/2}$, and the observation that $K_{max}$ for these $\Delta K$s is less than $K_{ISCC}$ (=13 MN/m$^{3/2}$), are characteristic of true corrosion fatigue as discussed previously. In this case, the relationship in this $\Delta K$ range is independent of frequency, which differs from that of Fig. 7.120 for the high-strength maraging steel. For that alloy, the slopes were the same for growth rates in the aggressive environment and the inert environment, but the growth rates increased with decrease in frequency.

The nonlinear form of the crack-growth curves in Fig. 7.121 at $\Delta K > 10$ MN/m$^{3/2}$ is characteristic of corrosion fatigue for $K_{max} > K_{ISCC}$ where stress corrosion becomes a major contributor to the crack growth. These sections of the fatigue crack growth curves, particularly for $10^{-2}$ and $10^{-3}$ Hz, are very similar in form to stress-corrosion crack growth versus K curves (Fig. 7.103) where the threshold stress is designated as $K_{ISCC}$. The threshold $\Delta K$ for a stress-corrosion contribution to crack growth in Fig. 7.121 is about 10 MN/m$^{3/2}$ at $10^{-3}$ Hz. This is designated as $\Delta K_{SCC}$ ($10^{-3}$ Hz), the threshold stress-intensity range for stress corrosion under cyclic stressing, to differentiate it from $\Delta K_{ISCC}$ determined under sustained stress. Since for this steel/environment system, $\Delta K_{ISCC} = K_{ISCC} = 13$ MN/m$^{3/2}$, the threshold stress under cyclic stress is lower than under sustained stress, which indicates that cyclic loading has some influence on the stress-corrosion mechanisms. $\Delta K_{SCC}$ increases slightly with increasing frequency between $10^{-3}$ and 1 Hz.

The increase in fatigue crack growth rate with decreasing frequency at $\Delta K$s, which include a stress-corrosion-crack-growth component, is attributed to increasing time per cycle for the stress-corrosion mechanism to act at the advancing crack. Conversely, at high frequency, negligible stress corrosion occurs, and the fatigue-crack growth behavior approaches that of the inert environment. The proposed mechanism of fatigue crack growth below $\Delta K_{SCC}$ is hydrogen embrittlement similar to that proposed for the maraging steel previously discussed. The rapid increases in crack-growth rates at $\Delta K_{SCC}$, followed by the plateau ranges of $\Delta K$ over which the changes in growth rates are significantly lower, indicate rapid changes in the factors controlling the crack-growth mechanisms. If $R = 0$, then the crack tip is opening and closing as the stress-intensity factor varies from $K = 0$ to $K = K_{max} = \Delta K$ at the applied frequency. These factors determine the size of the opening and the strain rate at the crack tip. If oxide-film cracking is a critical step in the mechanism and is enhanced by an increase in $K_{max}$ at the existing strain rate, an increase in area of bare metal can be accompanied by an increased anodic dissolution rate of the metal and increased hydrogen generation. The trend for the change in crack-growth rate to decrease and even for the growth rate to become constant in the plateau region indicates that transport processes, both individual species diffusion and bulk transport due to the "pumping" action of the opening and closing of the crack, are limiting the growth rate. Included are transport of the environment to the crack tip, transport of corrosion-product ions from the crack interface, and transport of oxygen or other species to the interface.

The corrosion fatigue crack growth behavior in saline environments is more complex for a low-carbon (0.16 wt% C) line pipe steel than for the high-strength 4340M. The growth rate versus $\Delta K$ behavior for the line pipe steel in 3.5% NaCl solution is shown in Fig. 7.122 when cathodically coupled to zinc (coupled potential = $-800 \pm 10$ mV (SHE)) and in Fig. 7.123 when at its free-corrosion potential ($-440 \pm 30$ mV (SHE) (Ref 169). Lines are shown for four frequencies at a stress intensity ratio of $R = 0.2$; the average growth rate in air from Fig. 7.119 is also shown for reference. (Data points on which these lines are based are shown in Ref 169.) Figure 7.122 shows that the behavior is similar to that of the high-strength 4340M steel shown in Fig. 7.121; at each frequency, the growth rate initially increases rapidly and linearly (region I), then deviates to a plateau range of $\Delta K$ of almost constant growth rate (region II), and on further increase in $\Delta K$ (region III), the rates approach those observed in air. True corrosion fatigue occurs in the linear region I and is associated with hydrogen-embrittlement mechanisms as described previously in relationship to the corrosion fatigue crack growth behavior of the 4340M steel (Fig. 7.121). The transition from region I to region II is associated with initiation of a stress-corrosion component, which adds to the true corrosion-fatigue component. The decreased change in growth rate in region II results from the increasing influence of time-dependent processes at the crack tip, such as diffusion of corrosion reactants and products or steps in the hydrogen adsorption/absorption process. If these processes become controlling, the crack-growth rate can become independent of $\Delta K$ as is observed at each frequency in Fig. 7.122. The crack-growth rates of the plateau ranges increase with decreasing fatigue cycle frequency similar to the 4340M steel. The increased rate is associated with the increased time-per-cycle during which straining at the crack tip exposes bare metal to the environment. This influence of strain rate is also responsible for a lower growth rate for a square-versus-sinusoidal-cycle profile, both at 0.1 Hz. The rise time of the stress to a maximum value for the square wave is much shorter than for the sinusoidal wave.

The line identified as region I has been interpreted as indicating that the data for each frequency merges, on decreasing ΔK, to the same linear relationship. Similar to interpretation of the fatigue behavior of the 4340M steel, the value of ΔK at which the data become linear identifies, at least approximately, a $\Delta K_{SCC}$ below which a stress-corrosion mechanism is no longer additive to the true corrosion-fatigue mechanism responsible for the linear relationship between growth rate and stress-intensity range. The data indicate that $\Delta K_{SCC}$ decreases with increase in frequency.

Figure 7.123 shows the corrosion-fatigue behavior of the line pipe steel in the 3.5% NaCl solution at the corrosion potential, –440 ± 30 mV (SHE) compared with –800 ± 10 mV (SHE) when cathodically coupled (Ref 169). At the higher free-corrosion potential, the slope of re-

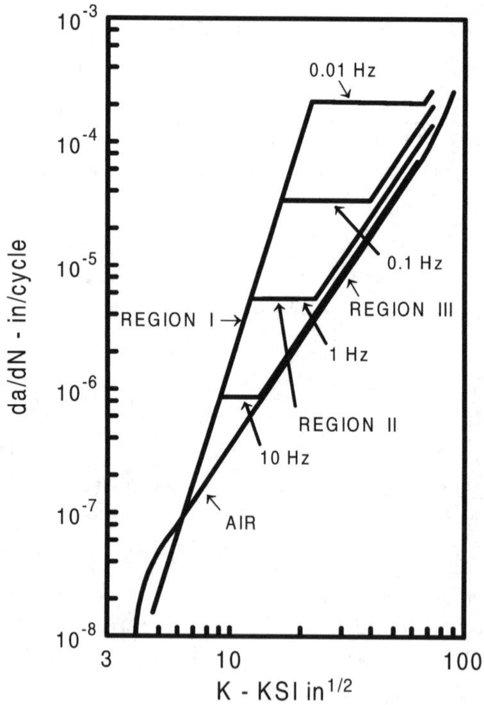


**Fig. 7.122** Corrosion-fatigue-crack-growth rate as a function of stress-intensity range for X-65 line pipe steel in air and in 3.5% NaCl solution under cathodic coupling to zinc. Cycled at indicated frequencies and R = 0.2. Coupled potential = –800 ± 10 mV (SHE). (Note: Original reference includes data on which these lines are based.) Source: Ref 169

gion I is lower, region II is ΔK dependent rather than essentially ΔK independent (plateau range) and merges into the air data without a clear region III. The observation that the fatigue crack growth rates are significantly greater when the steel is cathodically coupled is attributed to the greater hydrogen production at the lower potential, thereby resulting in increased hydrogen embrittlement.

A more complex effect of frequency on corrosion fatigue crack growth rate is observed for a Ti-6Al-4V alloy in aqueous 0.6 M NaCl, Fig. 7.124 (Ref 170). Growth rates are shown for three frequencies and for air as a reference environment; measurements were made at a stress-intensity-factor ratio of $R = K_{min}/K_{max} = 0.1$. The growth rate in air was independent of frequency. In the saline environment, a frequency-dependent transition is observed in the crack growth rate at which a stress-corrosion mechanism increases the rate above that of true corrosion fatigue, which is dominant at lower ΔKs. The ΔK of the initiation of the transition is identified in Fig. 7.124 as $\Delta K_{SCC}$ and is observed to increase with decrease in frequency. As discussed previously, the transition occurs when $K_{max}$ during a cycle exceeds $K_{ISCC}$ and will occur for a $K_{max} < K_{ISCC}$ if cyclic straining accelerates the sustained-

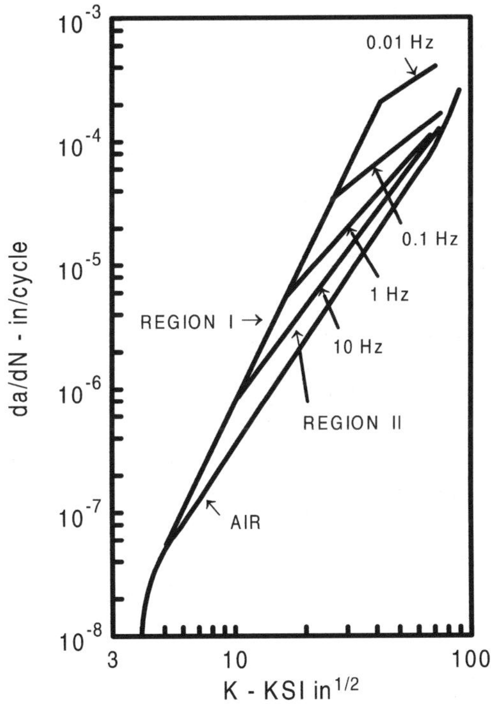


**Fig. 7.123** Corrosion-fatigue-crack-growth rate as a function of stress-intensity range for X-65 line pipe steel in air and at the free corrosion potential in 3.5% NaCl at indicated frequencies and R = 0.2. Corrosion potential = −440 ± 30 mV (SHE). (Note: Original reference includes data on which these lines are based.) Source: Ref 169

load stress corrosion mechanism. The problem of relating $K_{ISCC}$, which is determined in sustained-load tests, with a $\Delta K_{ISCC}$ defined for cyclic loading is discussed earlier. Using the relationship, $K_{max} = (\Delta K/(1 - R))$, the ΔK at which the sustained-load stress-corrosion mechanism should become active is approximated by $\Delta K = 0.9\ K_{ISCC}$. This value is indicated in Fig. 7.124 where it should be noted that the values of $\Delta K_{SCC}$ at the several frequencies are lower, indicating that the stress-corrosion mechanisms are influenced by the opening and closing of the fatigue crack.

It is also evident in Fig. 7.124 that the effect of frequency on the fatigue crack growth rate reverses on increasing ΔK above $\Delta K_{SCC}$. At a specific ΔK below $\Delta K_{SCC}$, the crack-growth rate increases with increase in frequency; the reverse frequency dependency is observed above $\Delta K_{SCC}$.

The SCC behavior of a high-strength aluminum alloy in several environments is shown in Fig. 7.113. The effect of cyclic stress-intensity range, ΔK, on the growth of fatigue cracks in the same high-strength aluminum alloy and in similar environments is shown in Fig. 7.125 (Ref 173). The similarity in the shapes of the two sets of curves should be

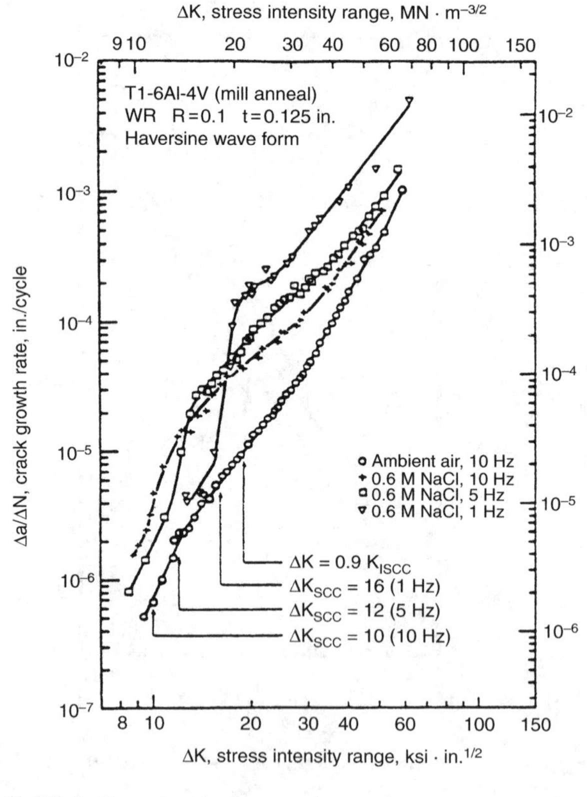


**Fig. 7.124** Corrosion-fatigue-crack-growth rate as a function of stress-intensity range for Ti-6Al-4V alloy in air and in 0.6 M NaCl at indicated frequencies and R = 0.1. Source: Ref 170

noted. In both cases, a range (plateau) of K or ΔK is observed over which the change of crack growth rate is small; also both sets of data initiate at critical values of K or ΔK, below which the growth rate is zero or becomes very small (i.e., a threshold value is required to initiate crack growth). As previously discussed, the SCC threshold is $K_{ISCC}$, and the corrosion-fatigue threshold is $\Delta K_{SCC}$. $\Delta K_{SCC}$ is interpreted as the ΔK that extends to a $K_{max} = \Delta K/(1 - R)$ during the cycle, which is less than $K_{ISCC}$ if the stress-corrosion mechanism is influenced by the cyclic loading. For this alloy and environments, the threshold ΔK is the same for distilled water and the several environments. Since the test data for the inert dry argon do not include lower values of ΔK, a comparison of threshold ΔKs in the inert and active environments cannot be made. In the intermediate stress-intensity range, ΔK = 10 ksi$\sqrt{\text{in.}}$, the crack-growth rate in distilled water is greater than that in argon by a factor of about four; the presence of 3.8 M KBr increases this factor to about 30. At higher ΔKs, the growth rates in distilled water and the halide solutions are approximately the same, but higher than for the inert dry argon.

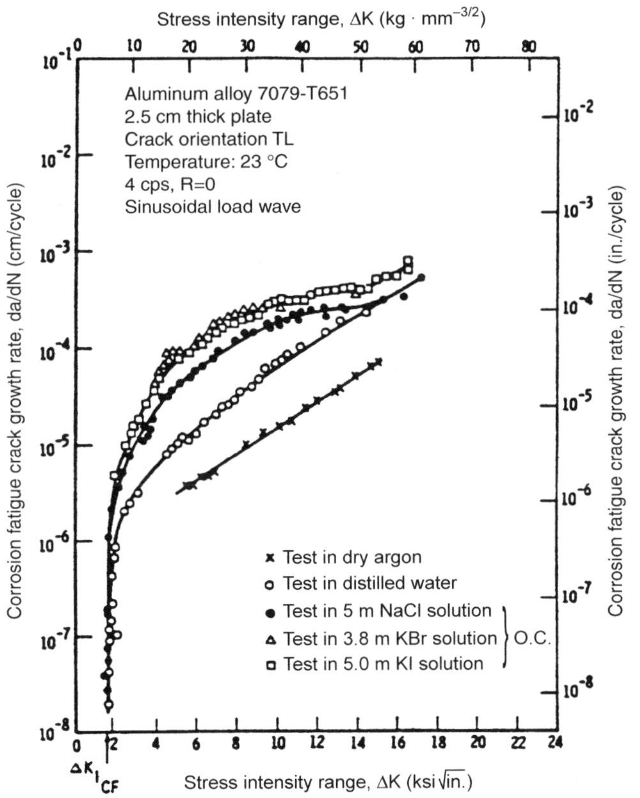


**Fig. 7.125** Corrosion-fatigue-crack-growth rate as a function of stress-intensity range for a high-strength aluminum alloy in dry argon and indicated halide solutions. Source: Ref 173

# References


1. Z. Smialowska, *Pitting Corrosion of Metals,* National Association of Corrosion Engineers, 1986
2. A.A. Stein, MIC in the Power Industry, *Microbiologically Influenced Corrosion,* G. Kobrin, Ed., NACE International, 1993, p 21–24
3. D.M. Hewette, "Pitting of Austenitic Stainless Steel, the Electrochemical Basis and Validity of Potentiostatic Measurements," master's thesis, University of Tennessee Library, Knoxville, TN, 1978
4. N. Sato, The Passivity of Metals and Passivating Films, *Passivity of Metals,* R.P. Frankenthal and J. Kruger, Ed., The Electrochemical Society, 1978, p 29–58
5. D.D. Macdonald, S.R. Biaggio, and H. Song, Steady-State Passive Films, *J. Electrochem. Soc.,* Vol 139, 1992, p 170–177
6. G. Okamoto, Passive Film of 18-8 Stainless Steel Structure and Its Function, *Corros. Sci.,* Vol 13, 1973, p 471–489
7. M. Pourbaix, *Atlas of Electrochemical Equilibria,* Pergamon Press, 1974, p 307–321
8. Z. Smialowska, Influence of Sulfide Inclusions on the Pitting Corrosion of Steels, *Corrosion,* Vol 28, 1972, p 388–396
9. G.S. Eklund, Initiation of Pitting at Sulfide Inclusions in Stainless Steel, *J. Electrochem. Soc.,* Vol 121, 1974, p 467–473
10. A.P. Bond, G.F. Bolling, H.A. Domian, and H. Bilon, Microsegregation and the Tendency for Pitting Corrosion in High-Purity Aluminum, *J. Electrochem. Soc.,* Vol 113, 1966, p 773–778
11. G.T. Burstein and P.C. Pistorius, Surface Roughness and the Metastable Pitting of Stainless Steel in Chloride Solutions, *Corrosion,* Vol 51, 1995, p 380–385
12. M. Janik-Czachor, AES Investigations of Passive Films on Iron and Iron Base Alloys, *Corrosion,* Vol 35, 1979, p 360–366
13. R.T. Foley, Localized Corrosion of Aluminum Alloys—A Review, *Corrosion,* Vol 42, 1986, p 277–288
14. G.C. Wood, W.H. Sutton, J.A. Richardson, T.N.K. Riley, and A. Malherbe, The Mechanism of Pitting of Aluminum and Its Alloys, *Localized Corrosion NACE 3,* B.F. Brown, J. Kruger, and R.W. Staehle, Ed., National Association of Corrosion Engineers, 1974, p 526–539
15. L. Tomcsanyi, K. Varga, I. Bartik, G. Horanyi, and E. Maleczki, Electrochemical Study of the Pitting Corrosion of Aluminum and Its Alloys II, Study of the Interaction of Chloride Ions with a Passive Film on Aluminum and Initiation of Corrosion, *Electrochim. Acta,* Vol 34, 1989, p 855–859

16. P. Zaya and M.B. Ives, A Comparison of Models for Local Breakdown of Passivity, *Passivity of Metals and Semiconductors,* M. Froment, Ed., Elsevier, New York, 1983, p 287–292
17. M.U. Macdonald and D.D. Macdonald, Theoretical Analysis of the Effects of Alloying Elements on Distribution Functions of Passive Breakdown, *J. Electrochem. Soc.,* Vol 136, 1989, p 961–967
18. D.A. Jones and B.E. Wilde, Galvanic Reactions during Localized Corrosion on Stainless Steel, *Corros. Sci.,* Vol 18, 1978, p 631–643
19. T. Suzuki, M. Yamabe, and Y. Kitamura, Composition of Anolyte within Pit Anode of Austenitic Stainless Steels in Chloride Solution, *Corrosion,* Vol 29, 1973, p 18–23
20. H.W. Pickering, A Critical Review of IR Drops and Electrode Potentials within Pits, Crevices and Cracks, *Advances in Localized Corrosion NACE 9,* H.S. Isaacs, U. Bertocci, J. Kruger, and S. Smialowska, Ed., National Association of Corrosion Engineers, 1990, p 77–84
21. J.R. Galvele, Transport Processes and the Mechanism of Pitting of Metals, *J. Electrochem. Soc.,* Vol 123, 1976, p 464–474
22. Z. Smialowska, Influence of Sulfide Inclusions on the Pitting Corrosion of Steels, *Corrosion,* Vol 28, 1972, p 388–396
23. P.E. Manning, D.J. Duquette, and W. Savage, The Effect of Test Method and Surface Condition on Pitting Potential of Single and Duplex Phase 304L Stainless Steel, *Corrosion,* Vol 35, 1979, p 151–157
24. H.S. Isaacs, Potential Scanning of Stainless Steel during Pitting Corrosion, *Localized Corrosion NACE 3,* R.W. Staehle, B.F. Brown, J. Kruger, and A. Agrawal, National Association of Corrosion Engineers, 1974, p 158–167
25. H.S. Isaacs, The Pitting of Iron in Dilute Chloride and Sulfate Solutions, *Advances in Localized Corrosion NACE 9,* H.S. Isaacs, U. Bertocci, K. Kruger, and S. Smialowska, Ed., National Association of Corrosion Engineers, 1990, p 221–226
26. R.M. Counce, "Potential Oscillations in the Pitting of 304 Stainless Steel," senior thesis, Department of Materials Science and Engineering, The University of Tennessee, Knoxville, TN, 1973
27. M.J. Johnson, Relative Critical Potentials for Pitting Corrosion of Some Stainless Steels, *Localized Corrosion,* STP 516, ASTM, 1972, p 262–272
28. H.P. Leckie and H.H. Uhlig, Environmental Factors Affecting the Critical Potential for Pitting in 18-8 Stainless Steel, *J. Electrochem. Soc.,* Vol 113, 1968, p 1262–1267
29. J.R. Galvele, Present State of Understanding of the Breakdown of Passivity and Repassivation, *Passivity of Metals,* R.P. Frankenthal and J. Kruger, Ed., The Electrochemical Society, 1978, p 285–327

30. I. Lee, "Polarization and Corrosion Behavior of 316L Stainless Steel and Alloy AL6X in Acid Chloride Environments," master's thesis, University of Tennessee Library, Knoxville, TN, 1986
31. R.J. Brigham, Pitting of Molybdenum Bearing Austenitic Stainless Steels, *Corrosion,* Vol 28, 1972, p 177–179
32. R.J. Brigham and E.W. Tozer, Temperature as a Pitting Criterion, *Corrosion,* Vol 29, 1973, p 33–36
33. R.J. Brigham and E.W. Tozer, Effect of Alloying Additions on the Pitting Resistance of 18% Chromium Austenitic Stainless Steel, *Corrosion,* Vol 30, 1974, p 161–166
34. P.E. Manning, Comparison of Several Accelerated Laboratory Tests for the Determination of Localized Corrosion Resistance of High-Performance Alloys, *Corrosion,* Vol 39, 1983, p 98–102
35. A.J. Sedriks, *Corrosion of Stainless Steels,* John Wiley & Sons, 1979
36. K. Osozawa, N. Okato, Y. Fukase, and K. Yokato, Effects of Alloying Elements on the Pitting Corrosion of Stainless Steel, *Proceedings of the Fifth International Congress on Metallic Corrosion,* National Association of Corrosion Engineers, 1974, p 270–274
37. I. Olefjord and L. Wegrelius, Surface Analysis of Passive State, *Corros. Sci.,* Vol 31, 1990, p 89–98
38. J. Horvath and H.H. Uhlig, Critical Potentials for Pitting Corrosion of Nickel, Chromium-Nickel, Chromium-Iron, and Related Stainless Steels, *J. Electrochem. Soc.,* Vol 115, 1968, p 791
39. K. Sugimoto and Y. Sawada, The Role of Molybdenum Additions to Austenitic Stainless Steels in the Inhibition of Pitting in Acid Chloride Solutions, *Corros. Sci.,* Vol 17, 1977, p 425–445
40. A.E. Yaniv, J.B. Lumsden, and R. Staehle, The Composition of Passive Films on Ferritic Stainless Steels, *J. Electrochem. Soc.,* Vol 124, 1977, p 490–496
41. A.R. Brooks, C.R. Clayton, K. Doss, and Y. Lu, On the Role of Cr in the Passivity of Stainless Steel, *J. Electrochem. Soc.,* Vol 133, 1986, p 2459–2464
42. C.R. Clayton and Y.C. Lu, A Bipolar Model of the Passivity of Stainless Steel: The Role of Molybdenum Addition, *J. Electrochem. Soc.,* Vol 133, 1986, p 2465–2473
43. R. Goetz and D. Landolt, The Influence of Chromium Content and Potential on the Surface Composition of Fe-Cr-Mo Alloys Studied by AES, *Electrochim. Acta,* Vol 29, 1984, p 667–676
44. I. Olefjord and B.O. Elfstrom, The Composition of the Surface during Passivation of Stainless Steels, *Corrosion,* Vol 38, 1982, p 46–52
45. K. Ogura and T. Ohama, Pit Formation in the Cathodic Polarization of Iron IV. Repair Mechanisms by Molybdate, Chromate and Tungstate, *Corrosion,* Vol 40, 1984, p 47–50

46. M.O. Spidel, Properties and Applications of High Nitrogen Steels, *High Nitrogen Steels—HNS 88,* J. Foct and A. Hendry, Ed., The Institute of Metals, London, 1989, p 92–96
47. J.E. Truman, Effects of Nitrogen Alloying on Corrosion Behavior of High Alloy Steels, *High Nitrogen Steels—HNS 88,* J. Foct and A. Hendry, Ed., The Institute of Metals, London, 1989, p 225–238
48. R.C. Newman and T. Shahrabi, The Effect of Alloyed Nitrogen or Dissolved Nitrate Ions on the Anodic Behavior of Austenitic Stainless Steel in Hydrochloric Acid, *Corros. Sci.,* Vol 27, 1987, p 827–838
49. R. Bandy and D. Van Royan, Properties of Nitrogen-Containing Stainless Alloy Designed for High Resistance to Pitting, *Corrosion,* Vol 41, 1985, p 228–233
50. K. Osozawa and N. Okato, Effect of Alloying Elements, Especially Nitrogen, on Initiation of Pitting in Stainless Steel, *Passivity and Its Breakdown in Iron and Iron-Base Alloys,* R.W. Staehle and H. Okada, Ed., National Association of Corrosion Engineers, 1976, p 135–139
51. *Guidelines for Selection of Marine Materials,* The International Nickel Co., New York, 1971, p13
52. B. Little, P. Wagner, and F. Mansfeld, Microbiologically Influenced Corrosion of Metals and Alloys, *Int. Mater. Rev.,* Vol 36, 1991, p 253–272
53. G.E. Coates, Effect of Some Surface Treatments on Corrosion of Stainless Steel, *Mater. Perform.,* Vol 29 (No. 8), 1990, p 61–65
54. L.J. Krstulovic, B. Kulusic, and M. Karsulin, The Influence of the Polarization Time on the Passivation of Iron in Sulphuric Acid, *Corros. Sci.,* Vol 21, 1981, p 95–100
55. S. Matsuda and H.H. Uhlig, Effect of pH, Sulfates, and Chlorides on Behavior of Sodium Chromate and Nitrite as Passivators for Steel, *J. Electrochem. Soc.,* Vol 111, 1964, p 156–161
56. M. Pourbaix, *Lectures on Electrochemical Corrosion,* Plenum Press, 1973, p 264
57. H.S. Campbell, A Review: Pitting Corrosion of Copper and Its Alloys, *Localized Corrosion NACE 3,* R.W. Staehle, B.F. Brown, J. Kruger, and A. Agrawal, Ed., National Association of Corrosion Engineers, 1974, p 625–638
58. V.F. Lucey, Mechanism of Pitting Corrosion of Copper in Supply Waters, *Br. Corros. J.,* Vol 2, 1967, p 175
59. H.P. Godard, Aluminum, *The Corrosion of Light Metals,* H.P. Godard, W.B. Jepson, M.R. Bothwell, and R.L. Kane, Ed., John Wiley & Sons, 1967, p 3–209
60. H. Kaesche, Pitting Corrosion of Aluminum and Intergranular Corrosion of Aluminum Alloys, *Localized Corrosion NACE 3,* B.F.

Brown, J. Kruger, and R.W. Staehle, Ed., National Association of Corrosion Engineers, 1974, p 516–525

61. J.R. Galvele, S.M. De Micheli, I.L. Muller, S.B. de Wexler, and I. Alanis, Critical Potentials for Localized Corrosion of Aluminum Alloys, *Localized Corrosion NACE 3,* B.F. Brown, J. Kruger, and R.W. Staehle, Ed., National Association of Corrosion Engineers, 1974, p 580–598
62. H. Bohni and H.H. Uhlig, Environmental Factors Affecting the Critical Pitting Potential of Aluminum, *J. Electrochem. Soc.,* Vol 116, 1969, p 906–910
63. A.J. Sedricks, J.A. Green, and J. Novak, Corrosion Processes and Solution Chemistry within Stress Corrosion Cracks in Aluminum Alloys, *Localized Corrosion NACE 3,* R.W. Staehle, B.F. Brown, J. Kruger, and A. Agrawal, Ed., National Association of Corrosion Engineers, 1974, p 569–575
64. T.R. Beck and R.C. Alkire, Occurrence of Salt Films during Initiation and Growth of Corrosion Pits, *J. Electrochem. Soc.,* Vol 126, 1979, p 1662–1666
65. J.W. Oldfield and W.H. Sutton, Crevice Corrosion of Stainless Steels I. A Mathematical Model, *Br. Corros. J.,* Vol 13, 1978, p 13–22
66. B.E. Wilde, On Pitting and Protection Potentials: Their Use and Possible Misuses for Predicting Localized Corrosion Resistance of Stainless Alloys in Halide Media, *Localized Corrosion NACE 3,* B.F. Brown, K. Kruger, and R.W. Staehle, Ed., National Association of Corrosion Engineers, 1974, p 342–352
67. "Guide for Crevice Corrosion Testing of Iron-Base and Nickel-Base Stainless Alloys in Seawater and Other Chloride-Containing Aqueous Environments," G 78, *Annual Book of ASTM Standards,* ASTM, 1992, p 463–470
68. J.H. Payer, W.E. Berry, and R. Parkins, Application of Slow Strain-Rate Technique to Stress Corrosion Cracking of Pipeline Steel, *Stress Corrosion Cracking—The Slow Strain Rate Technique,* STP 665, G.M. Ugiansky and J.H. Payer, Ed., ASTM, 1979, p 222–234
69. G.G. Geesey, Biofilm Formation, *A Practical Manual on Microbiologically Influenced Corrosion,* Gregory Kobrin, Ed., NACE International, 1993, p 11–13
70. K. Xu, S.C. Dexter, and G.W. Luther III, Voltammetric Microelectrodes for Biocorrosion Studies, *Corrosion,* Vol 54 (No. 10), 1998, p 814
71. Z. Lewandowski and P. Stoodley, Internal Mass Transport in Heterogeneous Biofilms—Recent Advances, Paper 222, *Corrosion 95* (Houston, TX), NACE International, 1995

72. R.E. Tatnall, Introduction, *A Practical Manual on Microbiologically Influenced Corrosion,* Gregory Kobrin, Ed., NACE International, 1993, p 1–9
73. R.G.J. Edyvean and S.C. Dexter, MIC in Marine Industries, *A Practical Manual on Microbiologically Influenced Corrosion,* Gregory Kobrin, Ed., NACE International, 1993, p 47–63
74. W.A. Hamilton, Sulfate-Reducing Bacteria and Anaerobic Corrosion, *Ann. Rev. Microbiol.,* Vol 39, 1985, p 195–217
75. J.A. Hardy,Utilization of Cathodic Hydrogen by Sulfate-Reducing Bacteria, *Br. Corros. J.,* Vol 18 (No. 4), 1983, p 190–193
76. R.A. Buchanan, P. Li, X. Zhang, N.D.E. Dowling, E.E. Stansbury, J.C. Danko, and D.C. White, "Electrochemical Studies of Microbiologically Influenced Corrosion," EPRI Report NP-7468, Project 2939-5, Electric Power Research Institute, Palo Alto, CA, 1991
77. P. Chandrasekaran and S.C. Dexter, Mechanism of Potential Ennoblement on Passive Metals by Seawater Biofilms, Paper 493, *Corrosion 93* (Houston, TX), NACE International, 1993
78. W.H. Dickinson and Z. Lewandowski, Manganese Biofouling of Stainless Steel: Deposition Rates and Influence on Corrosion Processes, Paper 291, *Corrosion 96* (Houston, TX), NACE International, 1996
79. R.E. Tatnall and D.H. Pope, Identification of MIC, *A Practical Manual on Microbiologically Influenced Corrosion,* Gregory Kobrin, Ed., NACE International, 1993, p 65–77
80. A.A. Stein and Y. Mussalli, Chemical Treatment for Mitigation of MIC, *A Practical Manual on Microbiologically Influenced Corrosion,* Gregory Kobrin, Ed., NACE International, 1993, p 113–121
81. S.J. Roswenberg and C.R. Irish, Solubility of Carbon in 18-Percent-Chromium—10-Percent-Nickel Austenite, *J. Res. Natl. Bur. Stand.,* Vol 48, 1952, p 40
82. C.J. Novak, Structure and Constitution of Wrought Austenitic Stainless Steels, *Handbook of Stainless Steels,* D. Peckner and I.M. Bernstein, Ed., McGraw-Hill, 1977, p 4-1 to 4-78
83. M. Hishida and H. Nakada, Critical Cooling Rate of 18Cr-8Ni Stainless Steel for Sensitization and Subsequent Intergranular Stress Corrosion Cracking, *Corrosion,* Vol 34, 1978, p 338–344
84. W.O. Binder, C.M. Brown, and R. Franks, Resistance to Sensitization of Austenitic Stainless Steels of 0.03% Max Carbon Content, *Trans. Am. Soc. for Met.,* Vol 41, 1949, p 1301
85. H.F. Ebling and M.A. Scheil, Time-Temperature-Sensitization Diagrams for Types 347, 304L, and 316 Stainless Steel, *Advances in the Technology of Stainless Steels and Related Alloys,* STP 369, ASTM, 1963, p 275–282

86. M. Henthorne, Intergranular Corrosion in Iron and Nickel-Base Alloys, *Localized Corrosion: Cause of Metal Failure,* STP 516, ASTM, 1972, p 6–119
87. R.M. Davison, T. DeBold, and M. Johnson, Corrosion of Stainless Steels, *Corrosion,* Vol 13, *Metals Handbook,* 9th ed., ASM International, 1987, p 547–565
88. R. Steigerwald, Metallurgically Influenced Corrosion, *Corrosion,* Vol 13, *Metals Handbook,* 9th ed., ASM International, 1987, p 123–135
89. F.K. Krysiak, Corrosion of Weldments, *Corrosion,* Vol 13, *Metals Handbook,* 9th ed., ASM International, 1987, p 344–368
90. J.J. Demo, *Structure, Constitution, and General Characteristics of Wrought Ferritic Stainless Steels,* ASTM, 1977
91. R.L. Cowan and C.S. Tedmon, Intergranular Corrosion of Iron-Nickel-Chromium Alloys, *Advances in Corrosion Science,* Vol 3, M.G. Fontana and R.W. Stahle, Ed., Plenum Press, 1973, p 293–400
92. R.A. Lula, Ed., *Duplex Stainless Steels,* American Society for Metals, 1983
93. S.J. Pawel, "The Sensitization Behavior of Cast Stainless Steels Subjected to Weld Repair," master's thesis, University of Tennessee Library, Knoxville, TN, 1983
94. M.A. Streicher, Relationship of Heat Treatment and Microstructure to Corrosion Resistance in Wrought Ni-Mo-Cr Alloys, *Corrosion,* Vol 19, 1963, p 272t–284t
95. R.B. Leonard, Thermal Stability of Hastelloy Alloy C-276, *Corrosion,* Vol 25, 1969, p 222–228
96. M.V. Hyatt and M.O. Spidel, *High Strength Aluminum Alloys, Stress-Corrosion Cracking in High Strength Steels and in Titanium and Aluminum Alloys,* B.F. Brown, Ed., Naval Research Laboratory, Washington DC, 1972, p 147–244
97. M.O. Speidel and M.V. Hyatt, Stress-Corrosion Cracking of High Strength Aluminum Alloys, *Advances in Corrosion Science and Technology,* M.G. Fontana and R.W. Staehle, Ed., Vol 2, Plenum Press, 1972, p 115–335
98. E.H. Hollingsworth and H.Y. Hunsicker, Corrosion of Aluminum and Aluminum Alloys, *Corrosion,* Vol 13, *Metals Handbook,* 9th ed., ASM International, 1987, p 583–609
99. J.R. Galvele and S.M. De Micheli, Mechanism of Intergranular Corrosion of Al-Cu Alloys, *Corros. Sci.,* Vol 10, 1970, p 795–807
100. L.R. Honnaker, Stainless Steels for Corrosion Resistance, *Chem. Eng. Prog.,* Vol 54, 1958, p 79–82
101. V. Cihal, Metallurgical Aspects of Knife-Line Attack on Stabilized Stainless Steels, *Localized Corrosion NACE 3,* R.W. Staehle, B.F.

Brown, J. Kruger, and A. Agrawal, Ed., National Association of Corrosion Engineers, 1974, p 502–514

102. "Standard Practices for Detecting Susceptibility to Intergranular Attack in Austenitic Stainless Steels," A 262-91, *Annual Book of ASTM Standards,* Vol 03.02, ASTM, 1992, p 1–18
103. K. Osozawa, K. Bohnenkamp, and H. Engell, Potentiostatic Study on the Intergranular Corrosion of an Austenitic Chromium-Nickel Stainless Steel, *Corros. Sci.,* Vol 6, 1966, p 421–433
104. W.L. Clarke, V.M. Romero, and J. Danko, *Detection of Sensitization in Stainless Steel Using Electrochemical Techniques,* GEAP-21382, NRC-5, National Technical Information Service, 1975
105. P. Muraleedharan, J.B. Gnanamoorthy, and K. Rao, EPR Method Versus ASTM A 262 Practice E for Measuring DOS in Austenitic Stainless Steels, *Corrosion,* Vol 45, 1989, p 142–149
106. A.P. Majidi and M.A. Streicher, Potentiodynamic Reactivation Method for Detecting Sensitization in AISI 304 and 304L Stainless Steels, *Corrosion,* Vol 40, 1984, p 393–408
107. G. Kobrin, Materials Selection, *Corrosion,* Vol 13, *Metals Handbook,* 9th ed., ASM International, 1987, p 321–337
108. B.F. Brown, Fundamentals of Stress Corrosion Cracking, *Stress-Corrosion Cracking in High Strength Steels and in Titanium and Aluminum Alloys,* B.F. Brown, Ed., U.S. Government Printing Office, Washington DC, 1972, p 1–16
109. M. Henthorne, Stress Corrosion, *Chem. Eng.,* 20 Sept 1971, p159–164
110. F.P. Ford, Current Understanding of the Mechanisms of Stress Corrosion and Corrosion Fatigue, *Environment-Sensitive Fracture: Evaluation and Comparison of Test Methods,* STP 821, S.W. Dean, E.N. Pugh, and G.M. Ugiansky, Ed., ASTM, 1984, p 32–51
111. R.N. Parkins, Stress Corrosion Spectrum, *Br. Corros. J.,* Vol 7, 1972, p 15–28
112. R.N. Parkins, Environmental Aspects of Stress Corrosion Cracking in Low Strength Ferritic Steels, *Stress Corrosion Cracking and Hydrogen Embrittlement of Iron Base Alloys NACE 5,* R.W. Staehle, J. Hochman, R.O. McCright, and J.E. Slater, Ed., National Association of Corrosion Engineers, 1972, p 601–619
113. M.G. Alvarez, C. Manfredi, M. Giordano, and J. Galvele, Anodic Rate Controlling Steps in Transgranular Stress Corrosion Cracking of Alpha-Brass in $NaNO_2$ Solutions, *Corros. Sci.,* Vol 24, 1984, p 769–780
114. R.N. Parkins and N.J.H. Holroyd, Stress Corrosion Cracking of 70/30 Brass in Acetate, Formate, Tartrate, and Hydroxide Solutions, *Corrosion,* Vol 38, 1988, p 245–255

115. F.P. Ford, Stress Corrosion Cracking, *Corrosion Processes,* R.N. Parkins, Ed., Applied Science Publishers, New York, 1982, p 271–309
116. R.N. Parkins, A Critical Evaluation of Current Environment-Sensitive Fracture Test Methods, *Environment-Sensitive Fracture: Evaluation and Comparison of Test Methods,* STP 821, ASTM, 1984
117. R.N. Parkins, P.W. Slattery, and B. Poulson, The Effects of Alloying Additions to Ferritic Steels upon Stress Corrosion Cracking Resistance, *Corrosion,* Vol 37, 1981, p 650
118. C.D. Kim and B.E. Wilde, A Review of the Constant Strain-Rate Stress Corrosion Cracking Test, *Stress Corrosion Cracking—The Slow Strain Rate Test,* STP 665, G.M. Uganisky and J.H. Payer, Ed., ASTM, 1979, p 97–112
119. R.B. Diegle and W.K. Boyd, The Role of Film Rupture during Slow Strain-Rate Stress Corrosion Cracking Testing, *Stress Corrosion Cracking—The Slow Strain Rate Technique,* STP 665, G.M. Ugainsky and J.H. Payer, Ed., ASTM, 1979, p 26–46
120. J.R. Scully and P.J. Moran, Influence of Strain on Hydrogen Assisted Cracking of Cathodically Polarized High-Strength Steel, *Environmentally Assisted Cracking: Science and Engineering,* STP 1049, ASTM, 1990, p 5–29
121. N.J. Holroyd and G.M. Scamans, Slow-Strain-Rate Stress Corrosion Testing of Aluminum Alloys, *Environment-Sensitive Fracture: Evaluation and Comparison of Test Methods,* STP 821, S.W. Dean, E.N. Pugh, and G.M. Ugiansky, Ed., ASTM, 1984, p 202–241
122. P. Lacomb and R.N. Parkins, Low Strength Steels, *Stress Corrosion Cracking and Hydrogen Embrittlement of Iron Base Alloys NACE 5,* R.W. Staehle, J. Hochman, R.D. McCright, and J.E. Slater, National Association of Corrosion Engineers, 1973, p 521–523
123. G. Sandoz, High Strength Steels, *Stress-Corrosion Cracking in High Strength Steels and in Titanium and Aluminum Alloys,* B.F. Brown, Ed., Naval Research Laboratory, Washington DC, 1972, p 79–145
124. C.S. Carter and M.V. Hyatt, Review of Stress Corrosion Cracking in Low Alloy Steels with Yield Strength below 150 KSI, *Stress Corrosion Cracking and Hydrogen Embrittlement of Iron Base Alloys NACE 5,* R.W. Staehle, J.H. Hochman, R.D. McCright, and J.E. Slater, Ed., National Association of Corrosion Engineers, 1973, p 524–600
125. R.N. Parkins, Stress Corrosion Cracking of Low-Strength Ferritic Steels, *The Theory of Stress Corrosion Cracking of Alloys,* J.C. Scully, Ed., NATO, Brussels, 1971, p 167–185

126. S. Ahmad, M.L. Mehta, S.K. Saraf, and I. Saraswat, Stress Corrosion Cracking of Sensitized 304 Austenitic Stainless Steel in Petroleum Refinery Environment, *Corrosion,* Vol 38, 1982, p 347–353
127. H.H. Horowitz, Chemical Studies of Polythionic Acid Stress-Corrosion Cracking, *Corros. Sci.,* Vol 23, 1983, p 353–362
128. S. Ahmad, M.L. Mehta, S.K. Saraf, and I. Saraswat, Anodic Polarization Characteristics of Sensitized 304 Austenitic Stainless Steel in Polythionic Acid Environment, *Corrosion,* Vol 39, 1983, p 330
129. R.F. Steigerwald, A.P. Bond, H.J. Dundas, and E. Lizlovz, The New Fe-Cr-Mo Ferritic Stainless Steels, *Corrosion,* Vol 33, 1977, p 279–295
130. B. Wallen and J. Olsson, Corrosion Resistance in Aqueous Media, *Handbook of Stainless Steels,* D. Peckner and I.M. Bernstein, Ed., McGraw Hill Book Co., 1977, p 16-1 to 16-89
131. D.R. McIntire and C.P. Dillon, *Guidelines for Preventing Stress Corrosion Cracking in the Process Industries,* MTI Publication 15, Materials Technology Institute of the Chemical Process Industry, Inc., March 1985, p 23
132. H.R. Copson, Effect of Composition on Stress Corrosion Cracking of Some Alloys Containing Nickel, *Physical Metallurgy of Stress Corrosion Fracture,* T. Rhodin, Ed., Interscience Publishers, Inc., New York, 1959
133. R.M. Latanision and R.W. Staehle, in *Fundamental Aspects of Stress Corrosion Cracking NACE 1,* R.W. Staehle, A.J. Forty, and D.Van Rooyan, Ed., National Association of Corrosion Engineers, 1969, p 214
134. K.F. Krysiak, Corrosion of Weldments, *Corrosion,* Vol 13, *Metals Handbook,* 9th ed., ASM International, 1987, p 344–374
135. H.D. Solomon and T.M. Devine, Duplex Stainless Steels—A Tale of Two Phases, *Duplex Stainless Steels,* R. A. Lula, Ed., American Society for Metals, 1983, p 693–756
136. D.O. Sprowls and R.H. Brown, Stress Corrosion Mechanisms for Aluminum Alloys, *Fundamental Aspects of Stress Corroson Cracking NACE 1,* R.W. Staehle, A.J. Forty, and D. Van Rooyan, Ed., National Association of Corrosion Engineers, 1969, p 466–509
137. A.F. Beck and P.R. Sperry, The Relationship between Structure and Susceptibility to Stress Corrosion in Aluminum-Magnesium Alloys, *Fundamental Aspects of Stress Corrosion Cracking NACE 1,* R.W. Staehle, A.J. Forty, and D. Van Rooyan, Ed., National Association of Corrosion Engineers, 1969, p 513–529
138. M.O. Speidel, Interaction of Dislocations with Precipitates in High Strength Aluminum Alloys and Susceptibility to Stress

Corrosion Cracking, *Fundamental Aspects of Stress Corrosion Cracking NACE 1,* R.W. Staehle, A.J. Forty, and D. Van Rooyan, Ed., National Association of Corrosion Engineers, 1969, p 561–579
139. G. Edmunds, E.A. Anderson, and R. Waring, Ammonia and Mercury Stress-Cracking Tests for Brass, *Symposium on Stress Corrosion Cracking of Metals,* ASTM, 1944, p 7–18
140. E.N. Pugh, J.V. Craig, and A. Sedriks, The Stress-Corrosion Cracking of Copper, Silver and Gold Alloys, *Fundamental Aspects of Stress Corrosion Cracking NACE 1,* R.W. Staehle, A.J. Forty, and D. Van Rooyen, Ed., National Association of Corrosion Engineers, 1969, p 118–158
141. R.C. Newman and G.T. Burstein, Early Stages of Film Growth on Alpha Brass Immersed in Ammoniacal Cu (II) Solutions, *Corrosion,* Vol 40, 1984, p 201–204
142. H.E. Johnson and J. Leja, On the Potential/pH Diagrams of the Cu-NH3-H2O and Zn-NH3-H2O Systems, *J. of Electrochem. Soc.,* Vol 112, 1965, p 638–641
143. E.N. Pugh, The Mechanisms of Stress Corrosion Cracking of Alpha-Brass in Aqueous Ammonia, *The Theory of Stress Corrosion Cracking in Alloys,* J.C. Scully, Ed., NATO, Brussels, 1971, p 418–441
144. A.J. Forty and P. Humble, Surface Films and Stress-Corrosion Cracking, *Environment-Sensitive Mechanical Behavior,* A.R.C. Westwood and N.S. Stoloff, Ed., Gordon and Breach, New York, 1966, p 403–420
145. E.N. Pugh, A Post Conference Evaluation of Our Understanding of the Failure Mechanisms, *Stress Corrosion Cracking and Hydrogen Embrittlement of Iron Base Alloys NACE 5,* R.W. Staehle, J.H. Hochman, R.D. McCright, and J.E. Slater, Ed., National Association of Corrosion Engineers, 1977, p 1977
146. T.P. Hoar and C.J.L. Booker, The Electrochemistry of the Stress Corrosion Cracking of Alpha Brass, *Corros. Sci.,* Vol 5, 1965, p 821–840
147. H.E. Johnson and J. Leja, Surface Chemical Factors in the Stress-Corrosion Cracking of Alpha Brass, *Corrosion,* Vol 22, 1966, p 178–189
148. E.N. Pugh, J.V. Craig, and W. Montaguer, Factors Influencing the Path of Stress-Corrosion Cracking in Alpha-Phase Copper Alloys Exposed to Aqueous Ammonia Environments, *Trans. ASM,* Vol 61, 1968, p 468–473
149. A. Parthasarathi and N.W. Polan, Stress Corrosion Cracking of Copper Alloys: The Effects of Corrosion Potential and Tarnishing Characteristics, *Corrosion,* Vol 43, 1987, p 747–755
150. T.C. Wilson, G. Edmunds, E.A. Anderson, and W. Peirce, Effect on Season Cracking of Alloy Additions to Cartridge Brass, *Sym-*

*posium on Stress-Corrosion Cracking of Metals,* ASTM, 1944, p 173–192

151. J.A.S. Green and E.N. Pugh, The Role of Tarnishing in the Stress-Corrosion Cracking of Alpha-Brass in Aqueous Citrate and Tartrate Solutions, *Metall. Trans.,* Vol 2, 1971, p 1379–1383
152. U. Bertocci and E.N. Pugh, Chemical and Electrochemical Aspects of SCC of Alpha-Brass in Aqueous Ammonia, *International Congress on Metallic Corrosion,* National Research Council of Canada, Toronto, 1984, p 144–152
153. A. Kawashima, A.K. Agrawal, and R. Staehle, Effect of Oxyanions and Chloride Ion on the Stress Corrosion Cracking Susceptiblity of Admiralty Brass in Nonammonical Aqueous Solutions, *Stress Corrosion Cracking—The Slow Strain-Rate Technique,* STP 665, G.M. Uglansky and J.H. Payer, Ed., ASTM, 1979, p 266–278
154. E.N. Pugh, Progress toward Understanding the Stress Corrosion Problem, *Corrosion,* Vol 41, 1985, p 517–526
155. B. Tompkins, Role of Mechanics in Corrosion Fatigue, *Met. Sci.,* The Metals Society, July 1979, p 387–395
156. J.C. Scully, Fractographic Aspects of Stress Corrosion Cracking, *The Theory of Stress Corrosion Cracking of Alloys,* J.C. Scully, Ed., NATO Scientific Affairs Division, Brussels, 1971, p 127–166
157. J.A. Beavers and E.N. Pugh, The Propagation of Transgranular Stress-Corrosion Cracks in Admiralty Metal, *Metall. Trans.,* Vol 11A, 1980, p 809–820
158. H.W. Liu, Stress-Corrosion Cracking and the Interaction between Crack-Tip Stress Field and Solute Atoms, *J. Basic Eng.,* 1970, p 633–638
159. M.O. Speidel, Current Understanding of Stress Corrosion Crack Growth in Aluminum Alloys, *The Theory of Stress Corrosion Cracking in Alloys,* S.C. Scully, Ed., NATO Scientific Affairs Division, Brussels, 1971, p 289–354
160. J.R. Galvele, A Stress Corrosion Mechanism Based on Surface Mobility, *Corros. Sci.,* Vol 27, 1987, p 1–33
161. B.F. Brown, The Application of Fracture Mechanics to Stress-Corrosion Cracking, *Metall. Reviews,* Vol 13, 1968, p 171–183
162. D. Broek, *Elementary Engineering Fracture Mechanics,* Sijthoff and Nordhoff International Publishers, Netherlands, 1978
163. D. Broek, *The Practical Use of Fracture Mechanics,* Kluwer Academic Publishers, Boston, 1989
164. H.L. Ewalds and R.J.H. Wanhill, *Fracture Mechanics,* Edward Arnold, London, 1984
165. M.O. Speidel, Stress Corrosion Crack Growth in Austenitic Stainless Steel, *Corrosion,* Vol 33, 1977, p 199–203

166. M.O. Speidel, Stress Corrosion Cracking of Stainless Steels in NaCl Solutions, *Metall. Trans.*, Vol 12A, 1981, p 779–789
167. P.M. Scott, Chemistry Effects in Corrosion Fatigue, *Corrosion Fatigue: Mechanics, Metallurgy, Electrochemistry, and Engineering,* STP 801, ASTM, 1983, p 319–350
168. S. Suresh, *Fatigue of Materials,* Cambridge University Press, New York, 1992
169. O. Vosikovsky, Fatigue-Crack Growth in an X-65 Linepipe Steel at Low Cyclic Frequencies in Aqueous Environments, *Int. J. Mater. Technol. (Trans. ASME),* Vol 97, 1975, p 298–304
170. D.B. Dawson and R.M. Pelloux, Corrosion Fatigue Crack Growth of Titanium Alloys in Aqueous Environments, *Metall. Trans.,* Vol 5, 1974, p 723–731
171. J.M. Barsom, Effect of Cyclic Stress Form on Corrosion Fatigue Crack Propagation below KISCC in a High Yield Strength Steel, *Corrosion Fatigue,* O.F. Devereux, A.J. McEvily, and R.W. Staehle, Ed., National Association of Corrosion Engineers, 1971, p 424–436
172. M.O. Speidel, Corrosion Fatigue in Fe-Ni-Cr Alloys, *Stress Corrosion Cracking and Hydrogen Embrittlemment in Iron Base Alloys NACE 5,* R.W. Staehle, J.H. Hochman, R.D. McCrigh, and J.E. Slater, Ed., National Association of Corrosion Engineers, 1973, p 1071–1080
173. M.O. Speidel, M.J. Blackburn, T.R. Beck, and J. Feene, Corrosion Fatigue and Stress Corrosion Crack Growth in High Strength Aluminum Alloys, Magnesium Alloys, and Titanium Alloys Exposed to Aqueous Solutions, *Corrosion Fatigue,* O.F. Devereux, A.J. McEvily, and R.W. Staehle, Ed., National Association of Corrosion Engineers, 1971, p 324–345

# Selected Sources of Information: Corrosion Properties of Materials and Corrosion Testing

## Corrosion Properties of Materials

- *Corrosion,* Vol 13, *Metals Handbook,* 9th ed., ASM International, 1987
- B.D. Craig and D. Anderson, Ed., *Handbook of Corrosion Data,* 2nd ed., ASM International, 1995
- N.E. Hammer, Ed., *Corrosion Data Survey—Metals Section,* 6th ed., NACE International, 1985
- E.D. During, *Corrosion Atlas,* 3rd ed., Elsevier Science Publishers, 1997
- K.M. Pruett, *Chemical Resistance Guide for Metals and Alloys,* Compass Publications, La Jolla, CA, 1995
- P.A. Schweitzer, Ed., *Corrosion Resistance Tables,* 4th ed., Marcel Dekker, 1995
- S.L. Chawla and R.K. Gupta, *Materials Selection for Corrosion Control,* ASM International, 1993

## Corrosion Testing

Selected ASTM Standards: *Annual Book of ASTM Standards,* American Society for Testing and Materials, 1916 Race Street, Philadelphia, PA, 19103

- A 143, Practice for Safeguarding against Embrittlement of Hot-Dip Galvanized Structural Steel Products and Procedure for Detecting Embrittlement
- A 262, Practices for Detecting Susceptibility to Intergranular Attack in Austenitic Stainless Steels
- A 763, Practices for Detecting Susceptibility to Intergranular Attack in Ferritic Stainless Steels
- B 117, Test Method of Salt Spray (Fog) Testing
- B 368, Method for Copper-Accelerated Acetic Acid-Salt Spray (Fog) Testing (CASS Test)
- B 380, Method of Corrosion Testing of Decorative Electrodeposited Coatings by the Corrodkote Procedure
- B 457, Method for Measurement of Impedance of Anodic Coatings on Aluminum
- B 537, Practice for Rating of Electroplated Panels Subjected to Atmospheric Exposure
- B 577, Test Methods for Hydrogen Embrittlement of Copper
- B 583, Test Methods for Porosity in Gold Coatings on Metal Substrates
- B 627, Test Method for Electrolytic Corrosion Testing (EC Test)
- B 651, Test Method for Measurement of Corrosion Sites in Nickel Plus Chromium or Copper Plus Nickel Plus Chromium Electroplated Surfaces with the Double-Beam Interference Microscope
- B 680, Test Method for Seal Quality of Anodic Coatings on Aluminum by Acid Dissolution
- B 732, Test Method for Evaluating the Corrosivity of Solder Fluxes for Copper Tubing Systems
- B 735, Test Method for Porosity in Gold Coatings on Metal Substrates by Nitric Acid Vapor
- B 741, Test Methods for Porosity in Gold Coatings on Metal Substrates by Paper Electrography
- B 808, Test Method for Monitoring of Atmospheric Corrosion Chambers by Quartz Crystal Microbalances
- B 810, Test Method for Calibration of Atmospheric Corrosion Test Chambers by Change in Mass of Copper Coupons
- B 825, Test Method for Coulometric Reduction of Surface Films on Metallic Test Samples

- B 826, Test Method for Monitoring Atmospheric Corrosion Tests by Electrical Resistance Probes
- C 692, Method of Evaluating the Influence of Wicking-Type Thermal Insulations on the Stress Corrosion Cracking Tendency of Austenitic Stainless Steel
- C 694, Test Method for Weight Loss (Mass Loss) of Sheet Steel during Immersion in Sulfuric Acid Solution
- C 876, Test Method for Half-Cell Potentials of Uncoated Reinforcing Steel in Concrete
- D 130, Test Method for Detection of Copper Corrosion from Petroleum Products by the Copper Strip Tarnish Test
- D 610, Test Method for Evaluating Degree of Rusting on Painted Steel Surfaces
- D 665, Test Method for Rust-Preventing Characteristics of Inhibited Mineral Oil in the Presence of Water
- D 849, Test Method for Copper Corrosion of Industrial Aromatic Hydrocarbons
- D 930, Test Method of Total Immersion Corrosion Test of Water-Soluble Aluminum Cleaners
- D 1141, Specification for Substitute Ocean Water
- D 1176, Test Method for Sampling and Preparing Aqueous Solutions of Engine Coolants or Antirusts for Testing Purposes
- D 1193, Specification for Reagent Water
- D 1280, Test Method of Total Immersion Corrosion Test for Soak Tank Metal Cleaners
- D 1374, Test Method of Aerated Total Immersion Corrosion Test for Metal Cleaners
- D 1384, Test Method for Corrosion Test for Engine Coolants in Glassware
- D 1414, Test Methods for Rubber O-Rings
- D 1654, Method for Evaluation of Painted or Coated Specimens Subjected to Corrosive Environments
- D 1743, Test Method for Corrosion Preventive Properties of Lubricating Greases
- D 1838, Test Method for Copper Strip Corrosion by Liquefied Petroleum (LP) Gases
- D 2059, Test Method for Resistance of Zippers to Salt Spray (Fog)
- D 2251, Test Method for Metal Corrosion by Halogenated Organic Solvents and Their Admixtures
- D 2570, Method for Simulated Service Corrosion Testing of Engine Coolants
- D 2649, Test Method for Corrosion Characteristics of Solid Film Lubricants

- D 2671, Test Methods for Heat-Shrinkable Tubing for Electrical Use
- D 2803, Test Method for Filiform Corrosion Resistance of Organic Coatings on Metal
- D 2809, Test Method for Cavitation Erosion-Corrosion Characteristics of Aluminum Pumps with Engine Coolants
- D 2847, Practice for Testing Engine Coolants in Car and Light Truck Service
- D 2933, Test Method for Corrosion Resistance of Coated Steel Specimens (Cyclic Method)
- D 3263, Test Methods for Corrosivity of Solvent Systems for Removing Water-Formed Deposits
- D 3310, Test Method for Determining Corrosivity of Adhesive Materials
- D 3316, Test Method for Stability of Perchloroethylene with Copper
- D 3482, Practice for Determining Electrolytic Corrosion of Copper by Adhesives
- D 3603, Test Method for Rust-Preventing Characteristics of Steam Turbine Oil in the Presence of Water (Horizontal Disk Method)
- D 4048, Test Method for Detection of Copper Corrosion from Lubricating Grease by the Copper Strip Tarnish Test
- D 4340, Test Method for Corrosion of Cast Aluminum Alloys in Engine Coolants under Heat-Rejecting Conditions
- D 4585, Practice for Testing Water Resistance of Coatings Using Controlled Condensation
- D 4627, Test Method for Iron Chip Corrosion for Water Soluble Metalworking Fluids
- D 4636, Test Method for Corrosiveness and Oxidation Stability of Hydraulic Oils, Aircraft Turbine Engine Lubricants, and Other Highly Refined Oils
- E 712, Practice for Laboratory Screening of Metallic Containment Materials for Use with Liquids in Solar Heating and Cooling Systems
- E 745, Practice for Simulated Service Testing for Corrosion of Metallic Containment Materials for Use With Heat-Transfer Fluids in Solar Heating and Cooling Systems
- E 937, Test Method for Corrosion of Steel by Sprayed Fire-Resistive Material Applied to Structural Members
- F 326, Test Method for Electronic Hydrogen Embrittlement Test for Cadmium Electroplating Processes
- F 359, Practice for Static Immersion Testing of Unstressed Materials in Nitrogen Tetroxide ($N_2O_4$)
- F 363, Method for Corrosion Testing of Gaskets

- F 482, Test Method for Corrosion of Aircraft Metals by Total Immersion in Maintenance Chemicals
- F 483, Test Method for Total Immersion Corrosion Test for Aircraft Maintenance Chemicals
- F 519, Method for Mechanical Hydrogen Embrittlement Testing of Plating Processes and Aircraft Maintenance Chemicals
- F 746, Test Method for Pitting or Crevice Corrosion of Metallic Surgical Implant
- F 897, Test Method for Measuring Fretting Corrosion of Osteosynthesis Plates and Screws
- F 945, Test Method for Stress-Corrosion of Titanium Alloys by Aircraft Engine Cleaning Materials
- F 1089, Test Method for Corrosion of Surgical Instruments
- F 1110, Test Method for Sandwich Corrosion Test
- F 1111, Test Method for Corrosion of Low-Embrittling Cadmium Plate by Aircraft Maintenance Chemicals
- F 1113, Test Method for Electrochemical Measurement of Diffusible Hydrogen in Steels (Barnacle Electrode)
- G 1, Practice for Preparing, Cleaning, and Evaluating Corrosion Test Specimens
- G 2, Test Method for Corrosion Testing of Products of Zirconium, Hafnium, and Their Alloys in Water at 680 °F or in Steam at 750 °F
- G 3, Practice for Conventions Applicable to Electrochemical Measurements in Corrosion Testing
- G 4, Method for Conducting Corrosion Coupon Tests in Plant Equipment
- G 5, Standard Reference Test Method for Making Potentiostatic and Potentiodynamic Anodic Polarization Measurements
- G 8, Test Methods for Cathodic Disbonding of Pipeline Coatings
- G 15, Terminology Relating to Corrosion and Corrosion Testing
- G 16, Guide for Applying Statistics to Analysis of Corrosion Data
- G 19, Test Method for Disbonding Characteristics of Pipeline Coating by Direct Soil Burial
- G 28, Test Methods of Detecting Susceptibility to Intergranular Attack in Wrought, Nickel-Rich, Chromium-Bearing Alloys
- G 30, Practice for Making and Using U-Bend Stress-Corrosion Test Specimens
- G 31, Practice for Laboratory Immersion Corrosion Testing of Metals
- G 32, Test Method for Cavitation Erosion Using Vibratory Apparatus
- G 33, Practice for Recording Data from Atmospheric Corrosion Tests of Metallic-Coated Steel Specimens

- G 34, Test Method for Exfoliation Corrosion Susceptibility in 2XXX and 7XXX Series Aluminum Alloys (EXCO Test)
- G 35, Practice for Determining the Susceptibility of Stainless Steels and Related Nickel-Chromium-Iron Alloys to Stress-Corrosion Cracking in Polythionic Acids
- G 36, Practice for Evaluating Stress-Corrosion Cracking Tests in a Boiling Magnesium Chloride Solution
- G 37, Practice for Use of Mattsson's Solution of pH 7.2 to Evaluate the Stress-Corrosion Cracking Susceptibility of Copper-Zinc Alloys
- G 38, Practices for Making and Using C-Ring Stress-Corrosion Test Specimens
- G 39, Practice for Preparation and Use of Bent-Beam Stress-Corrosion Test Specimens
- G 41, Practice for Determining Cracking Susceptibility of Metals Exposed under Stress to a Hot Salt Environment
- G 42, Test Methods for Cathodic Disbonding of Pipeline Coatings Subjected to Elevated Temperatures
- G 44, Practice for Evaluating Stress Corrosion Cracking Resistance of Metals and Alloys by Alternate Immersion in 3.5% Sodium Chloride Solution
- G 46, Practice for Examination and Evaluation of Pitting Corrosion
- G 47, Test Method for Determining Susceptibility to Stress-Corrosion Cracking of High-Strength Aluminum Alloy Products
- G 48, Test Method for Pitting and Crevice Corrosion Resistance of Stainless Steels and Related Alloys by the Use of Ferric Chloride Solution
- G 49, Practice for Preparation and Use of Direct Tension Stress Corrosion Test Specimens
- G 50, Practice for Conducting Atmospheric Corrosion Tests on Metals
- G 51, Test Method for pH of Soil for Use in Corrosion Testing
- G 52, Practice for Exposing and Evaluating Metals and Alloys in Surface Seawater
- G 53, Practice for Operating Light- and Water-Exposure Apparatus (Fluorescent UV-Condensation Type) for Exposure of Nonmetallic Materials
- G 54, Practice for Simple Static Oxidation Testing
- G 57, Method for Field Measurement of Soil Resistivity Using the Wenner Four-Electrode Method
- G 58, Practice for the Preparation of Stress Corrosion Test Specimens for Weldments
- G 59, Practice for Conducting Potentiodynamic Polarization Resistance Measurements

- G 60, Method for Conducting Cyclic Humidity Tests
- G 61, Test Method for Conducting Cyclic Potentiodynamic Polarization Measurements for Localized Corrosion Susceptibility of Iron-, Nickel-, or Cobalt-Based Alloys
- G 62, Test Methods for Holiday Detection in Pipeline Coatings
- G 64, Classification of the Resistance of Stress-Corrosion Cracking of Heat-Treatable Aluminum Alloys
- G 66, Method for Visual Assessment of Exfoliation Corrosion Susceptibility of 5XXX Series Aluminum Alloys (ASSET Test)
- G 67, Test Method for Determining the Susceptibility to Intergranular Corrosion of 5XXX Series Aluminum Alloys by Mass Loss after Exposure to Nitric Acid (NAMLT Test)
- G 68, Practice for Liquid Sodium Corrosion Testing of Metals and Alloys
- G 69, Practice for Measurement of Corrosion Potentials of Aluminum Alloys
- G 71, Guide for Conducting and Evaluating Galvanic Corrosion Tests in Electrolytes
- G 78, Guide for Crevice Corrosion Testing of Iron-Base and Nickel-Base Stainless Alloys in Seawater and Other Chloride-Containing Aqueous Environments
- G 79, Practice for Evaluation of Metals Exposed to Carburization Environments
- G 80, E 1, Test Method for Specific Cathodic Disbonding of Pipeline Coatings
- G 82, Guide for Development and Use of a Galvanic Series for Predicting Galvanic Corrosion Performance
- G 84, Practice for Measurement of Time-of-Wetness on Surfaces Exposed to Wetting Conditions as in Atmospheric Corrosion Testing
- G 85, Practice for Modified Salt Spray (Fog) Testing
- G 87, Method for Conducting Moist $SO_2$ Tests
- G 95, E 1, Test Method for Cathodic Disbondment Test of Pipeline Coatings (Attached Cell Method)
- G 91, Test Method for Monitoring Atmospheric $SO_2$ Using the Sulfation Plate Technique
- G 92, Practice for Characterization of Atmospheric Test Sites
- G 96, Practice for On-Line Monitoring of Corrosion in Plant Equipment (Electrical and Electrochemical Methods)
- G 97, Test Method for Laboratory Evaluation of Magnesium Sacrificial Anode Test Specimens for Underground Applications
- G 100, Method for Conducting Cyclic Galvanostaircase Polarization
- G 101, Guide for Estimating the Atmospheric Corrosion Resistance of Low-Alloy Steels

- G 102, Practice for Calculation of Corrosion Rates and Related Information from Electrochemical Measurements
- G 103, Test Method for Performing a Stress-Corrosion Cracking Test of Low Copper Containing Al-Zn-Mg Alloys in Boiling 6% Sodium Chloride Solution
- G 104, Test Method for Assessing Galvanic Corrosion Caused by the Atmosphere
- G 106, Practice for Verification of Algorithm and Equipment for Electrochemical Impedance Measurements
- G 107, Guide for Formats for Collection and Compilation of Corrosion Data for Metals for Computerized Database Input

# Index

## A

# B

# D

# K



# L

# N

# O

# P

## Q



## R



## S

## U



## V



## W